Natural Particulars

**Dibner Institute for the History
of Science and Technology**

Dibner Institute Studies in the History of Science and Technology
Jed Buchwald, general editor

Anthony Grafton and Nancy Siraisi, editors, *Natural Particulars: Nature and the Disciplines in Renaissance Europe*

Natural Particulars

Nature and the Disciplines in Renaissance Europe

———

edited by Anthony Grafton and Nancy Siraisi

The MIT Press
Cambridge, Massachusetts
London, England

This book was set in Bembo by Graphic Composition, Inc. and printed and bound in
the United States of America

Library of Congress Cataloging-in-Publication Data

Natural particulars : nature and the disciplines in Renaissance Europe / edited by
  Anthony Grafton and Nancy Siraisi.
      p.   cm. — (Dibner Institute studies in the history of science and technology)
  Based on papers from a workshop held at Dibner Institute for the History of
Science and Technology, May 5–6, 1995.
  Includes bibliographical references and index.
  ISBN 0-262-07193-2 (hc. : alk. paper)
  1. Nature Congresses. 2. Life sciences—Europe—History Congresses. 3. Science,
Renaissance Congresses. I. Grafton, Anthony. II. Siraisi, Nancy G. III. Series.
QH81.N3   1999
509′.4′09031—dc21   99-30932
                                                                              CIP

Contents

PREFACE

The essays collected in this volume discuss forms of learned response to the natural world in fourteenth- to early-seventeenth-century Europe. Its title and subtitle reflect our continuing conviction that one cannot study the history of understanding of nature in this period without mastering the learned disciplines of the time, however alien these may now seem. The subtitle also evidently reveals that we think there is a valid use for the term "Renaissance" in this context (not that we would dispute the broader merits of "early modern" as a chronological or social history category). In the learned disciplines treated here, the classical heritage always played a central role, whether it was being defended or attacked, preserved or revised, and whether those doing so were humanists or scholastics. By focusing on questions of revival, transmission, and appropriation, we treat our protagonists from a standpoint that they themselves might accept as legitimate, while recognizing both the traditional and the innovative qualities of their work. In this heuristic sense, at least, we feel certain that both the term and the notion of the Renaissance retain their validity.

We would like to acknowledge the help of many friends and colleagues, without whom this volume would not have been possible. The genesis of the collection was the workshop on "Renaissance Natural Philosophy and the Disciplines," held at the Dibner Institute for the History of Science and Technology at the Massachusetts Institute of Technology on 5–6 May 1995. Jed Buchwald and Evelyn Simha of the Dibner Institute invited us to organize a meeting on this theme, and carried out all the necessary technical preparations for it with miraculous speed and efficiency, leaving us to enjoy some memorable exchanges of ideas in peace and comfort. We are also grateful to Jed Buchwald, in his capacity as general editor of the Dibner Series in the History of Science and Technology, for arranging for the publication of these essays, which represent revised and enlarged forms of the papers presented at the workshop. Our greatest debt is, of course, to our authors, who have borne patiently with our requests for revisions and with various delays. Special thanks to Katharine Park, who graciously allowed us to use her title for the collection as a whole. Thanks are also due, finally, to the anonymous referees who read the volume for the press, to Matthew Abbate and Larry Cohen of the MIT Press for their help, and to Alice Falk for exceptionally careful and thoughtful copyediting.

CONTRIBUTORS

Michael J. B. Allen is co-editor of *Renaissance Quarterly,* professor of English at UCLA, and an authority on Renaissance Platonism. His books include *Icastes: Marsilio Ficino's Interpretation of Plato's Sophist* (1989), *Nuptial Arithmetic* (1994), *Plato's Third Eye: Studies in Marsilio Ficino's Metaphysics and its Sources* (1995), and *Synoptic Art: Marsilio Ficino on the History of Platonic Interpretation* (forthcoming).

Ann Blair is associate professor of history and of history and literature at Harvard University and the author of *The Theater of Nature: Jean Bodin and Renaissance Science* (1997). Her current project is a study of the development of encyclopedic reference works in the sixteenth and seventeenth centuries.

Brian P. Copenhaver is professor of history and philosophy and provost of the College of Letters and Science at UCLA. His *Hermetica* appeared in 1992, his *Renaissance Philosophy* (with Charles Schmitt) in 1993. He is now working on the *De inventoribus rerum* of Polydore Virgil and the *900 Conclusions* of Giovanni Pico della Mirandola.

Chiara Crisciani is associate professor in the Department of Philosophy of the University of Pavia. She has devoted her research mainly to epistemological aspects of scholastic medicine and the development of Latin alchemy, in the context of the history of philosophical and scientific thought of the late Middle Ages. She is currently studying the relationship and the differences between alchemy and medicine in the pre-Paracelsian period as well as new trends in Italian alchemy in humanistic and early Renaissance cultural contexts.

Luc Deitz spent a number of years as a research fellow at the Warburg Institute (University of London) before being appointed curator of manuscripts and rare books at the Bibliothèque nationale de Luxembourg. His publications include a critical edition, with German translation and commentary, of books 1–4 and 7 of Scaliger's *Poetices libri septem,* as well as a number of articles on Renaissance poetics and philosophy, and on the history of classical scholarship.

Paula Findlen teaches Renaissance history and history of science and medicine at Stanford University. She is the author of *Possessing Nature: Museums, Collecting, and Scientific Culture in Early Modern Italy* (1994) and *A Fragmentary Past: Museums and the Renaissance* (forthcoming). Currently she is completing a project on women and science in early modern Italy.

Anthony Grafton teaches history and history of science at Princeton University. His publications include *Joseph Scaliger* (1983–1993), *Defenders of the Text* (1991), and *The Footnote: A Curious History* (1997).

James Hankins is professor of history at Harvard University. He has published widely on Renaissance intellectual history, including *Plato in the Italian Renaissance* (1990).

Thomas DaCosta Kaufmann is a professor in the Department of Art and Archaeology, Princeton University. His recent books include *The Mastery of Nature: Aspects of Art, Science, and Humanism in the Renaissance* (1993; Japanese translation, 1995) and *Court, Cloister, and City: The Art and Culture of Central Europe, 1450–1800* (1995; German translation, 1998). He is at present at work on books on the geography of art and on the historiography of art before Winckelmann.

John Monfasani is professor of history at the University at Albany, State University of New York, and executive director of the Renaissance Society of America. His research has focused on Renaissance humanism, rhetoric, philosophy, and Greek Renaissance émigrés. His publications include *George of Trebizond: A Biography and a Study of His Rhetoric and Logic* (1976), *Collectanea Trapezuntiana* (1984), and two volumes of collected articles: *Language and Learning in the Italian Renaissance* and *Byzantine Scholars in Renaissance Italy.* He is presently preparing a study of the Plato-Aristotle controversy of the fifteenth century as well as a study of the Greek Renaissance émigré Theodore Gaza.

Daniela Mugnai Carrara has been a fellow of the Harvard University Center for Italian Renaissance Studies at Villa I Tatti, Florence; a Jean Monnet Fellow at the European University Institute, Florence; and a CNR (Italy)/NATO Fellow at the Wellcome Unit for the History of Medicine at Oxford. She discovered the unknown manuscript of Nicolò Leoniceno's library in the Biblioteca Bertoliana of Vicenza and published it in *La biblioteca di Nicolò Leoniceno. Fra Aristotele e Galeno: cultura e libri di un medico umanista* (1991). Her main field of research is the history of the relationships between philosophy and medicine.

William R. Newman   is professor in the Department of History and Philosophy of Science at Indiana University. He has published *The "Summa perfectionis" of pseudo-Geber* (1991) and *Gehennical Fire: The Lives of George Starkey, an American Alchemist in the Scientific Revolution* (1994). His research interests include the history of alchemy and early chemistry, premodern matter-theory, early attempts to produce artificial life, and the history of the occult sciences as a whole.

Vivian Nutton   is professor of the history of medicine at University College London; since 1996 he has headed the Academic Unit of the Wellcome Institute for the History of Medicine. He has published on many aspects of the history of medicine from the Egyptians onwards, but his work has centered on Galen (his edition of Galen's *On Prognosis* appeared in 1979 and his *editio princeps* of Galen's *On My Own Opinions* is in press) and on the Renaissance. A selection of his papers, *From Democedes to Harvey,* was published in 1988; and along with his Wellcome colleagues, he has written *The Western Medical Tradition* (1995).

Katharine Park   is Zemurray Stone Radcliffe Professor of the History of Science and Women's Studies at Harvard University. She writes on the history of medicine and the life sciences in late medieval and Renaissance Europe. She is the author of *Doctors and Medicine in Early Renaissance Florence* (1985) and *Wonders and the Order of Nature, 1150–1750* (1998, with Lorraine Daston).

Nancy Siraisi   is professor of history at Hunter College and the Graduate School of the City University of New York. She has devoted her research primarily to late medieval and Renaissance medicine in its intellectual and social context. Her most recent book is *The Clock and the Mirror: Girolamo Cardano and Renaissance Medicine* (1997).

Natural Particulars

Introduction
Anthony Grafton and Nancy Siraisi

I

Recent work in the history of science has cast new light on many shadowy areas of past thought and practice. Historians have offered new interpretations of the earliest cuneiform texts in which Mesopotamian diviners used the stars to forecast the future and of the austere modern spaces in which molecular biologists analyze genes to decipher the language of life itself. Innovative books have called back to new life such understudied spaces for scientific work as the museum, the seminar, and the arsenal. New editions, translations, and commentaries have offered close readings of scientific works ranging from antiquity to recent times, and on subjects from anatomy and physiology to alchemy and astrology. Popular but undercultivated fields like the history of cartography—long the province of well-intentioned amateurs—have been settled, plowed, and tended by highly qualified specialists. Even the sales of books in history of science remain brisk, in defiance of the trends in scholarly publishing in general and history—especially European history—in particular.

These signs of growth, however, are accompanied by much more worrying signs of underlying disagreement—disagreement not only about the interpretation of individual works and careers, which will and should be with us always, but also about the basic subject matter and methods of the field. A divisive, partisan rhetoric has crept into critical discussions. Advocates of new methods have tried, in recent years, to blur the borders between the sciences now recognized and other fields of thought and action. They argue, reasonably enough, that knowledge about the natural world is always conditioned by the historical and cultural conditions within which it is framed. By precept and example they demand that historians pay attention not only to the classic texts in which the heroes of science published their results, but also to the particular contexts—social, political, and economic—in which they created them. Yet at times, they advance these claims in ways that seem to reduce the great men once primarily studied by historians of science to glib careerists whose efforts to master natural phenomena were motivated solely by a desire for prominence or money. Defenders of tradition, by contrast, insist that rigorous interpretation of texts must precede and underpin the imaginative re-creation of places of scientific work and the transactions that took place in them. Emphasizing the intellectual vitality and power of early, as well as recent, scientific work, they insist that only those who have substantial aptitude for and training in the modern sciences can study the history of science in a profound way.

More than once, the rhetoric of these debates has become divisive, even ferocious. Some of the more provocative new-style historians of science seem to argue that study of the traditional classics of science is little more than a distraction from their true task: showing that early scientists were engaged in an enterprise characteristically different, in vital ways, from what goes on in a modern university laboratory. Some of the more traditionalist scholars charge the innovators with lacking any interest in the substantial achievements of early scientists—of the methodological and substantive discoveries traditionally associated with the Scientific Revolution, for example. At times representatives of each side have accused the other of errors that vitiate their work, and even of deliberate falsification. The pages of the journals that encourage such polemics offer visions of scholarship looking less like *The School of Athens* than like a lurid nineteenth-century panorama of a battlefield, the ground littered with the bleeding bodies and fallen standards of two grappling armies. And the divisions that find expression in savage review essays are also embodied in divisions of other kinds, as conferences systematically exclude one set of voices or the other and journals dedicate themselves to the pursuit of those violent polemics that editors affect to dread but really love.

The present collection of essays is chiefly designed to present a wide range of new work on the study of nature in early modern Europe. But both it and the conference at which the papers were first read and discussed also had a larger purpose: to suggest a different point of view about the entire conflict. To put it very briefly: we believe that neither the revisionist nor the traditionalist approach sketched above does justice to the rich and complex scholarship of the last century on the study of nature in Renaissance Europe. Both approaches, we would argue, grew up simultaneously, in many branches of study. Neither of them is really so radical or traditional as their more extreme proponents seem to think. Properly applied, in fact, they do not refute but reinforce one another—as many of the essays that follow clearly show. The study of Renaissance approaches to nature, in other words, can be rewarding both as an introduction to a rich and fascinating set of texts, individuals, and historical developments and as an object lesson in the principle always insisted on by the great historian of historical thought Arnaldo

Momigliano: that the tradition of historical scholarship and writing has richer resources in it than current polemics reveal.

II

Almost half a century ago, Erwin Panofsky published his vigorous defense of the scientific originality of the Renaissance in "Artist, Scientist, Genius."[1] Few today would question the essential correctness of his view that natural philosophy and other nonmathematical scientific disciplines underwent crucial transformations in the fifteenth and sixteenth centuries. Indeed, salient developments of the Renaissance—including enhanced attention to description and depiction, both verbal and visual; the accumulation of fresh data (geographical, anthropological, zoological, botanical, anatomical); and the emergence of new social structures and environments in which the study of nature was pursued (botanic gardens, anatomy theaters, courts, museums, collecting, and artistic endeavors)—clearly transformed the study of the natural world.

But in the decades since Panofsky wrote, the historiography of Renaissance approaches to the world of nature has itself undergone dramatic transformation. Two generations of scholarship have left us considerably better informed about the sources, scope, and varieties of Renaissance thinking about nature than he could be at midcentury. Even more important, perspectives have radically shifted, boundaries have dissolved, new themes and new methodologies have emerged. Conceptual and chronological frontiers once apparently secure—"medieval science" and "the Scientific Revolution," for example—have changed out of recognition or, in the view of some scholars, disappeared altogether. The seminal studies of Michel Foucault and Frances Yates, even if not fully persuasive in every aspect, have made it impossible for historians ever again to ignore the role of various forms of magical thinking and practice in the Renaissance understanding of the natural world. The focus of inquiry has shifted away, to a considerable extent, from analysis of the content or transmission of individual major scientific texts of the period toward the identification of broader social, political, and cultural factors that shaped learning and practice. No one can any longer doubt that these new perspectives have immeasurably enriched understanding. Yet the need for textual studies has not diminished. On the contrary, in the case of many texts—some major and highly influential ones as well as a much broader range of "routine" writing about natural philosophy, medicine, and so on—in which Renaissance writers embodied their view of nature, even the most basic work remains to be done. New textual studies continue to make fundamental contributions to the field. Increasingly, moreover, such studies are themselves now informed by newer historiographic insights, as students of the classical tradition try to reconstruct not only the fortunes of individual authors but also the larger social, institutional, and architectural worlds—the schools, universities, academies, libraries, and museums—in which they were copied, studied, and applied.

Our goal in assembling the collection of essays in this volume is to present new work that exemplifies both the central themes and the important methodologies in the history of Renaissance natural philosophy and non-mathematical sciences as it is practiced today. The topics covered—Aristotelianism, Platonism, so-called new philosophies of nature, alchemy, medicine, and natural history—may seem disparate, but they reflect intellectual and disciplinary realities of the period. Four main themes, moreover, run through every section and unify the collection as a whole. The first is the exploration of specific examples of the appropriation, manipulation, and reworking of older traditions of knowledge—not only classical but also medieval—involved in so much of Renaissance innovation in theory and practice. This theme emerges, for example, with particular clarity in two essays that both deal directly with the continuing significance of Aristotelianism—and in so doing connect traditional ways of doing natural philosophy with the developing sixteenth-century taste for encyclopedic reference works of many different kinds. These two studies concern Aristotelian texts such as the *Problemata* and the *History of Animals,* long available in medieval Latin translation, which gained new types of diffusion, significance, and function in the fifteenth and sixteenth centuries. John Monfasani reveals the extensive modifications a humanist translator introduced as he rendered such works from Greek into Latin, and traces the publishing history that allowed his versions to become dominant in early editions. Ann Blair interprets the content and publication history of the *Problemata* and works based on it, in both Latin and vernacular versions, as a form of nascent "popular" science. Similarly, two essays on Renaissance Platonism by Michael Allen and James Hankins illuminate, respectively, aspects of Marsilio Ficino's original philosophical manipulations of Platonic concepts and the continued importance of the medieval Platonic tradition, characterized by study of the *Timaeus,* in early Renaissance Italy.

A second theme—illustrated by almost every essay—is the enormous impact of Renaissance humanism, and the new forms of philological scholarship and Greek learning associated with it, on the knowledge and practice of sciences of nature and mankind. During the fifteenth and sixteenth centuries, access to an enlarged range of ancient texts, ideas, or observations and the use of new humanistic translations markedly affected every one of the natural philosophical traditions and scientific disciplines discussed in this volume, with the possible exception of alchemy. Nor was this effect limited to the realm of rhetoric. Enhanced and sometimes critical attention to Greek descriptive science played an essential part in some of the most striking sixteenth-century developments in scientific activity, notably the expansion—and heightened epistemological status—of observation and description in natural history and some aspects of medicine.

A third theme is the influence of disciplinary settings in shaping the way in which Renaissance individuals were able to study nature. The changing map of the disciplines, as new ones rose and old ones altered in status or content, brought about radical shifts in the alignment and interaction of the sciences. Thus, alchemy, natural history, and medicine all intersected repeatedly in this period. To the extent that alchemy was concerned, at least theoretically, with the healing of the human body (as Chiara Crisciani's essay in this volume demonstrates), its content overlapped with that of medicine. Moreover, both disciplines combined an authoritative textual basis with a tradition of practice involving material manipulations. Natural history, too, overlapped with medicine. Interest in plants (and some animals and minerals) had to a large extent grown out of, and was often still concerned with, their therapeutic uses. But the three occupied very different disciplinary spaces. Medicine enjoyed the prestige of a higher discipline established for centuries in the university curriculum. This endorsement was never acquired by alchemy, which always remained officially outside the university, though it was often—like medicine—welcomed at courts. Natural history, by contrast with both, provides a salient example of a science in the process of carving out a disciplinary position for itself, as two essays illustrate in complementary ways. Katharine Park draws attention to early interest in the subject among scholastic physicians who produced texts that bring together medical observation (of mineral springs) with fascination with "marvels of nature"; Paula Findlen traces the emergence of a scientific community.

Finally, a fourth theme that weaves through the collection as a whole is that of the material and practical means of the dissemination of knowledge. Translation and adaptation, manuscript diffusion, publishing strategies, the popularization of scientific information, papal patronage, the spread of humanist medicine from Italy to remote parts of Europe, the exchange of letters among like-minded botanists—all these modes of dissemination crop up repeatedly in the essays in this volume. None of the authors would claim that a particular system of dissemination fully explains the success of a given text or approach; all of them would insist that the available forms of transmission and publication of knowledge, from the formal lecture to the easily copied marginal note, played a central role in shaping natural knowledge in early modern Europe.

III

At the end of the twentieth century, scholarship on the history of Renaissance natural philosophy and life sciences builds on a long history of previous endeavors. One important result has been to clarify the relation between the intellectual world of the Renaissance and that of the Middle Ages. The sharp contrast that Panofsky confidently drew between the Renaissance and all that had gone before was in part a reaction against the way that historians of medieval science had tended to dismiss the Renaissance entirely. The period between the 1920s and the 1960s was in some respects a golden age for the history of medieval science. It was certainly a time in which the subject entered the mainstream of historical inquiry in a way it had never done before. The new attention paid to medieval science came largely in response to—or reaction against—the influence of the ideas of Pierre Duhem (1861–1916). He had argued that the origins of developments in the physical and mathematical disciplines regarded as central to the Scientific Revolution lay in the late Middle Ages. George Sarton (1884–1956), one of the chief founders of the discipline of the history of science in the United States, also focused his attention on the Middle Ages. Meanwhile, the wide-ranging manuscript and bibliographical research of Lynn Thorndike (1882–1965) brought to light a vast corpus of largely unpublished and unstudied medieval writing on magical, astrological, and medical topics. In the middle years of the century, a younger generation of notable medievalist historians of science explored in detail the breadth and sophistication of fourteenth-century scholastic natural science and philosophy at Paris and Oxford. They concentrated on areas connected to the disciplines traditionally seen as central in standard accounts of the Scientific Revolution: physical science, cosmology, astronomy, and approaches to mathematization. The resulting impressive body of editions, translations, and studies revealed the medieval tradition of Archimedes and the range of the thought of Oresme, the Oxford Calculators, and many other scholastic authors as never before.[2]

To a hitherto entirely unprecedented extent, this body of work enriched understanding of medieval approaches to the world of nature. It also laid an indispensable foundation for the study of the history of erudite traditions over the following centuries. The immediate effect was, however, to direct the attention of historians of science toward thirteenth- and fourteenth-century scholasticism and away from the world of humanism. In some instances, moreover, the influence of humanism was explicitly decried as hostile to scientific ideas and activities. Thorndike, for example, militantly repudiated the entire concept of a Burckhardtian Renaissance and dismissed the impact of humanism on science as negligible or even harmful.[3] Some influential scholars framed their work in strong theses that more or less explicitly defined the inquiry as a search for the origins of Western science, which they located in the Christian Middle Ages. Thus, in a widely read interpretive survey and in specialized studies Alistair Crombie claimed to have identified an experimental tradition in the thirteenth century. He held that this tradition combined with Greek philosophy, as transformed by thirteenth-century scholastic analysis of Aristotle's *Posterior Analytics,* and with characteristics peculiar to Christianity to produce a specifically Western scientific outlook. Crombie subsequently modified but never entirely abandoned these views.[4] In the light of these attitudes and interests, neither humanism nor fifteenth- and sixteenth-century cultural developments more generally seemed topics of central interest.

Few would now maintain the strong form of the thesis of continuity between scholastic science and a clearly delimited Scientific Revolution. As the author of a recent survey of *The Beginnings of Western Science* puts it, "the more extreme claims made on behalf of medieval science and its anticipation of early modern developments are not merely exaggerated, but false."[5] Attention instead has shifted away from tracing linear development in selected sciences toward an endeavor to map the whole range of medieval and Renaissance natural knowledge. Indeed, rather than a unified history of "science," we now have a picture of many different and sometimes overlapping sciences, skills, and disciplines, developing at different rates and in different ways. At the same time, awareness has grown of the extent to which common intellectual assumptions and methods characterized the disciplines studied in the world of the medieval universities.[6] Over the last thirty years, in addition, the history of all periods of premodern science has been profoundly affected by a general shift toward attention to life sciences, to social and intellectual contexts of scientific knowledge, to connections between sciences and systems of belief, and to the cultural relations of ideas and practices about nature. The influences ultimately responsible for this shift are too varied to be summarized here, though the work of Foucault should probably be included among them. Newer work also tends to have a different geographic and institutional (or rather extrainstitutional) scope; attention is paid somewhat less to Oxford and Paris than formerly, and more to southern Europe and the world outside the universities. All these trends, by now incorporated in works too numerous to mention in an introductory essay, may be exemplified by a journal devoted to "nature, sciences, and medieval societies" launched in 1993. An editorial preface asserts the intention to bridge "sciences of nature and social history" and "the history of scientific thought and cultural anthropology."[7] Themes of successive annual issues have included the body, sciences at a royal court (that of Frederick II), alchemy, and the theater of nature. From this perspective, the development of physical and mathematical sciences appears as only one of many strands—though obviously a major and central one—in the tapestry of natural knowledge woven between the thirteenth and the seventeenth centuries.

These bodies of scholarship—at once textual and social, internalist and revisionist—ensure that we are now in a much better position than Panofsky was to appreciate the continued vitality of medieval, as well as classical, forms of erudition, not only throughout the fifteenth and sixteenth centuries, but even into the early seventeenth century. Notable studies have revealed the strength and diversity of Renaissance Aristotelianism. At the most fundamental level, new bibliographical reference tools have charted the full extent of the vast output of Aristotle editions and commentaries in the first two centuries of printing.[8] In addition, some leading scholars of medieval Aristotelianism have carried their research forward into the following period. Thus, studies by Edward Grant, whose earlier work concerned medieval physics and cosmology, trace the enduring appeal into the seventeenth century of traditional topics and methodology in debates about space, place, and the structure of the universe.[9] The diversity and vitality of Renaissance Aristotelianism are stressed in the work of Charles Schmitt and Edward Mahoney. Indeed, as Schmitt remarked, it seems preferable to speak of Renaissance Aristotelianisms in the plural, since approaches to the Philosopher between the fifteenth and the early seventeenth century were striking in their variety. Alongside the scholastic treatments of cosmology traced by Grant, humanists concentrated their attention on editing, retranslating, and commenting upon the Greek text, on integrating the full range of ancient Greek commentary into their own work, and on weeding out spuria. A lively Averroist tradition persisted even as biologists and anatomists developed a new interest in Aristotle's works on animals and in the botanical writings of his pupil Theophrastus. Both the radical naturalism of Pomponazzi and the interpretations of the Jesuit Coimbra commentators found shelter under the capacious Aristotelian umbrella. Moreover, these different versions of Aristotelianism often did not exist in isolation from each other but rather mingled and cross-fertilized. For example, Nicoletto Vernia (d. 1499), who was formed entirely in the Latin scholastic tradition, and his pupil Agostino Nifo (ca. 1470–1538), who learned Greek in middle age, both began their careers as Averroists. When they subsequently abandoned Averroës for religious as well as philosophical reasons, they developed a preference for Aristotle's Greek commentators, one of whom, Themistius, they knew in the humanist translation of Ermolao Barbaro. Subsequently, some Platonic themes also appeared in Nifo's work. Indeed, much Renaissance Aristotelianism had an eclectic cast, drawing freely on other traditions. Thus, as Brian Copenhaver and Giancarlo Zanier have pointed out, even the thought of Pomponazzi, the most radical of Aristotelians, reveals influences from Ficinian Platonism and ideas about natural magic.[10]

Like Aristotelian philosophy, Hippocratic-Galenic medicine was an intellectual tradition of long standing in the Latin West. The large body of Latin medical literature that took shape between the late eleventh and the early fourteenth century conveyed ideas and therapeutic practices derived from Hippocratic-Galenic medicine, but they were transmuted by translation and reworked by the interpretations of Arabic and Latin writers for a new social and cultural context. Medicine was endowed with a more or less standardized curriculum of authoritative texts, taught in association with scholastic Aristotelianism. When universities arose, it was accepted as one of the three higher disciplines. Scholastic physicians of the thirteenth to early fifteenth centuries subjected Hippocratic, Galenic, and Arabic works to a continuing barrage of exegesis, elaboration, and problem solving in works of traditional form, *quaestiones* and commentaries, but the medicine of the late Middle Ages also had a strongly practical side. Medicine offered the example and potential of a discipline that rested on ancient textual authority. Yet in some respects—for example, in central aspects of physiology—it offered a challenge to Aristotle. It was securely entrenched in the universities but was also practice-based, requiring attention to particulars of patients, diseases, plants, animals, and so on. The structure, methodology, and much of the scientific content of this medical system continued to underlie sixteenth-century medical teaching and learning. The Latin medical literature of the eleventh to early fifteenth centuries was widely disseminated in numerous editions during the first seventy-five years of printing. Some texts—for example, the medieval Latin translations of the principal works of the major Arabic medical authors Avicenna and Rasis, and the *Conciliator* (of medical and philosophical opinions) of the celebrated physician, philosopher, and astrologer Pietro d'Abano (d. 1315)—had a much longer printing history.[11] In sixteenth-century universities, justly celebrated new practices, new sites for teaching and learning, and newly edited and translated texts that greatly amplified knowledge of ancient medicine were more often additions to than substitutions for existing curricular arrangements.

The modern study of medieval medicine still rests in part on foundations established by medical historians of the late nineteenth and early twentieth century, most notably Karl Sudhoff (1853–1938). These scholars were primarily engaged in tracing the history of the scientific content of their own professional discipline. But over the last thirty years the history of medicine in all periods (not just premodern medicine) has become a subject of interest to social and cultural historians. At the same time, the turn in the history of medieval science toward life sciences and the social aspects of knowledge has also ensured a more prominent place for medicine.[12] Fundamental contributions of recent years have included editions and studies of major texts and their transmission as well as much archival work on social aspects of medicine.[13] The connections between medical ideas and natural philosophy have also begun to be explored. Medicine has taken its place as a key element in medieval intellectual culture as well as social practice.[14] Emergent historical themes—for example, the body, sexuality, and gender—have also stimulated new work.[15] In particular, recent studies have revealed the early involvement of medical practitioners in various peripheral activities relating to the care or investigation of the body, such as the early history of autopsy and endeavors to prolong life.[16] These activities seem highly relevant to the Renaissance expansion of the scope of medicine to intersect with or give birth to other branches of knowledge.

Medicine and alchemy intersect in various ways, for alchemy as it developed in the late medieval West was a science of life as well as of matter. But as noted above, an early established, enduring, and significant difference between the two disciplines lay in their levels of institutionalization. That alchemy was practiced outside the university and had potentially illicit aspects perhaps constrained its development in certain ways, but it may also have fostered conceptual freedom.[17] Revived interest in the history of alchemy is yet another manifestation of the broader approach to the history of medieval science characteristic of recent years. Indeed, in the view of one of the principal modern historians of the subject, "the historiography of alchemy is still in a pioneering state."[18] Studies published in the last decade substantially revise the traditional account of the development of the discipline between the reception of Arabic alchemical texts in the twelfth century and the seventeenth century. It is now apparent that the sixteenth-century reception and transformation of alchemical tradition by Paracelsus and his followers and other sixteenth- and seventeenth-century alchemists was one stage in a long history, the end of a series of earlier intellectual and textual transformations. The major text known as the *Summa perfectionis,* now securely identified as a product of the thirteenth-century Latin West, gave a central place to mercury and sulfur and developed the concept of *minima naturalia.* Its author, who wrote under the name of Geber, may have been a Franciscan friar.[19] Analysis of some of the alchemical writings attributed to Raymond Lull has contributed much new information about the fourteenth- and fifteenth-century development of an alchemy suffused with both mysticism and vitalism. In the concept of an "elixir" that would heal and preserve human bodies as well as play a part in transmutatory alchemy, ideas regarding medicine, alchemy, and the prolongation of life came together and mingled.[20] Indeed, the more alchemical processes were described in a sexualized or physiological language, the more medicine and alchemy drew on a common body of terms and ideas.[21]

In sum, much of the institutional structure and organization of knowledge within which new information, new ideas, and new disciplines burst forth in the fifteenth and sixteenth centuries had been established in the thirteenth century. The set of disciplines taught in the universities was put in place during the first century of their existence. Moreover, standard texts and curricula (for example, the division of medical teaching into separate courses in *theoria* and *practica*) also took shape in the thirteenth or early fourteenth centuries. The same may be said of teaching methods: the practice of teaching by commentary on authoritative texts and the arguing of *quaestiones* were enduring aspects of early university instruction that continued to shape the intellectual environment of the fifteenth and sixteenth centuries.[22] In due course, scholastic commentary would be replaced by humanistic commentary, and new observations would be brought to bear on old *quaestio* topics.

The establishment of natural philosophy and medicine as university disciplines ensured the continuity and respectability of systematic teaching about the natural world. In the Italian schools, the close relation between these disciplines was expressed institutionally, in the formation of student universities of "arts and medicine" and doctoral colleges of philosophy and medicine (that is, in both cases, liberal arts, natural philosophy, and medicine), as well as biographically, in the careers of numerous masters who at different times of their lives taught logic, philosophy, and medicine.[23] At the same time, other branches of natural knowledge flourished outside or on the periphery of institutionalized university teaching. And many signs show that academic and nonacademic branches of knowledge freely intersected and influenced one another. Thus, by the thirteenth century, surgery, like medicine, was equipped with substantial and authoritative specialized literature in Latin. Learned surgeons, who wrote in Latin, imitated writers on medicine in claiming for aspects of their discipline the status of *scientia* (in the Aristotelian sense of a subject in which syllogistic reasoning from generally accepted premises could lead to universal truths). Surgery was indeed occasionally taught in a university setting in Italy. More important, central concepts—about therapy as well as the nature of science—were shared between the university world and this quintessentially manual discipline. By the fourteenth and fifteenth centuries, as many scholars have shown, much scientific and technical literature with origins in the learned tradition was circulating outside the academy in vernacular form.[24] Furthermore, it was not participation in manual activities that distinguished surgeons, alchemists, and magical practitioners from the inhabitants of the university world. Whether based in or outside the universities, medical practitioners and astrologers, too, were inextricably involved with manual, practical, and technical skills.

The natural knowledge of the later Middle Ages, then, was already diverse and open to many varieties of learning and experience. Indeed, the principal trend that has been noted in the medicine of the fifteenth century is growing interest in individual cases, diseases, remedies, and events of daily life.[25] It has further been argued that a line of descent can be traced from fascination with marvels and wonders to the attention to particulars that characterizes Renaissance descriptive sciences and ultimately to the emergence of the concept of objective factual information.[26] But the line is not equally straight and full in all fields. Danielle Jacquart has drawn attention both to the increasing interest in experience of fourteenth- and fifteenth-century medical authors and to their awareness of its often problematic nature.[27] By contrast, if one may judge by such striking thirteenth-century precedents as Frederick II's *De arte venandi cum avibus* and portions of the works on animals of Albertus Magnus, writers on zoology seem to have had more confidence. Nevertheless, as Katharine Park points out in this volume, the ability of the authors of fourteenth- and fifteenth-century works on natural thermal springs to make additions to knowledge based on their own observation represents a significant step toward a new and broader natural history.

The early history of anatomical dissection—yet another area of medieval scientific endeavor that has become the subject of new attention and some revisionism—may serve as a final example, one that encapsulates much of the complicated recent history that lies behind Renaissance approaches to the world of nature. The most celebrated early practitioner of human dissection, Mondino de' Liuzzi (d. 1326), professor of practical medicine in the University of Bologna, flourished in an academic environment and produced scholastic commentaries.[28] For the most part, to be sure, dissections were undertaken infrequently and used to illustrate medical teaching based on compendia that very inadequately represented Galen's anatomical knowledge. The results bore little resemblance to the conduct, depiction, and narrative of anatomy made possible in the sixteenth century by the reception of Galen's full anatomical works, by changing artistic values, by printing, and by the investigative research of Vesalius and others. These developments are now once again a focus of attention for historians of medicine who are able to build on a substantial groundwork of technical analysis provided by historians of an earlier generation. As one might expect, the new work on Renaissance anatomy emphasizes cultural, social, and intellectual connections.[29] Academic anatomical dissections of the human cadaver were indeed only one of several practices originating in the fourteenth century or earlier that involved opening the body. Other purposes included funerary embalming, forensic investigation, and private autopsy. Yet it remains noteworthy that human dissection was institutionalized in the high scholastic period as an accepted part of the most advanced type of medical education, reviving a practice that had lapsed for a thousand years. The confidently invasive attitude toward the human body that this innovation seems to imply is well exemplified by the proem to Mondino de' Liuzzi's anatomical textbook, which prefaces instructions for dissecting the corpse of a criminal with a ringing statement about the nobility of mankind.[30]

IV

Just as Panofsky set too sharp a break between medieval and Renaissance ways of organizing and pursuing the study of nature, he also assumed too readily that Renaissance ways of studying the classics departed radically from those that had flourished in the cathedral schools and universities of medieval Europe. At the outset of his essay, he restated a principle that he had developed in the 1930s, working in collaboration with Fritz Saxl at the Warburg Institute—the interdisciplinary research institution for the study of the classical tradition founded by the art historian Aby Warburg in Hamburg, which moved to London in the 1930s with its incomparable stock of books and much of its incomparable group of affiliated scholars intact. Using classical mythology as their case in point, Panofsky and Saxl argued that medieval culture had been characterized by "the principle of disjunction." Medieval scholars knew a vast amount about the names and characters of the ancient gods, whose adventures they interpreted at length—usually as allegorical accounts of ethical principles or early human achievements. Medieval artists knew a vast amount about the forms with which Greek and Roman artists had represented the gods in sculpture and painting. But the persistent separation between the world of the intellectuals and that of the artists—and the radical lack of any sense of anachronism characteristic of all medieval thinkers—made it impossible for anyone to combine classical form with classical content in an internally consistent way.

In the Renaissance, by contrast, scholars came to see the ancient past from "a fixed historical distance"—just as artists came to see the physical world from a fixed distance. The "historical perspective" attained by men like Petrarch and Valla enabled them to see the differences between their world and the ancient one; to fuse the ancient forms once more with the beings they had represented; and, by doing so, to revive ancient culture as a whole. For Saxl and Panofsky, the fusion of the star maps transmitted over the centuries in the Islamic world with the classical images of the constellations transmitted in Latin manuscripts of the Aratea—a synthesis that took place in the schools and courts of Renaissance Italy—represented the first characteristically modern conquest of the historical world. In his essay on the Renaissance, Panofsky offered only one example: the juxtaposition of Palladio's Villa Maser with the Pantheon. But he also argued that the historical scholarship of the Renaissance humanists proved basic to the Scientific Revolution. Only the techniques of philology enabled intellectuals, for the first time since antiquity, to understand in detail the classic works of ancient science, to dismantle the frame of medieval commentaries that hung about and distorted many of them, and to identify both their strengths and their weaknesses.[31]

The tradition of research into the classical tradition that Panofsky and Saxl helped to found has flourished in the decades since. One of the greatest of the many great Renaissance scholars who came to maturity in the 1930s, Eugenio Garin, has devoted much of his career to developing similar theses about the historical revolution caused by Renaissance humanism. But the tradition has also grown in directions that these scholarly pioneers could not have predicted. Another great German-Jewish scholar of the same generation, Paul Oskar Kristeller, spent much of his remarkable career arguing that the revival of ancient learning and philosophy in the Renaissance took many forms, only a few of them governed by the philological historicism that Panofsky saw as typical of the period. Relying on a vast amount of new evidence, Kristeller argued that students of the philosophical classics often drew as heavily on the traditions of medieval as of humanistic learning. More paradoxically still, students of Plato—the preeminent scholarly rediscovery of the Italian Renaissance—often read his dialogues in a highly anachronistic way, through the interpretative screen provided by the treatises and commentaries of late antique Neoplatonists like Plotinus—whom Ficino not only studied intensively but also translated into Latin.[32] Similar arguments were advanced by other scholars of the same generation, like Raymond Klibansky, who organized a great collective project to edit the documents that attest to the continuing life of Plato in the Middle Ages.[33]

Thanks above all to Kristeller and his students, from Charles Schmitt to James Hankins, it has now become clear that there was far more continuity than Panofsky believed between medieval and Renaissance efforts to understand the classics of ancient thought about the natural world. Texts long available in medieval schools and libraries—like the translation of the *Timaeus* by Calcidius—continued to be studied by humanists from Petrarch onward, as Hankins shows in his essay in this volume. The cosmology of the *Timaeus,* he demonstrates, offered a radical challenge to the Aristotelian views of eternity, time, and nature that flourished in the medieval universities: indeed, the *Timaeus* inspired Johannes Kepler to undertake his first sketch of a radical new cosmology, the *Mysterium cosmographicum* of 1596. But this challenge represented, in some respects, less a classical revival than the revenge of the Platonist school of Chartres of the twelfth century against the new Aristotelianism of the universities.

True, recent work on the classical tradition has also emphasized, as Panofsky did, the growth of new philological techniques—and has connected these with the rise of the new technology of printing and the transformation of education that took place in the new secular schools that urban elites created at the urging of Italian and northern humanists. Hankins shows not only that the Latin *Timaeus* of the Middle Ages continued to be read but also that it came to be flanked by the Greek *Timaeus,* which Ficino, Pico, and many others read in the original, with close attention to the details of wording and argument. But more recent research has also qualified Panofsky's thesis in crucial ways. Kristeller, Garin, D. P. Walker, Frances Yates, and others have shown, for example, that the scholars of the Renaissance did not rely on the canon of pure ancient authorities that the historical scholars of eighteenth- and nineteenth-century Germany identified as authentic and reliable. The Greek dialogues ascribed to the ancient Egyptian sage Hermes Trismegistus—works really written in the first through third centuries C.E., though evidently based in part on earlier Egyptian materials—were translated by Ficino, equipped with commentaries by him and other influential scholars, and widely interpreted as the sources of Plato's dialogues.[34] The Jewish tradition of Cabalistic Bible interpretation—transmitted in texts that claimed even older origins than the Hermetic corpus—fascinated Pico, who saw Cabalistic techniques as far more ancient and profound than the philological techniques of interpretation that he learned from the humanists of his time. Brian Copenhaver's article in this volume examines this process in detail, showing how medieval traditions of reading meaning into the forms of Hebrew letters played a crucial role in Pico's development of one of his most ambitious intellectual projects.[35]

A second vital point also emerged most clearly from the work of Kristeller, Klibansky, and Walker. Renaissance students of the ancient world took not only the objects they studied but many of the methods they applied to them from what they saw as a coherent, unbroken classical tradition. The late antique Neo-Platonists, like Proclus, and the fathers of the church offered what they took as profound interpretations of Plato's dialogues and doctrines: and these later readings, which nineteenth- and early-twentieth-century scholars tended to dismiss as schematic or fanciful, often proved as vital as Plato's own work, or more so, in shaping the cosmologies of Renaissance thinkers. Both Michael Allen and Luc Deitz document this point in detail, showing how the most committed Platonists of fifteenth- and sixteenth-century Italy took vital elements of their systems from Numenius and Proclus.

These technical and internal strands of historical analysis do not represent the only legacy of the Warburg Institute. Warburg himself was as fascinated by postclassical adaptations and misuses of ancient texts and symbols as by philological efforts to understand them as they really were. He was passionately interested, for example, in the way the Florentine merchants of the later Middle Ages and the Renaissance imagined the goddess Fortuna—who became, in their eyes, a wind goddess who could fill or refuse to fill the sails of their richly loaded trading ships. Under his direction and Saxl's, the Warburg Institute devoted as much attention to the cheap German pamphlets and popular images that brought knowledge of the ancient star gods to a wide public as to the grand Italian frescoes in which their ancient forms were restored with archaeological faithfulness.[36] Renaissance and later responses to the classical tradition have proved to be as varied, and in part as wild, when studied through the Warburg Institute's multiple lenses, as those of their medieval predecessors—a point emphasized at the Warburg Institute, in recent years, by Charles Schmitt and Jill Kraye, who have richly documented the continuing use of and respect given to texts now generally dismissed by classical scholars as pseudo-Aristotelian. In dealing with the Aristotelian *Problemata* both John Monfasani and Ann Blair take one such text and its fate as their subject.

Finally, the scholars associated with the Warburg also made clear a point Panofsky omitted from his essay—that the choice of classical texts to study, and of approaches to take to them, is far from neutral. Renaissance scholars' decisions about which texts to analyze and which analytical methods to apply to them often reflected less a calm, rational study of the whole classical corpus than a particular political situation—such as the collapse of the Florentine republic, which, as Felix Gilbert, Delio Cantimori, and others showed, inspired Machiavelli and his associates in the circle of the Rucellai family to steep themselves in the study of Roman politics, using the histories of Polybius and Livy. The creation of a canon of texts could often have profound methodological and intellectual consequences, as, for example, when humanists chose Tacitus rather than Livy as their model for political analysis of past societies.[37] Paula Findlen's essay and Thomas Kaufmann's comment show that even texts as apparently innocent as the works on natural history of Dioscorides and Pliny could serve powerful political purposes—within both the narrower context of disciplinary politics and the wider one of courtly patronage.

The world of scientific knowledge and ideas described by the essays in this volume is nonetheless very different from that of the twelfth to early fifteenth centuries. It was transformed by changes in the political, social, and religious sphere, as well as the philosophical and scientific; by the invention of printing and the discovery of the New World; and, just as radically, by the challenges to the intellectual authority of texts posed by artists like Leonardo da Vinci. This volume offers only a partial introduction to the many kinds of research currently being done in Renaissance approaches to the natural world, and it examines only some of the many ways in which natural science and philosophy were transformed.[38] But it does illustrate, in a powerful sense, the fruitfulness of the disciplinary strategy that Panofsky thought characteristic of the Renaissance itself. For it shows that decompartmentalization—the breaking down of divisions between the Middle Ages and the Renaissance, textual analysis and social history, high and low culture, the university and the court—remains an effective way to attack the culture of the period, to reveal the continuing value of older traditions of analysis, and to shift the attention of historians from contemporary skirmishes to the lines where the real intellectual battles of the fifteenth and sixteenth centuries were fought.

NOTES

1. Erwin Panofsky, "Artist, Scientist, Genius: Notes on the 'Renaissance-Dämmerung'," in *The Renaissance: Six Essays* (New York: Harper, 1953), pp. 121–182.

2. The bibliography of the history of natural philosophy and branches of scientific knowledge in the High Middle Ages and Renaissance (ca. 1100–ca. 1600) that has grown up since the mid-twentieth century is far too extensive to be listed here. In this and the following notes, we confine examples to a highly selective list of authors and, in most cases,

to one work for each author: see Marshall Clagett, *The Science of Mechanics in the Middle Ages* (Madison: University of Wisconsin Press, 1959); Anneliese Maier, *An der Grenze von Scholastik und Naturwissenschaft,* 2nd ed. (Rome; Edizioni di Storia e letteratura, 1952); John Murdoch, "'Mathesis in philosophiam scholasticam introducta': The Rise and Development of the Application of Mathematics in Fourteenth-Century Philosophy and Theology," in *Arts libéraux et philosophie au moyen âge* (Montreal: Institut d'études mediévales; Paris: Vrin, 1969), pp. 215–252, Edith Dudley Sylla, "Medieval Concepts of the Latitude of Forms: The Oxford Calculators," *Archives d'histoire doctrinaire et littéraire du moyen âge* 30 (1973): 223–283; and Curtis Wilson, *William Heytesbury and the Rise of Mathematical Physics* (Madison: University of Wisconsin Press, 1956). An important feature of the work on medieval science between the 1950s and early 1970s was the publication of scholarly text editions: for example, Marshall Clagett, ed., *Archimedes in the Middle Ages,* 5 vols. in 10 (vol. 1, Madison; University of Wisconsin Press, 1964; vols. 2–5, Philadelphia: American Philosophical Society, 1976–1984); Edward Grant, ed. and trans., *Nicole Oresme and the Kinematics of Circular Motion* (Madison: University of Wisconsin Press, 1971); David C. Lindberg, ed. and trans., *John Pecham and the Science of Optics* (Madison: University of Wisconsin Press, 1970).

3. To quote one of his milder remarks on the subject; "Strange, is it not, that these medieval and scholastic centuries which were ever seeking after something up-to-date, should have been stigmatized as benighted and behind the times by subsequent historians, while the humanist reaction that followed, with its turning back to Rome and Greece, should have been hailed as the beginning of the modern mind and times!" Lynn Thorndike, *A History of Magic and Experimental Science* (New York: Columbia University Press, 1923–1958) 3:262–263.

4. A. C. Crombie, *Augustine to Galileo: The History of Science, A.D. 400–1650* (London: Falcon Press, 1952), second edition published as *Medieval and Early Modern Science,* 2 vols. (Garden City, N.Y.: Doubleday, 1959); and idem, *Robert Grosseteste and the Origins of Experimental Science, 1100–1700* (Oxford: Clarendon Press, 1953). For a detailed retrospective evaluation, see Bruce Eastwood, "On the Continuity of Western Science from the Middle Ages: A. C. Crombie's *Augustine to Galileo,*" *Isis* 83 (1992): 84–99.

5. David C. Lindberg, *The Beginnings of Western Science: The European Scientific Tradition in Philosophical, Religious, and Institutional Context, 600 B.C. to A.D. 1450* (Chicago: University of Chicago Press, 1992), p. 360.

6. For an important statement of common assumptions and methods as regards natural philosophy and theology, see John Murdoch, "From Social into Intellectual Factors: An Aspect of the Unitary Character of Medieval Learning," in *The Cultural Context of Medieval Learning,* ed. Murdoch and Edith Dudley Sylla (Dordrecht: Reidel, 1975), pp. 271–348.

7. Agostino Paravicini Bagliani, "Perchè *Micrologus?*" unpaginated preface to *Micrologus: Natura, scienze e società medievali* 1 (1993).

8. F. Edward Cranz and Charles B. Schmitt, *A Bibliography of Aristotle Editions, 1501–1600,* 2nd ed. (Baden-Baden: Koerner, 1984); Charles H. Lohr, *Latin Aristotle Commentaries,* vol. 2, *Renaissance Authors* (Florence: Olschki, 1988).

9. Edward Grant, *Much Ado about Nothing: Theories of Space and Vacuum from the Middle Ages to the Scientific Revolution* (Cambridge: Cambridge University Press, 1981); idem, *Planets, Stars, and Orbs: The Medieval Cosmos, 1200–1687* (Cambridge: Cambridge University Press, 1994).

10. See Charles B. Schmitt, *Aristotle and the Renaissance* (Cambridge, Mass.: published for Oberlin College by Harvard University Press, 1983), with remark alluded to at p. 10; Edward P. Mahoney, "Philosophy and Science, I: Nicoletto Vernia and Agostino Nifo," in *Scienza e filosofia all' Università di Padova nel Ouattrocento,* ed. Antonino Poppi (Trieste: Lint, 1983), pp. 135–202; Andrew Cunningham, "Fabricius and the 'Aristotle Project' in Anatomical Teaching and Research at Padua," in *The Medical Renaissance of the Sixteenth Century,* ed. Andrew Wear, Roger K. French, and Ian M. Lonie (Cambridge: Cambridge University Press, pp. 195–222; Brian Copenhaver, "Did Science Have a Renaissance?" *Isis* 83 (1992): 387–407; Giancarlo Zanier, *Ricerche sulla diffusione e fortuna del "De incantationibus" di Pomponazzi* (Florence: La Nuova Italia Editrice, 1975).

11. See Nancy G. Siraisi, *Avicenna in Renaissance Italy: The "Canon" and Medical Teaching in Italian Universities after 1500* (Princeton: Princeton University Press, 1987).

12. For a thoughtful and persuasive evaluation of the relation between the two disciplines, see John Harley Warner, "The History of Science and the Sciences of Medicine," *Osiris* 10 (1995): 164–193.

13. Charles Burnett and Danielle Jacquart, eds., *Constantine the African and Ali ibn al-Abbas al-Magusi* (Leiden: Brill, 1994); L. Garcia-Ballester, J. A. Paniagua, and Michael R. McVaugh, general editors, *Arnaldi de Villanova Opera medica omnia,* 7 vols. in 9 to date (Granada: Seminarium Historiae Medicae Granatensis, 1975–); Michael R. McVaugh, ed., *Guigonis de Caulhiaco (Guy de Chauliac) Inventarium sive Chirurgia Magna,* 2 vols. (Leiden: Brill, 1996–1997). Mention should also be made of the various studies by Alain Touwaide on the Byzantine Dioscorides manuscripts and tradition and of the ongoing research by Monica Green into the manuscript tradition of texts associated with the name of Trotula. A notable example of the integration of archival research into social aspects of medicine with its intellectual history is provided by Michael R. McVaugh, *Medicine before the Plague: Practitioners and Their Patients in the Crown of Aragon, 1285–1345* (Cambridge: Cambridge University Press, 1993).

14. Jole Agrimi and Chiara Crisciani, *Edocere medicos: Medicina scolastica nei secoli XIII–XV* (Naples: Guernini, 1988); Mark Jordan, "Exegesis and Argument in Salernitan Teaching on the Soul," in *Renaissance Medical Learning: Evolution of a Tradition,* ed. Michael R. McVaugh and Nancy G. Siraisi, *Osiris,* 2nd ser., 6 (Philadelphia: History of Science Society, 1990), pp. 42–61.

15. Danielle Jacquart and Claude Thomasset, *Sexualité et savoir médicale au Moyen Age* (Paris: Presses universitaires de France, 1985); Joan Cadden, *The Meanings of Sex Difference in the Middle Ages* (Cambridge: Cambridge University Press, 1993).

16. Luke Demaitre, "The Care and Extension of Old Age in Medieval Medicine," in *Aging and the Aged in Medieval Europe,* ed. Michael M. Sheehan (Toronto: Pontifical Institute of Mediaeval Studies, 1990), pp. 3–22, Katharine Park, "The Criminal and the Saintly Body: Autopsy and Dissection in Renaissance Italy," *Renaissance Quarterly* 47 (1994):

1–33; Agostino Paravicini Bagliani, *Medicina e scienza della natura alla corte dei papi del Due-cento* (Spoleto: Centro italiano di studi sull'Alto Medioevo, 1991).

17. Chiara Crisciani, "Aspetti della trasmissione del sapere nell'alchimia latina: Un'immagine di formazione, uno stile di commento," *Micrologus* 3; *Le crisi di alchimia* (1995): 149–210.

18. William R. Newman, *Gehennical Fire: The Lives of George Starkey, an American Alchemist in the Scientific Revolution* (Cambridge, Mass.: Harvard University Press, 1994), p. 92.

19. Geber, *The "Summa perfectionis" of Pseudo-Geber*, ed. and trans. William R. Newman (Leiden: Brill, 1991).

20. Michela Pereira, *The Alchemical Corpus Attributed to Raymond Lull* (London: Warburg Institute, University of London, 1989).

21. Barbara Obrist, "Alchemie und Medizin im XIII. Jahrhundert," *Archives internationales d'histoire des sciences* 43 (1993): 209–246; Chiara Crisciani, "Medici e alchimia nel secolo XIV: date e problemi di una ricerca," in *Atti del congresso internazionale su medicina medievale e scuola medica salernitana* (Salerno: Centro Studi Medicina "Civitas Hippocratica," 1994), pp. 102–118.

22. Brian Lawn, *The Rise and Decline of the Scholastic "Quaestio Disputata"; With Special Emphasis on Its Use in the Teaching of Medicine and Science* (Leiden: Brill, 1993).

23. Nancy Siraisi, *Taddeo Alderotti and His Pupils: Two Generations of Italian Medical Learning* (Princeton: Princeton University Press, 1981), offers some examples.

24. See, for example, Monica Green, "Obstetrical and Gynecological Texts in Middle English," *Studies in the Age of Chaucer* 14 (1992): 53–88. Extensive work has been done on German vernacular medicine by Gundolf Keil and many others.

25. See, for example, Danielle Jacquart, "Theory, Everyday Practice, and Three Fifteenth-Century Physicians," in McVaugh and Siraisi, *Renaissance Medical Learning*, pp. 140–160.

26. See Lorraine Daston and Katharine Park, *Wonders and the Order of Nature, 1150–1750* (New York: Zone Books, 1998).

27. Danielle Jacquart, *La médecine médiévale dans le cadre parisien* (Paris: Fayard, 1998), pp. 415–432.

28. On Mondino, see now Romana Martorelli Vico's introduction to her edition of Mondini de Leuciis, *Expositio super capitulum De generatione embrionis Canonis Avicennae cum quibusdam quaestionibus*, Fonti per la Storia d'Italia (Rome: Istituto Storico Italiano per il Medio Evo, 1993), and the introductory material in Mondino de' Liuzzi, *Anothomia*, ed. Piero P. Giorgi and Gian Franco Pasini (Bologna: Istituto per la storia dell' Università di Bologna, 1992).

29. Two noteworthy studies are Andrea Carlino, *La fabbrica del corpo: Libri e dissezione nel Rinascimento* (Turin: Einaudi, 1994), translated as *Books of the Body* (Chicago: University of

Chicago Press, 1999); and Giovanna Ferrari, *L'esperienza del passato: Alessandro Benedetti filologo e medico umanista* (Florence: Olschki, 1996). A special issue of the *Journal of the History of Medicine and Allied Sciences* has also recently been devoted to early anatomy; see vol. 50, no. 1 (January 1995).

30. Mondino, *Anothomia,* pp. 100–104.

31. See Erwin Panofsky and Fritz Saxl, "Classical Mythology in Mediaeval Art," *Metropolitan Museum Studies* 4 (1933): 228–280.

32. See the classic studies by Paul Oskar Kristeller collected as *Renaissance Thought and Its Sources,* ed. Michael Mooney (New York: Columbia University Press, 1979).

33. Raymond Klibansky, *The Continuity of the Platonic Tradition during the Middle Ages* (London: Warburg Institute, 1939).

34. See the classic survey of Frances A. Yates, *Giordano Bruno and the Hermetic Tradition* (Chicago: University of Chicago Press, 1964). The large and lively controversial literature on Yates (and Hermes Trismegistus) is best approached through the very well-informed introduction in *Hermetica; The Greek "Corpus Hermeticum" and the Latin "Asclepius," in a New English Translation,* ed. trans. Brian P. Copenhaver (Cambridge: Cambridge University Press, 1992). A powerful dissenting view is that of Paola Zambelli, *L'ambigua natura della magia* (Milan: Il Saggiatore, 1991).

35. See the great work of Chaim Wirszubski, *Pico della Mirandola's Encounter with Jewish Mysticism* (Cambridge, Mass.: Harvard University Press, 1989). New work on the diffusion of the Cabala in Christian circles is owed above all to Moshe Idel; see, e.g., his foreword to Johann Reuchlin, *The Art of the Kabbalah: De arte cabalistica,* trans. Martin and Sarah Goodman (Lincoln: University of Nebraska Press, 1993).

36. Aby Warburg, *Ausgewählte Schriften und Würdigungen,* ed. Dieter Wuttke, 2nd ed. (Baden-Baden: Koerner, 1980); Fritz Saxl, *Lectures,* 2 vols. (London: Warburg Institute, University of London, 1957).

37. See, e.g., *Tacitus and the Tacitean Tradition,* ed. T. J. Luce and A. J. Woodman (Princeton: Princeton University Press, 1993).

38. For other views of Renaissance approaches to the natural world, see two other recent collections: *Renaissance and Revolution; Humanists, Scholars, Craftsmen, and Natural Philosophers in Early Modern Europe,* ed. J. V. Field and Frank A. J. L. James (Cambridge: Cambridge University Press, 1993), and *Reading the Book of Nature: The Other Side of the Scientific Revolution,* ed. Allen G. Debus and Michael T. Walton (Kirksville, Mo.: Sixteenth Century Journal, 1998).

I

NATURAL PHILOSOPHIES

1

NUMBER, SHAPE, AND MEANING IN PICO'S CHRISTIAN
CABALA: THE UPRIGHT *TSADE,* THE CLOSED *MEM,* AND
THE GAPING JAWS OF AZAZEL

Brian P. Copenhaver

I

The last of the nine hundred *Conclusions* that Giovanni Pico della Mirandola
planned to defend in Rome in 1487 is also the last of seventy-two Cabalist
conclusions that Pico meant to represent his own opinion, as apart from the
"teaching of the wise Hebrew Cabalists" that he interpreted in forty-seven
other conclusions presented earlier. In this last of all his theses, Pico claimed
that "just as the true astrology teaches us to read in the book of God, in the
same way Cabala teaches us to read in the book of the Law."[1] In what way
does Pico's Christian Cabala teach us to read the book of the Law—the
Torah? And how is that way like the reading of nature—the book of God—
taught by the true astrology? I wish to frame an approach to these questions
and to offer a few answers. Some will be conjectural—my speculations about
Pico's esoteric intentions. Whatever its value, my understanding of Pico's
Cabala depends above all on the superb book by Chaim Wirszubski, completed by Moshe Idel and published in 1989, twelve years after Wirszubski's
death.[2]

Jacob Burckhardt, publishing his essay on the Renaissance in 1860,
ended his memorable chapter on "the discovery of the world and of man"
with an even more striking passage from the first part of Pico's famous *Oration,* written to introduce the *Conclusions.* There, in Burckhardt's words, Pico
told how God had "made man . . . to know the laws of the universe, to love
its beauty, to admire its greatness." Later, in the part of his somber chapter on
"morality and religion" that deals with "the influence of ancient superstition," Burckhardt again praised Pico for resolving the problem of universal law in favor of human freedom by rejecting astrology. "Pico . . . made
an epoch in the subject by his famous refutation," wrote Burckhardt. "His
main achievement was to set forth . . . a positive Christian doctrine of the freedom of the will and the government of the universe. . . . The first result
of his book was that the astrologers ceased to publish their doctrines." Having inflated the contemporary effect of Pico's *Disputations against Astrology,*

Burckhardt promoted Pico's modern fame by making him a hero of a progressive Renaissance and a forerunner, at least, of the advancement of science.[3]

To love science, Pico, and the Renaissance all at once became harder for later historians, but in 1927, when his *Individuum und Kosmos* appeared in the Studies of the Warburg Library, Ernst Cassirer was no less emphatic than Burckhardt about the impact of Pico's *Disputations:* "With one blow, he destroys the sphere of influence of astrology," he claimed. And yet this achievement "must at first seem a strange historical anomaly, for [Pico's] . . . philosophy of nature as well as his philosophy of religion is strongly influenced by magical and cabbalistic thought." Can the right-thinking scourge of the stargazers have been a woolly-brained wizard as well? To resolve this puzzle, Cassirer looked for "independent inner forces . . . based not on Pico's view of nature, but on his total *ethical* view. . . . His treatise attacking astrology builds upon this foundation of ethical humanity," wherein the natural world lies open to a free human spirit without finally confining it, where "turning towards the . . . cosmos always implies the ability not to be bound to any one part." This moral freedom to explore nature, argues Cassirer, has its consequences in method and epistemology, requiring Pico to "distinguish the form of mathematical-physical causality from astrological causality. The latter is based upon . . . occult qualities; the former is satisfied with . . . experience. . . . Phenomena are to be understood by their *own* principles. . . . Whatever the heavens contain of *real* influences . . . consists simply of *light* and *heat.* . . . [Pico] has discovered . . . nothing less than the concept of *vera causa,* which Kepler and Newton will later embrace."[4]

In 1934, in the fourth volume of his monumental *History of Magic and Experimental Science,* Lynn Thorndike contradicted Cassirer and Burckhardt: "The importance of Pico della Mirandola . . . has often been grossly exaggerated," he complained. "The darling of enthusiasts for the so-called Italian Renaissance, his reputation must decline with its. . . . Pico's numerous theses involving astrology, divination, and demonology, . . . magic, Orphism and the cabala . . . suffice to show where he stood with respect to magic and experimental science." In the same volume that spends two chapters dismissing Pico, Thorndike gave only one to Marsilio Ficino, titling it "Ficino the Philosophaster." Hence, his reaction to Pico's earlier career, his dismissal of the "loose eulogies of the sublimity of his mind," is unsurprising. Noting that "Pico's later attack on astrology to some extent represented a right about face from his earlier attitude," Thorndike acknowledged in the *Disputations* his "wide, if not exhaustive, acquaintance with the past literature." Then, having criticized its first book "as a hasty, superficial, rhetorical, and inadequate performance," he described the eleven other parts of this unfinished work as "rambling and ineffective."[5]

Another product of the 1930s was Eugenio Garin's early work on Pico in the book that remains the starting point for any serious treatment of this important thinker. Up to a point Garin followed Burckhardt and Cassirer in crediting the anti-astrological Pico with "a notable methodological contribution to the astronomical sciences . . . [and] an attempt at a natural philosophy . . . foreshadowing, as some have claimed, the science of Galileo, Kepler and Newton." But he was less inclined to detach Pico's occultism from the rest of his achievement or to disconnect it altogether from the *Disputations*. What the friend of magic and the enemy of astrology had in common was a rational philology.[6] "Both cabbala and magic are closely related to humanistic 'philology,'" wrote Garin in a later work, locating "the deepest meaning" of the humanist enterprise in "the search for a contact with 'nature,' the latter . . . understood, in a polemical sense, as . . . opposed to the crystallization of a tradition." Garin also described Pico as having reinterpreted "the very ancient doctrine that the universe was a grand book of great originality, . . . [and as having thereby] demonstrated how historical-philological research coincided with the investigation of nature."[7] I find Garin's approach to Pico's thought more plausible than Thorndike's medievalist polemic or the panegyrics of Burckhardt and Cassirer, all of which abandon the possibility that his investigations of Cabala as much as his criticisms of astrology might bear on the history of science.

Pico's Cabala, in fact, was more original than his *Disputations*. His Cabalist conclusions reveal—or rather conceal—a remarkable effort to find new tools for understanding nature, seen as God's creation. On this point, Frances Yates extended Garin's insights in her famous book of 1964, *Giordano Bruno and the Hermetic Tradition*, which contains a long chapter titled "Pico della Mirandola and Cabalist Magic." To Pico and also to Ficino, Yates ascribed a "Hermetic attitude toward the cosmos and toward man's relation to the cosmos," calling it "the chief stimulus of that new turning toward the world and operating on the world which, appearing first as Renaissance magic, was to turn into seventeenth-century science." She also attributed the start of a "Hermetic-Cabalist tradition" to Pico and stressed the statue-magic of the Hermetic *Asclepius* as the source of Pico's exaltation of the magus who uncovers the world's secrets in order to apply them actively. Inasmuch as Pico's (and Ficino's) natural magic was a theory meant to have useful effects, Yates was right—in some general sense—to emphasize the influence of their refined philosophizing on the status of magic and thence on the development of natural-philosophical (though not yet "scientific," in our terms) theory as a basis for practical application.[8]

Yates made much—correctly so—of Pico's having started his *Oration* with the ringing words of the *Asclepius:* "Man is a great miracle."[9] But she made nothing of other facts about Pico's use of the *Hermetica,* particularly in relation to his Cabala. Pico mentions Hermes or the Hermetic *Asclepius* only twice in the *Oration:* once in his grand opening, and again in a roster of authorities on the "ancient theology" that says nothing about magic. A third piece of Hermetic material comes uncited from the *Asclepius,* in a passage that urges the cleansing of the lower soul by moral philosophy, the purging of "the whole sensual part into which the lure of the body settles and ties the soul up like a noose round its neck."[10] Human effort has its place in the regimen of the *Oration* for "emulating a Cherub's life on earth"; it is the beginning—but not the end—of an *askēsis* that leads through moral philosophy, dialectic, and natural philosophy to theological contemplation and mystical union. The transit from world-engaging to world-escaping *gnōsis* that Pico makes into a spiritual curriculum is more orderly than the jarring movements of the Hermetic treatises from cosmic optimism to bleak metaphysical pessimism, but it is no less remote from any post-Baconian sense of material progress through scientific inquiry.[11]

Yates recognized such contradictions in the Hermetic texts, but if she was conscious of the scarcity of magic in the Greek *Hermetica* translated by Ficino, it did not deter her from treating Pico's natural magic as both "Hermetic" and "Cabalist." Actually, in the part of the *Oration* devoted to magic Pico mentions many authorities—twenty or more, from Apollonius to Zoroaster. But Hermes is not among them, and for good reason.[12] Yates claimed that the magic in the *Oration* comes from the *Asclepius,* but the demonolatry of its celebrated god-making passages is just what Pico took pains to disavow by distinguishing natural magic from demonic. Better suited to his pious and eirenic aims were Plotinus, Porphyry, Iamblichus, Proclus, and the Chaldaean *Oracles,* sources from which it was possible to extract—as Ficino would in the third of his *Three Books on Life* and elsewhere—a *philosophical* theory of *natural* magic.[13] The small usefulness to Pico of the *Hermetica* is also apparent in the *Conclusions,* of which only ten of nine hundred are ascribed to Mercurius Trismegistus. One of these deals with divination, but none with any other natural magic and only one expressly with Cabala. Both for magic and for Cabala, the *Orphica* interested Pico more than the *Hermetica.* "Nothing is more effective in natural magic than the hymns of Orpheus," he wrote, adding that "just as David's hymns are wonderfully helpful to the work of Cabala, so are the hymns of Orpheus to the work of truly licit and natural magic."[14]

Thus, there is more evidence of an "Orphic-Cabalist" than of a "Hermetic-Cabalist" Pico, but there are better places than either of these ancient gentile theologies to begin an inquiry about his Cabala and natural philosophy. Garin claims, for example, that Pico was much taken with the "ancient doctrine that the universe was a grand book of great originality." Agreeing with Garin, I believe this explains why Pico put the very last of his nine hundred *Conclusions* in so conspicuous a place: he wanted to draw attention to a belief of great importance to him and to his teacher, Yohanan Alemanno—that "just as the true astrology teaches us to read in the book of God, in the same way Cabala teaches us to read in the book of the Law."[15]

Our strongest memory of the injunction to read nature as a book written by God comes, of course, from Galileo, who in one place pointed to the "hundred texts from Sacred Scripture that teach us how the glory and grandeur of Almighty God are . . . divinely read in the open book of heaven" and in another described philosophy as "written in this greatest book of all, I mean the universe, that always stands open before the eyes but cannot be understood unless one first learns to understand the language and read the characters (*caratteri*) in which it is written, . . . the language of mathematics, . . . [whose] characters are triangles, circles, and other geometrical figures (*figure*) without which it is impossible for humans to understand a word of it."[16] Galileo's success as a grammarian of this language inspired Descartes and his successors: so effective were they in combining the figurative force of geometry with the analytic power of algebra to form a new mathematics capable of extending Galileo's physical insights that we may forget how dubious this project seemed to some contemporaries, especially those who shared the common Aristotelian disdain for mathematics as an instrument of physics. One such adversary of the new science, for example, astounded Descartes with the accusation that his coordinate geometry was really a kind of magic. Since this charge focused on the use of shapes in physical explanation, Descartes replied that if his critic were right, then "a key, a sword, a wheel and all other objects whose effects depend on shape are . . . tools of magic."

Whether Descartes knew it or not, he was involved in a very old quarrel about the reputedly magical effects of pictorial shapes carved on natural objects, especially astrological talismans. By finding Aristotelian principles that might account for such effects, Aquinas had fueled a debate that still burned in the sixteenth and seventeenth centuries. Fearing that a quantitative mechanics based on shape would destroy the qualitative physics of the schools, Descartes's opponent tried to discredit him by treating shape as something that would interest only a magician.[17] But this attempt to smear Descartes would apply better to Pico, as his enemies understood. Pedro Garcia, one of the six bishops appointed in 1487 by Pope Innocent VIII to examine the *Conclusions,* later wrote a lengthy refutation of the material from Saint Thomas that supports Pico's views on the role of shape (*figura*) in natural magic.[18] Bishop Garcia could not know, of course, that Pico's wish to use shape and number to understand nature, a desire reflected in the Cabalist *Conclusions,* would one day find a more threatening expression in the Cartesian repudiation of traditional natural philosophy.

II

The celebrated yet elusive speech called the *Oration on the Dignity of Man*—though not so called by its author—introduced the even less accessible *Conclusions,* whose crude Latin, scholastic form, seeming incoherence, and recondite sources have delayed the study or even the editorial inquiry that so important a work by so famous a figure usually invites. As far as I know, the recent Italian translation with Latin text by Albano Biondi is the first complete version published in any modern language. The Latin edited by Kieszkowski is unusable, so this puzzling document still awaits a critical edition.[19] In these circumstances, it will be useful to survey the form and content of the *Conclusions* before descending to particulars. Although Pico meant to defend nine hundred theses, a sample of thirty-two will suffice to show that natural philosophy, cosmology, and mathematics were prominent among the topics that he planned to dispute. The general subject of natural philosophy; the relation of physics to metaphysics, of theory to practice, of mathematics, psychology, and magic to natural philosophy; the nature of number; the cause of motion; the substance of the heavens; the ensoulment of heavenly bodies; the presence of intelligences in the heavens; the action of higher things on lower; motion, matter, body, vacuum, elements, qualities, stars, spheres, sun, moon, planets, heat, light, sensation, vision, optics, perspective: all these interest Pico as he reads the book of nature through the book of Scripture.[20]

The twenty-one conclusions following come from the first 402, the group that Pico does not claim as his own (the name in parenthesis is the source of the conclusion; the number is Pico's number, by section):

1. "The moving body is the subject of natural knowledge." (Albertus: 14)

2. "Besides the thing that generates or the thing that removes the obstacle, no mover moves heavy bodies and light bodies." (Thomas: 37)

3. "Anything can move itself from virtual act to formal act." (Scotus: 18)

4. "If a vacuum exists and anything in it is moved, it will be moved instantaneously." (Giles of Rome: 11)

5. "The heavens are a simple body, not a composite of matter and form." (Averroës: 9)

6. "In the heavens is matter of the same type as the matter of lower bodies." (Avicenna: 3)

7. "Bodies from the heavens bestow nothing formally on lower things except heat." (Isaac of Narbonne: 4)

8. "The heavens make lower things hot through their light that falls upon them." (Abu Marwan: 3)

9. "If the heavens were without soul, they would be less noble than any ensouled body, but to say this is wicked in philosophy." (Theophrastus: 1)

10. "Besides the soul that moves it as efficient cause, there is a special intelligence attending any part of the heavens and moving it as final cause; in substance this intelligence is completely distinct from such a soul." (Alexander of Aphrodisias: 2)

11. "In Themistius the agent intellect that only illuminates is the same thing as Metatron in Cabala, I believe." (Themistius: 2)

12. "Knowledge of the soul is midway between natural knowledge and divine." (Themistius: 3)

13. "Soul, when it descends, does not descend as a whole." (Plotinus: 2)

14. "The soul is the source of motion and the mistress of matter." (Adelard of Bath?: 7)

15. "The craftsman of the world is the hypercosmic soul." (Porphyry: 2)

16. "In the stars of the heavens there is no power as such to do evil." (Iamblichus: 7)

17. "There are four troops of younger gods: the first dwells between the first heaven and the beginning of air; the second from there to the middle of the air; the third from there to the earth." (Proclus: 36)

18. "Through the secret of the straight, reflected, and refracted ray in the knowledge of perspective, we are reminded that nature is threefold—having mind, soul, and body." (Pythagoras: 7)

19. "Whatever there is from the Moon upward is pure light, and that is the substance of the cosmic spheres." (Chaldaean Theologians: 6)

20. "All that is moved is bodily; all that moves is not bodily." (Hermes Trismegistus: 2)

21. "When the light of the mirror that does not shine is made like that of the mirror that shines, then, as David says, night will be like day." (Cabalists: 20)[21]

Eleven more conclusions are from the 498 that Pico labeled as representing his "own opinion."

22. "Even though their language differs, Averroës and Avicenna cannot disagree fundamentally on whether the natural philosopher accepts from the metaphysician that body is composite." (Conciliating Conclusions: 16)

23. "The way that lower things are made hot by higher things, as given by Aristotle, seems by no means correct." (Dissenting Philosophical Conclusions: 42)

24. "Given any practicable object, the operation that practices it is nobler than the one that contemplates it, other things being equal." (Innovating Philosophical Conclusions: 46)

25. "The sense posited in nature by Alkindi, [Roger] Bacon, William of Paris, and others, but especially by all the Magicians, is nothing other than the sense in the vehicle [of the soul] that the Platonists propose." (Platonic Conclusions: 45)

26. " . . . to turn more to sense than to intellect belongs to the soul not as soul but as falling." (Conclusions from the *Liber de causis:* 9)

27. "Just as Aristotle's claim about the ancients—that they went wrong in thinking about nature because they treated physical objects mathematically—would be true if the ancients had understood mathematicals materially and not formally, likewise it is entirely true that the moderns who dispute mathematically about natural objects demolish the foundations of natural philosophy." (Mathematical Conclusions: 5)

28. "Whether all things are written out in the heavens and shown through signs to anyone who knows how to read." (Questions to be Answered through Number: 74)

29. "In the same place by the roots of the earth they can mean nothing but vegetal life, in conformity with the sayings of Empedocles, who attributes metempsychosis even to plants." (Zoroastrian and Chaldaean Conclusions: 4)

30. "Magic is the practical part of natural knowledge." (Magical Conclusions: 3)

31. "Through the ogdoad of maritime hymns the property of bodily nature is marked out for us." (Orphic Conclusions: 12)

32. "Just as the true astrology teaches us to read in the book of God, in the same way Cabala teaches us to read in the book of the Law." (Cabalist Conclusions: 72)[22]

This final proposal to read both the heavens and the Law as divine writing takes us to Psalm 19, a textually vexed piece of Scripture that treats nature as endowed with speech. Its Vulgate version can be rendered as follows:

> The heavens tell out the glory of God, and the firmament announces his handiwork.
> One day casts his word to the next, and night discloses knowledge to night.
> There is no language and no speech in which their voices are not heard.

> Their sound has gone out into all the land, and their words into the ends
> of the earth.

Jacques Lefèvre d'Etaples, a contemporary of Pico, spotted the problem in
verse 3, whose Hebrew original (probably a gloss) says "There is no speech
and there are no words, their voice is not heard" instead of "There is no language and no speech in which their voices are not heard." Hence, a modern
English version of the Psalm:

> The heavens tell out the glory of God, heaven's vault makes known his
> handiwork.
> One day speaks to another, night to night imparts knowledge,
> And this without any speech or language or sound of any voice.
> Their sign shines forth on all the earth, their message to the ends of the
> world.

According to the Vulgate, God speaks so powerfully in the Heavens that his
celestial voice fills all other discourse; but closer attention to the Hebrew suggests that God's heavenly speech speaks without speaking.[23] This latter, paradoxical reading of Psalm 19, open to anyone who looked at the Hebrew,
was known to the author of the *Conclusions,* as one learns from a fragment of
his commentary on it, though the fragment lacks the greater part of the comment that would have dealt with the Hebrew text. From what remains, it
seems that Pico preferred the usual Latin version, wherein "the firmament
everywhere expresses the works of God and the heavens make visible his
glory."[24]

The specifically magical power of divine speech in Hebrew appears at
several points in Pico's "Twenty-six Magical Conclusions according to His
Own Opinion":

19. Speech and words (*uoces et uerba*) are effective for doing magic because it is
the speech (*uox*) of God through which nature first works magic.

20. Any speech (*uox*) has power in magic insofar as it is formed by the speech
(*uoce*) of God.

21. Speech (*uoces*) that does not signify (*non significatiuae*) can do more in magic
that speech that signifies (*significatiuae*), and a person of depth can understand the
basis of this conclusion from the preceding conclusion.

22. No names (*nomina*) taken as signifying (*significatiua*), and insofar as they are
names (*nomina*) individually and in themselves, can have power for doing magic
unless they are Hebrew or closely derived from it.[25]

These four conclusions follow a dense description of magic whose aphoristic form masks its coherence. Pico begins by admitting that the Church is

right to ban the common magic of his time because it depends on demons. His own magic, by contrast, is lawful. He calls it "natural magic" or the "practical part of natural knowledge," claiming for it both theoretical foundations and pragmatic results. It works by uniting and thereby activating forces otherwise dispersed through heaven and earth, and God's grace is its ultimate origin.[26]

Pico links his magic to Cabala, another source of the wonders whose first author is God, but then he takes pains to distinguish the miracles of Christ from the effects of magic and Cabala both. Christ's miracles prove his divinity not because unusual things happened but because they were made to happen in a certain way. Having asserted this difference between a wondrous event and its divine agency, Pico reached one of the conclusions condemned by the Church; "no knowledge gives us more certainty of Christ's divinity than magic and Cabala." Moreover, no magic works without Cabala, whose effects reach beyond magic if its use is pure and direct.[27] Pico also claims that language is magically effective because "it is the speech of God through which nature first works magic." The divine words that spoke nature into being are the primal text behind any righteous language whose effect is magical. Recognizing God's speech as the basis of magical language, Pico then makes the odd claim (discussed below) that "speech that does not signify can do more in magic than speech that signifies." Moreover, names taken as names and as signifying have no magic unless they are Hebrew or derived from Hebrew.[28] However, "characters and figures" as well as names, words, and sounds have magical power. In fact, "characters and figures can do more in a work of magic than any material quality." Characters do for the magus what numbers do for the Cabalist. Pico also mentions a "magical arithmetic" and suggests that numbers have a role in magic.[29]

III

Like the magical group, the Cabalist conclusions that Pico called his own begin by making distinctions, first between speculative and practical Cabala, then among four types of the speculative kind. Speculative Cabala is about the *Sefirot,* the ten emanations of supernal divinity, often represented as a "tree" (figure 1.1). Pico next says that the first division of speculative Cabala is "revolving the alphabet" and that the three other parts form "the triple Merchiava . . . dealing with divine, intermediate, and sensible natures."[30] What Pico means by "revolving the alphabet" would have been clear to any reader of late medieval commentaries on the *Sefer Yezirah* or *Book of Formation,* a work of the sixth century or earlier that derives the structure of the

Crown

Intelligence   Wisdom

Power   Greatness

                        Beauty

Majesty   Endurance

                Righteous

                Kingdom

Figure 1.1
The ten *Sefirot.*

BuYu BaYa BiYi BeYe BoYo   בֹּי   בֶּי   בִּי   בָּי   בְּי

BuHu BaHa BiHi BeHe BoHo   בֹּה   בֶּה   בִּה   בָּה   בְּה

BuVu BaVa BiVi BeVe BoVo   בֹּו   בֶּו   בִּו   בָּו   בְּו

Figure 1.2
Revolving the alphabet.

world from the letters of the Hebrew alphabet. Seeing the letters as the roots of the universe, Jewish mystics contemplated their permutations and combinations with one another and with the names of God in rituals of meditative ecstasy, as in figure 1.2 which shows variations on the letter *bet* (ב) with the sequence of vowels and with the consonants of the divine name *YHWH* (יהוה). The term *Merchiava* is Pico's transliteration of מֶרְכָּבָה (*merkabah*) or "chariot," naming a line of esoteric interpretation that had long been applied to the first chapter of Ezekiel. For Maimonides, *Ma'aseh Merkabah* or the Work of the Chariot had meant metaphysics as a study higher than physics, which was called the Work of the Beginning or *Ma'aseh Bereshit* and focused on the opening of Genesis. By a threefold *Merkabah,* Pico meant a division of the ten *Sefirot* (see figure 1.1) into triads corresponding to higher, middle, and lower levels of physical and metaphysical structure. Abraham Abulafia, a mystic of the thirteenth century who greatly influenced the original parts of Pico's Cabala, wrote a commentary on Maimonides that connects the metaphysical Work of the Chariot with ecstatic alphabetic meditations based on the *Sefer Yezirah*. This is the main burden of Pico's speculative Cabala: revolving the letters of the Torah and contemplating the world of the *Sefirot* in a quest for ecstasy.[31]

Pico's third Cabalist thesis defines practical Cabala as the effort "to make practical the whole of formal metaphysics and of lower theology." If the metaphysics of forms is about the *Sefirot* and if lower theology is about divine attributes inferred from nature, then Pico's view of Cabala was like that of Abulafia, who was greatly concerned with techniques of alphabetic manipulation as an ecstatic practice whose theoretical principles implied a sophisticated metaphysics and theology. Since Abulafia's technique of revolving or combining letters assumed this speculative context, we may understand that Pico's initial (but fluid) division of Cabala into a practice dealing with names and a theory about *Sefirot* fits the same framework. We may also infer that what Pico did not mean by practical Cabala was the practice that Abulafia had condemned in the medieval magicians called *Ba'alei Shemot,* the Masters of Names who used the names of God to manipulate angelic and demonic powers for good or ill.[32]

In the *Oration,* Pico had distinguished the written Torah that God gave to Moses from its secret interpretation, originally unwritten but later compiled in seventy books of theology, metaphysics, and natural philosophy. Coming upon Pico's account of this Cabala as confirming trinitarian and other Christian doctrine, a Christian reader of the *Oration* would have had few clues to Pico's real meaning at this point and small hope of understanding the conclusions on Cabala. Although medieval Latin texts written by converted Jews—of which Raymond Martini's *Pugio fidei* was the most famous—had long since exploited Talmudic and Midrashic materials, their purposes were narrowly and aggressively apologetic and they were not informative about Cabala. Having been instructed by Elijah Delmedigo, Flavius Mithridates, Yohanan Alemanno, and other experts on Cabala, Pico was the first person born a Christian to take it seriously and apply it in creative ways; but his *Conclusions* could only elude most contemporaries—which suited his esoteric aims if not his eventual need to justify the ways of Cabala to Rome.[33]

For that purpose, the *Apology* that Pico wrote to defend the condemned conclusions stresses Cabala's distance from demonic magic and admits its strangeness to Christian eyes. Faced with hostile ignorance, Pico presented Cabala as mainly the same secret tradition of scriptural interpretation described in the *Oration* and now deployed through his *Conclusions* "for the confirmation of our faith against the Jews."[34] But since any mystery kept hidden might be treated as a Cabala, the Jews had used the term in this broader but secondary sense, in two applications especially: one was a method of combining letters resembling the Art of Raymund Lull; another was the study of supralunar things, which is the apex of natural magic. Explaining that his conclusions refer to both types of secondary Cabala, Pico divides the latter again, allowing that the study of higher things might or might not be part of natural magic. Then he says that this secondary study, detachable from natural magic, helps us know Christ's divinity. With or without natural magic, this good Cabala is entirely distinct from still another, evil type, related to Pico's Cabala as necromancy is to natural magic. Evil Cabalists use "the secret names of God . . . to bind demons and do wonders," claiming that Christ worked his miracles in the same way. The *Apology* repeats Pico's disavowal of the demonic magic that the Church forbids, approving only a magic that can be called natural because it works by joining natural forces together in order to make them active, in the way that the placement of a leonine stone and a leonine plant under the influence of the constellation Leo will concentrate forces otherwise dispersed through this series.[35] Pico then locates "characters and figures" within the safe limits of natural action because they share the active power attributed by Pythagorean philosophers to mathematical entities. Unlike the unmathematical Aristotelians, the Pythagoreans argued that "just as mathematicals are more formal than physical entities, so are they more active."[36]

IV

Pico's *Apology* made explicit what his *Conclusions* implied by connecting Cabala with natural magic: good practical Cabala cannot be a demonic magic of divine names. But Pico also professed a magic of "characters and figures" whose linguistic peculiarities were highlighted by their association with Cabala. Characters are conventional signs—sometimes of astrological, alchemical, or angelic significance, sometimes without apparent meaning—used since ancient times to mark texts, talismans, or other materials intended for magical use (figure 1.3).[37] Unlike numerals, characters are not used for counting, and unlike letters they do not form words. Their efficacy in magic comes from their shape, without regard for ordinary meaning, verbal or numerical. Pico put shaped signs without such meaning on the safe side of the divide between natural and demonic magic, the same division that distinguished classes of magical objects for earlier thinkers. While an amulet is a natural object—a stone, a plant, any piece of matter—worn on the body to ward off harm, natural or supernatural, a talisman is an amulet marked with

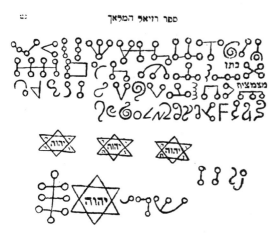

Figure 1.3
Characters from the *Sefer Raziel*.

a sign meant to fortify its power. Augustine had condemned simple amulets because he thought them, even unmarked, to be evidence of pacts with demons. But Albert and other Christian authorities accepted medical amulets as natural objects involving no commerce with demons. On this looser construction, an amulet that bears no sign sends no message to the beings who can receive messages—other persons, whether human or superhuman.[38]

Communication with nonhuman persons—God, angels, and demons— is the Church's preserve. The Church approves certain rites but not others, distinguishing piety from superstition, illicit talismans from relics and other lawful tokens of grace. Outside the bounds of liturgy and law, to talk with good or bad spirits is to sin against religion. Marsilio Ficino knew this, yet he also believed that stones marked with astrological signs were medically useful. Inspired by Proclus and Aquinas, he tried to solve the problem by treating astrological signs on gems as causes of membership in a particular ontological class. Thus, the zodiacal sign of Leo on a talisman groups it with the animal lion, the constellation Leo, and higher spiritual powers belonging to the same set. Ficino treated the work of the sign as metaphysical rather than linguistic and therefore innocent. Just as a natural form makes a thing one of its kind, a member of one species rather than another, so the artificial sign of Leo puts the stone in a leonine series, drawing power not from communication with persons but from inclusion in a class. This hedge of Ficino's against demonic magic was a metaphysical construct based on the Neoplatonic concept of a *taxis* or series and the Peripatetic notion of a substantial or specific form.[39]

Pico's hedge against demonic magic was a more elaborate semiotic construct embedded in a bolder conception of magic. His practical Cabala is about signs that are powerful and yet devoid of the meaning that would make them messages to demons. How this conception could apply to "characters and figures" is clear enough, since they differ from letters and numbers in not bearing the meaning that these signs carry. But if practical Cabala is to be a natural magic, and if natural magic must avoid the meaningful speech that demons might hear, then not only characters but also the Hebrew letters through which Cabala operates must be efficacious without being meaningful in the ordinary way. The source of efficacy for Hebrew letters is obvious: their power is primal because God inspired Moses to write them as representing the first divine words. So holy are the letters that medieval Cabalists saw them as the shapes or forms of God himself, and Johann Reuchlin would later ask: "If the words by which the almighty creator made heaven and earth are made into a talisman, . . . would anyone think them powerless?"[40] As elements of the pre-Adamic language of creation, Hebrew letters have even higher status than things made later by God but before any human artifice. Yet these powerful signs, given in the act of creation and hence prior to the world, are letters that make up words. How can they be like characters, without ordinary meaning but effective?

In terms well-known in Pico's time and long before, Thomas Aquinas had traced the extraordinary meaningfulness of biblical language to the fact that

> the author of Holy Scripture is God, who has the power to attach not only speech (*uoces*) . . . but also things themselves (*res ipsas*) to meaning (*significandum*). Hence, while speech has meaning in all fields of knowledge, it belongs especially to [scriptural] knowledge that the things meant by speech also mean something themselves. This first kind of meaning, then, wherein speech means things, pertains to the first kind of understanding [of Scripture], which is historical or literal. But that meaning wherein the things meant by speech also mean other things is called the spiritual understanding; it is built upon the literal and takes its place (*supponit*).[41]

This summary of scholastic hermeneutics shows why Christians of Pico's day and before treated the relations among things, meanings, and words found in the Bible as a special semiotic field. In the section of his *Apology* that deals with the words of consecration in the Mass, Pico left a clue to his Cabalist version of this scriptural semiotics. To defend a suspect eucharistic conclusion, he also called on another medieval theory, the subtle semantics of signification and supposition.

Supposition deals with relations between words and the concepts represented by them, distinguishing terms that are linguistically identical but different in reference and propositional context. If they occur in different propositions, terms with the same lexical meaning may have different suppositions, in which case they will be said to "suppose for" different concepts. Having claimed that the words of consecration are to be taken "in a material rather than a signifying sense," Pico explained in his *Apology* that when the priest speaks Christ's words—"Hoc est enim corpus meum"—the priest's words "suppose" *materially* (in the manner of a recitation or *recitative,* as Pico said) for other words whose original supposition was signifying or *personal.* When Christ first said it, the word *corpus* supposed *personally* or by signification for the speaker's concept of his body, but when the priest says *corpus,* it supposes *materially* or by recitation for Christ's original word. (Note that *personal* and *material* here are technical terms remote from ordinary usage. The distinction is roughly like that between *use* and *mention* in contemporary philosophical terminology.) Like the words of institution as the priest says them in the Mass, the words of creation as we read them in the Torah will be nonsignifying or *material* in Pico's sense, words supposing for the words of another who spoke them as *personal* or signifying.[42] William of Occam summarized the distinction as follows: "Supposition is personal (*personalis*) when a term supposes (*supponit*) for what it signifies (*significato*) and is used as signifying (*significatiue*). . . . Supposition is material (*materialis*) when a term does not suppose as signifying (*significatiue*) but supposes for something spoken (*uoce*) or written (*scripto*)."[43] In this way, having identified God's speech as the source of magical language, Pico's magical conclusions call nonsignifying speech more powerful than speech that signifies, meaning that a human speaker's material use of God's words as nonsignifying will have greater effect than personal use of his own words as signifying. Pico's analysis in the *Apology* of eucharistic language shows that speech without signification in this technical sense can be speech of great power.

Elsewhere in the *Apology,* referring to Plato's *Cratylus,* to Stoic ideas about language, and to Origen on certain Hebrew words such as *hosanna, hallelujah,* and *sabaoth,* Pico offers another defense for the use of names (*nomina*) in natural magic, arguing that these special terms, mainly onomatopoeic, are effective "not as signifying by convention but as natural objects in themselves . . . unless perhaps there are some in which the signification is natural."[44] The point of disconnecting certain signs from ordinary signification becomes clearer in light of Pico's adaptation of Cabalist techniques to Christian purposes, which depended on his exploration of the Hebrew language in its written form. "If any language is primal and not contingent, many interpretations show clearly that it is Hebrew": the unique status of this primordial language convinced Pico of its special power. "One who grasps [its] . . . structure deeply and by the roots," he wrote, "and knows how to keep that [structure] fitted to the fields of knowledge will have a pattern and a rule for the complete discovery of anything that can be known."[45]

V

The elements of this amazing "pattern and rule" are the twenty-seven signs of the Hebrew alphabet (figure 1.4). Twenty-two are distinct letters, but five are special final forms for certain letters when they appear at the end of a word. Each of the twenty-two letter-signs is also a number-sign or numeral. The first sign in the alphabet is the letter *alef* (א) and the numeral 1, the second is the letter *bet* (ב) and the numeral 2, and so on through the tenth sign for the letter *yod* (י) and our compound numeral 10, after which some compound numerals like 11 require two letters—*yod, alef* (יא)—while others like 20 correspond to one letter, *kaf* (כ). The last letter in the alphabet, *taw* (ת), is a single sign whose numerical value is 400. Because all Hebrew letters are also numerals, every Hebrew word has numerical value. The holiest word in Hebrew is the chief name of God, the Tetragrammaton or four-lettered name, so sacred that it must not be spoken (figures 1.5 and 1.6). The four letters of this name—*yod* (י), *he* (ה), *waw* (ו), and *he* (ה) again—are also the numerals 10, 5, 6, and 5, summing to 26, a simple example of the technique called *gematria*. A more complex calculation makes a running total of these letters as they occur in the sacred name, beginning with the value of the first, then of the first plus the second, and so on, giving 10 for *yod* (י), 15 for *yod* (י) plus *he* (ה), 21 for *yod* (י) plus *he* (ה) plus *waw* (ו), and 26 for *yod* (י) plus *he* (ה) plus *waw* (ו) plus *he* (ה). In a further step, the sum of these four subtotals produces 72, a number of great power.[46]

An arithmetic representation of the process would look like figure 1.6. But the shape of the arrangement in figure 1.7 suggests more to a Trinitarian Cabalist. The same structure, expressed in the figural numbers of Pythagorean arithmetic, is the famous tetractys, a shape that shows how the number 10 is the sum of the first four numbers, 1, 2, 3, and 4. Pico alludes to this procedure in his fifty-sixth Cabalist conclusion, writing that "one who knows how to extend the quaternary into the denary has a way, if he is expert in Cabala, to deduce the name of seventy-two letters from the ineffable name." An inverse triangulation of the same Hebrew numerals gives the same result in a different way, as in figure 1.8. In the very next conclusion, Pico urges the application of "formal arithmetic" to God's unutterable name, and

Letter	Final	Number	Name	Sound
א		1	*alef*	'
ב		2	*bet*	b
ג		3	*gimel*	g
ד		4	*dalet*	d
ה		5	*he*	h
ו		6	*waw*	v
ז		7	*zayin*	z
ח		8	*heth*	ch
ט		9	*tet*	t
י		10	*yod*	y
כ	ך	20 (500)	*kaf*	k
ל		30	*lamed*	l
מ	ם	40 (600)	*mem*	m
נ	ן	50 (700)	*nun*	n
ס		60	*samek*	s
ע		70	*ayin*	'
פ	ף	80 (800)	*pe*	p
צ	ץ	90 (900)	*tsade*	ts
ק		100	*qof*	k
ר		200	*resh*	r
ש		300	*shin*	sh
ת		400	*taw*	t

Figure 1.4
Hebrew letters and numerals.

yod ʾ   he ה   waw ו   he ה

  10   5   6   5   =   26

Figure 1.5
The letters of the Tetragrammaton.

ʾ   10   yod

ה + ʾ   15   yod + he

ו + ה + ʾ   21   yod + he + waw

ה + ו + ה + ʾ   <u>26</u>   yod + he + waw + he

                          72

Figure 1.6
72 in the Tetragrammaton.

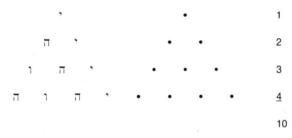

ʾ				•			1
ה	ʾ			•	•		2
ו	ה	ʾ		•	•	•	3
ה	ו	ה	ʾ	•	•	•	•	<u>4</u>
							10

Figure 1.7
The Tetragrammaton and the tetractys.

ʾ ʾ ʾ ʾ	10 + 10 + 10 + 10	= 40	• • • •
ה ה ה	5 + 5 + 5	= 15	• • •
ו ו	6 + 6	= 12	• •
ה	5	= <u>5</u>	•
		72	

Figure 1.8
72 in the Tetragrammaton, inverted.

in the conclusion immediately preceding he says that Cabala is well suited to "Pythagorean arithmetic." To "extend the quaternary into the denary" is an operation of Pythagorean arithmetic that transforms the quaternary of the first four numbers into a denary not just by adding them arithmetically but by displaying them figurally in triangular form. Understanding Pico's formal numbers as figural also makes sense of a mathematical conclusion on Joachim of Fiore, who anticipated Pico in looking for meaning in the shapes of the signs that he puzzled over in his *Book of Figures:* "In his prophecies Joachim used no other method," Pico claimed, "than that of formal numbers." In this context, the terms *formal arithmetic* and *Pythagorean arithmetic* probably refer to the same procedure, made even more powerful when a figural number is composed of Hebrew signs which, as letters, make up the name of God; as shapes, reflect the very form of divinity; and, as numerals, yield the wondrous quantity of 72.[47]

Pico's conclusions on the Orphic hymns show why 72 is a number to conjure with. Since the hymns have the same relation to natural magic that the Psalms of David have to Cabala, we may expect that Orphic notions will have correspondences in Cabala, as in fact they do—the monstrous Typhon with the Satanic Samael, the Orphic Night with the Cabalist Infinite, and so on. So close is this connection that the hymns can do "nothing without the work of Cabala, whose work it is to make practical every formal, continuous, and discrete quantity."[48] Discrete quantity is the subject of the arithmetic of integers, in which the operations of addition and multiplication combine distinct numerical values. Continuous quantity is the subject of geometry, which can represent the results of other operations, such as the taking of roots, that simple integers sometimes cannot express. Between the geometry of continuous quantity and the arithmetic of discrete quantity is a third domain of formal quantity, by which Pico seems to have meant a representation of number that combines the figurative force of geometry with the precise distinctions of arithmetic. If Pythagorean figural numbers such as the tetractys represent formal quantity, then "Pythagorean arithmetic" and "formal arithmetic" are the same, as the Cabalist conclusions imply.[49]

In fact, Pico uses "the form of the Pythagorean quaternary" to interpret the number of the Orphic hymns, which, as he tells us, "is as great as the number by which the triple God created time." This number is 86, which in Pico's day and long after was the number of the hymns. Figures 1.9 and 1.10 display Wirszubski's construction of that number as the sum of the five numerals that make up the name of God first mentioned in the Bible, *Elohim* (אֱלֹהִים), less sacred than the Tetragrammaton and therefore pronounceable: *alef* (א) 1 + *lamed* (ל) 30 + *he* (ה) 5 + *yod* (י) 10 + *mem* (ם) 40 = 86.

Elohim	אֱלֹהִים	
alef	א	1
lamed	ל	30
he	ה	5
yod	י	10
mem	ם	40
		86

Figure 1.9
86 in *Elohim*.

אֶהְיֶה	1 + 5 + 10 + 5 =	21	Ehyeh
אֲדֹנָי	1 + 4 + 50 + 10 =	65	Adonai
יהוה	10 + 5 + 6 + 5 =	26	YHWH
		112	

= (26) + 86 =   אֱלֹהִים   +   (יהוה)

Elohim   +   (YHWH)

Figure 1.10
86 in the three great quaternaries.

But with five letters, *Elohim* (אֱלֹהִים) is not a quaternary; moreover, Pico's number must be that of "the triple God." The solution is to find the right triad of quaternary divine names, *Ehyeh, Adonai,* and the Tetragrammaton, and to calculate their numbers, as in figures 1.9 and 1.10. The sum of the sums of these three names is 112, but 26 must be subtracted from this amount because that much of it is silenced by the unsayable name—leaving 86, which is at once the number of the five-lettered name *Elohim* (אֱלֹהִים), of the three four-lettered names with the Tetragrammaton suppressed, and finally of the frequent appellation *YHWH Elohim* (יהוה אֱלֹהִים) with the holiest name again mute and subtracted. Pico called אֶהְיֶה (*Ehyeh*), יהוה (*YHWH*), and אֲדֹנָי (*Adonai*) "the three great quaternary names of God that occur in the secrets of the Cabalists and through a wondrous appropriation to the three persons of the Trinity ought to be attributed . . . to the Father, . . . Son . . . [and] Holy Spirit."[50]

VI

As Pico was so intent on a Cabalist interpretation of the number of the Orphic hymns, he would not have neglected the number of his own statements about Cabala. These seventy-two aphorisms are not just the last of his theses. They are the climax of the project that Pico announced in the *Oration* and displayed so extravagantly in the *Conclusions*. His aim was a contemplative therapy for the soul grounded in philosophy and leading to final ecstatic union in the One. The *Oration* prescribes an ascent through moral philosophy, dialectic, natural philosophy, and theology toward the

> friendship of the single-spirited wherein all spirits do not just accord in one mind that is above every mind but, in some way beyond telling, end up as altogether one. This is the friendship that the Pythagoreans say is the end of all philosophy, this is the peace . . . by [which] even humans ascending into heaven become angels. Let us wish this peace for our friends, . . . for our soul. . . . She will desire to die in herself that she may live in her bridegroom, in whose sight the death of his saints is precious beyond doubt—death, I say, if the fullness of life should be called death; the sages have said that to meditate upon it is the study of philosophy.[51]

The erotic charge on this passage recalls the lines of the *Commento,* again clarified by Wirszubski, in which Pico explains how the death of the kiss is the ecstatic goal of Cabala, "when the soul in intellectual rapture unites so much with the separated substances that it is raised beyond the body and abandons it entirely." The first part of the thirteenth Cabalist conclusion describes the rapturous death recommended in the *Commento* and the *Oration:* "One who works in Cabala without mixing in the extraneous and stays at the work a long time will die the death of the kiss." A steady dose of pure Cabala will kill, but this good death is the heavenly goal of the contemplative program of the *Oration.* The other death described in the second part of the thirteenth conclusion is hellish and horrifying. If the Cabalist "goes wrong in the work or comes to it unpurified, he will be devoured by Azazel through the property of judgment."[52]

The name "Azazel" (עֲזָאזֵל) occurs in only one passage of the Bible, in Leviticus 16:7–28, where the Lord gives Moses instructions for Aaron about the scapegoat ceremony for the Day of Atonement. Aaron is to "cast lots over the two goats, one to be for the Lord and the other for Azazel": the Lord's goat is a sin-offering (חַטָּאת), Azazel's an offering for atonement (לְכַפֵּר). The Book of Enoch made Azazel a leader of the rebel angels, whose teachings to mankind include "charms and spells." In this vein, the *Zohar* mentions angels called Uzza and Azael whom the Almighty cast down from heaven for questioning his creation of man. They took on bodies, had sex with women, and taught "magic, witchcraft, and sorcery." An error that might provoke such a demon to devour the Cabalist would be a mistake in his specialty of magic. When Pico adds that "the property of judgment" will be the means of Azazel's attack on the erring Cabalist, he refers to the *Sefirot,* in particular to the fifth *Sefirah* called Power or Judgment (see figure 1.1). In the standard representation or tree of the *Sefirot,* the left side *Sefirah* of Judgment becomes a force of limitation or evil, in contrast to the opposite *Sefirah* of Greatness or Love on the right, the source from which good flows freely. Such conceptions led to the elaboration of an entire world of *Sefirot* on the left, including the eighth and ninth, which correspond to Samael and Lilith: one the usual name in Cabala for Satan, the other his female counterpart. Pico fears Azazel as part of this evil world of unclean forces, whence the demon draws his mastery of magic.[53] Going wrong with the magical alphabet of Cabala would send the devouring demon a written invitation.

The desire for countermagic against demonic powers and physical ills inspired Cabalists to make magic out of certain requirements of traditional Judaism—the wearing of *tefillin* or phylacteries, for example—although such magical uses were controversial. In this manner, the *Zohar* speaks of the mezuzah as standing "continually by the door . . . [where] a demon [is] waiting . . . on the left-hand side. When a man raises his eyes, he sees the mystery of his Master's name, and mentions it, and then the demon cannot harm him." Texts from Scripture (Deut. 6:4–9; 11:13–21, written in twenty-two lines) inside the mezuzah represent the *Sefirot* and secure their blessing, while the three-lettered divine name *Shaddai* (שַׁדַּי) visible on the outside wards off the demon lurking at the doorpost. Well before Pico's time, the mezuzah had come to be decorated with names of angels and magical signs, including the hexagram known first in Arabic texts as the Seal of Solomon but by the fourteenth century in Hebrew sources as the Magen David (מָגֵן דָּוִד), or Shield of David. Other words of power also strengthened the six-pointed star. One was the divine name of seventy-two letters; another was the angelic name Taftafiyyah (טפטפיה), which was a name of the great angel Metatron (מטטרון).[54]

Christians of the later Middle Ages knew the mezuzah as a magical device, sometimes reviling it, sometimes adapting it for their own uses. In the latter case, they might arrange Christian symbols—such as the name "Jesus" in a Hebrew form like ישו (*Yeshu*)—in patterns inspired by Jewish talismans.

One type of Jewish design that could have instructed Christians was the famous childbirth talisman (figure 1.11) depicted in the *Sefer Raziel,* a manual
of angel magic whose various parts circulated in Latin after the thirteenth
century. Two equilateral triangles interlaced and inscribed in nested circles
make up the basic structure of the talisman. Four words outside all the circles
list the rivers of paradise (Gen. 2:10–14), while words written between the
two circular bands are proper names (Adam, Eve, Lilith, etc.) and a Psalm
verse (91:11) invoking angelic protection. Inside the inmost circular band but
outside the central hexagon where the two triangles intersect is the divine
name of forty-two letters written in fourteen groups of three. The bottom
half of the hexagon itself contains the divine name *Quph* (קוף) in its six permutations; words from Exodus 11:8 in the top half repel the host of
demons.[55]

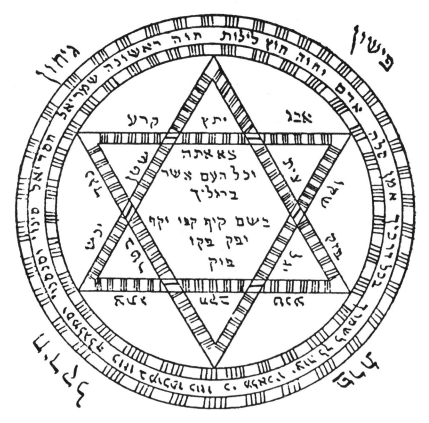

Figure 1.11
Talisman from the *Sefer Raziel.*

The holy name *Shaddai* (שַׁדַּי), often rendered "Almighty," made a good choice for a talisman meant to call down heavenly power to protect its wearer. Gematria told the Cabalist that the numerical value of שַׁדַּי was 314, the same as the number of Metatron, an angel so mighty that he needed seventy names, of which one was Taftafiyyah, taken from the first two letters of three verses of Psalm 119.[56] In a conclusion from Themistius, Pico explains that the Agent Intellect of the Greek commentators on Aristotle corresponds to Metatron in Cabala. He makes the same point cryptically in a Chaldaean conclusion, alluding there to one of Metatron's other names, which was *na 'ar* (נַעַר) or "boy." The *Commento* explains further that a human who dies to the sensible world can be reborn in the intelligible world as an angel, as the Cabalists say Enoch was changed into "Matatron, an angel of divinity."[57] This mighty spirit was named "boy" because he serves God, but his service brings him so near the Holy One that heretics made him into a second deity. The metaphysical aspect of Metatron's theological eminence is his identification with the Agent Intellect, understood by Maimonides as the channel through which God sends prophecy to chosen mortals. Hence, when Abulafia sought union with God through the discipline of alphabetic combination, Metatron as Agent Intellect became an intermediate goal of his ecstatic meditations. The kiss is the first of six erotic images that Abulafia uses to capture the prophetic process; the others are intercourse, semen, impregnation, a son, and new birth.[58] The new birth of immortality is the death that comes of complete union with the Mind of God, the death that Pico also compared to a kiss. Metatron, the angel of ecstatic death, appears in Pico's tenth Cabalist conclusion as the figure called "Pallas by Orpheus, Paternal Mind by Zoroaster, Son of God by Mercury, Wisdom by Pythagoras, Intelligible Sphere by Parmenides," all of them good names for a hypostasis close to the central abyss of the Godhead yet not entirely within it.[59]

Metatron plays a key role in Pico's speculative Cabala. As the chief of angels and a foe worthy of Samael and Azazel, he should also be a force in practical Cabala. But in discussing practical Cabala, Pico says nothing openly about him, which is not surprising; he believed that the greatest truths of Cabala were its deepest secrets. In the *Apology* he argued that the simple story of the Bible made God's mercy and justice plain enough to guide the unlettered faithful toward good behavior and correct worship, "but to divulge publicly the more secret mysteries hidden under the bark of the Law and the crude surface of words . . . was this any different than . . . casting pearls before swine? Thus, to keep these things secret from the mob for communication to the perfect . . . was a matter not of human judgment but of divine command."[60] Imputing such esoteric motives to Pico's silence about

Metatron (and other topics), I think it possible that he meant the whole collection of seventy-two Cabalist conclusions as a figured number, the makings
of a talisman to attract this greatest of angels. Angelic talismans like those of
the *Sefer Raziel* (see figures 1.3 and 1.11) and well-known in Pico's time invoked Metatron against Samael with a hexagram made of the seventy-two-
lettered name of God. This name derives from Exodus 14:19–21, where each
of three verses contains exactly seventy-two letters (figure 1.12). Introducing
the miracle of the Red Sea, they tell how the angel of God protected the Is-
raelite column. Noticing the symmetry of these verses, interpreters arranged
the seventy-two letters in three horizontal boustrophedon lines and then read
each vertical group of three letters as the name of an angel—Vehuiah, Ieliel,
Sitael, and so on, totaling seventy-two names in all. The composition of an
angelic Shield of David, the Jewish Orpheus, may have been Pico's aim in setting out his Christian Cabala in seventy-two theses, as he had already used
Cabala to find the names of God in the number of the Orphic hymns. He
could also have found encouragement for talismanic magic in the teaching of
Yohanan Alemanno, who wrote that "the secret of the world of the letters is

Exodus 14: 19–21—the first three of the seventy-two angel names

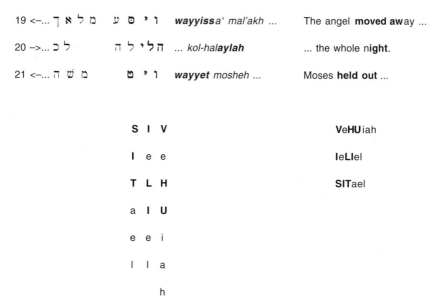

19 <–... ך א ל מ   ע ס י ו   *wayyissa' mal'akh* ...   The angel **moved aw**ay ...

20 –>... כ ל   ה ל י ל ה   ... *kol-halaylah*   ... the whole **night**.

21 <–... ה ש מ   ט י ו   *wayyet mosheh* ...   Moses **held out** ...

S	I	V
I	e	e
T	L	H
a	I	U
e	e	i
I	I	a
	h	

VeHUiah

IeLIel

SITael

Figure 1.12
Angel names in Exodus 14.

that they are the forms and seals which receive supernal spiritual emanations as are the seals which receive astral emanations."[61]

VII

If Pico wanted his seventy-two conclusions to form an angelic talisman for repelling Azazel by summoning Metatron, one of two things follows: either he crossed the line dividing characters that operate by shape from letters that carry verbal meaning; or else he had in mind a talisman using no letters or words that might be taken to invoke angels or demons. Indeed, the first choice might be the error in Cabala that would invite the hungry Azazel, but Pico could avoid it by making his talisman a Pythagorean figured number in Cabalist form. The common sight of the hexagram on Jewish talismans would show how to proceed. Moreover, Pico knew how to "extend the quaternary into the denary . . . [in order] to deduce the name of seventy-two letters from the ineffable name."[62] The resulting calculation (figures 1.7, 1.8, and 1.13) produces two forms of the tetractys. Both are denary triangles, but they point in opposite directions—one up, the other down—and superimposed they form the hexagram of the Shield of David. As a device for doing angel magic, this figure derives its power from its hexagonal form, a good hexadic fit for the seventy-two angelic names. Its attraction for Pico would have been that it uses only a shape without actually expressing any names of angels.

Nonetheless, without making himself guilty of theurgy, Pico might have taken the further step of putting Hebrew letters in a figured talisman, as long as they addressed the triune God and not his angelic ministers. In the twenty-second magical conclusion, he implied that individual names can be used with magical effect if they are Hebrew or derived from Hebrew. In this case, unlike other speech (*uox*) that loses power unless its supposition is non-signifying, Hebrew names (*nomina*) may retain both their efficacy and their

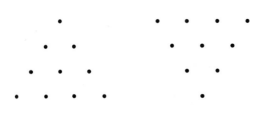 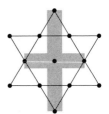

Figure 1.13
Figured numbers and the Shield of David.

signifying supposition. God's Hebrew names, as in the completely hypothetical ingredients for a talisman in figure 1.14, are words of great power that
truly signify what they stand for; even the letters of the names are the shapes
of God. Here, in place of the three-lettered name *Shaddai* (שַׁדַּי) seen on the
Shield of David, the three letters in its central band—*yod* (י), *shin* (ש), and *waw*
(ו)—correspond to Pico's fourteenth and fifteenth Cabalist conclusions,
where he substituted a *shin* (ש) for the *he* (ה) of the Tetragrammaton and inserted it between *yod* (י) and *waw* (ו) to spell a Hebrew form of the name Jesus (יְשׁוּ) in three letters.[63] At the bottom of the putative hexagram in figure
1.14 is a single letter *mem* (מ), whose shape and location in semiotic and hy-
postatic space were of great interest to Pico.

Several of his conclusions highlight one of Cabala's most striking features, its focus on particles of language that add little or nothing to any ordinary reading, loading heavy measures of meaning on the smallest bits of
text, even single letters. What justifies taking so much from so little is the
primary rule of Cabalist hermeneutics: nothing in the Torah lacks meaning.[64] Pico pushed this principle to its limit in his forty-first Cabalist con-

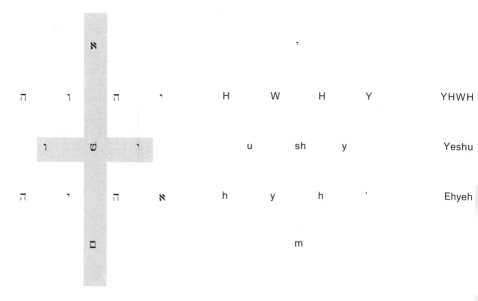

Figure 1.14
YHWH, Yeshu, Ehyeh, and *'asham* in a stellate talisman.

clusion, which deals with the shape of one letter in Isaiah 9:6–7, a famous
Messianic passage:

> For to us a child is born, to us a son is given;
> and the government is upon his shoulder,
> and his name is called:
> Wonderful Counselor, Mighty Hero,
> Eternal Father, Prince of Peace,
> *for the increase of the realm*
> and for peace without end.

The phrase "for the increase of the realm" at the beginning of verse 7 renders the Hebrew words *lemarbeh hammisrah* (לְמַרְבֵּה הַמִּשְׂרָה) in the Masoretic
text, but the apparatus points to a problem with the second letter of the first
word. That letter is *mem,* written not in its usual "open" form (מ) but in the
"closed" form (ם) required at the end of a word. In this passage *mem* is not final, so its shape is wrong. A textual critic would look for a scribal mistake in
the group *lemarbeh* (לְמַרְבֵּה), traditionally read as a combination of the preposition *le* (לְ) with the noun *marbeh* (מַרְבֵּה), meaning "for the increase." A different reading treats the first two letters, *lamed* and *mem* (לם), as a group
meaning "to him" or "for him"; this makes the next three letters the separate word *rabbah* (רַבָּה), the feminine singular of the adjective *rav* (רַב), which
means "great," suggesting that something great in the feminine gender comes
to the Hero of the previous verse.[65]

The riddle of the closed *mem* challenged Jewish exegetes, who saw the
anomalous letter as signifying a closed or unrevealed doctrine, as the hidden
date of the end of time, as the walls of Jerusalem closed again after the cap-
tivity, and so on. Some Cabalists connected the *mem* with *Binah* (Intelligence), the third *Sefirah,* regarded in this case by the *Zohar* as the barren male
or "son" who governs the time of exile before the coming of the Messiah.
This (and other) material in the *Zohar* shows traces of Christian influence, but
Christians who were aware of the closed *mem* took a different position, as
Nicholas of Lyra explained in the fourteenth century: "The closed letter *mem*
is placed here in the middle of an expression contrary to its nature . . . to de-
note that Christ, of whom Scripture speaks here, was born contrary to the
course of nature from a closed virgin." Although this Marian reading stressed
the extraordinary character of the Messiah's birth, Pico's claim in his forty-
first Cabalist conclusion is Trinitarian: Cabala can tell us "through the mystery of the closed *mem* why Christ sent a Paraclete after him." For Pico, the
Prince of Peace to whom greatness comes is Christ, the second person of
the Trinity, whom the Spirit follows as Paraclete. The Spirit first appears in

the Bible in the second verse of Genesis as *ruach elohim* (רוּחַ אֱלֹהִים), the Spirit of God. But the Hebrew word for "Spirit," *ruach* (רוּחַ), is a feminine noun, which in the phrase "great Spirit" would be governed by the form *rabbah* (רַבָּה) required by reading the *mem* in Isaiah 9:7 as final, a closed (ם) rather than an open (מ) *mem*.[66]

Pico gives the first key to this Trinitarian cipher in his twelfth thesis "according to the mathematics of Pythagoras," which says that "in numbering forms we should not exceed the quadragenary."[67] The quadragenary or 40 is the value of *mem* (מ) as a numeral (figure 1.15). But why should 40 limit the numbering of forms?

In the *Sefer Yezirah,* the letters *shin* (שׁ), *alef* (א), and *mem* (מ) are called "mothers." God chose them "from among the elementals . . . and . . . set them in his great name," which Cabalists took to mean that *mem* (מ) produced the *yod* (י), *shin* (שׁ) produced the *he* (ה), and *alef* (א) produced the *waw* (ו) of the Tetragrammaton, giving them the glory of kinship to the three letters that make up the unutterable name.[68] Because it is *shin* (שׁ) that makes the *he* (ה) of *YHWH* (יהוה), the mothers could have told Pico which letter to substitute for *he* (ה) in converting *YHWH* (יהוה) into *Yeshu* (ישׁו). Moreover, in the transposed order מ–שׁ–א, the mothers of the Tetragrammaton spell out אָשָׁם (*'asham*), the "guilt-offering" described in Leviticus and Numbers as a lamb or a sheep or a goat, celebrated by Isaiah as the "lamb brought to the slaughter . . . for the transgression of the people" and, for Christian believers, recalled in John's Gospel as "the lamb of God . . . who takes away the sin of the world": "Agnus dei . . . qui tollit peccatam mundi." These words of the Bap-

---

mem   מ = 40

the mothers: *shin* שׁ   *alef* א   *mem* מ

mem   מ -> *yod*   י

shin   שׁ -> *he*   ה   *yod*   י + *he* ה + *waw*   ו + *he* ה -> יהוה (*YHWH*)

alef   א -> *waw*   ו

אָשָׁם   (*'asham*)   guilt-offering

Figure 1.15
Mem, the mothers, YHWH, and *'asham.*

tist in the Vulgate had entered the Mass as a triple expiation long before the time of Pico, who would also have known the *agnus dei* as a common amulet: a piece of wax taken from Easter candles, imprinted with a lamb bearing a cross, and worn for protection against lightning, disease, and other kinds of harm.[69] If Pico thought of the name *Yeshu* (יֵשׁוּ) as the horizontal triad at the center of a talisman (see figure 1.14), he could scarcely have found a better choice than *'asham* (אָשָׁם) for the intersecting vertical, the two together making a cross within the Shield of David, proclaiming that Jesus is the Lamb of God, surrounding this name with the holiest names of the old covenant, and marking the bottom or limit of this cruciform star with a closed *mem* (ם).

Cabalists themselves used the mothers figurally for their own sacred purposes, to form three horizontal bands tying the *Sefirot* together in the Sefirotic tree with a *mem* (מ) marking the bottom band. Figure 1.16 depicts

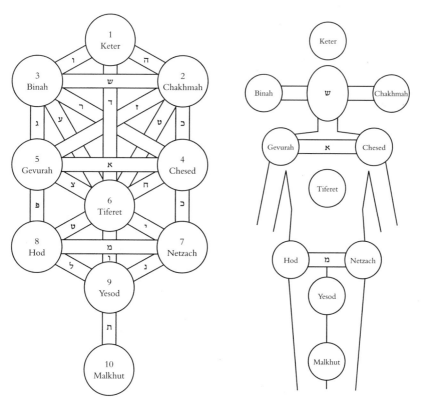

Figure 1.16
Mem as a boundary in the *Sefirot* and Adam Kadmon.

another common Sefirotic diagram, in which the anatomy of the supernal man, Adam Kadmon, represents the disposition of the *Sefirot,* and this more concrete figure shows how *mem* might limit a higher world of forms standing above a lower world of generation. Pico emphasized such a limit in the fifth Cabalist conclusion, declaring that "any Hebrew Cabalist . . . is forced inevitably to concede of the Trinity and each divine person—Father, Son, and Holy Spirit—what the Catholic faith of Christians asserts, that precisely, no more, no less, and without variation."[70]

This same strict injunction to limit occurs earlier in the fourteenth Chaldaean conclusion, which applies "the arithmetic of the higher Merchiava" to a saying of Zoroaster limiting sacrifice to "three days and no more." The "higher Merchiava" is the supreme triad of the *Sefirot,* corresponding to the three Plotinian hypostases. Pico probably understood the third *Sefirah, Binah* (בִּינָה, "Intelligence"), in the form *Ben Yah* (בֶּן־יָהּ, "Son of God"), and surely he knew the section of Augustine's *City of God* that attacked Porphyry for having misunderstood this Trinity because he had been deceived by the *Chaldaean Oracles.*[71] Regulating sacrifice by the number 3 sets the correct trinitarian limit on divine worship, a theological restriction that occurs again in the first conclusion based on Plato (figure 1.17). Speaking of the series of double and triple numbers that harmonize the world soul in *Timaeus* 35B, Pico wrote that from "the triple numbers . . . placed . . . in a triangle signifying the soul," we can tell "how far one may progress in numbering forms through . . . the first forming form." The "first forming form" is unity, from which Plato derived two geometrical progressions that later commentators saw as branches descending from the apex of a Greek *lambda* (Λ) and making two sides of a triangle. Unity followed by the next three powers of 2 makes the progression 1, 2, 4, and 8 of "double numbers," as Pico called them; his "triple numbers" form the analogous series 1, 3, 9, and 27. Three powers of 2 and three powers of 3 make up the two legs of Plato's psychic *lambda,* but 3s make the better trinitarian boundaries. The series of 3s stops at 27, 3 multiplied by itself three times. For Pico's Cabala, however, greater interest attaches to the sum of the series of triple numbers (1 + 3 + 9 + 27 = 40), showing yet again how *mem* (מ) as a numeral stands at the edge of divinity. The shape of the *mem* in Isaiah 9:7 is closed because it limits the extent of the Godhead; ם as a Platonic 40 closes the numbering of forms; thus, the open מ taken from the mothers to produce the closed ם in the word *'asham* (אָשָׁם) becomes the final element in a Christian Cabalist talisman.[72]

Whether the *mem* is open or closed, its normal value is 40, but 600 is an alternative figure for the closed *mem* (read as the second of the five final letters placed at the end of the alphabet after *taw* [ת] = 400; see figure 1.4).

Λ *lambda*

1

2   3

4   9

8   27

$3^0 = 1$, $3^1 = 3$, $3^2 = 9$, $3^3 = 27$ ... 3^n

$2^0 = 1$, $2^1 = 2$, $2^2 = 4$, $2^3 = 8$ ... 2^n

$1 + 3 + 9 + 27$ $= 40$ $= \text{מ}$

$1 + 2 + 4 + 8$ $= 15$ $= \text{יה}$

15×40 $= 600 = \text{ם}$

Figure 1.17
Mem and Plato's triple numbers.

600 is also the product of 40 (מ) multiplied by 15 (יה), the latter being the divine name *Yah* as well as the sum of the double numbers of the Platonic
lambda. Since the six-hundredth of Pico's *Conclusions* is also the first Platonic
thesis on the double and triple numbers that sum to 15 and 40, the coincidence confirms a Cabalist reading of this conclusion and also argues that Pico
meant us to find numerology in the organization of his work.[73] If so, we face
the most tempting question of all about the structure of the *Conclusions:* why
are there just nine-hundred of them, the nine-hundredth being the proposal
"to read in the book of God, in the same way Cabala teaches us to read in
the book of the Law"?

Pico clearly regarded 900 as a meaningful number. As the product of
3^2 or 9 multiplied by 10^2 or 100, its numerological attractions are obvious,
especially in light of the twenty-third magical conclusion that highlights the
power of threes and tens as formal numbers and as "numbers of numbers
in magical arithmetic." While making final plans for his thwarted Roman
dispute, Pico sent news to a recent visitor, Girolamo Benivieni, that the

Conclusions "had grown after you left to 900 and were moving toward 1,000 except that I called a halt, content to take my stand at the former number—a mystical one, naturally." At this late stage of his project, Pico made 900 the symbol of the soul's mantic excitement when stirred inwardly by the nine Muses. That excitement, as he told Benivieni, still moved him on the day after his friend's visit, driving his thoughts to a passage from John's Gospel in which Jesus foretold his departure to the Father and the subsequent arrival of the Paraclete. This prophecy of peace and enlightenment brought by the Holy Spirit is a promise of philosophical peace for Pico, who in his transport therefore added to the *Oration* "certain things about peace that bestow praise on philosophy." Culturally this peace is the concord of seemingly discordant systems.[74] Spiritually it is the Cabalist death of the kiss.[75] But as a spiritual goal it is also the peace sought in the cultural school of the Muses, in the learned labor that Ficino (in the *Platonic Theology* and in comments on Plato's *Ion* and *Phaedrus*) had described as a Bacchic ascent through eight celestial orbs and the ninth sphere of the World-Soul toward Apollonian illumination and Jovian rapture. "Jupiter seizes hold of Apollo," wrote Ficino, "Apollo illuminates the Muses, the Muses arouse . . . the poets, the poets inspire their interpreters, and the interpreters move their listeners."[76]

A similar Orphic mythology recurs in the *paideia* of four grades—the soul's march through moral philosophy, dialectic, natural philosophy, and theology—that shapes the first half of the *Oration*. "Who does not wish to be inspired by those Socratic frenzies that Plato sang in the *Phaedrus*?" asked Pico.

> We shall be stirred by [them] . . . if moral philosophy also . . . tempers the force of the passions . . . and dialectic makes reason progress in due measure so that together, when we have been charged with the frenzy of the Muses, our ears will drink their fill of heavenly harmony. Bacchus, leader of the Muses, will then make us drunk with the flowing plenty of God's house by showing us, as we philosophize in his mysteries—which are the visible signs of nature—the invisible works of God. If, like Moses, we remain faithful in the whole of this house, theology most holy will approach and ensoul us in a double frenzy. Lifted to the summit of this loftiest watchtower, . . . we shall be . . . prophets of Phoebus . . . and like burning seraphs . . . no longer ourselves but the Very One who made us.

The lessons of the Muses and a fourfold cherubic discipline have led the soul to the seraphic blaze of mystical union. On the way, it was the Bacchic mystery of natural philosophy—seeing the unseen in nature's signs—that prepared the soul for theology and final joining in the Godhead.[77] Is this what the mystical 900 measured for Pico, standing, as it does, at the end of his seventy-two Cabalist theses?

In this part of the *Oration,* as in the *Conclusions* and elsewhere, Pico urges us to seek what we cannot see.[78] Even without such encouragement, a Cabalist will have wanted to know more about the 900 and its alphabetic equivalent (see figure 1.4), which is the final *tsade* (ץ)—the very last of the five finals—whose normal form is צ (= 90). And a Christian Cabalist, even one unaware of the inspirations descending to mortals from the triad of Jupiter, Apollo, and the Muses, will have noticed the Trinity in the Gospel passage brought to Pico's mind by the 900. Jesus, having warned that death will take him to the Father, promises that the Paraclete will come to keep his peace. The foresight is prophetic and the promise is Messianic, fulfilling the words of Isaiah about the Prince of Peace and the Spirit of God that reveal "through the closed *mem* why Christ sent a Paraclete after him."[79]

A late Midrashic work, the *Otiot de-R. Akiva,* shows how to make Christian Cabala out of the *tsade,* in the manner of Pico's interpretation of the closed *mem.* This text, explicated by Yehuda Liebes, influenced the *Zohar* in a passage where *tsade* joins the other letters of the alphabet in beseeching the Almighty to make use of it in creating the world. The Creator acknowledges that the *tsade* is righteous (*tsaddik,* צדיק) but insists that "you must be concealed . . . lest you give the world cause for offence," alluding to Jesus as the *tsade* but in an unflattering (and sexually insulting) manner: Jesus is the impotent "son" whose sterility will be healed only when the Messiah comes, and in the meantime this secret will be a dangerous scandal to the gentiles. Some of the older texts from the *Otiot de-R. Akiva* underlying this Zoharic passage are plainly hostile: "Why does *tsadi* have two heads? Because it is Jesus of Nazareth who seized two heads—one of Israel and one of Edom [Christianity]— . . . and led people astray. And because Israel saw him thus, they . . . crucified him." But other such Midrashim are more suitable for Pico's Christian Cabala: "Why does *tsadi* have two forms? . . . And why does it have two heads? This is the Messiah, son of David, as it is written: 'a shoot shall grow out of the stump of Jesse.' Why is he called the Messiah? Because he is the head of all. And why is the bottom part . . . bent? Because it is burdened and sick from Israel's sins. . . . And why is the other *tsadi* . . . straight? Because God testifies to Israel through his prophets that he is a true branch: 'See a time is coming . . . when I will raise up . . . [a true branch of David's line].'"[80] These suggestive Christological Midrashim had found their way into Cabalist interpretations probably accessible to Pico, though whether Pico actually knew of them remains uncertain. In any case, the interpretation of the number of the 900 *Conclusions* as a Messianic cryptogram seems

Figure 1.18
The upright cruciform *tsade*.

credible to me; it may even be that the shape of the upright tsade (ץ) is a
shadow of Christ crucified (figure 1.18).[81]

    That Pico should wish the meaning of his whole Cabalist project—all
the nine-hundred *Conclusions* culminating in the seventy-two theses on Cabala—to be seen in the compass of just one letter is consistent with the
hermeneutics that he pioneered for Christian use and expressed in the first
set of conclusions "according to the teaching of the wise Hebrew Cabalists":
there he declared that "there are no letters whose forms, ligatures, separa-
tions, twisting, direction, defect, excess, smallness, greatness, crowning, closing, opening, and order do not reveal the secrets of the ten *Sefirot*."[82] This
principle, which justifies Pico's scrutiny of the shape of the closed *mem* (ם),
will also account for seeing so much in the figure of the upright *tsade* (ץ).
Shape is what letters and numerals have in common with figures and characters. The shape of Hebrew letters is a mean joining the riskier extremes of
characters used in magic and numbers used in Cabala, as in Pico's twenty-
fifth magical conclusion, which claimed that "just as characters belong to the
work of magic, numbers belong to the work of Cabala; between them the
use of letters is a mean, applicable to the extremes by modification." Modifying the shapes and values of the *mem* and *tsade*—from open (מ) to closed (ם),
from bent (צ) to upright (ץ)—showed Pico the way to Cabalist speculation.
Since the Cabalist is the partner of the good magus, shape should play a be-
nign role in natural magic, as Pico claimed in the *Apology* when discussing
the natural action of characters.[83] Letters and numbers are like characters
when shape rather than ordinary meaning is their active feature, as with
Pico's speculative Cabala on the closed *mem*. If my speculations about Messianic talismans in the seventy-two Cabalist theses are true to his purposes,
then the same principle will hold for Pico's practical Cabala and will support
his claim that its magic is a good, natural magic.

VIII

The distinction between shape and message, between figure and language, lets Pico tell good magic from bad and good Cabala from bad. Will the same distinction, applied to astrology, rescue its notorious amulets and talismans, producing the "true astrology" of Pico's final conclusion? Keeping the hungry Azazel at bay is a matter of knowing how to write and read like a good practical Cabalist—with shapes rather than messages—when the risk of theurgy arises, as it will in addressing Metatron. The shape and structure alone of the six-pointed shield (see figure 1.13), without regard to semantic content, forms a starry cross that maps the 72s of the divine name, of the roster of angels from Exodus, and of Pico's culminating conclusions. The very last conclusion, sealing the 900 with a final 72, declares that "just as the true astrology teaches us to read in the book of God, in the same way Cabala teaches us to read in the book of the Law."[84] The good Cabalist who deciphers shapes in the Torah shows the true astrologer how to decode signs in the stars. Together they find power in figures drawn by God, both in the lights of the heavens and in the letters of the Law. Having asked "whether all things are written out in the heavens and shown through signs to anyone who knows how to read,"[85] Pico gave an esoteric answer, certainly, yet an emphatically positive one. In the constellation of his seventy-two conclusions are clues that make the talismanic star of David visible as a figure effective in the world of nature but revealed in the book of Scripture. Having comprehended the seventy-second conclusion, the true astrologer will reinforce the work of the good magus who, as Pico tells us, can do nothing without the help of the Cabalist. And since the seventy-second Cabalist conclusion counts as 900 in the whole set, the true astrologer will also scan the heavens for the seal of the upright *tsade,* of the Messianic Christ and Creator who left his signature on the celestial scroll when he hung the stars in the sky. This trigon conjunction of arcana—astrology, magic, and Cabala—depends on Pico's analysis of letters, numbers, and shapes as they are to be read wherever God has put them.

As heirs of Galileo, we've learned to read our signs differently, depending on where we find them. Since Galileo's time, the book of nature has taught us to fashion systems of mathematical science whose remarkable powers of explanation have promoted a secularism which, in turn, when confronted with the book of Scripture or with its ghosts of conscience past, has caused us to devise hermeneutic systems with remarkable powers of evasion, whether moral or theological. To remain a believer, Galileo needed a hermeneutic that would decouple the reading of Scripture from the reading of nature, especially when Scripture speaks about nature. Where Galileo wanted detachment, Pico looked for coincidence, believing that a reading of Scripture and nature through Cabala would disclose the magic and astrology that the natural philosopher should regard as true and the Christian should respect as good. Although Pico's natural philosophy was not Galileo's new science, the historian of philosophy or science who wants to understand Pico's time or Galileo's must engage the natural philosophy of that period on its own terms, terms that include Pico's learned, refined, and original speculations about number, shape, and meaning in Cabala.

NOTES

1. Pico, *Conclusiones* 5.11.72 (B 140): "Sicut uera astrologia docet nos legere in libro Dei, ita cabala docet nos legere in libro Legis." Citations are to the Latin text in Giovanni Pico della Mirandola, *Conclusiones nongentae: Le novecento tesi dell'anno 1486,* ed. and trans. Albano Biondi (Florence: Olschki, 1995), with Biondi's page numbers in parentheses, preceded by the section, subsection, and item number of the conclusions cited. Biondi's Latin text of the *Conclusiones,* though neither a critical edition nor free of errors, should now be preferred to *Conclusiones sive theses DCCCC Romae anno 1486 publice disputandae sed non admissae,* ed. Bohdan Kieszkowski (Geneva: Droz, 1973). I have followed Biondi, but with a few corrections and with the punctuation modified. The first edition of the *Conclusiones* was printed in Rome by Eucharius Silber (Franck) in 1486. A widely cited sixteenth-century text, with the associated *Oration* and *Apology,* appears in Giovanni Pico della Mirandola and Gian Francesco Pico, *Opera omnia (1557–73)* (Hildesheim: Olms, 1969), 1:63–240, 313–331. *De hominis dignitate, Heptaplus, De ente et uno e scritti vari,* ed. Eugenio Garin (Florence: Vallecchi, 1942), is the best modern edition of these works and of the *Commento.* See also *Disputationes adversus astrologiam divinatricem,* ed. E. Garin (Florence: Vallecchi, 1946–1952). For a recent summary of Pico's thought, see Brian P. Copenhaver and Charles B. Schmitt, *Renaissance Philosophy,* vol. 3 of *A History of Western Philosophy* (Oxford: Oxford University Press, 1992), pp. 163–176. W. G. Craven, *Giovanni Pico della Mirandola, Symbol of His Age: Modern Interpretations of a Renaissance Philosopher* (Geneva: Droz, 1981), is a review of the historiography. *L'Opera e il pensiero di Giovanni Pico della Mirandola nella storia dell'umanesimo: Convegno internazionale, Mirandola, 15–18 settembre 1963,* 2 vols. (Florence: Istituto Nazionale di Studi sul Rinascimento, 1965), collects the papers of a major international conference on Pico.

I read the first version of this paper at the 1994 conference in Mirandola celebrating the five-hundredth anniversary of Pico's death. A preliminary sketch without documentation, "L'Occulto in Pico: Il mem chiuso e le fauci spalancante da Azazel: La magia cabalistica di Giovanni Pico," appeared in the proceedings of that conference: *Giovanni Pico della Mirandola: Convegno internazionale di studi nel cinquecentismo anniversario della morte (1494–1994), Mirandola, 4–8 ottobre 1994,* ed. Gian Carlo Garfagnini (Florence: Olschki, 1997), 1:213–236. Having read versions of the paper half-a-dozen times at conferences since then, I'm grateful to many people for their advice and criticisms, but especially to Michael Allen and Moshe Idel.

2. Chaim Wirszubski, *Pico della Mirandola's Encounter with Jewish Mysticism* (Cambridge, Mass.: Harvard University Press, 1989).

3. Jacob Burckhardt, *The Civilization of the Renaissance in Italy,* trans. S. G. C. Middlemore (London: Penguin, 1990), pp. 229, 328; Copenhaver, "Astrology and Magic," in Charles B. Schmitt et al., *The Cambridge History of Renaissance Philosophy* (Cambridge: Cambridge University Press, 1988), pp. 267–270.

4. Ernst Cassirer, *The Individual and the Cosmos in Renaissance Philosophy,* trans. M. Domandi (New York: Harper and Row, 1963), pp. 86, 115–117.

5. Lynn Thorndike, *A History of Magic and Experimental Science* (New York: Columbia University Press, 1923–1958), 4:485, 496, 530, 532, 540, 562–573.

6. Eugenio Garin, *Giovanni Pico della Mirandola: Vita e dottrina* (Florence: Le Monnier, 1937), pp. 146–149, 180; cf. *Astrology in the Renaissance: The Zodiac of Life,* trans. Carolyn Jackson, June Allen, and Clara Robertson (London: Routledge and Kegan Paul, 1983), pp. 1–14, for a more recent view from Garin. Contemporary with Garin's early work was another Italian study with more detail on Pico's Cabala: Eugenio Anagnine, *G. Pico della Mirandola: Sincretismo religioso-filosofico, 1463–1494* (Bari: Laterza, 1937), but on this topic all previous studies are superseded by Wirszubski, above, note 2.

7. Eugenio Garin, *Italian Humanism: Philosophy and Civic Life in the Renaissance,* trans. Peter Munz (Oxford: Blackwell, 1965), pp. 106–111; idem, "Interpretations of the Renaissance," in *Science and Civic Life in the Italian Renaissance,* trans. Peter Munz (New York: Doubleday, 1969), pp. 11–12.

8. Frances A. Yates, *Giordano Bruno and the Hermetic Tradition* (Chicago: University of Chicago Press, 1964), pp. 85–86, 90, 103, 106–107, 110; idem, "The Hermetic Tradition in Renaissance Science," in *Art, Science, and History in the Renaissance,* ed. Charles S. Singleton (Baltimore: Johns Hopkins University Press, 1968), pp. 257, 259, 263, 270, 272. For the controversy about the "Yates thesis," see Copenhaver, "Natural Magic, Hermetism, and Occultism in Early Modern Science," in *Reappraisals of the Scientific Revolution,* ed. David Lindberg and Robert Westman (Cambridge: Cambridge University Press, 1990), pp. 261—301.

9. Pico, *Oratio,* ed. Garin, p. 102: "Cui sententiae illud Mercurii adstipulatur: Magnum, o Asclepi, miraculum est homo"; *Asclep.* 6; Yates, *Bruno,* p. 90.

10. Pico *Oratio,* ed. Garin, pp. 102, 114 n. 2, 144: "totam sensualem partem in quam sedet corporis illecebra quae animam obtorto, ut aiunt, detinet collo"; *Asclep.* 12: "Res enim dulcis . . . in hac corporali uita . . . quae res animam obtorto, ut aiunt, detinet collo."

11. Pico, *Oratio,* ed. Garin, pp. 112–114; *Hermetica: The Greek "Corpus Hermeticum" and the Latin "Asclepius" in a New English Translation,* ed. and trans. Brian P. Copenhaver (Cambridge: Cambridge University Press, 1992), pp. xxxix, lii, 102–103, 139–140, 144–145, 152–153, 185–186, 199, 206, 223, 234, below, notes 51, 52, 57, 75–77.

12. Copenhaver, *Hermetica,* pp. xxxv–xl, 149, 203–204, 226.

13. Pico, *Oratio,* ed. Garin, p. 148; Copenhaver, *Hermetica,* pp. 81, 89–91, 238–240, 254–257; idem, "Natural Magic," pp. 270–275; idem, "Hermes Theologus: The Sienese Mercury and Ficino's Hermetic Demons," in *Humanity and Divinity in Renaissance and Reformation: Essays in Honor of Charles Trinkaus,* ed. John O'Malley et al. (Leiden: Brill, 1993), pp. 149–182; idem, "Lorenzo de' Medici, Marsilio Ficino, and the Domesticated Hermes," in *Lorenzo il Magnifico e il suo mondo: Convegno internazionale di studi, Firenze, 9–13 giugno 1992,* ed. G. C. Garfagnini (Florence: Olschki, 1994), pp. 225–257.

14. Pico, *Conclusiones* 4.8.1–10; 5.10.2, 4; 5.11.10 (B 54, 120, 122, 126): 4.8.7: "Sex uiis futura homini Deus denunciat: per somnia, portenta, aues, intestina, spiritus et Sybillam"; 4.8.10: "Decem ultores de quibus dixit secundum Mercurium praecedens conclusio uidebit profundus contemplator correspondere malae coordinationi denariae in Cabala et praefectis illius de quibus ego in cabalisticis conclusionibus nihil posui quia est secretum"; 5.10.2: "Nihil efficientius hymnis Orphei in naturali magia si debita musica, animi intentio et ceterae circumstantiae qua[s] norunt sapientes fuerint adhibitae"; 5.10.4: "Sicut hymni Dauid operi Cabalae mirabiliter deseruiunt, ita hymni Orphei oper[i] uere licitae et naturalis magiae." Yates, *Bruno,* pp. 109–110, points out that in copying the list of twelve "tormentors" in *Corpus Hermeticum* 13.7 from Ficino's Latin *Pimander (Mercurii Trismegisti liber de potestate et sapientia Dei: Corpus Hermeticum I–XIV, Versione latina di Marsilio Ficino,* ed. Sebastiano Gentile [Florence: SPES, 1989], fol. 50r), Pico "left out two" in the preceding ninth conclusion "to make a comparison with Cabala in . . . [the] tenth"; this is true, but his compression of the twelve tormentors to ten also anticipated later critics who reduced the twelve to match "the arrival of the decad" in *C. H.* 13.10–12; Copenhaver, *Hermetica,* pp. 51–52, 187–192. On the "evil tenfold grouping" in Cabala, see Wirszubski, *Jewish Mysticism;* p. 188; on the *Orphica, Hermetica,* and Cabala, pp. 195–199. See also below, notes 48, 50.

15. Moshe Idel, "The Magic and Neoplatonic Interpretations of the Kabbalah in the Renaissance," in *Jewish Thought in the Sixteenth Century, ed.* B. D. Cooperman (Cambridge, Mass.: Harvard University Press, 1983), pp. 200–202; above, notes 1, 7. On the chronology of Pico's acquaintance with Alemanno, see also Wirszubski, *Jewish Mysticism,* pp. 256–257, referring to the *Commento,* ed. Garin, pp. 535, 595.

16. Galileo Galilei, *Opere,* ed. Franz Brunetti (Turin: Unione Tipografico, 1964), 1:573, 631–632.

17. Aquinas, *Summa contra gentiles* 3.105; Tommaso Campanella, *Magia e grazia: Theologicorum liber XIV,* ed. R. Amerio (Rome: Istituto di Studi Filosofici, 1957), pp. 192–194; Marin Mersenne, *Quaestiones in Genesim* (Paris, 1623), col. 1151; idem, *Correspondance,* ed. C. De Waard et al. (Paris: Beauchesne, etc., 1932–1988), 2:443–445; René Descartes, *Oeuvres de Descartes,* ed. C. Adam and P. Tannery (Paris: Vrin, 1964–1976), 2:73–74, 363–368; 3:211–212, 367, 371–372, 420–421, 460–464, 485–517, 523–524, 528, 535, 558–559, 598–599; 4:77–78, 85–89; 5:125–128; 6:239, 245; 7:586, 596; 8.2:15–16, 22–23, 142, 150–152, 174, 179; Stephen Gaukroger, *Descartes: An Intellectual Biography* (Oxford: Clarendon, 1995), pp. 354–361; Copenhaver, "Astrology and Magic," pp. 282–283, 295; below, note 39.

18. *Determinationes magistrales reuerendi patris domini Petri Garsiae episcopi Usselensis contra conclusiones apologeticas Ioannis Pici Mirandulani* . . . (Rome: Eucharius Silber, 1489), sigs.

hv–viii, lv–m, miiii–v; Henri Crouzel, *Une Controverse sur Origène à la renaissance: Jean Pic de la Mirandole et Pierre Garcia* (Paris: Vrin, 1977), pp. 23–33.

19. Above, note 1.

20. For a comprehensive account of many of these topics as they arose in medieval natural philosophy, see Edward Grant, *Planets, Stars, and Orbs: The Medieval Cosmos, 1200–1687* (Cambridge: Cambridge University Press, 1994).

21. Pico, *Conclusiones* 1.1.14 (B 8): "Corpus mobile est subiectum scientiae naturalis"; 1.2.37 (B 12): "Grauia et leuia a nullo alio motore mouentur quam uel a generante uel a remouente prohibens"; 1.4.18 (B 16): "Aliquid potest mouere seipsum de actu uirtuali ad actum formalem"; 1.6.11 (B 20): "Dato uacuo, si aliquid in [eo] moueatur, in instanti mouebitur"; 2.1.9 (B 20) "Coelum est corpus simplex, non compositum ex materia et forma"; 2.2.3 (B 24): "In coelo est materia eiusdem rationis cum materia inferiorum"; 2.4.4 (B 28): "Corpora celestia non largiuntur formaliter inferioribus nisi caliditatem"; 2.5.3 (B 28): "Coelum calefacit inferiora per lumen suum super ea cadens"; 3.1.1 (B 30): "Si coelum inanimatum esset, esset quocunque animato corpore ignobilius, quod dicere impium est in philosophia"; 3.4.2 (B 34): "Cuilibet coelo, praeter animam quae mouet eum efficaciter, propria assistit intelligentia quae illum mouet ut finis ab anima tali secundum substantiam omnino distincta"; 3.5.2 (B 34): "Intellectus agens illuminans tantum credo sit illud apud Themistium quod est Metatron in Cabala"; 3.5.3 (B 34): "Scientia de anima est media inter scientias naturales et diuinas"; 4.1.2 (B 36): "Non tota descendit anima quum descendit"; 4.2.7 (B 38): "Anima est fons motus et gubernatrix materiae"; 4.3.2 (B 38): "Opifex mundi est supermundana anima"; 4.4.7 (B 40): "Nulla est uis celestium astrorum quantum est in se malefica"; 4.5.36 (B 48): "Quatuor sunt exercitus iuniorum deorum. Primus habitat a primo coelo usque ad principium aeris, secundus inde usque ad dimidium aeris, tertius inde usque ad terram"; 4.6.7 (B 52): "Per secretum radii recti, reflexi et refracti in scientia perspectiuae triplicis naturae admonemur: intellectualis, animalis et corporalis"; 4.6.7 (B 54): "Quicquid est [a] luna supra purum est lumen, et illud est substantia orbium mundanorum"; 4.8.2 (B 54): "Omne motum corporeum, omne mouens incorporeum"; 4.9.20 (B 58): "Cum fiet lux speculi non lucentis sicut speculi lucentis, erit nox sicut dies, ut dicit Dauid."

22. Pico, *Conclusiones* 5.1.16 (B 64): "An corpus compositum accipiat physicus a metaphysico discordare fundamentaliter non possunt Auenrois et Auicenna etsi uerbis discrep[e]nt"; 5.2.42 (B 72): "Motus a[b] Aristotele datus quomodo calefiant inferiora a superioribus nullo modo rectus apparet"; 5.3.46 (B 84): "Dato quocunque obiecto practicabili, nobilior est operatio quae eum practicat quam quae eum contemplatur, si cetera sint paria"; 5.5.45 (B 102): "Sensus naturae quem ponunt Alchindus, Bacon, Guillielmus Parisiensis et quidam alii, maxime autem omnes Magi, nihil est aliud quam sensus uehiculi quem ponunt Platonici"; 5.6.9 (B 106): " . . . quod declinare plus ad sensum quam ad intellectum non est animae ut anima est sed ut cadens est"; 5.7a.5 (B 106): "Sicut dictum Aristotelis de antiquis—dicentis quod ideo errarunt in physica contemplatione quia mathematice res physicas tractarunt—uerum esset si illi materialiter mathematica non formaliter accepissent, ita est uerissimum modernos qui de naturalibus mathematice disputant naturalis philosophiae fundamenta destruere"; 5.7b.74 (B 114): "Utrum in coelo sint descripta et significata omnia cuilibet scienti legere"; 5.8.4 (B 114): "Ibidem per

radices terrae nihil aliud intelligere possunt quam uitam uegetalem, conuenienter ad dicta Empedoclis, qui ponit transanimationem etiam in plantis"; 5.9.3 (B 116): "Magia est pars practica scientiae naturalis"; 5.10.12 (B 112): "Per octonarium numerum hymnorum maritimorum corporalis naturae nobis propriet[as] design[atur]"; 5.11.72 (B 140): "Sicut uera astrologia docet nos legere in libro Dei, ita cabala docet nos legere in libro Legis."

23. Vulg. Ps. 19:1–4:

Caeli enarrant gloriam Dei,
Et opera manuum eius annuntiat firmamentum.
Dies diei eructat uerbum,
Et nox nocti indicat scientiam.
Non sunt loquelae, neque sermones,
Quorum non audiantur uoces eorum,
Et in fines orbis terrae uerba eorum.

Cf. Ps. 19:1–4 in *Revised English Bible* (Oxford: Oxford University Press, 1989), p. 474; Ps. 19:4 in *Biblia Hebraica Stuttgartensia* (Stuttgart: Deutsche Bibelstiftung, 1967/1977), p. 1101: אֵין־אֹמֶר וְאֵין דְּבָרִים בְּלִי נִשְׁמָע קוֹלָם Jacques Lefèvre d'Etaples, *Quincuplex Psalterium: Facsimilé de l'édition de 1513* (Geneva: Droz, 1979), fol. 28v; C. A. and E. G. Briggs, *A Critical and Exegetical Commentary on the Book of Psalms* (Edinburgh: Clark, 1906), 1:164–166, 172.

24. *Ioannis Pici Mirandulae, etc. expositio posterior naturalis XVIII Psalmi, Coeli enarrant gloriam Dei, etc.,* in Biblioteca Comunale di Ferrara, MS cl. 2, 26, fols. 65r–86r, esp. 66v: *"Non sunt loquellae etc.:* Hebrei qui legunt *Nec auditur uox eorum,* ut supra est indicatum, se ad sensibilia interpretantur quod non ita, *Coeli Dei gloriam enarrant,* ut sensibili uoce quae audiatur deum laudent. Eadem autem uerba uidentur et hunc sensum recipere, *Non sunt loquellae neque sermones,* idest multas dabis linguas, multos sermones, et ita erit quod non audietur apud illos, scilicet uox eorum, et hunc sensum Arabes sequuntur. Potest et sic exponi, *Non sunt loquellae neque sermones* caelorum, et idem magis explicans subdit: *Quorum non audiatur uox eorum,* id est quod caelorum uoces non audiuntur, et erit sensus Hebreorum. Sed quae sequuntur, *In omnem terram, etc.* uidentur magis coherere primo sensui quam isti; possunt referri et ad hoc quod *Firmamentum* ubique enunciet opera Dei et *Coeli* eius gloriam manifestent." This and other Psalm commentaries by Pico are studied by Eugenio Garin, "Il commento ai Salmi: Le parte inedite," in *La cultura filosofica del rinascimento, ricerche e documenti* (Milan: Bompiani, 1994), pp. 249–253.

25. Pico, *Conclusiones* 5.9.19–22 (B 118–20): 19. "Ideo uoces et uerba in magico opere efficientiam habent quia illud in quo primum magicam exercet natura uox est Dei"; 20. "Quaelibet uox uirtute[m] habet in magia in quantum Dei uoce formatur"; 21. "Non significatiuae uoces plus possunt in magia quam significatiuae, et rationem conclusionis intelligere potest qui est profundus ex praecedenti conclusione"; 22. "Nulla nomina ut significatiua, et inquantum nomina sunt singula et per se, sumpta in magico opere uirtutem habere possunt nisi sint Hebraica uel inde proxime deriuata."

26. Pico, *Conclusiones* 5.9.1–6, 11, 13 (B 116–118). For the background, see Eugenio Garin, "Magic and Astrology in the Civilization of the Renaissance," in *Science and Civic Life,* pp. 145–165; idem, *Astrology in the Renaissance,* pp. 45–52, 90–103; D. P. Walker, *Spiritual and Demonic Magic from Ficino to Campanella,* Studies of the Warburg Institute 22

(London: Warburg Institute, University of London, 1958), pp. 30–59, 75–84; Paola Zambelli, "Platone, Ficino e la magia," in *L'ambigua natura della magia: Filosofi, streghe, riti nel Rinascimento* (Milan: Il Saggiatore, 1991), pp. 29–52; idem, "Il problema della magia naturale nel Rinascimento," in ibid., pp. 121–152; Copenhaver, "Magic and Astrology"; idem, "Natural Magic"; idem, "Hermes Trismegistus, Proclus, and the Question of a Philosophy of Magic in the Renaissance," in *Hermeticism and the Renaissance: Intellectual History and the Occult in Early Modern Europe,* ed. Ingrid Merkel and Allen G. Debus (Washington, D.C.: Folger Books, 1988), pp. 79–110.

27. Pico, *Conclusiones* 5.9.6–9, 15, 18, 26 (B 118–120): 9. "Nulla est scientia quae nos magis certificet de diuinitate Christi quam magia et cabala." See Wirszubski, *Jewish Mysticism,* pp. 123–124, 194–195; Idel, "Neoplatonic Interpretations," p. 233 n. 65.

28. Above, note 25; below, note 45; Wirszubski, *Jewish Mysticism,* p. 189; Gershom Scholem, "Der Name Gottes und die Sprachtheorie der Kabbala," in *Judaica,* vol. 3, *Studien zur jüdischen Mystik* (Frankfurt/Main: Suhrkamp, 1973), pp. 21–33, 67–70.

29. Pico, *Conclusiones* 5.7a.5, 9.23–25 (B 106, 120): 24. "Ex secretioris philosophiae principiis necesse est confiteri plus posse characteres et figuras in opere magico quam possit quaecunque qualitas materialis"; Wirszubski, *Jewish Mysticism,* pp. 83, 140–141, 144.

30. Pico, *Conclusiones* 5.11.1–2 (B 126): 1. "Quicquid dicant ceteri Cabalistae, ego prima diuisione scientiam Cabalae in scientiam Sephirot et Semot, tanquam in practicam et speculatiuam distinguerem", 2. " . . . ego partem speculatiuam Cabalae quadruplicem diuiderem. . . . Prima est scientia quam ego uoco alphabetariae reuolutionis. . . . Secunda, tertia et quarta pars est triplex Merchiaua . . . de diuinis, de mediis et sensibilibus naturis"; Wirszubski, *Jewish Mysticism,* pp. 135–136, 139; Idel, "Neoplatonic Interpretations," pp. 197–200; Gershom Scholem, *Kabbalah* (Jerusalem: Keter, 1974), pp. 106–109. The order of "practicam et speculatiuam" in 5.11.1 is unexpected, either because Pico assumed that the reader ought to know that the Cabala of names was practical and the Cabala of *Sefirot* theoretical or because he wanted to reverse these categories. Paolo Fornaciari (*Conclusione cabalistiche* [Milan: I Cabiri, 1994], p. 35 n.1) reads the Latin as given and understands the Cabala of names as "that which works on the names of God with the effect of gaining mystical, metaphysical knowledge and thus is speculative." In any case, theory and practice are so interdependent in Pico's Cabala that the distinction becomes blurred.

31. Wirszubski, *Jewish Mysticism,* pp. 136–138, 261; Scholem, *Kabbalah,* pp. 5–6, 10–30, 53–55, 105, 180–182; Moshe Idel, *The Mystical Experience in Abraham Abulafia,* trans. J. Chipman (Albany: State University of New York Press, 1988), pp. 7–41; idem, *Kabbalah: New Perspectives* (New Haven: Yale University Press, 1988), pp. xii, 49–56, 97–103, 106, 127, 146–149, 169–170, 188, 200–209, 214–215, 235–238; Colette Sirat, *A History of Jewish Philosophy in the Middle Ages* (Cambridge: Cambridge University Press, 1985), pp. 262–266.

32. Pico, *Conclusiones* 5.11.3 (B 126): "Scientia quae est pars practica Cabalae practicat totam metaphysicam formalem et theologiam inferiorem." Wirszubski, *Jewish Mysticism,* pp. 139–143, 254; Scholem, *Kabbalah,* pp. 182–189, 310–311; Idel, *Kabbalah,* pp. 41, 47, 101, 202–203; above, note 30.

33. Pico, *Oratio*, ed. Garin, pp. 154–158; Gershom Scholem, "Considérations sur l'histoire des débuts de la Kabbale chrétienne," in *Kabbalistes chrétiens, ed.* A. Faivre and F. Tristan (Paris: Albin Michel, 1979), pp. 19–26, 34–36; P. O. Kristeller, "Giovanni Pico della Mirandola and His Sources," in *L'Opera e il pensiero di Giovanni Pico*, 1:74–75; François Secret, *Les Kabbalistes chrétiens de la renaissance* (Paris: Dunod, 1964), pp. 1–32, 38–43; David B. Ruderman, "The Italian Renaissance and Jewish Thought," in *Renaissance Humanism: Foundations, Forms, and Legacy,* ed. A. Rabil (Philadelphia: University of Pennsylvania Press, 1988), 1:397–404, 414–415, below, note 61.

34. Pico, *Opera (1557–73),* pp. 170–171, 175–176, 180: "Praeter legem quam Deus dedit Moysi in monte, . . . reuelatam quoque fuisse . . . ueram legis expositionem cum manifestatione omnium mysteriorum et secretorum. . . . De hac uero mandatum ei a Deo ne ipsam scriberet sed sapientibus solum . . . communicaret, quos idem Moyses . . . elegerat . . . eisque itidem praeciperet ne eam scriberet sed successoribus suis uiua uoce reuelarent. . . . Ex quo modo tradendi istam scientiam per successiuam scilicet receptionem . . . dicta est ipsa scientia scientia Cabalae . . . quod scientia receptionis. . . . Fuerunt autem postea haec mysteria literis mandata, . . . et illi dicti sunt libri Cabalae, in quibus libris multa—imo pene omnia—inueniuntur cosona fidei nostrae. . . . Ex istis libris . . . ego multas posui conclusiones ad confirmationem fidei nostrae contra Iudaeos."

35. Pico, *Opera (1557–73),* pp. 180–181: "Verum quia iste modus tradendi per successionem qui dicitur Cabalisticus uidetur conuenire unicuique rei secretae et mysticae, hinc est quod . . . unumquodque scibile quod per uiam occultam alicunde habeatur dicatur haberi per uiam Cabalae. In uniuersali autem duas scientias hoc etiam nomine honorificarunt, unam quae dicitur ars combinandi et est simile quid sicut apud nos dicitur ars Raymundi . . . Aliam quae est de uirtutibus rerum superiorum quae sunt supra lunam, et est pars magiae naturalis suprema. . . . Ista . . . quae uno modo potest capi ut pars magiae naturalis, alio modo ut res distincta ab ea, est illa de qua loquor in praesenti conclusione, dicens quod adiuuat nos in cognitione diuinitatis Christi. . . . Verum, sicut cum olim magi tantum dicerentur sapientes, necromantes deinde et diabolici uiri sapientis sibi falso nomen uendicantes magos se uocauerunt, ita et quidam apud Hebraeos . . . nihil a necromantibus differentes dixerunt se habere secreta Dei nomina et uirtutes quibus daemones ligarent et miracula facerent"; p. 171: "Refellam omnem magiam prohibitam ab Ecclesia . . . protestans me solum loqui de magia naturali, et . . . declarans . . . quod per istam magicam nihil operamur nisi solum actuando uel uniendo uirtutes naturales."

36. Pico, *Opera (1557–73),* pp. 171–172: "Praedictam autem . . . restrictionem intentionis meae in conclusionibus magicis ad magiam naturalem intendo esse applicandam cuilibet conclusioni particulari, et ita cum dico de actiuitate characterum et figurarum in opere magico, loquor de uera actiuitate sua et naturali. . . . Et licet secundum Peripateticos longe minoris sint actiuitatis et uirtutis quam sint qualitates materiales, tamen dicerent Pythagorici . . . quod sicut mathematica sunt formaliora physicis, ita etiam actualiora."

37. T. Schrire, *Hebrew Magic Amulets: Their Decipherment and Interpretation* (New York: Behrman House, 1966), pp. 46–47; John G. Gager, ed., *Curse Tablets and Binding Spells from the Ancient World* (Oxford: Oxford University Press, 1992), pp. 6–12, 56–58, 65–71, 107–110, 169–171, 181, 237–239. Figure 1.3 comes from a modern printed edition of the *Sefer Raziel,* first published in Amsterdam in 1701, but the magical texts underlying this

compilation circulated in the Middle Ages and were certainly known to Pico's contemporaries; Cusanus cites a book "qui Salamoni inscribitur et vocatur Sepher Raziel" in the first part of his *Sermo I, De nominibus* (Cusanus, *Opera omnia, vol. 17, Sermones I.1,* ed. R. Haubst et al. [Hamburg, 1970], p. 6). Latin versions, one associated with the patronage of Alfonso the Wise (1221–1284), survive in manuscripts from the fifteenth and sixteenth centuries; see François Secret, "Sur quelques traductions du *Sefer Razi'el*," *Revue des études juives* 128 (1969): 223–245; Idel, "Neoplatonic Interpretations," pp. 193–194; cf. Wirszubski, *Jewish Mysticism,* pp. 130, 150.

38. Augustine, *De doctrina Christiana* 2.23.36, 29.45; Albertus, *De mineralibus* 1.1.5–6, 9; 2.1.3–4, 3.6; [Albertus,] *Speculum astronomiae* 11, 16; Aquinas, *Summa contra gentiles* 3.99, 104–105; *Summa theologiae* 2–2.96.2; *De occultis operibus naturae* 14, 17–20. Walker, *Spiritual and Demonic Magic,* pp. 42–44; Copenhaver, "Scholastic Philosophy and Renaissance Magic in the *De vita* of Marsilio Ficino," *Renaissance Quarterly* 37 (1984); 523–554.

39. Copenhaver, "Scholastic Philosophy," pp. 533–545; idem, "Hermes Trismegistus," pp. 84–91; idem, "Renaissance Magic and Neoplatonic Philosophy: 'Ennead' 4.3–5 in Ficino's 'De vita coelitus comparanda,'" in *Marsilio Ficino e il ritorno di Platone: Studi e documenti,* ed. G. C. Garfagnini (Florence: Olschki, 1986), 2:351–369; idem, "Iamblichus, Synesius, and the Chaldaean *Oracles* in Marsilio Ficino's *De vita libri tres:* Hermetic Magic or Neoplatonic Magic?" in *Supplementum Festivum: Studies in Honor of Paul Oskar Kristeller,* ed. James Hankins, John Monfasani, and Frederick Purnell, Jr. (Binghamton, N.Y.: Medieval and Renaissance Texts and Studies, 1987), pp. 441–455; Michael Allen, "Summoning Plotinus: Ficino, Smoke, and the Strangled Chickens," in *Plato's Third Eye: Studies in Marsilio Ficino's Metaphysics and Its Sources* (Aldershot: Variorum, 1995), pp. 63–88.

40. Johann Reuchlin, *On the Art of the Kabbalah: De arte cabalistica,* trans. Martin and Sarah Goodman (Lincoln: University of Nebraska Press, 1993), p. 348: " . . . at nunc uerba quibus creator omnipotens fecit coelum et terram alligata, num putauerit inquiunt aliquis nihil posse?" See Idel, *Kabbalah,* pp. 188–190.

41. Aquinas, *Summa theologiae* 1.1.10: "Respondeo dicendum quod auctor sacrae scripturae est Deus, in cuius potestate est ut non solum uoces ad significandum accommodet . . . sed etiam res ipsas. Et ideo, cum in omnibus scientiis uoces significent, hoc habet proprium ista scientia quod ipsae res significatae per uoces etiam significant aliquid. Illa ergo prima significatio, qua uoces significant res, pertinet ad primum sensum, qui est sensus historicus uel litteralis. Illa uero significatio qua res significatae per uoces iterum res alias significant dicitur sensus spiritualis; qui super litteralem fundatur et eum supponit." Thomas's position, though influential, was not uncontroversial: see G. R. Evans, *The Language and Logic of the Bible: The Road to Reformation* (Cambridge: Cambridge University Press, 1985), pp. 39–50.

42. Pico, *Opera (1557–73),* pp. 231–232, 240: "Non intendo ista uerba fuisse materialiter accipienda cum proferebantur a Christo sed solum quando proferuntur a sacerdote euangelico consecrante. . . . Oratio prolata a tali sacerdote supponit pro oratione prolata a Christo significatiue sumpta. . . . Id quod recitatur a sacerdote est illa oratio, 'Hoc est enim corpus meum,' materialiter sumpta pro simili oratione et ex similibus uocibus composita a Christo significatiue prolata." The conclusion (*Conclusiones* 5.4.10 [B 90]) under discussion here is the tenth that Pico called "rather different from the common mode of

speaking of the theologians." Note that his terms are *significatiue, materialiter,* and *recitatiue,* but not *personaliter,* as in the passage from Occam (below, note 43). On supposition and signification, see Copenhaver and Schmitt, *Renaissance Philosophy,* pp. 99–101.

43. William of Occam, *Summa totius logicae* 1.100.63: "Suppositio personalis est quando terminus supponit pro suo significato et significatiue. . . . Suppositio materialis est quando terminus non supponit significatiue sed supponit pro uoce uel pro scripto."

44. Pico, *Opera (1557–73),* p. 175: "Similiter de nominibus quod habeant aliquam ac-tiuitatem naturalem . . . non habent ut significatiua sunt ad placitum sed ut sunt in se quaedam res naturales . . . nisi forte essent aliqua quibus significatio esset naturalis, sicut Stoici dicunt de omnibus nominibus, quibus ut aduersantur Peripatetici, ita Plato in *Cratilo* assentitur. . . . Origenes autem de Hebraicis . . . dicit quod quaedam nomina . . . in sacris literis sicut Osanna, Sabaoth, halleluia et similia fuerunt . . . non mutata in aliam linguam in qua non retinuissent suam naturalem significationem et consequenter uir-tutem." For Pico the key text here is Origen, *Contra Celsum* 1.24, which summarizes the disagreement among Platonists, Aristotelians, and Stoics on meaning, onomatopoeia, and etymology; A. A. Long, *Hellenistic Philosophy: Stoics, Epicureans, Sceptics,* 2nd ed. (Berke-ley: University of California Press, 1986), pp. 131–135; Long and D. N. Sedley, *The Hel-lenistic Philosophers,* vol. 1, *Translations of the Principal Sources with Philosophical Commentary* (Cambridge: Cambridge University Press, 1987), pp. 192–195. Greeks and Romans at-tributed magical power to words from Egyptian, Hebrew, and other unfamiliar languages, as noted by Plotinus (*Enn.* 5.8.6), Iamblichus (*De myst.* 2.11; 7.4.254–256), and *Corpus Hermeticum* 16.2. Because Ficino had translated only the first fourteen treatises of the Greek *Corpus Hermeticum,* Pico may not have known *C.H.* 16, and he cites neither Plo-tinus nor Iamblichus on this issue (Pico, *Conclusiones* 4.1.1–15; 4.4.1–9 [B 36, 40–42]); see also John Ferguson, *The Religions of the Roman Empire* (London: Thames and Hudson, 1970), pp. 167–168, 177.

45. Pico, *Conclusiones* 5.2.80, 3.55 (B 76, 84): 80. "Si qua est lingua prima et non casu-alis, illam esse hebraicam multis patet coniecturis"; 55. "Qui ordinem hebraicae linguae profunde et radicaliter tenuerit atque illum proportionaliter in scientiis seruare no-uerit, cuiuscumque scibilis perfecte inueniendi normam et regulam habebit" above, notes 25, 28.

46. Reuchlin, *De arte cabalistica,* p. 264; Scholem, *Kabbalah,* pp. 337–343. Since much turns on there being just seventy-two Cabalist theses at the end of nine-hundred conclu-sions, it is important to note that the *title* of the final set of Cabalist theses reads "LXXI" rather than "LXXII," even though the number of actual items in this group is seventy-two. I agree with Paola Zambelli (*L'apprendista stregone: Astrologia, cabala e arte lulliana in Pico della Mirandola e seguaci* [Venice: Marsilio, 1995], pp. 38–39) that the discrepancy is probably just a mistake; cf. Giovanni di Napoli, *Giovanni Pico della Mirandola e la proble-matica dottrinale del suo tempo* (Paris: Desclee, 1965), pp. 85–87; Fornaciari, *Conclusione ca-bilistiche,* pp. 15–16.

47. Pico, *Conclusiones* 4.6.3 (B 52): "Ubi unitas punctualis [c]adit in alteritatem binarii, ibi est primo triangulum"; 5.7a.9–10 (B 108): 9. "Per arithmeticam non materialem sed formalem habetur optima uia ad prophetiam naturalem"; 10. "Joachim in prophetiis suis alia uia non processit quam per numeros formales"; 5.11.55–57 (B 136): 55. "Quod

dicunt Cabalistae, lumen repositum in septuplo lucere plus quam lumen relictum, mirabiliter conuenit arithmeticae pythagoricae"; 56. "Qui sciuerit explicare quaternarium in denarium habebit modum, si sit peritus cabalae, deducendi ex nomine ineffabili nomen LXXII litterarum"; 57. "Per praecedentem conclusionem potest intelligens in arithmetica formali intelligere quod operari per Scemamphoras est proprium rationali naturae"; Reuchlin, *De arte cabalistica,* p. 272. Louis Valcke ("Des *Conclusiones* aux *Disputationes:* Numérologie et mathématiques chez Jean Pic de la Mirandole," *Laval théologique et philosophique* 41 [1985]: 43–56) distinguishes the Pythagorean numerology of the *Conclusiones* from what he sees as the more properly mathematical posture of Pico's later work. For figured numbers, see Paul-Henri Michel, *De Pythagore à Euclide: Contribution à l'histoire des mathématiques préeuclidiennes* (Paris: Les Belles Lettres, 1950), pp. 295–325; and for figured numbers in Ficino, see Michael J. B. Allen, *Nuptial Arithmetic: Marsilio Ficino's Commentary on the Fatal Number in Book VIII of Plato's "Republic"* (Berkeley: University of California Press, 1994), pp. 97–100. "Forms" in the hylemorphic sense of Aristotelian metaphysics—the forms of physical objects—were identified by Iamblichus, *On Pythagoreanism* (excerpted by Psellus), as "physical numbers," and Iamblichus included the tetractys among "physical numbers of the formal type" (φυσικοὶ ... ἀριθμοὶ κατὰ τὸ εἶδος), as explained in Dominic J. O'Meara, *Pythagoras Revived: Mathematics and Philosophy in Late Antiquity* (Oxford: Clarendon, 1989), pp. 53–70, 217–219. See also Theon of Smyrna, *De utilitate mathematicae* 93.17–99.16, on the tetractys as the building block of the world. On Joachim of Fiore, see Marjorie Reeves, *Joachim of Fiore and the Prophetic Future* (London: SPCK, 1976), pp. 5–25, 83–95; and on the number 72, see below, note 61; also note 40, above.

48. Pico, *Conclusiones* 5.10.2, 4, 13, 15, 21 (B 120–124): 13. "Idem est Typhon apud Orpheum et Z[a]mael in Cabala"; 15. "Idem est Nox apud Orpheum et Ensoph in Cabala"; 21. "Opus praecedentium hymnorum nullum est sine opere Cabalae cuius est proprium practicare omnem quantitatem formalem, continuam e[t] discretam"; Wirszubski, *Jewish Mysticism,* pp. 141, 148, 195–196; above, note 14.

49. Above, note 47, though Wirszubski, *Jewish Mysticism,* pp. 141–142, 196, offers a different understanding of "quantitas formalis, continua e[t] discreta"; he concludes that Pico had in mind *middah* (*proprietas* in Latin) as a synonym of *Sefirah,* referring to the divine attributes.

50. Pico, *Conclusiones* 5.10.5 (B 122): "Tantus est numerus hymnorum Orphei quantus est numerus cum quo Deus triplex creauit saeculum sub quaternarii pythagorici forma numeratus"; 5.11.6 (B 126): "Tria magna Dei nomina quaternari[a] quae sunt in secretis Cabalistarum per mirabilem appropriationem tribus personis trinitatis ita debere attribui ut nomen [אהיה] sit patris, nomen [יהוה] sit filii, nomen [אדני] sit spiritus sancti intelligere potest qui in scientia cabalae fuerit profundus"; Reuchlin, *De arte cabalistica,* p. 340; Wirszubski, *Jewish Mysticism,* p. 197. At the end of the eighteenth century, an edition of Thomas Taylor's translation still counted eighty-six Orphic hymns: Taylor, *The Mystical Initiations or Hymns of Orpheus* (London: T. Payne and Son, 1787), pp. 224–227.

51. Pico, *Oratio,* ed. Garin, pp. 118–120: " . . . unanimi amicitia qua omnes animi in una mente quae est super omnem mentem non concordent adeo sed ineffabili quodam modo unum penitus euadant. Haec est illa amicitia quam totius philosophiae finem esse

Pythagorici dicunt, haec illa pax . . . per eam ipsi homines ascendentes in caelum angeli fierent; hanc pacem amicis . . . optemus, . . . animae nostrae. . . . In se ipsa cupiet mori ut uiuat in sponso, in cuius conspectu preciosa profecto mors sanctorum eius, mors, inquam, illa, si dici mors debet plenitudo uitae, cuius meditationem esse studium philosophiae dixerunt sapientes"; below, notes 52, 74–77.

52. Pico, *Commento,* ed. Garin, pp. 118–120: "Puo dunque per la prima morte, che è separazione solo dell'anima dal corpo . . . vedere lo amante l'amata Venere celeste, . . . ma chi più intrinsecamente ancora la vuole possedere e, non contento del vederla e udirla, essere degnato de' suoi intimi amplessi e anelanti baci, bisogna che per la seconda morte dal corpo per totale separazione si separi . . . e ambedue una sola anima chiamare si possono. E nota che la più perfetta e intima unione che possa l'amante avere della celeste amata si denota per l'unione di bacio, e perchè e' sapienti cabalisti vogliono molti degli antiqui padri in tale ratto d'intelleto essere morti, troverai appresso di loro essere morti di binsica [בנשיקה], . . . cioè morte di bacio, è quando l'anima nel ratto intelletuale tanto alle cose separate si unisce che dal corpo elevata in tutto l'abbandona"; *Conclusiones* 4.9.44, 5.11. 11, 13 (B 60, 128): 11. "Modus quo ration[ales] animae per archangelum Deo sacrificantur, qui a Cabalistis non exprimitur, non est nisi per separationem animae a corpore, non corporis ab anima nisi per accidens, ut contigit in morte osculi, de quo scribitur, 'praeciosa in conspectu domini mors sanctorum eius'"; 13 "Qui operatur in Cabala sine admixtione extranei, si diu erit in opere, morietur ex binsica, et si errabit in opere aut non purificatus accesserit, deuorabitur ab Azazele per proprietatem iudicii"; Ps. 116:15; Wirszubski, *Jewish Mysticism,* pp. 50, 153–154, 158–160; Idel, *Kabbalah,* p. xvi; idem, *Abulafia,* pp. 180–184; Fornaciari, *Conclusione cabilistiche,* p. 41 n. 10; above, note 51.

53. Israel Tishby and Fischel Lachower, eds., *The Wisdom of the Zohar,* trans. D. Goldstein (Oxford: Oxford University Press, 1989), 2:449, 453, 459–464, 511–512, 529–530, 623–624, 631; Scholem, *Kabbalah,* pp. 53–56, 116–118, 320–326, 356–361, 385–388; Idel, *Kabbalah,* p. 167. On the term "property" (מִדָּה), see Wirszubski, *Jewish Mysticism,* pp. 38–39, 141–142.

54. Tishby and Lachower, *Zohar,* 2:531, 917–921, 1161–1167; Scholem, *Kabbalah,* pp. 362–366; *Encyclopedia Judaica,* 2:906–915; 11:687–697, 1474–1477.

55. Schrire, *Amulets,* pp. 62, 71, 97–99, 128, plates 42–43; Schrire reproduces and explains the childbirth amulet from the 1701 edition of the *Sefer Raziel,* on which see above, note 37.

56. Schrire, *Amulets,* pp. 68, 105–109; Idel, *Kabbalah,* pp. 33, 117, 230.

57. Pico, *Conclusiones* 3.5.2, 5.8.13 (B 34, 116); for Themistius, see above, note 21; the thirteenth Chaldaean conclusion is "Per puerum apud interpretes nihil aliud intellig[ibile] quam intellectum"; Pico, *Commento,* ed. Garin, p. 554: "nè altrimenti el detto si debbe intendere de' sapienti cabbalisti quando o Enoch in Matatron, angelo della divinità,o universalmente alcuno altro uomo in angelo dicono trasformarsi." For a related passage in the *Oration,* see Eugenio Garin, "La prima redazione dell' 'Oratio de hominis dignitate,'" in *Cultura Filosofica,* p. 235: "Nam et Hebraeorum theologia secretior nunc Enoch sanctum in angelum diuinitatis quem uocant ✦✦✦✦, nunc in alia alios numina reformant." Cf. Wirszubski, *Jewish Mysticism,* pp. 193, 199–200, who makes sense of all this by substitut-

ing מטטרון for the asterisks and arguing that the tenth Cabalist conclusion (*Conclusiones* 5.11.10 [B 128]) should read: "Illud quod apud Cabalistas dicitur מטטרון, illud est sine dubio quod ab Orpheo Pallas, a Zoroastre [p]aterna mens, a Mercurio Dei filius, a Pythagora sapientia, a Parmenide sphaera intelligibilis nominatur"; see also Idel, *Kabbalah,* pp. 60, 67.

58. Tishby and Lachower, *Zohar,* 2:626–629; Scholem, *Kabbalah,* pp. 377–381; idem, "Débuts de la kabbale," p. 25; Sirat, *Jewish Philosophy,* pp. 192–195, 262–266; Idel, *Abulafia,* pp. 27, 40, 116–119, 125–128, 137, 179–205; idem, *Kabbalah,* pp. 33, 39–40, 60–61, 67, 149, 230; Gedaliahu Stroumsa, "Form(s) of God: Some Notes on Metatron and Christ," *Harvard Theological Review* 76 (1983): 269—288; Daniel Abrams, "The Boundaries of Divine Ontology: The Inclusion or Exclusion of Metatron in the Godhead," *Harvard Theological Review* 87 (1994): 300–305.

59. Above, note 57.

60. Pico, *Opera (1557–73),* pp. 122–123: " . . . at mysteria secretiora et sub cortice legis rudique uerborum praetextu latitantia . . . plebi palam facere quid erat aliud quam . . . inter porcos spargere margaritas? Ergo clam uulgo habere perfectis communicanda . . . non humani consilii sed diuini praecepti fuit"; *Oratio,* ed. Garin, p. 156; *Conclusiones* 4.8.10, 5.9.24, 10.1, 11.11, 63 (B 54, 120, 128, 138); above, note 34; below, notes 74–77.

61. Alemanno's *Collectanea* are cited in Idel, "Neoplatonic Interpretations," p. 235 n. 95; see also pp. 202–205, 211; Reuchlin, *De arte cabalistica,* pp. 258, 264, 270–274, 310; Scholem, *Kabbalah,* p. 366; Schrire, *Amulets,* pp. 98–99, 105–109; see above, note 40. Fornaciari, *Conclusione cabalistiche,* p. 61 n. 40, points out the relevance of the long list of names in Exodus 34:6 to the name of seventy-two letters. On various uses of 72, see Idel, *Kabbalah,* pp. 46, 122–125, 169, 261; idem, *Abulafia,* pp. 15, 22–23, 35–38, 105–107; above, note 33.

62. Above, notes 47–49, 52, 54–55.

63. Pico, *Conclusiones* 5.11.14, 15, 43 (B 128, 134): 14 "Per litteram Scin quae mediat in nomine Jesu significatur nobis cabalistice quod tum perfecte quieuit tanquam in sua perfectione mundus cum Iod coniunctus est cum Vau, quod [f]actum est in C[h]risto, qui fuit uerus Dei filius et homo"; 15 "Per nomen Iod he u[au] he, quod est nomen ineffabile, quod dicunt Cabalistae futurum esse nomen Messiae euidenter cognoscitur futurum eum Deum dei filium per spiritum sanctum hominem factum, et post eum ad perfectionem humani generis super homines paraclytum descensurum"; 43 "Per mysterium duarum litterarum Vau et Iod scitur quomodo ipse Messias ut Deus fuit principium sui ipsius ut homo"; Wirszubski, *Jewish Mysticism,* pp. 165–166, 216; above, notes 25, 41–43, 54.

64. Pico, *Conclusiones* 5.11.20, 32, 34, 41, 60 (B 130, 132, 134, 138); Tishby and Lachower, *Zohar,* 2:1079–1080; Scholem, *Kabbalah,* pp. 169–170.

65. The apparatus in *Biblia Hebraica Stuttgartensia,* p. 688, notes the reading לְמוֹ (לוֹ =) רַבָּה for לְסַרְבֵּה, but the form of the preposition לְ here is unusual; M. Friedländer, ed., *The Commentary of Ibn Ezra on Isaiah* (London: N. Trübner, 1873), p. 52. Cf. J. A. Alexander, *Commentary on the Prophecies of Isaiah* (Grand Rapids, Mich.: Zondervan, 1975), pp. 207–209;

G. B. Gray, *A Critical and Exegetical Commentary on the Book of Isaiah* (Edinburgh: Clark, 1906), 1:176–177.

66. Pico, *Conclusiones* 5.11.41 (B 134): "Sciri potest in Cabala per mysterium mem clausi cur post se Christus miserit paracletum"; Nicolaus de Lyra, *Postilla super totam Bibliam: Strassburg, 1492* (Frankfurt/Main: Minerva, 1971), vol. 2, sig. CCiiir: " . . . *lemarbe . . .* significat ad multiplicandum et ponitur ibi littera *mem* clausa in medio dictionis contra naturam suam quia semper ponitur in fine dictionis et nunquam in medio nec in principio et hic ad denotandum quod Christus de quo scriptura hic loquitur natus fuit contra cursum naturae de uirgine clausa"; Yehuda Liebes, "Christian Influences on the Zohar," in *Studies in the Zohar,* trans. A. Schwartz et al. (Albany: State University of New York Press, 1993), pp. 145–150; Friedländer, *Ibn Ezra,* p. 52; Alexander, *Commentary on Isaiah,* p. 208. On threes, see below, notes 71, 72.

67. Pico, *Conclusiones* 4.6.12 (B 52): "In formis numerandis non debemus extendere quadragenarium."

68. *Sefer Yezirah* 1.13, 2.1, 3.1: בירר שלש אותיות מן הפשוטות בסוד שלש אמות אמ"ש וקבעם בשמו הגדול; text and translation in Aryeh Kaplan, *Sefer Yetzirah: The Book of Creation* (York Beach, Me.: Samuel Weiser, 1993), pp. 80–81, 95, 139.

69. Lev. 5:6, 6:10; Num. 6:12, 18:9; Isa. 53:2–12, esp. 10; Vulg. John 1:29; Acts 8:26–39. See Alexander, *Commentary on Isaiah,* pp. 303–304; Thorndike, *Magic and Experimental Science,* 1:737–738; 2:352–353; Keith Thomas, *Religion and the Decline of Magic* (New York: Scribner's, 1971), pp. 30–31; Richard Kieckhefer, *Magic in the Middle Ages* (Cambridge: Cambridge University Press, 1989), pp. 75–80. The *Liber redemptionis,* a commentary (probably by Abulafia) on the *Guide* of Maimonides in a Latin translation done for Pico by Flavius Mithradates, describes the transposition of אמש into אשם. Flavius also interpolated *agnus dei* (שֶׂה אֵל) into another work by Abulafia, *On the "Secrets of the Law"*: see Wirszubski, *Jewish Mysticism,* pp. 84–85, 89, 91, 97, 113.

70. Pico, *Conclusiones* 5.11.5 (B 126): "Quilibet hebraeus Cabalista secundum principia et dicta scientiae Cabalae cogitur ineuitabiliter concedere de trinitate et qualibet persona diuina, patre, filio et spiritu sancto, illud praecise sine additione uel diminutione aut uariatione, quod ponit fides catholica christianorum"; for the diagrams, see Kaplan, *Sefer Yetzirah,* pp. 29, 151.

71. Cf. *Orac. Chald.* 73 (Des Places):

Ἐν τούτοις ἱερὸς πρῶτος δρόμος, ἐν δ᾽ ἄρα μέσσῳ
ἠέριος, τρίτος ἄλλος ὃς ἐν πυρὶ τὴν χθόνα θάλπει.
Ἀρχαῖς γὰρ τρισὶ ταῖσδε λάβροις δουλεύει ἅπαντα.

In them the first holy stage; in the middle then
the airy one; third the other that warms the earth in fire.
All things together serve these three mighty masters.

For other threes in the *Oracles,* see 2, 22–23, 26–29, 31, 44, 48. See Exod. 19:15 for another limitation by the number three; also Augustine, *De ciuitate dei* 10.23–30; Wirszubski, *Jewish Mysticism,* pp. 193–194, 197; Idel, "Neoplatonic Interpretations," p. 198.

72. Pico, *Conclusiones* 5.5.1 (B 94): "Per numeros triplares qui a Platone in Timaeo ponuntur in triangulo animam significante admonemur quousque in formis numerandis sit progrediendum per naturam illius quod est prima forma formans. Per numeros uero duplares ibidem positos admonemur quatenus, positis duobus extremis terminis, coordinanda sunt media per naturam eius quod est medium in uniuerso"; Plato, *Timaeus* 35B; Francis M. Cornford, *Plato's Cosmology: The Timaeus of Plato Translated with a Running Commentary* (1937; reprint, New York: Liberal Arts Press, 1957), pp. 66–68; Allen, *Nuptial Arithmetic,* pp. 43–47, 62–63, 68, 71, 74–75, 111–112.

73. Pico, *Conclusiones* 5.5.1 (B 94). For the closed *mem* read as 40, see Wirszubski, *Jewish Mysticism,* pp. 71, 74; and Idel, *Abulafia,* p. 101, for the same letter read as 600.

74. Pico, *Conclusiones* 5.9.23 (B 120): "Quilibet numer[i] praeter ternarium et denarium sunt materiales in magia; isti formales sunt et in magica arithmetica sunt numeri numerorum"; Pico to Girolamo Benivieni, 12 November 1486 (in L. Dorez, "Lettres inédites de Jean Pic de la Mirandole," *Giornale storico della letteratura italiana* 25 [1895]; 358): "Postquam abisti, ad 900a excreuerunt progrediebanturque, nisi receptui cecinissem, ad mille. Sed placuit in eo numero, utpote mistico, pedem sistere. Est enim (si uera est nostra de numeris doctrina) symbolum animae in se ipsam oestro Musarum percitae recurrentis. Accessit et orationi id quod ad te mitto. Cum enim statutum sit mihi ut nulla pretereat dies quin aliquid legam ex euangelica doctrina, incidit in manus illud Christi: 'Pacem meam do uobis, pacem meam do uobis, pacem relinquo uobis.' Illico subit[a] quadam animi concitatione de pace quaedam ad philosophiae laudes facientia tanta celeritate dictaui ut notarii manum praecurrerem saepe et inuerterem." In Vulg. John 14:15–31, Jesus says "I give . . ." only once, not twice as in the trinity of promises remembered by Pico; cf. Garin, *Pico: Vita e dottrina,* p. 73; Wirszubski, *Jewish Mysticism,* p. 4 n. 6.

75. Below, note 77, with cross-references.

76. Ficino, *Platonis opera omnia* (1484; reprint Venice, 1491), fol. 60r, cited and translated in Michael Allen, "The Soul as Rhapsode: Marsilio Ficino's Interpretation of Plato's *Ion,"* in *Plato's Third Eye,* XV–144–5: "Iupiter rapit Apollinem; Apollo illuminat Musas; Musae suscitant et exagitant lenes et insuperabiles uatum animas; uates inspirati interpretes suos inspirant; interpretes autem auditores mouent"; Allen, *The Platonism of Marsilio Ficino: A Study of His Phaedrus Commentary, Its Sources, and Genesis* (Berkeley: University of California Press, 1984), pp. 132–135.

77. Pico, *Oration,* ed. Garin, pp. 122–124: "Quis non Socraticis illis furoribus a Platone in *Phaedro* decantatis sic afflari non uelit? . . . Agemur Socraticis furoribus . . . si et per moralem affectuum uires ita per debitas competentias ad modulos fuerint intentae . . . et per dialecticam ratio ad numerum se progrediendo mouerit, Musarum perciti furore caelestem armoniam auribus combibemus. Tum Musarum dux Bacchus in suis mysteriis, idest uisibilibus naturae signis, inuisibilia Dei philosophantibus nobis ostendens, inebriabit nos ab ubertate domus Dei, in qua tota si uti Moses erimus fideles, accedens sacratissima theologia duplici furore nos animabit. Nam in illius eminentissimam sublimati speculam . . . Phoebei uates, huius alati erimus amatores, et ineffabili demum caritate, quasi oestro perciti, quasi Saraphini ardentes extra nos positi, numine pleni, iam non ipsi nos sed ille erimus ipse qui fecit nos"; above, notes 11, 51–52, 57.

78. Above, notes 60, 77.

79. Above, notes 65–66, 74.

80. Quotations from the *Zohar* and from the *Otiot de-R. Akiva* are in Liebes, "Christian Influences," pp. 154–158; Isa. 11:1; Jer. 23:5, 33:15; Wirszubski, *Jewish Mysticism,* pp. 164–165 n. 8; above, note 66.

81. Above, note 40.

82. Pico, *Conclusiones* 4.9.33 (B 60): "Nullae sunt litterae in tota lege quae in formis, coniunctionibus, separationibus, tortuositate, directione, defectu, superabundantia, minoritate, maioritate, coronatione, clausura, apertura et ordine decem numerationum secreta non manifestent."

83. Pico, *Conclusiones* 5.9.25 (B 120): "Sicut caracteres sunt proprii operi magico, ita numeri sunt proprii operi Cabalae, medio existente inter utrosque et appropriabili per declinationem ad extrema usu litterarum."

84. Above, note 1.

85. Pico, *Conclusiones* 5.7b.74 (B 114): "Utrum in coelo sint descripta et significata omnia cuilibet scienti legere."

2

The Study of the *Timaeus* in Early Renaissance Italy

James Hankins

In 1363 an anonymous Italian commentator on Plato's *Timaeus,* justifying his own work on the dialogue, claimed that "none of his predecessors, after Calcidius, had taken the trouble to expound or comment upon Plato, perhaps because of his strange manner of speaking."[1] An odd remark to make, at least from the perspective of modern scholarship on medieval philosophy. It now seems clear that Plato's *Timaeus,* in the partial translation of Calcidius, was among the more frequently studied texts of the High Middle Ages. In addition to major lemmatic commentaries by Bernard of Chartres and William of Conches, we have dozens of heavily glossed manuscripts, mostly northern French and German, ranging from the late eleventh to the early thirteenth centuries. What the "1363 commentator" (as we may as well christen him) could not have known was that he stood on the threshold of a great revival of Platonic scholarship—centered in Italy and culminating in the work of Marsilio Ficino a century later—in which Calcidius' *Timaeus* would once again be frequently copied, taught, glossed, and cited.[2]

Indeed, as Paul Dutton has recently shown, the late fourteenth and fifteenth centuries were second only to the twelfth century as a great age of Calcidian studies.[3] This revival of interest in the Calcidian and Chartrean tradition of Timaean study in early Renaissance Italy has gone largely unnoticed by historians of philosophy, part of a general tendency to ignore Renaissance revivals of medieval philosophy. The aim of the present study is to explore the nature and extent of this revival and to offer some speculations regarding its impact on the later development of Renaissance natural philosophy.

To begin with the manuscript tradition of Calcidius, of the 198 known manuscripts containing Calcidius' translation of or commentary on the *Timaeus,* at least 40 (not counting excerpts) were written in the later fourteenth or fifteenth centuries, and at least 28 in Italy. One can, moreover, document the presence of another 23 manuscripts written in Europe before 1350 that were present in Italian collections or were studied by Italian scholars during the fifteenth century. So something like a quarter of the surviving manuscripts either were written in Italy or were present in Italy during the early Renaissance.[4]

The presence of Calcidius manuscripts can be documented in most of the important public and princely libraries of early Renaissance Italy. Coluccio Salutati, chancellor of Florence and the mentor of a whole generation of humanists, had a copy made for his large library; this codex, now in the Vatican library, later entered the collection of Pope Nicholas V. The papal collections already possessed by 1436 another copy, which had almost certainly been the property of the humanist cardinal Giordano Orsini. The great collector Niccolò Niccoli, whose books later formed the nucleus of the Library of San Marco, Florence's public library, owned no fewer than four copies of Calcidius. The Visconti library in Milan owned three copies, including one formerly owned and annotated by Petrarch. Cardinal Bessarion's great library, later the nucleus of the Biblioteca Marciana in Venice, contained a copy; other codices from the Veneto include two owned by the fourteenth-century Trevisan collector Oliviero Forzetta,[5] one owned by Doge Pietro Mocenigo, and one owned by the Zabarella family, later in the possession of the philosopher Giacomo Zabarella. Among the famous collectors of the later fifteenth century, copies were made or purchased for the libraries of the condottiere Federico of Urbino; Alessandro Sforza, signore of Pesaro; Cardinal Agostino Patrizi Piccolomini (an associate of Pius II and Pomponio Leto); the Florentine merchant Guglielmo Sachetti; Andrea Matteo III d'Acquaviva, duke of Atri; and Lorenzo de' Medici, *il Magnifico*.

We can also identify numerous copies of Calcidius owned by humanists and philosophers of quattrocento Italy. All the major Platonic philosophers of the fifteenth century possessed copies of the text. In addition to Bessarion's copy, already mentioned, Nicholas of Cusa owned two copies, Niccolò Leoniceno had one copy, and Marsilio Ficino possessed a copy written and annotated in his own hand. Pierleone da Spoleto, Lorenzo de' Medici's physician and an associate of Ficino, annotated a twelfth-century manuscript of the text.[6] Pico della Mirandola annotated, or rather scribbled on, a manuscript of Calcidius, now in Naples—of which more anon.

The humanists of the early Renaissance also had ready access to the dialogue. I have already mentioned the copies owned by Petrarch, Salutati, and Niccoli. Leonardo Bruni, the most important translator of Plato before Ficino, studied Calcidius in a manuscript owned by his great friend Niccoli, probably in the first decade of the fifteenth century. Pier Paolo Vergerio, a older contemporary and friend of Bruni, studied the *Timaeus* while still a young man; there survive excerpts from the dialogue written by him in 1388. Two famous humanist schoolmasters of the early quattrocento, Guarino

Veronese and Gasparino Barzizza, owned copies that they may have used in their teaching. Two associates of Bessarion's circle in Rome, Guillaume Fichet and Nicolaus Modrussiensis, also had the text in their libraries. Nicolaus Modrussiensis's copy may have served as one of the witnesses for the *editio princeps* of Calcidius, printed in Paris in 1520.[7]

Renaissance Italians knew not only Calcidius' interpretation of the *Timaeus* but also the Chartrian commentary tradition on the dialogue.[8] Many years ago Raymond Klibansky pointed out that the "Contius" and the "Policrates" referred to by Marsilio Ficino in a letter describing his Platonic sources could be identified, respectively, with William of Conches and with the *Policraticus* of John of Salisbury.[9] More recently, Sebastiano Gentile has identified silent borrowings from William's *Glosae super Platonem* and the *Policraticus* of John of Salisbury in Ficino's letters and in an early theological work, *Di Dio et anima*.[10] It would have been easy enough for Ficino to have consulted William's *Glosae super Platonem,* as there were at least two copies in quattrocento Florence, one formerly owned by Niccoli.[11] Other, non-Florentine copies of the text were owned by a certain Leonardus (not Bruni), who seems to have read it with some attention, and by Cardinal Bessarion.[12]

In addition to William's glosses on the *Timaeus,* Chartres produced at least one other important set of anonymous glosses, recently identified by Paul Dutton as the work of Bernard of Chartres.[13] Like William's, Bernard's glosses not only circulated as an integral lemmatic commentary but also were copied freely into the margins of many manuscripts of the Latin *Timaeus* written during the twelfth century and afterward. Bernard was known to the grammarians of the Veneto in the late fourteenth century, as is evidenced by the glossary commentary of Antonius de Romagno (discussed below). Glossed manuscripts influenced by Bernard were also known in quattrocento Florence: Salutati's manuscript, later owned by Nicholas V, had glosses descending from the Bernardine tradition, as did two of the manuscripts owned by Niccolò Niccoli and a manuscript, now in London, copied by the Florentine scribe Piero Strozzi.[14] Ficino, too, may well have known these Bernardine glosses, as he once or twice refers in his own notes on Calcidius to "post-Calcidian commentaries."[15]

This brings us to the subject of Renaissance glosses on the Calcidian *Timaeus.* We have over a dozen manuscripts surviving from the early Renaissance that contain extensive glosses, notes, or study materials of various kinds assembled by Italian humanists and philosophers. From the fourteenth century we have notes by Petrarch, the large lemmatic commentary by the "1363 commentator," and a glossary commentary by Antonius de Romagno, a grammarian and humanist from the Veneto. From the fifteenth century we have several anonymous sets of glosses: one composed by a Venetian humanist;[16] one by a midcentury humanist whom we shall christen the "Recanati Master"; glosses written for Andrea Matteo III d'Acquaviva;[17] and a set of glosses, surviving in three copies, probably composed in Padua shortly after the middle of the quattrocento. Identifiable annotators of Calcidius include Marsilio Ficino's associate Pierleone da Spoleto, who added a few glosses to his twelfth-century copy of the *Timaeus;* Ficino himself; and his student and colleague Giovanni Pico della Mirandola.

Petrarch's glosses can be dealt with summarily, as they have been the object of a recent study by Sebastiano Gentile, who is preparing a new edition of them.[18] The glosses, consisting mostly of *notabilia,* short comments, and cross-references, were clearly written for his private use rather than for teaching purposes. Gentile supposes them to have been written in two distinct stages, an earlier stage (around 1335–1338 or even earlier) and a later stage (after 1355). In the first stage the *Timaeus* seems to have been among the books Petrarch read in preparation for writing his *Rerum memorandarum liber* (1343–1345), while during the second stage Petrarch's reading of the dialogue was closely connected with the composition of his invective *De sui ipsius et multorum aliorum ignorantia* (1367). It was in his reading of the *Timaeus* and its creation myth that Petrarch was able to find confirmation of Augustine's view, so important for the history of Renaissance Platonism, that Plato was the closest of the ancient philosophers to Christianity. He was therefore an ideal authority to be used in the polemic against the godless Aristotelianism of the universities—a polemic continued by humanists and philosophers down to the end of the sixteenth century.[19]

The other two fourteenth-century encounters with the *Timaeus* preserved in manuscripts can both be associated with the northern Italian grammatical tradition.[20] Both of these commentaries were intended for classroom use, and manifest the typical interest of the northern Italian grammarians in exhibiting the logical structure of the text, in the solution of *dubia* and in the illustration of parallel texts. The older of the two, composed by the "1363 commentator," consists mostly of paraphrases of the text itself. Sometimes he also explains technical matters, using Boethius, Macrobius, Cicero, Apuleius, and (silently) Calcidius. These authors are in fact mentioned in his preface, where he speaks of his desire to render intelligible the opinions of these "and other Platonists" by explicating the "holy opinions of Plato." The preface also speaks acerbically of "the many learned theologians and philosophers who nowadays are proud to cite Plato without having ever read him—or without fully understanding him if perchance they have read him." This remark (and the classing of Cicero as a Platonist) suggests that the 1363 commentator may have been a follower of Petrarch, for the latter made similar criticisms of contemporary scholastics with regard to their knowledge of Plato.[21]

Like Petrarch's, the 1363 commentator's study of the *Timaeus* seems to have taken place largely in isolation from the previous medieval glossary tradition associated with Chartres and Paris. There is no trace of Bernard's or William's influence, and the commentator shows his independence of the tradition in such matters as his extensive treatment of the historical myth of Atlantis, neglected by earlier commentators, and his idiosyncratic division of the text.[22] The same cannot be said of Antonius de Romagno de Feltro, a grammarian and humanist (fl. 1388–1408) who worked primarily in the Veneto.[23] With Antonius' glossary commentary on the *Timaeus* we are once again back in the mainstream of Calcidian and Chartrian interpretation. Antonius' commentary, in fact, is often dependent on Bernard of Chartres, sometimes verbally.[24] More often, however, Antonius reworks Bernard's material, condensing and combining it with his own observations or with matter drawn from Calcidius, Boethius, and Macrobius, as well as from other staples of the late medieval grammarian such as Servius' commentaries on Virgil.[25]

Antonius' interest in Plato's moral and political thought is seen most clearly in his *accessus,* in which he emphasizes the derivation of "popular" or customary justice—what the medieval tradition called "positive justice"—treated in the *Republic,* from "natural justice," treated in the *Timaeus.* The model polity constructed by Socrates is parallel to the model universe laid out by Plato in the *Timaeus.* By observing the justice implicit in God's creation of the natural world, human beings will be better able to imitate it in their own lives; indeed, God's just order in Nature shows us what human justice was like in the time before the Fall.[26] The actual moral lessons Antonius draws in his reading of the *Timaeus* seem rather more homely stuff, however, as when he allegorically compares the normal motion of the firmament to the good, while the opposite motion of the planets reminds him of the wicked, returning to their sins like the proverbial dog to his vomit.[27]

Antonius is at his most original when treating Plato's elemental theory. Plato's theory posits a substrate, *hyle* or matter, metaphysically prior to the traditional four elements, and *ydeas* or exemplars, which serve as models for the informed matter created by the Demiurge. Antonius correctly saw Plato's account as a challenge to Aristotle's elemental theory, and he boldly identifies himself with Plato's conception (writing explicitly "et hoc est mea sententia" next to Plato's summary at 53A). He defends against Aristotle (who is not named) Plato's view that *hyle* can have some kind of metaphysical status as a *tertium quid* between being and non-being.[28]

Like a true Platonist, Antonius echoes Plato's belief (52B) that our normal intuitions of reality are like dreams; behind them is "the marvelous nature of the philosophers" consisting of idea and matter: "Truly we experience or think what dreamers think, since we think whatever exists exists in a material place, and we think nothing exists except what is in heaven or on earth, or in water or in air. Which is false, because before these things existed, *hyle* and *ydea* existed in their marvelous nature of sorts according to the philosophers."[29] Not very sophisticated, perhaps, but Antonius still has the honor of being the first Renaissance thinker to defend the Platonic theory of matter against the overwhelming hegemony of Aristotelian physics, a defense that would not become common until the time of Marsilio Ficino.

Calcidius' Plato continued to be read by the grammarians of the Veneto even in the second half of the fifteenth century. From the third quarter of the fifteenth century we have a highly traditional glossary commentary, surviving in three manuscripts, which is likely to be the work of a Paduan arts master.[30] All three manuscripts of the commentary also contain Latin translations of Plato's *Gorgias, Phaedo,* and *Crito* made by Leonardo Bruni (1370–1444), showing that the medieval *Timaeus* had now entered the orbit of the humanistic book.[31] But in compiling his glosses the Paduan master showed no interest in enlarging his understanding of Plato by consulting the new humanist translations made by Bruni, Uberto, and Pier Candido Decembrio, and others. His commentary is largely a pastiche of Calcidius, with echoes of Bernard and other medieval glossators.[32] The glossator provides textual summaries and *divisiones,* gives explanations of philosophical terms, and indicates a few parallel passages from Macrobius (probably relying on intermediate sources). The glossator may have had some kind of university training, for he shows a familiarity with scholastic terminology and his interests seem to be more philosophical than literary. In one of the few glosses not dependent on Calcidius, we get some sense of his philosophical profile:

> There is one good thing alone which is only good and nothing else. This is the first good which is good in that it exists. There is also a second good which also is called good in that it exists, but in a certain other sense, namely, because that which itself is good flows from his will whose Being is good. Whence every white thing is good. Therefore white both exists and is good. But it is good in that it exists as flowing from his will who is good, not as something said to be white in that it exists, but it is only said to be white because he is not white who wished it to be white. Thus there-

fore the nature of every single thing is capable of beatitude and receives a similitude of some kind of its artificer.[33]

This gloss, significantly, is placed next to a famous passage of the *Timaeus* that emphasizes—in contrast to later medieval Christian theology—the necessary and unique character of the created universe. It was also a locus classicus for the Plotinian interpretation of Plato's metaphysics, which orders the realm of Being into descending hypostases of Mind, Soul, and Body, causally dependent on the One beyond Being. Plotinian metaphysics, too, was at odds with the major theological traditions of the later Middle Ages, which identified God with Being itself. The glossator here seemingly wishes to impose an Augustinian reading of Plato's text by emphasizing the identity of Being and Goodness in the First Good and the derivative character of second goods, whose existence depends on the will of the First Good (making them *eo ipso* contingent goods). It is their derivative character that enables them to receive a similitude of their Maker and makes them capable of beatitude. All of this, of course, is foreign to the thought of Plato (and Calcidius).

It is only toward the middle of the fifteenth century that we begin to find glosses on the *Timaeus* markedly different in character from the kind of annotations associated with the medieval grammatical tradition. From that period we have a hitherto unknown set of glosses, possibly Roman in origin, that seem to be the work of a humanist teacher; we shall label him the "Recanati Master" after the present location of the manuscript that preserves his glosses.[34] The Recanati Master shows no signs of having had a formal university training in arts. Nor does he divide the text, search for *dubitationes,* or cite parallel authorities in the manner of the professional grammarian. Rather, his chief aim seems to have been to summarize and clarify the text, and to provide an aid for the compilation of copybooks. He draws attention especially to passages illustrating Plato's piety and the harmony of his cosmological thought with Christianity. Sometimes the search for harmony pushes him to the edge of syncretism, as when he identifies Plato's "children of the gods" with Christian "sancti uiri" (40E), Plato's "lower gods" with angels (42D), and Plato's vague remarks about the lower gods "receiving back the mortal things [they] have created" with the "resurrectio universalis" (41D). His longest gloss attempts to compare the concepts of time and eternity (37D), but one does not feel oneself in the presence of a powerful philosophical intelligence. If the Recanati Master was indeed a Roman humanist, his was not the Rome of Lorenzo Valla or Cardinal Bessarion.

By the second half of the fifteenth century, Florence had emerged as a center of philosophical study to rival Rome and the Paduan Studio; and, thanks to Marsilio Ficino, it had become the most important center for the study of Plato since the closing of the ancient Platonic Academy. Ficino's study of the *Timaeus* began at a very early age and continued up to the last few years of his life. His very first work, the lost *Institutiones ad Platonicam disciplinam,* is supposed to have been largely a commentary on the *Timaeus,* and we have a manuscript containing Calcidius' commentary written by Ficino in 1454, when he was about twenty-one years old.[35] The codex is festooned with notes in Ficino's own hand, which may well have been added at a later date.[36] The notes seem to have been intended for use in his private teaching and are strikingly different in character from those on any of the other annotated manuscripts discussed so far.

Ficino's glosses on Calcidius are no mere rearrangements of traditional materials and topics, but give the impression of a fresh reading of the text by a critical philosophical intelligence. The glosses (as usual) indicate *notabilia,* summarize arguments, identify sources,[37] and offer further illustrations and explanations of Calcidius' text. The Florentine studies Calcidius' idiosyncratic vocabulary, and he frequently notes philosophical, astronomical, musical, and mathematical terms, sometimes comparing them with the terminology used in his own time. As one would expect of the self-styled *doctor animarum,* he is deeply interested in what Calcidius has to say about the World-Soul and soul in general; he recognizes some unorthodox tendencies in Calcidius' thought but insists that Plato should not be accused of vulgar error in the matter of the transmigration of souls.[38] Yet he is attracted to Numenius' solution to the problem of evil, that evil is caused by a lower soul latent in matter which is mastered by a higher, rational soul imposed by God on matter. His own solution offers slight refinements on Numenius:

> There are two World-Souls according to Plato. I think that in matter there are two souls: one educed from the potency of matter itself, which is vegetative and has motion, which always existed in it, which is subcorporeal and agitates matter without reason using every irrational motion, which we call evil, that is, "rash." The other is the soul God created when he wished to adorn the cosmos, which has reason and thus clarifies the cosmos through the order of motion.[39]

Ficino also studies carefully Calcidius' remarks on astronomy, noting differences of opinion among ancient authorities about the order of the planets. He glosses approvingly Calcidius' theory that planetary retrogradations are optical illusions.[40] He takes particular notice of the passage where Calcidius says that from the point of view of soul, it is the sun, not the earth, that must be considered the center of the universe: the warm, vital, beating heart, not the dark and motionless uterus.[41] This is an image that recurs in Ficino's short treatise *De sole,* one of the texts that inspired Copernicus' heliocentrism.[42] Ficino also read closely Calcidius' discussion of Plato's theory of vision, but—surprisingly for so devoted a Platonist—seems to have disagreed with Plato in this matter.[43]

The part of Calcidius' commentary that most occupies Ficino, however, is the long section at the end on matter and elemental theory. Ficino uses Calcidius as a sourcebook for collecting the opinions of the ancient philosophers on elemental theory. Indeed, at one point in his marginal notes, Ficino puts together a kind of *placita philosophorum* on the nature of matter and the primary elements, in which he tries to collate ancient opinions with the ancient schools as described in Diogenes Laertius' *Lives of the Philosophers* (in the translation of Ambrogio Traversari).[44] Some of this material would ultimately find its way into Ficino's mature commentary on the *Timaeus* (1484, 1493–1496) where he would use it in his campaign to undermine Aristotle's elemental theory; we shall return to this shortly.

As the notes cited show, Ficino's annotations on Calcidius often have a strong pedagogical flavor, and we can now confirm that they were actually used by his most famous pupil, Giovanni Pico della Mirandola.[45] Despite Pico's tendency to play the *ingratus discipulus,* Ficino supported his student's studies generously by lending him for long periods rare codices from his own library (as we know from Ficino's patient letters asking for their return).[46] Such would seem to be the case with Ficino's manuscript of Calcidius. In Pico's own copy of Calcidius a vocabulary list and a number of Ficinian notes are copied word for word, while other notes of Pico are clearly dependent on Ficino's.[47] Pico also used Ficino's text to fill lacunae in his own and entered variants from the Ambrosiana codex in the margins of his own manuscript. In themselves, Pico's notes are not of great interest, as they consist mostly of short *notabilia.* Occasionally, however, we can catch sight of Pico's concordist juices flowing, as when he remarks sharply, "Plato seems to have said other things in agreement with Aristotle, aside from these matters about the substance of souls" (*Alia, praeter ea quae de substantia animarum, videtur Aristoteli Plato consentanea dicere*), next to a passage where Calcidius summarizes various Middle Platonic objections to Aristotle's definition of the soul as an entelechy.[48]

Ficino's and Pico's notes on Calcidius show signs of a fresh approach to the Calcidian *Timaeus,* but the real watershed that separates traditional from early modern study of the *Timaeus* is formed by Ficino's mature *Compendium in Timaeum.* First published in 1484 and republished in enlarged form in 1496, this commentary is probably the least well-known of Ficino's major works, but it certainly deserves further study.[49] It is important for at least three reasons.

In the first place, Ficino's commentary brings a whole new range of ancient sources to bear on the problem of understanding Plato's text, sources that the dialogue's sixteenth-century commentators continued to invoke.[50] Traditional Calcidian and Chartrian grammatical commentaries had used for this purpose a relatively narrow range of Latin sources, mostly Middle Platonic in inspiration: Calcidius himself, Boethius, Apuleius, and Macrobius. Ficino, thanks to his knowledge of Greek and to the matchless resources of the Medici and San Marco libraries, was able to consult a number of fresh ancient sources of more varied philosophical hue: the eclectic Middle Platonist Galen, Plutarch's *De fato* and *De animae procreatione in Timaeo,* Theon of Smyrna's *Expositio rerum mathematicarum ad legendum Platonem utilium,* the academic skeptic Cicero,[51] the eclectic Diogenes Laertius, the first part of Proclus' Neoplatonic *Timaeus* commentary, and the neo-Pythagorean Περὶ ψυχᾶς κόσμω καὶ φύσιος, a work falsely attributed to Timaeus of Locri, the main interlocutor of Plato's *Timaeus.*[52] The latter work, composed in pseudo-Doric dialect, is now considered a first-century C.E. neo-Pythagorean forgery based on the Platonic *Timaeus.* Ficino, gullible in such matters, believed the work to be a genuine opusculum of Timaeus of Locri and the main source for Plato's dialogue. This in turn led him to characterize the *Timaeus* as one of Plato's "Pythagorean" works, that is, a work that reports Pythagorean doctrine but does not necessarily represent Plato's own settled views.[53]

Ficino's tendency to distance the doctrine of the *Timaeus* from Plato's own doctrine brings us to the second reason for the importance of the *Compendium in Timaeum.* It was a central contention of Ficino's own philosophical writing, and the justification for his life's work, that Plato's philosophy offered a more adequate basis for Christian theology than did Aristotle's. Whether this was in fact the case would become a major subject of debate during the sixteenth century among Italian and northern European philosophers. The *Timaeus* was of course a central text for this debate, as it afforded grist for all the main philosophical mills: Platonists, Aristotelians, and concordists. Platonists naturally felt that Plato's doctrine of creation brought him closer to Christian truth than was Aristotle, who had believed in the eternity of the world. But the matter was hardly simple. In fact, the *Timaeus,* read closely, contained a whole syllabus of errors that the partisans of Aristotle could point to gleefully. Plato's seeming belief in the eternal recurrence of the ages; in an historical chronology that conflicted with biblical chronology, in polytheism; in the indirect creation or subcreation of the physical world by lower divinities (a heterodox solution to the problem of evil), in the transmigration of souls, in salvation through the rational control of appetites (a doctrine close to Pelagianism), in the uniqueness of the universe (arguably a limitation on God's absolute power), in the extradeical existence of the Ideas, in the eternity of the "receptacle," preexisting the act of creation and thus challenging creation ex nihilo—all of these Timaean doctrines threw down a formidable challenge to would-be Christianizing interpreters of Plato. Ficino's *Compendium in Timaeum* offered exegetical solutions to many of these problems (some more plausible than others), and thus became the point of departure for many sixteenth- and seventeenth-century partisans of Plato's cosmology.

Which brings us to a third reason for the importance of Ficino's commentary to the later tradition of Timaean study. Though Ficino's own position in the Plato-Aristotle controversy was officially concordist—he believed that Aristotle as well as Plato should be part of the formation of Christian thinkers and theologians[54]—he did not, in fact, hesitate to criticize Aristotle in order to demonstrate the superiority of Plato. His criticisms of Aristotle's elemental theory in the *Timaeus* compendium came to assume particular importance in the sixteenth century. As we have seen, Calcidius and some later medieval commentators were conscious that Plato's theory of matter implicitly challenged Aristotle's, and Antonius de Romagno at the end of the fourteenth century explicitly sided with Plato against Aristotle on this subject. But it was not until Ficino that the West possessed a commentator of sufficient learning and philosophical acumen to use Plato's elemental theory to mount a full assault on Aristotelian physics.

Several chapters in Ficino's commentary, in fact, constitute a sharp critique of Aristotle's theories of quintessence and elemental motion. While not questioning the existence of a real distinction between celestial and terrestrial matter, Ficino nevertheless tends to reduce the contrast between the two realms by arguing that the four terrestrial elements of earth, water, air, and fire are also present virtually in the celestial realm, as they are present beyond the celestial realm as ideas in the mind of the *opifex mundi*. He uses the Neoplatonic theory of correspondence, in other words, to emphasize the continuities rather than the discontinuities in the created order, especially the continuity between terrestrial fire and celestial fire; indeed, according to Ficino, the heavenly world consists predominantly of celestial fire.[55] Likewise, he argues that Aristotle's contrast between the rectilinear motion of the terrestrial elements and the circular motion of quintessence is overly schematic and superficial. For Ficino, all elemental motion is naturally circular, rectilinear motion being essentially a derivative, a consequence of the displacement of the elements from their natural location. The apparent rectilinear motion of fire, for example, can be interpreted as the unnatural motion of an element seeking to return to its natural place; once it regains its place, it resumes its natural circular motion, as can be seen (Ficino says) from the circular motion of comets.[56]

We see in Ficino's *Compendium,* in other words, a number of the characteristic features of the countercultural science of the later sixteenth century. There is the desire for greater unity and simplicity in scientific explanation, expressed in the form of a theory of the elements and of elemental motions that embraces both terrestrial and celestial spheres—a direct challenge to Aristotle's two-sphere universe. There is the concern, seen clearly in Galileo, that the Aristotelian natural philosophy of the universities provides an inadequate basis for Christian theology, as well as the argument that biblical authority is more consonant with the new alternative science than with traditional Aristotelian science.[57] And there is also (what we do not, unfortunately, have space to consider here) the desire to find mathematical structures underlying appearances, the "mathematization of the cosmos,"[58] along with the affirmation of the Timaean principle that "the unobservables we postulate to account for properties of observables need not themselves possess those same properties"[58]—a principle that lies behind Galileo's distinction between primary and secondary qualities. All of these principles are taken up by the Timaean commentators and Platonic philosophers of the sixteenth century. Platonist criticism of Aristotelian natural philosophy thus not only comes to constitute one of several cultural solvents of hegemonic Aristotelianism but also provides several key principles adopted by the scientific counterculture of the later sixteenth century.

The persistent antagonism of Platonic philosophers toward school Aristotelianism—an antagonism sometimes revealed even by officially "concordist" philosophers such as Ficino himself and his followers, the Florentine philosophers Francesco da Diacceto and Francesco II de' Vieri—thus emerges as an attitude of some importance for understanding the history of early modern science. If one asks the reasons for the sense of rivalry between Platonists and Aristotelians, one need look no further than the institutional and disciplinary context for an explanation. The *Timaeus,* a central text in the twelfth-century schools of northern France, had been ejected from the curriculum of the universities some time in the first half of the thirteenth century. Why this happened is not known, but it probably had much to do with the mythical and literary characteristics of the work, as well as with issues of orthodoxy.[59] In any case the work disappeared from university curricula, to become the intellectual property of the lowly grammarians and eventually of the not-so-lowly humanists. The humanists, as ever advocates of philosophical pluralism, beginning with Petrarch used the *Timaeus* and the works of Plato generally as a stick with which to beat the integral Aristotelians. They kept alive the Chartrian tradition of glossary commentaries that emphasized the concord of Plato with Christianity. In the mid-fifteenth century, some Byzantine philosophers and their Italian acolytes, in order to demonstrate the value of the Greek and Byzantine heritage, argued that Plato's theology approximated Christian truth more nearly than Aristotle's.[60]

In the fullness of time humanist educators produced a philosopher, Marsilio Ficino, capable of moving beyond the Parthian shots of a Petrarch or a Bruni, a philosopher who was able to turn the cultural prejudices of the early humanists into a real philosophical movement. Yet the participants in this movement long retained their sense of being outsiders, of belonging to an esoteric sect. The Platonic dialogues remained rarely, if ever, read by professional university philosophers or theologians. Even in the mid-sixteenth century, Lodovico Boccadiferro, a philosophy professor at Bologna who wrote a commentary on the *Timaeus,* could complain that Plato was neglected in the universities, "read by no one, either publicly or privately": "First, because he himself, treating of natural subjects, mixes with them divinity and mathematics, thus not preserving a distinct order, but treating everything in a sort of disordered and confused way. Another reason is the deeply rooted practice of reading Aristotle; another is ignorance of Greek."[61] It was only in the second half of the sixteenth century, under pressure from secular rulers, that Plato's dialogues reentered the university as texts to be read by professional philosophers.[62]

Yet thanks to the centuries-old pattern of antagonism between Aristotelianism and Platonism, the "Platonic professors" of the later sixteenth century retained the sense of being outsiders in a university milieu dominated by Aristotelians. Hence they came to make common cause with the new cosmologists in the struggle of the latter against hegemonic Aristotelianism. Hence Ficinian Platonism became a powerful resource drawn on by the new science of the later sixteenth and seventeenth centuries in its search for new scientific principles and a new scientific vision.

APPENDIX 1   MANUSCRIPTS OF CALCIDIUS' TRANSLATION AND COMMENTARY ON PLATO'S *TIMAEUS* IN EARLY RENAISSANCE ITALY

This list includes all manuscripts of Calcidius' translation of or commentary on the *Timaeus* known to me that were written in early Renaissance Italy, documentably present in Italian collections of the early Renaissance, or owned by figures of the early Italian

Renaissance. The items marked with an asterisk (★) I have inspected *in situ*. The following abbreviations have been employed:

Berti = Ernesto Berti, *Il Critone latino di Leonardo Bruni e Rinuccio Aretino,* Studi dell'Accademia Toscana di scienze e lettere "La Colombaria" 61 (Florence: Olschki, 1985).

Dutton, *Bernard* = *The "Glosae super Platonem" of Bernard of Chartres,* ed. Paul Edward Dutton, Studies and Texts 107 (Toronto: Pontifical Institute of Mediaeval Studies, 1991).

Dutton, "Material Remains" = Paul Edward Dutton, "Material Remains of the Study of the *Timaeus* in the Later Middle Ages," *L'enseignement de la philosophie au XIIIe siècle: Autour du "Guide de l'étudiant" du ms. Ripoll 109,* ed. Claude Lafleur and Joanne Carrier (Turnhout: Brepols, 1996), pp. 203–230.

Dutton, "The Uncovering" = Paul Edward Dutton, "The Uncovering of the *Glosae super Platonem* of Bernard of Chartres," *Mediaeval Studies* 46 (1984): 192–221.

Gentile, *Ritorno* = *Marsilio Ficino e il ritorno di Platone, Mostra di manoscritti, stampe e documenti, 17 maggio–16 giugno 1984,* ed. Sebastiano Gentile, Sandra Niccoli, and Paolo Viti (Florence: Le Lettere, 1984).

Hankins, *Plato* = James Hankins, *Plato in the Italian Renaissance,* 2 vols., Columbia Studies in the Classical Tradition 17. 1–2 (Leiden: Brill, 1990).

Iter = Paul Oskar Kristeller, comp., *Iter Italicum; A Finding List of Uncatalogued or Incompletely Catalogued Humanistic Manuscripts of the Renaissance in Italian and Other Libraries,* 7 vols. (London: Warburg Institute; Leiden: Brill, 1963–1996).

Jeauneau, *Glosae* = Guillaume de Conches, *Glosae super Platonem,* ed. Edouard Jeauneau, Textes philosophiques du Moyen Age 13 (Paris: Vrin, 1965).

Jeauneau, *Lectio* = Edouard Jeauneau, *Lectio philosophorum: Recherches sur l'Ecole de Chartres* (Amsterdam: Hakkert, 1973).

Klibansky = Raymond Klibansky, *The Continuity of the Platonic Tradition during the Middle Ages* (London: Warburg Institute, 1939; reprinted with supplement, separately paginated, Munich: Kraus, 1981).

Mazzatinti = G. Mazzatinti and A. Sorbelli, eds., *Inventari dei manoscritti delle biblioteche d'Italia,* 107 vols. to date (Forli: Olschki, 1890–).

Pellegrin, *Vaticane* = E. Pellegrin et al., *Les manuscrits classiques latins de la Bibliothèque Vaticane,* 3 vols. (Paris: CNRS, 1975–1991).

Pellegrin, *Visconti* = Elisabeth Pellegrin, *La bibliothèque des Visconti et des Sforza, ducs de Milan, au XVe siècle* (Paris: CNRS, 1955).

Pellegrin, *Visconti, Supplément* = Elisabeth Pellegrin, *La bibliothèque des Visconti et des Sforza, ducs de Milan, Supplément,* ed. Tammaro de Marinis (Florence: Olschki, 1969).

Ullman = Berthold Louis Ullman and Philip Stadter, *The Public Library of Renaissance Florence* (Padua: Antenore, 1972).

Waszink = *Timaeus a Calcidio translatus commentarioque instructus,* ed. J. H. Waszink, 2nd ed., Corpus Platonicum Medii Aevi: Plato Latinus 4 (London: Warburg Institute; Leiden: Brill, 1975).

★1. Arezzo, Biblioteca della Città, MS 431. Written in central Italy, s. XV^1, two semigothic hands. A few annotations and corrections in a second, perhaps Florentine hand, s. XV^2. Copied from no. 38, below. Mazzatinti 6:240; Waszink, p. cix.

★2. Assisi, Biblioteca Comunale, MS 573. Italy, s. XII ex. (fols. 79r–86v); one gloss at the end (fol. 86vb), also s. XII ex. From the library of the Franciscan convent in Assisi; listed in its inventory of 1381 and presumably continuously present in the convent from that time. Cesare Cenci, *Bibliotheca manuscripta ad sacrum conventum Assisiensem* (Assisi: Casa editrice Francescana, 1981), 1:260–261, no. 432; 1:266, no. 454.

★3. Bergamo, Biblioteca Civica, "Angelo Mai," MS MA 350 (*olim* Delta VI 35). Paduan decoration, s. XV 3/4. One of three manuscripts containing the glosses of the "Paduan master"; also contains Bruni's translations of Plato's *Gorgias, Phaedo,* and *Crito. Iter* 1:8, 5:481–482; Berti, p. 151; Hankins, *Plato,* 2:671; see above, p. 82.

★4. Berne, Bürgerbibliothek, MS 681. Northern French, s. XII. Owned by Guillaume Fichet ("Ficheti theologi doctoris"), a French scholastic theologian with an interest in Italian humanism, connected with the circles of Cardinal Francesco Piccolomini, Bessarion, and Sixtus IV. Waszink, p. cx; on Fichet, see Paul Oskar Kristeller, "An Unknown Humanist Sermon on St. Stephan by Guillaume Fichet," in *Mélanges Tisserant,* Studi e Testi 236 (Vatican City: Biblioteca Apostolica Vaticana, 1964), 6:459–497.

5. Copenhagen, Kongelige Biblioteket, MS Gl. kgl S 208 fol. Written in Italy in 1470. Dr. Marianne Pade kindly informs me that the codex does not contain any glosses. Waszink, p. cxv.

★6. El Escorial, Biblioteca de El Escorial MS S III 5. French? s. XII in. Contains an owner's note written in a fifteenth-century Italian script, s. XV: "Francisci Sabadini codex hic est." G. Antolín, *Catálogo de los códices latinos de la Real Biblioteca del Escorial* (Madrid: Imprenta Helenica, 1910–1923), 4:57–58.

7. Ferrara, Biblioteca Comunale Ariostea, MS II 389. Italy, s. XV. *Iter* 2:503; Waszink, p. clxxxvii.

★8. Florence, Biblioteca Mediceo-Laurenziana, MS Plut. LXXXIV, 24. Italy, s. XV 4/4. A collection of Latin Platonica, written for Lorenzo de' Medici, his arms; illuminated by Attavante; derived from no. 13, below. Waszink, p. cxvi; Gentile, *Ritorno,* pp. 7–8, no. 6.

★9. ———, MS Plut. LXXXIX sup. 51. Italian, s. XII. Notes of the twelfth and fourteenth centuries. At least three fourteenth-century Italian hands, monastic rather than scholastic in character (according to Gabriella Pomaro), though one annotator cites Averroës's commentary on Aristotle's *Metaphysics,* book 12. The manuscript is the source for at least three other fifteenth-century copies, nos. 25, 32, and 33, below. Waszink, p. cxvi.

★10. Florence, Biblioteca Nazionale Centrale, MS Conv. soppr. E 8, 1398. S. XII. Contains the *Timaeus* in Calcidius' translation with the commentary of William of Conches. Some later glosses of the thirteenth century, mostly interlinear. From the monastery of SS. Annunziata. Waszink, p. cxiv; Jeauneau, *Glosae,* pp. 31–32 (s. XIII¹); Gentile, *Ritorno,* pp. 8–9, no. 7.

★11. ———, MS Conv. soppr. J 2, 49. French origin. Three sets of glosses from the twelfth, thirteenth, and fourteenth centuries; some of the twelfth-century glosses are excerpted from Bernard of Chartres's commentary. Owned by Niccolò Niccoli and left by him to the San Marco library. Waszink, p. cxiv; Ullman, pp. 71, 200; Dutton, *Bernard,* pp. 267–274.

★12. ————, MS Conv. soppr. J 2, 50. Glosses of the thirteenth century showing the influence both of William and of Bernard. Owned by Niccoli; formerly in the library of San Marco. Eugenio Garin, *Studi sul platonismo medioevale* (Florence: Le Monnier, 1958), p. 53n; Waszink, p. cxiv; Ullman, p. 200; Dutton, "Material Remains," p. 210.

★13. ————, MS Conv. soppr. J 4, 28. S. XI. Notes and corrections in the hand of Niccoli and Leonardo Bruni, mostly consisting of corrections of the text and short observations on the translation; e.g., at fol. 2v, Bruni writes: "Hec clausula cum greco textu male convenit"; at fol. 3r, "*rhapsodias:* non convenit cum greco" (= *Tim.* 21B, ed. Waszink, p. 12, 18, *memoriae*). From the library of San Marco. Waszink, p. cxiv; Ullman, pp. 71, 200.

★14. ————, MS Conv. soppr. J 9, 40. S. XII. Corrections and notabilia in a twelfth-century hand. One note only by Niccolò Niccoli (fol. 64v): "Hic plures chartae desunt." From the library of San Marco. Waszink, p. cxiv; Ullman, p. 200.

★15. London, British Library, MS Add. 22815. S. XII, except for one page of s. XV. Several sets of glosses, one set drawn from Bernard, another from William. One page of the *Timaeus* was written by the Florentine scribe Piero Strozzi, indicating a Florentine provenance. No. 46, below—also written by Piero Strozzi—was copied from this MS. Waszink, p. cvii; Edouard Jeauneau, "Extraits des *Glosae super Platonem* de Guillaume de Conches dans un manuscrit de Londres," *Journal of the Warburg and Courtauld Institutes* 40 (1977): 212–222; Dutton, *Bernard,* p. 278; Dutton, "Material Remains," p. 207.

★16. ————, MS Harl. 2652. German, s. XI. Owned but not annotated by Nicholas of Cusa. Waszink, p. cxiv; "Kritisches Verzeichnis der Londoner-Handschriften aus dem Besitz des Nicolaus von Kues," *Mitteilungen und Forschungsbeiträge des Cusanus-Gesellschaft* 3 (1963): 16–100, at 48–51 (reference from Dr. Martin Davies); Cyril E. Wright, *Fontes Harleiani: A Study of the Sources of the Harleian Collection of Manuscripts Preserved in the Department of Manuscripts in the British Museum* (London: British Museum, 1972), pp. 120–121.

★17. Milan, Biblioteca Ambrosiana, MS S 14 sup. Written by Marsilio Ficino in 1454; copied from no. 13, above, or an intermediary. Contains Calcidius' commentary on the *Timaeus* with Ficino's annotations, Bruni's translation of the *Gorgias,* Apuleius' *De deo Socratis,* and an excerpt from book 8 of Augustine's *De trinitate.* Waszink, p. cviii; Hankins, *Plato,* 2:700; see above, p. 84.

★18. Munich, Bayerische Staatsbibliothek, CLM 225. Written in 1478–1480 at Amberg for Hartmann Schedel by Heinrich Stolberger from no. 36, below. Contains the glosses of the Paduan Master as well as Bruni's translations of Plato's *Gorgias, Phaedo,* and *Crito.* Waszink, p. cxvii; *Iter* 3:613; Berti, pp. 152–154; Hankins, *Plato,* 2:702; above, p. 82, with note 30. The correspondence between Schedel, Stolberger, and Baptista Augustensis regarding the copying of this manuscript is published in Richard Stauber, *Die Schedelsche Bibliothek* (Freiburg i. Br.: Herder, 1908; reprint, Nieuwkoop: De Graaf, 1969), pp. 67–68, 242–244.

★19. Naples, Biblioteca Nazionale, MS V A 11. French, late caroline and gothic minuscules, s. XII. Glosses in a twelfth-century hand. Owned by Gasparino Barzizza, who annotated another text in the MS, Macrobius' commentary on the *Somnium Scipionis.* Later owned by Gasparino's son Guiniforte Barzizza, Janus Parrhasius, and Antonio Seripandi.

Waszink, p. cxviii; Lucia Gualdo Rosa et al., *"Molto più prezioso dell'oro": Codici di casa Barzizza alla Biblioteca Nazionale di Napoli* (Naples: Luciano, 1996), pp. 38–40, no. 20, with plate.

*20. ———, MS VIII E 29. Italy, s. XV 4/4, 300 x 200 mm., III + I + 177 (modern numeration) + III leaves, initials painted in blue and red, humanist cursive bookhand. Copied from a Niccoli codex (no. 14, above), with variants from Ficino's codex (no. 17, above). Owned and annotated by Giovanni Pico della Mirandola. Farnese binding; perhaps acquired by Pope Paul III from Cardinal Marino Grimani. Waszink, p. cxviii; François Fossier, *La bibliothèque Farnese: Etude des manuscrits latins et en langue vernaculaire* (Rome: Ecole Française, 1982), p. 379; above, p. 85. Probably to be identified with no. 469 in the inventory of Pico's library, published by Pearl Kibre, *The Library of Pico della Mirandola* (New York: Columbia University Press, 1936), p. 182. See figures 2.1–2.6.

*21. ———, MS VIII E 30. Northern Italian, s. XV med., semigothic bookhand. Closely related to Vat. lat. 1544 and Canon. class. lat. 175. Waszink, p. cxix.

*22. ———, MS VIII F 11. German, s. XII. Annotations in Greek and Latin by an Italian hand of the fifteenth century, very likely that of Guarino Veronese. The annotator addresses a reader named Franciscus on fols. 31r (twice) and 32v; the style of annotation is similar to that used by Guarino in his notes on Plato's *Republic* (where he also addresses "Franciscus," i.e., Francesco Barbaro). The Greek hand is similar to that used by students of Manuel Chrysoloras; in one place (fol. 28v), the annotator provides the Greek text behind Calcidius' quotation in Latin from Pythagoras' *Aurei versi,* ll. 70–71 (= Waszink, p. 177, 2–4). Waszink, p. cxix.

*23. Naples, Biblioteca Oratoriana dei Girolamini, MS C. F. 3–10 (formerly XVI. XVIII). Written by the scribe Donnus Vitus in 1507, probably in Naples, for Matteo III d'Acquaviva, duke of Atri. With glosses copied by the scribe. Waszink believed the manuscript to be perhaps copied from Vatican Library, MS Reg. lat. 1308, which if true would indicate an Italian Renaissance stage in the provenance of that MS as well. Waszink, p. cxx; A. Putaturo Murano and A. Perriccioli Saggese, *Codici miniati della Biblioteca oratoriana dei Girolamini* (Naples: Edizioni scientifiche italiane, 1995), pp. 119–121; above, p. 80, with note 17.

24. Oxford, Bodleian Library, MS Canon. class. lat. 175. Italy, s. XV. Closely related to Naples VIII E 30 and the *editio princeps* of 1520. Waszink, pp. cxix, clxx.

25. ———, MS Canon. class. lat. 176. Italy, s. XV ex. Waszink, p. cxx.

*26. Paris, Bibliothèque Nationale, MS Par. lat. 6280. France, s. XI. Acquired by Petrarch in Avignon; annotated by him. Later in the library of the Visconti dukes of Milan. Pierre de Nolhac, *Pétrarque et l'humanisme,* new ed. (Paris: H. Champion, 1907), 1:127–150; Pellegrin, *Visconti,* p. 98; Waszink, p. cxx; Armando Petrucci, *La scrittura di Francesco Petrarca* (Vatican City: Biblioteca Apostolica Vaticana, 1967), p. 126, no. 41; Klibansky, supplement, p. 70; S. Gentile, "Le Postille del Petrarca al *Timeo* latino," in *Il Petrarca latino e le origini dell'umanesimo: Atti del Convegno internazionale, Firenze 9–22 maggio 1991,* 2 vols., published as *Quaderni Petrarcheschi* 9–10 (1992–1993), 9:129–139; Dutton "Material Remains," p. 212; above, p. 80.

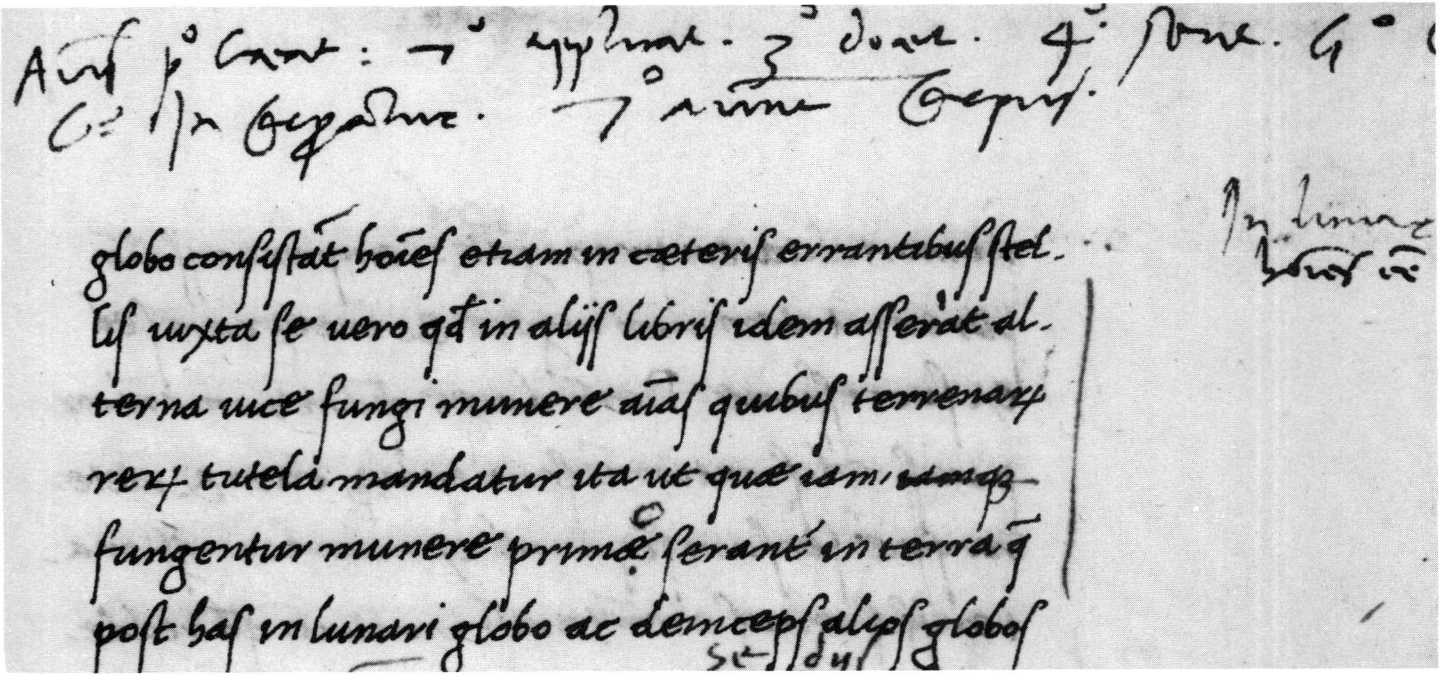

Figure 2.1

All plates are from Naples, Biblioteca Nazionale "Vittorio Emanuele III," Ms VIII E 29; reproduced by permission of the Ministero per i Beni Culturali e Ambientali of the Republic of Italy. Fol. 111r. With Pico's characteristic marginal sign. Top margin: "Animas primo creat, secundo applicat, tertio docet, quarto serat; quinto cadunt, sexto incorporantur, septimo animant corpus." Right margin: "In luna et in alia sydera homines esse." Cf. Calcidius, ed. Waszink, p. 220, 10.

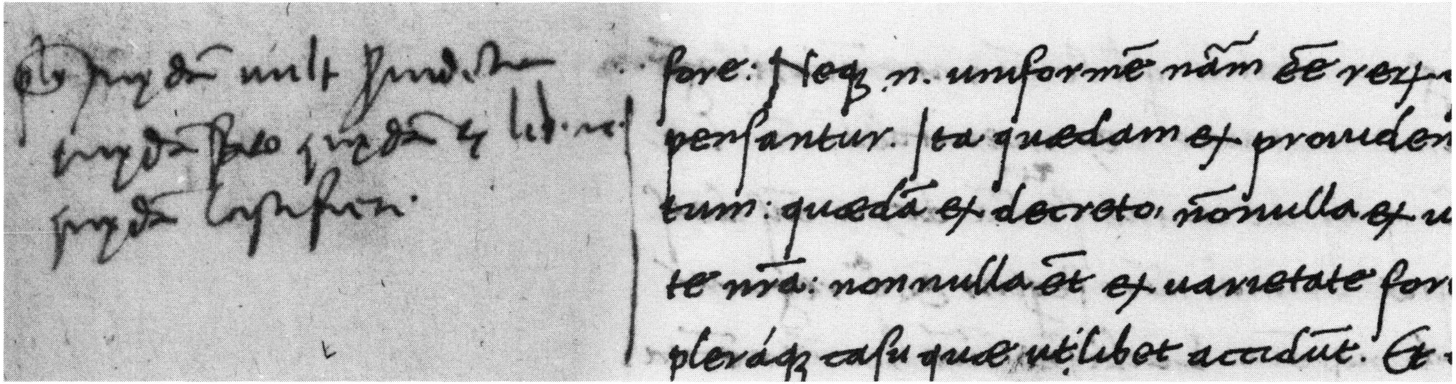

Figure 2.2

Fol. 92v. Left margin: "Plato quaedam uult providentia, quaedam fato, quaedam ex libero arbitrio, quaedam casu fieri." Cf. Calcidius, ed. Waszink, p. 183, 15.

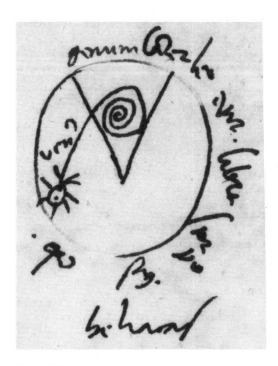

Figure 2.3

Fol. 69r. Pico's version of a Calcidian diagram explaining how the motions of the Same and the Different generate the spiral motion which in turn generates the illusion of contrary motion in the planets. Cf. Calcidius, ed. Waszink, p. 162, 5.

★27. ———, Par. lat. 6281. Southern France or northern Italy, s. XII in. Coeval marginal and interlinear glosses by a grammarian. From the Visconti library. Pellegrin, *Visconti,* p. 99.

★28. ———, Par. lat. 6282. France, ss. XI–XII. Coeval interlinear and marginal glosses on the *Timaeus* and Calcidius; none by Cusanus. Owner's note of N<icolaus> Cusan<us>. Waszink, p. cxxxi (s. XI med.).

★29. ———, Par. lat. 6283. France, s. XIII ex. No glosses but an *accessus* ("Capitulum universalis summe libri Platonis qui appellatur Thimeus") on fol. 1ra, *inc.* Osii Cordubensis episcopi rogatu. This *accessus* contains slight echoes of Bernard of Chartres's commentary. The manuscript also contains Cicero's partial translation of the *Timaeus* (*De essentia mundi*). Entered the Visconti library from the collection of Giacomo dalle Eredità and perhaps of Pasquino Capelli. Pellegrin, *Visconti,* p. 80; eadem, *Visconti, Supplément,* p. 14; Waszink, p. cxxi.

★30. ———, Par. lat. 7188. Normandy or England, s. XII in. No notes. "Ex bibliotecha Guilelmi Sacheti dono dedit Ael. Des Fontaines." Waszink, p. cxxi.

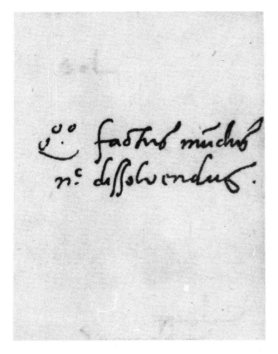

Figure 2.4
Fol. 25r. Pico's calligraphic hand. "Quomodo factus mundus nec dissolvendus." Cf. Calcidius, ed. Waszink, p. 73, 5.

★31. ———, Par. lat. 8677. Italy, s. XV 3/4, round humanistic bookhand, Florentine decoration. Marginal notabilia. A note on the flyleaf reads: "Visto per me Francisco da Luza, 1469"; this is followed by the owner's note: "domini Nicolai Leoniceni." Waszink, p. cxxi.

★32. Perugia, Biblioteca Comunale Augustea, 717 (*olim* J 111). Italy, written in 1500; copied from a codex derived from no. 9, above. In addition to Calcidius' version of the *Timaeus* on fols. 132v–151r, as indicated by Waszink, the manuscript also contains an excerpt from Calcidius' commentary (= ed. Waszink, pp. 204, 5 *Principio dicamus cuncta quae sunt*–210, 2 *plurimum distat*). Waszink, p. cxxii; *Iter* 2:60.

★33. Philadelphia, University of Pennsylvania, Van Pelt Library, MS Lat. 13. Florence s. XV ex. *Notabilia,* some in Greek; copied from a derivative of no. 9. Also contains cosmographical texts by pseudo-Aristotle (*De mundo*), Philo Judaeus, and Cleomedes. Waszink, p. cxxii; Norman Zacour et al., *Catalogue of Manuscripts in the Libraries of the University of Pennsylvania to 1800* (Philadelphia: University of Pennsylvania Press, 1965), p. 4.

★34. Recanati, Biblioteca della Casa Leopardi, MS 2 VIII F 7 (Libr. I rep. sup. C 119; *olim* LVII). Italy (Rome?), s. XV med., mbr., 103 leaves. Not s. XII, despite *Iter* and Waszink.

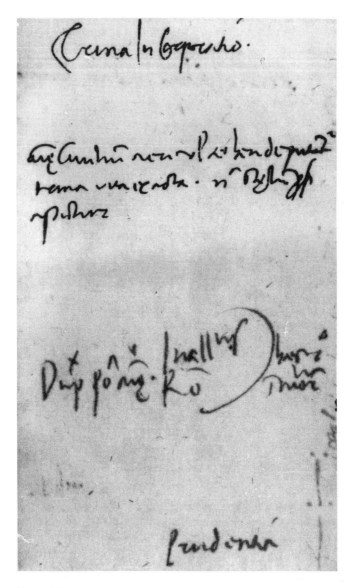

Figure 2.5

Fol. 89v. Left margin: "Trina incorporatio. animae ciuilium aeri vel aetheri deputantur trina vita exacta nec in caelum proficiscuntur. Duplex forma anime: intellectus, ratio: haec immortalia." Cf. Calcidius, ed. Waszink, p. 177, 10.

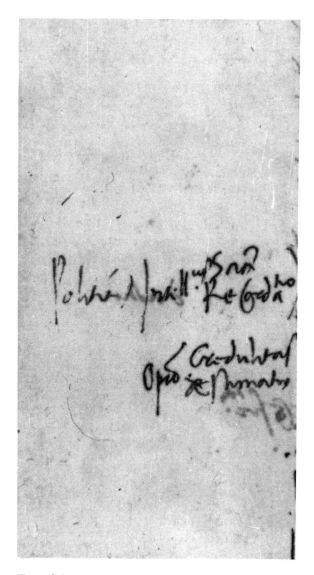

Figure 2.6
Fol. 167v. Pico notes Calcidius' version of the "Divided Line," Plato's account of the levels of cognition in *Republic* 6. 533D–534A. Cf. Calcidius, ed. Waszink, p. 334, 20.

$$
\text{Politia:} \begin{cases} \text{Intellectus} & \begin{cases} \text{Scientia} \\ \text{Recordatio} \end{cases} \\ \text{Opinio} & \begin{cases} \text{Credulitas} \\ \text{Aestimatio} \end{cases} \end{cases}
$$

Round humanistic bookhand, glosses in the hand of the scribe, large decorated initial (Paduan?). *Iter* 2:558–59; Waszink, p. clxxxvii; above, p. 83.

35. Reims, Bibliothèque municipale, MS 862. Italy, s. XV in. Written in Italy for Cardinal Guillaume Fillastre, who sent it to the chapter library of Reims Cathedral. Contains also Bruni's translations of Plato's *Gorgias* and *Phaedo*. Waszink, p. cxxiv; *Iter* 3:341; Hankins, *Plato,* 2:714. Fillastre's letter of transmission to the Chapter of Reims is edited in ibid. 2:496–497.

36. Stuttgart, Württembergische Landesbibliothek, MS Theol. et philos. fol. 58. Written by Baptista Augustensis in Padua in 1470; contains the glosses of the Paduan Master as well as Bruni's translations of Plato's *Gorgias, Phaedo,* and *Crito.* Waszink, p. cxxv; *Iter* 3:701; Berti, pp. 152–162; Hankins, *Plato,* 2:719; Dutton, "Material Remains," p. 225.

★37. Vatican City, Biblioteca Apostolica Vaticana, MS Arch s. Petri H 51. S. XII. Numerous twelfth-century glosses. Owned by Cardinal Giordano Orsini; entered the Vatican collection by 1436. Waszink, p. cxii; Jeauneau, *Lectio,* pp. 195–200; Giovanni Mercati, *Codices Latini Pico Grimani Pio* (Vatican City: Biblioteca Apostolica Vaticana, 1938), p. 147.

★38. ———, MS Barb. lat. 21. Two parts: the part containing the *Timaeus* is French, s. XII; that containing Calcidius' commentary is "perhaps Italian" (Pellegrin), s. XI. Glosses of the eleventh, thirteenth, fourteenth, fifteenth, and sixteenth centuries on the *Timaeus* and on Calcidius. One of the fifteenth-century hands has been identified by José Ruysschaert as that of Pierleone da Spoleto, physician of Lorenzo de' Medici and a member of Ficino's circle. The second part of the codex comes from the Franciscan convent in Siena, and is the source for nos. 1, above, and 47, below, according to Waszink. José Ruysschaert, "Nouvelles recherches au sujet de la bibliothèque de Pier Leoni, médecin de Laurent le Magnifique," *Académie royale de Belgique, Bulletin de la classe des lettres et des sciences morales et politiques,* 5th ser., 46 (1960): 50, no. 3; Waszink, p. cix; Pellegrin, *Vaticane,* 1:72; Dutton, "Material Remains," p. 207; James Hankins, "Pierleone da Spoleto on Plato's Psychogony (Glosses on the *Timaeus* in Barb. lat. 21)," in *Roma, magistra mundi: Itineraria culturae medievalis. Mélanges offerts au Père L. E. Boyle à l'occasion de son 75e anniversaire* (Louvain-la-Neuve: F.I.D.E.M., 1998), 3:337–348.

★39. ———, MS Chis. E V 156. Italy, s. XV, Italian gothica formata. Owned by Agostino Patrizi Piccolomini, bishop of Pienza (el. 1484). R. Avesani, "Per la biblioteca d'Agostino Patrizi Piccolomini," in *Mélanges Tisserant,* 6:43; Waszink, p. cxii; Pellegrin, *Vaticane,* 1:261 (ss. XIV or XV).

★40. ———, MS Chis. E VI 194. Italy, s. XV, round humanistic script, vine-stem initials. Owned and probably written for Alessandro Sforza, signore of Pesaro (1409–1473). The codex was copied from Petrarch's codex, no. 26, above. Waszink, p. cxii (where the owner is falsely identified as Cardinal Ascanio Sforza); Pellegrin, *Visconti, Supplément,* p. 62; Pellegrin, *Vaticane,* 1:262.

★41. ———, MS Ottob. lat. 1516. S. XII or XIII in., brought to Italy by the fourteenth century. Scattered *notabilia* in an Italian hand of the sixteenth century. Waszink, p. cxx; Pellegrin, *Vaticane,* 1:603.

★42. ———, MS Reg. lat. 1114. Italian, s. XIV or XV. Short glosses and notabilia in two hands, one a Venetian hand similar to (but not identical with) that of Bernardo Bembo. From the Jesuit house in Venice. Waszink, p. cxxiv; Pellegrin, *Vaticane,* 2.1:145.

★43. ———, MS Urb. lat. 203. Written for the condottiere and book collector Federico d'Urbino. Derived from no. 13, above. Waszink, p. cxxvi; Pellegrin, *Vaticane,* 2.2:525.

★44. ———, MS Vat. lat. 1544. Written in Italy ca. 1470 by Nicolaus Antonii de Montelpero, Lombard script. Rubricated *notabilia* and diagrams in a second hand. Owned by Nicolaus Modrussiensis, a Dalmatian humanist who worked mostly in Italy. This MS may have been one of those used by the editors of the *editio princeps* of Calcidius printed in Paris in 1520. Waszink, p. cxxvii; Pellegrin, *Vaticane,* 3.1:116.

★45. ———, MS Vat. lat. 2063. Italy, s. XIV. Contains a glossary commentary virtually identical with that found in Vienna, Österreichische Nationalbibliothek, MS 278; about a quarter of the glosses are derived from Bernard of Chartres's commentary. Sent to Coluccio Salutati by Giovanni Conversini da Ravenna (1343–1408); owned by Salutati and later by Pope Nicholas V. Waszink, p. cxxvii; Tullio Gregory, *Platonismo medievale: Studi e ricerche,* Studi storici 26–27 (Rome: Istituto storico italiano per il Medio Evo, 1958), pp. 88–89; Claudio Leonardi, *Codices Vaticani latini, Codices 2060–2117* (Vatican City: Biblioteca Apostolica Vaticana, 1987), pp. 7–9; Pellegrin, *Vaticane,* 3.1:510–511; Antonio Manfredi, *I codici latini di Niccolò V* (Vatican City: Biblioteca Apostolica Vaticana, 1994), p. 374, no. 596; Dutton, *Bernard,* pp. 292–295.

★46. ———, MS Vat. lat. 3348. Florence, s. XV med., copied by Piero Strozzi from no. 15, above. Also contains all of Bruni's translations from Plato. Waszink, p. cxxvii; Hankins, *Plato,* 2:729, no. 339.

★47. ———, MS Vat. lat. 4037. Italy, s. XV ex., copied from no. 38, above. Contains also the Latin works of Cardinal Bessarion, written by the same scribe, and Lilio Tifernate's translation of ps.-Timaeus Locrus. Waszink, p. cxxvii; John Monfasani, "Bessarion Latinus," *Rinascimento* 21 (1981): 167, 173–175, 185–188, 196–204.

★48. Venice, Biblioteca Marciana, MS Zan. lat. 469 (coll. 1856). Italy, s. XIV ex., written by four Italian gothic hands, I + 144 + I leaves. The glosses on fols. 1r–14v are in the hand of Antonius de Romagno de Feltro (signed, fols. 14v, 58r, 95r). Fol. 58r has the note, in Antonius' hand: "Excusetur scriptor si in locis compluribus liber iste corruptus invenietur. Sumpsit enim ab exemplari cuius summa emendatio erat esse corruptissimum." On fol. 144v there are a few notes regarding Antonius' family history. The codex later belonged to Cardinal Bessarion. Waszink, p. cxvii; see above, p. 81, and below, appendix 2.A.

★49. ———, MS Marc. lat. VI 137 (coll. 2853). Italy, s. XV. Scattered notes in a fifteenth-century hand. Owned by the philosopher Giacomo Zabarella (1533–1589). Waszink, p. cxviii.

★50. ———, MS Marc. lat. XIV 54 (coll. 4328), fol. 101r–v. Eight excerpts from the *Timaeus,* called "Alegabilia dicta collecta ex Thymeo Platonis," in the hand of Pier Paolo Vergerio (1370–1444) and dated "in Justinopoli, anno domini 1388, die VII a Septembris." The excerpts correspond to *Timaeus* 42A (Waszink, p. 37, 1–20), 42E (ibid., p. 38,

6–9), 44A (ibid., p. 40, 1–16), 44D (ibid., pp. 40, 19–41, 13), 47A–D (ibid., pp. 44, 4–45, 8), 48C–E (ibid., p. 46, 2–9), 51E (ibid., p. 50, 9–10). The variants suggest that the codex may have been copied from Petrarch's manuscript or a close relative. *Epistolario di Pier Paolo Vergerio,* ed. Leonardo Smith (Rome: Istituto storico italiano per il medio evo, 1934), pp. lxxiv, 4, 12, 492, with a plate of fol. 101v; Waszink, p. clxxxviii; *Iter* 2:264.

51. Vienna, Österreichische Nationalbibliothek, MS 2269. S. XIII in. Owned by "Petrus Mocenico dei gratia dux Venetiarum" (doge 1474–1476). Waszink, p. cxxvi.

Index of Owners, Scribes, and Glossators

Acquaviva, Matteo III d', 23

Antonius de Romagno de Feltro, 48

Assisi, Franciscan convent, 2

Baptista Augustensis, 36

Barzizza, Gasparino, 19

Barzizza, Guiniforte, 19

Bernard of Chartres, 11, 12, 15, 29, 45

Bessarion, Cardinal, 48

Bruni, Leonardo, 13

Capelli, Pasquino (?), 29

Conversini, Giovanni, da Ravenna, 45

Federico d'Urbino, 43

Fichet, Guillaume, 4

Ficino, Marsilio, 17

Fillastre, Cardinal Guillaume, 35

Florence, Monastery of SS. Annunziata, 10

Francisco de Luza, 31

Giacomo dalle Eredità, 29

Grimani, Cardinal Marino (?), 20

Guarino Veronese (?), 22

Leoniceno, Niccolò, 31

Medici library, Florence, 8, 9

Medici, Lorenzo de', 8

Mocenigo, Piero, 51

Niccoli, Niccolò, 11, 12, 13, 14

Nicholas of Cusa, 16, 28

Nicholas V, Pope, 45

Nicolaus Antonii de Montelpero, 44

Nicolaus Modrussiensis, 44

Orsini, Cardinal Giordano, 37

APPENDIX 2   EXTRACTS FROM RENAISSANCE GLOSSES ON THE *TIMAEUS*

A.   Antonius de Romagno

Source: Venice, Biblioteca Marciana, MS Zan. lat. 469 (1856). See appendix 1, no. 48. The edges of the folios have been cropped by the binder, resulting in the loss of parts of the text; conjectural restorations of the text are indicated by angle brackets.

Antonius de Romagno's *Accessus* to the *Timaeus*

/fol. 1r/ Quoniam quidem in huius operis principio quid inquiratur notare debemus, quis sit modus tractandi prius prelibare temptemus. Plato Socratem magistrum suum imitatur ne dispositissimis verbis suis dissentire videatur. Nanque Socrates cum de republica

tractare disposuerit, <ad> eius statum pertractandum virtutes necessarias repperit cete-
risque virtutibus in eiusdem reipublice statu et regimine justiciam prevalere cognovit,
excellen<tia> eius eminentie persuasionibus suis ciuitatum rectores admovit. Est autem iusticia qua unicuique quod iustum est redditur servata communi utilitate <et
conven>ientia, cuius adhibita reipublice status liberaliter disponitur. Volens igitur
Socrates de republica tractare, de iusticia populari, id est consuetudinaria, in quodam volumine suo diu diserere non permisit, cuius tractatum reipublice statui commodissimum
in decem libros diuisit. Sed quia scilicet in nulla ciuitate reipublice statum bene regi per-
cepit, cuiusdam reipublice effigiem sibi ad tractandum suscepit. Quandam enim civi-
tatem, non ut realiter esset, sed sola mentis consideratione sapienter edificauit in qua
quomodo prelati, quomodo subditi et quodlibet genus hominum unum sequi deberet pre
ordin<e quolibet>, quomodo respublica in qualibet ciuitate regenda esset manifestauit.
Et quemadmodum Socrates sibi quandam ciuitatem effigiauit cuius similitudine reipublice statum tenendum esse uoluit, ita Plato mundi artificem introducens quendam ar-
chetipum mundum in mente sua disposuisse dicit, ad cuius similitudinem mundus iste
sensilis factus fuerit.

Quoniam vero Socrates de iusticia populari, idest consuetudinaria, tantum trac-
tauerat et ideo de iusticia ad plenum non dixerat, quia unde procederet justicia popularis
querebatur et ideo oportebat ut ad justiciam naturalem que mater est eiusdem aperiendam
ascenderet, voluit Plato quod Socrates magister suus intactum reliquerat suscipere, idest
de justicia naturali disserere. Est autem iusticia naturalis per quam homines quod iustum
est faciunt, non seruili timore sed filiali amore, que unde sit, inde in <an>gelis nulli du-
bium est. Et quia de naturali iusticia prius tractare disposuit ut dei iusticiam demonstraret,
oportuit <demonstrare quod> deus hunc mundum et omnia que in mundo sunt, dando
unicuique quod suum est, iuste creauit. Asserit nanque quod quemadmodum <deus> in
sua dispositione nichil ordinauit iniuste, sed omnia dando unicuique quod suum est iuste
formauit, ita unusquisque homo suo in opere quod iustum est debet facere, unicuique
quod suum est dando.

Duabus de causis Platoni visum est oportunum dei iusticiam demonstrare, tum quia
tempus ostendere voluit in quo naturalis iusticia hominum maxime valuit, scilicet quando
statim facto mundo primum homo in seculum floruit, vel quia dei iusticiam ad argumen-
tum hominibus induxit. Et quia hoc negocium erat valde difficile, nolluit Plato sibi as-
cribere hoc opus more sapientum causa euitande arrogantie, et ne videretur magistro suo
se velle preferre, quia susceperat id agendum quod magister suus intactum reliquerat, ergo
Thimeo cuidam suo discipulo hoc opus asscripsit, quia illum magis quam se ipsum laudari
voluit.

Hunc autem librum P<latonis> Calcidius de greco in latinum transtulit, quem per-
suasione Osii pape transferre sibi commisisse librum suum valde commendabilem reddit,
cum vir tante auctoritatis hoc opus sibi iniunxerit, quod nunquam sibi iniunxisset nisi illum hoc posse perficere cognovisset.

Intentio huius libri est de naturali iusticia tractare, persuadendo hominibus secundum eam vivere. Materia est ipsa iusticia, causa <vero> intentionis est utilitas et fructus
iusticie, <ideoque> ethice supponitur. Sed secundum hoc quod tractat de inuisibilibus,
idest de consonantiis, supponitur phisice.

Videndum est quod Calcidius premittit quendam prologum more recte scriben-
tium in quo reddit lectores attentos, dociles et beni<volos>, in quo quidem prologo dum
Osii pape captat benivolentiam, eius commendat sapientiam.

Antonius de Romagno's Dependence on Bernard of Chartres

1. Bernard of Chartres, ed. Dutton, pp. 183–184, 11. 276–281. *Sed animal* (37D).

Vult accedere ad genituram temporis, ut ostendat quod sicut mundus intelligibilis est aeuo coaequaeuus, ita hic sensibilis tempori, et sicut hic mundus est imago illius, ita tempus est imago aeui. Continuatio. Immortalis genuit sensilem mundum, cuius natura non aequatur aeuo, *sed natura animalis,* id est *animal, quod generale,* id est intelligibilis mundus, *aeuo,* id est aeternitati, *exaequatur.*

Antonius de Romagno, MS cit., fol. 8r, right margin.

Ut ostenderet quod sicut mundus intelligibilis est coequeuus aeuo, ita hic sensilis tempori, et sicut hic mundus est ymago illius, ita tempus est ymago eui; coequaevum immortale genuit sensilem [im]mortalem etiam, cuius tamen natura non equatur euo, sed tempori.

2. Bernard of Chartres, ed. Dutton, p. 146, 11. 21–42. *Unus, duo, tres* (17A).

Nam subtracto quarto, remanent partes quae coniunctae faciunt primum perfectum numerum, id est sex, et ideo a perfecto incipit, ut notet perfectionem operis. Vel ideo quartum uoluit abesse, quia tractaturus erat de anima, quae ex tribus primis consonantiis primo loco figuratur constare, scilicet diatessaron, diapente, diapason. Vel ideo quia in his tribus numeris magna uis perpenditur, unitas enim fons est omnium numerorum: binarius et ternarius primi sunt qui in se ipsos et alter in alterum multiplicati firmam faciunt conexionem, sicut bis bini bis, ter terni ter, bis bini ter, ter terni bis. Quae tam firma et solida conexio praesenti operi de mundi genitura agenti bene conuenit, quod per tres auditores notatur. Si uero Socratem cum tribus consideres, quattuor sunt, in quo numero omnes musicas consonantias uel proportiones inuenies. Duo enim ad unum duplus est, scilicet diapason; tres ad duo sesquialter, id est diapente; quattuor ad tres sesquitercius, id est diatessaron; ad unum idem quattuor quadruplus, id est bis diapason. Quibus simphoniis mundi fabricam constructam esse docebit. Non sine causa ergo quartus auditor subtractus est. Hunc quartum dicunt fuisse in re Platonem, qui pro magistri reuerentia se subtraxit, ne uideretur se illi praeferre, si suppleret quod magister non poterat. Sed totum in significantia melius uidetur esse dictum.

Antonius de Romagno, MS cit., fol. 2r. *Unus, duo, tres.*

Quartus defuit non sine causa quia a senario numero incipit qui est perfectus, idest per partes eius. Realiter, ut quidam dicunt, quartum querebat scilicet Eudigedium,[1] sed invenire non potuit: vel quia quartum <fuis>se Platonem, sed arrogantiam vitando se ponere nolluit, ne videretur preferri magistro suo, videlicet Socrati.

Vel sub integumento *unus duo tres* posuit et non [sine] quartum,[2] ut perfectionem in suo opere designaret, ponendo quasdam numeri partes quibus aggregatis perfectus numerus constituitur, scilicet senarium. Et dicitur perfectus quia diuiditur in partes equaliter quibus aggregatis ipse redditur perfectus numerus. Diuiditur enim in duos ternarios et tres binarios et sex unitates equaliter, quibus adiunctis ipse perfectus existit, nam unum et duo et tres senarium perficiunt, nec plus nec minus reddunt.

Vel ut firmam ac solidam complexionem rerum denotaret, que firma et solida rerum complexio presenti operi, scilicet agenti de genitura mundi, bene conueniret, et hoc facit ponendo principium numerorum, videlicet unitatem et ipsos primos et cubicos numeros: qui primi et cubici dicuntur quia in se ipsis duplicati, vel <quia> unus in alterum firmam et solidam rerum complexionem denotant. Harum proportione notata ad ipsa elementa quibus omnia colligantur <referuntur>. Unde Boetius in libro De consolatione:

Tu numeris elementa ligas,★ in seipsis duplicati, ut bis bini bis et ter terni ter; unus in alterum, ut bis bini ter et ter terni bis.

Vel ideo ut se ostenderet principaliter velle tractare de anima que ex tribus primis consonantiis figuraliter dicitur constare, scilicet dyatesseron, dyapente et diapason.

Vel idcirco potius ut notaret se disserere de exemplari iusticia, hoc est naturali, unde Macrobius:★★ *Quattuor sunt quaternarum genera virtutum,* idest que quattuor modis tractantur. Nam alia *politica,* idest que tractatur civili modo, et alia *purgatoria,* idest illa <qua> homines cum adhuc sunt in mundo, mori iam mundo cupiunt; alia *animi iam purgati,* scilicet qua solum rimantur celestia et nil terrenum desid<erant>, ut illa que fuit in Paulo Heremita; alia *exemplaris* que est divina essentia. Et de hac in hoc libro intendit Plato.

Notes

[1]*scil.* Euthydemum

[2]quare *MS*

★Boethius, *Cons., metrum* 3.9.10

★★Macrobius, *In somn. Cic.* 1.8.5.

Antonius de Romagno on Matter and the Elements

/fol. 13v/ Ostendit superius multis rationibus et similitudinibus hilen propria forma carere. Nunc autem vult dicere nec terram nec aerem nec aliquid ex visibilibus creaturis nec etiam pura elementa esse matrem facti, sed quod factum visibiles et invisibiles creaturas comprehendit. . . . Matrem vero totius creature ait esse quandam informem capacitatem positam intra nullam et aliquam substantiam, que cum sit inter hec duo tam discreta, idest nullam et aliquam, hec[1] non nulla erit quia erit aliqua, idest informis, nec aliqua erit formatorum. Sicut cum inter album et nigrum medium ponimus, illud nec album nec nigrum dicimus, sed medium. Ait etiam illam nec plane intelligibilem propter formas receptas [i.e., because it has not received forms], nec plane sensibilem, idest sensibus subiacere, sed etiam talem que videtur intelligi per ea que variantur in ipsa. Et concludit dicendo, ergo, quia nulla propria species sensibilis est[2] <ei> tribuenda et hec sola inter elementa pura est, ipsa tantum mater totius facti dici debet.

Diceret aliquis: non est verum quod hec elementa omnium non sint mater. Quod removet Plato dicendo: Que sunt partes illius immutabiles rei, si illa que sine parte est in se pars dici potest?

/fol. 14r/ Aliquis fortasse diceret quod essentia hyles verteretur in essentiam idee et econtrario, sed hoc est falsum, quia numquam essentia unius in alterius essentiam vertitur.

Esse et fuisse (52D): idest erunt et fuerunt ante sensibilem mundum hec tria, scilicet existens idest idea, et locus idest hyles, et generatum idest pura quattuor elementa potentialiter generata, idest formata hile.

Quia hile semper erit, igitur omniformis videtur, et nutricula totius generationis.

Notes

[1]Nec *MS*

[2]est sensibilis *MS*

(Other passages from this commentary are cited in notes 27 and 29.)

B. The Recanati Master

Source: Recanati, Biblioteca Leopardiana, MS 2 VIII F 7 (Libr. I rep., sup. C 119). See appendix 1, no. 34. As this manuscript is difficult of access and photographs are not at present permitted, it seems desirable to give here a complete edition of the glosses.

20A   Timeus Locrensis. (10v)

22B   Quare nullus e Grecia senex. (16v)

27C   Nota quod etiam antiquorum consuetudo fuit in cuiuslibet operis principio divinitatis auxilium inuocare. (30v)
Nota de quibus intendit tractare. (30v)

27D   Inuocatio auxilii predicti. (30v)
Hic incipit explanare suam intentionem de quibus intendit edicere, diuidens librum. (31r)

28B   *Omne igitur celum uel mundus.* Subaudi: factus est ex aliqua legittima causa, quandoquidem supra dictum est quod nil fit cuius ortum <ex> causa legittima non procedat. (32r)

28C   Omnium deus est a<u>ctor nec a nobis inuenitur nec digne laudatur. (33r)

29D   Nota quare deus mundum fecerit. (36r)

29E   Nota quod dei voluntas omnis boni est causa nulliusque mali, quin immo cuncta quae ad malum tendere uidentur ex nature mobilitate ad bonum reducit suo ordine. (36v)

30B   Mundus est animatus anima intellectuali. (37v)

30C   Hic probat quod mundus perfectus creatus est et quod est tantum unus et non plures. (38r)

31B   Ignis et terra sunt fundamenta mundi. (39v)

31C   Hic probat quod necesse fuit ut quattuor essent elementa. (40r)

32A   Hic probat quod sicut est de toto mundo, ita de ipsius partibus: existunt enim ex quattuor elementis. (41r)

32B   Hic est conclusio: quod constat ex quattuor elementis. (41v)

32C   Preter dei uoluntatem mundus dissolui non potest. (42r)

33A   Ex diuina prouidentia mundus regitur. (42v)
[next to a diagram:] Per istud cognoscimus quomodo elementa coniungantur. (43v)

33B   Mundus est rotundus et a centro ad omnes partes equalis. (44r)

33C   Hic probat quod licet mundus sit admodum animalis, tamen membris non indiget. (44r)

34A   Sufficit mundo motus circularis. (45r)

34B   Deus fecit mundum summe perfectum, ita quod nihilo preter eum <indiget>. (46r)

34C   Deus constituit animam nobilitate preesse corporee nature. (46v)

36D   Orbes contrario motu rotant ut septem planete a firmamento opposite uertuntur. (50v)

37A   Deus omnia intuetur atque prescit. (52r)

37C   Ex beneficiis nobis a deo impensis noster intellectus hillarescit et ad sui conditoris complacentiam studet. (53r)

37D   Facto euo deus tempus constituit et eius partes, utputa annum, diem, mensem, septimam et horam, sed euo et tempori habitus est qua re euum est continuum non habens partes, tempus autem partium diuersitatem sortitur. Amplius euum est mensura simplicium substantiarum et in eorum partibilitatem, tempus uero mensura est corruptibilium. Item euum creatum est cum anima mundi, sed tempus cum motu celi qui est mensura ipsius et ipsum eius atque creatum celo. (53v)

38B   Hic sensibilis mundus similis intellectuali. (55r)

38C   Nota quare facte sunt stelle erratice, equidem ut temporis partes narrentur. (55v)

38D   Nota locationem septem planetarum. (56r)
        Explicare ornatum celi difficile est ipso opere. (56v)

39B   Achantus arbor est que folia habet in speram versa. (58r)
        Nota quod sol illuminat omnia que infra celum sunt. (58v)

39C   A<d> septem planetarum aliarum<que> stellarum discursus non notat uulgus, ex quibus diuersitas euentuum fit. (59r)

39D   Cum sol aliaque sydera compleuerit cursum, iterum circuunt ut annus tempusque continuetur. (59v)

39E   Non obstante quod mundus intelligibilis esset, sensibilem mundum fecit deus. (60v).
        Cum opifex deus aliquid facere uult, ideas respicit. (60v)
        Nota quattuor quae fecit deus. (61r)

40A   Nota ornamentum celi. (61v)

40B   Sydera que sunt in octavo orbe semper sunt in eodem statu. (62v)
        Nota quod terra est mater terrenorum custosque poli diei et noctis. (63r)

40C   Circulatio stellarum. (63r)
        Occultatio eadem debet. (64r)

40D   Diuinarum potestatum ratio assignari non potest, ideo credendum est divinis hominibus et coticibus. (64v)

40E   Nota quod antiqui posuerunt filios deum quibus credi debent. Sed apud nos sancti uiri intelligi debent huiusmodi. (65r)
        Enumerat hos filios deum. (65v)

41B   Nota de innouatione mundi. (67r)

41C   Loquitur in persona dei. (67r)

41D   Ad beatam uitam. (68v)
        Resurrectio uniuersalis. (69r)
        Anime electe ad eternam uitam. (69v)

42A   Vincendo carnis uicia habebimus uitam beatam. (70r)

42D   Curam hominum angelis dedit deus. (72v)

44C   Infernus. (78v)

44D   Formatio humani corporis. (78v)
        Nota de capite. (79r)

45B   Oculi et quare dati sunt. (80v)

45C   Quomodo creatur uisio. (81v)

45D   Quare non uidemus de nocte. (82v)

45E   De imaginibus que apparent in speculis et in liquidis. (83r)

46D   Nota quod licet corpora recipiant peregrinas impressiones, tamen non prospiciunt, sed sola anima sentit ipsaque habet rationem, intellectum et prudentiam. (85r)

47A   Nota de uisu. (86v)

47B   Maximum beneficium est uisus. (87v)

      Nota bene quare datus est uisus hominibus. (87v)

47C   Quare datus est auditus et sermo hominibus. (88r)

48A   Hic nota de mundo sensibili. (89v)

48D   Inuocatio dei. (91r)

48E   Tres mundos ponit: intelligibilem et exemplarem et sensibilem. (91v)

49B   Nota stabilitas est in naturalibus corporibus. (92v)

49C   Generatio et corruptio elementorum. (93r)

49D   Nota stabilitatem habent que in mundo sunt. (94r)

50A   Similitudo. (94r)

50B   Prima materia omnia recipit in se. (95v)

50C   Hic epilogat iam dicta, scilicet unum quod gignitur, aliud in quo gignitur atque <aliud> ad cuius similitudinem gignitur. (96v)

C.   The Paduan Master

Source: Bergamo, Biblioteca Civica "Angelo Mai," MS MA 350. See appendix 1, no. 3.

The Paduan Master's Dependence on Calcidius and "Chartrian" Sources

1. (Glossing Calcidius' preface, ed. Waszink, p. 5, 1)

Osius Hispaniensis episcopus hunc librum, uidens latinis profuturum quia ex eo pendent omnes sententie philosophorum, rogauit Calcidium archidiaconum suum ut hunc transferret. Qui cum transtulisset hanc partem misit ei, cum hac epistola in qua commendat amicitiam, uirtuti eam conferendo quia sicut uirtus uel res quasi impossibilem ad possibilem redigit, sic et amicitia.

Compare to Bernard of Chartres in Dutton, *Bernard,* p. 142, 9–10, and to the glosses in a twelfth-century manuscript from Mont-Saint-Michel (Avranches, Bibliothèque Municipale MS 226), edited in Jeauneau, *Lectio,* p. 226.

2. (Glossing *Tim.* 17A, *Unus duo tres*)

A numero incipit quia numerus est exemplar et origo omnium et ipse est tractaturus de archetipo mundo qui est exemplar totius sensibilis mundi.

Plato de iusticia sub quorundam philosophorum personis, id est Socratis, Thymei et Critie et Hermocratis egit, primo in persona Socratis de politica, id est de ea que ciuitatis dicitur positiua. Quod utique decem libris illis fecit quos de republica composuit. Ad quem tamen tractatum ex consequenti descendit quia principaliter de iusticia quesitum est, quam Trasimachus orator diffinierat eam esse que huic prodesset qui plurimum posset. Contra quem Socrates diffiniuit eam esse pocius que his prodesset qui minimum possent. Que ut non ex unius hominis ingenio sed illustriori urbis exemplo spectaretur, quandam urbis depinxit ymaginem que iustis institutis et moribus regeretur, a quibus, si quando degeneraret, improspera atque [in]exitiabilis huic morum mutacio fieret.

Compare to Calcidius' commentary, ed. Waszink, p. 59, 3–13.

3. (Glossing *Tim.* 22C, *Sed est uera*)

Sed uera est quia calor sole natus quasi filius solis, paulatim crescendo et humori prevalendo exhustionem immittit; sic emisso, humori paulatim crescendo et calori prevalendo diluvium gignit.

This naturalistic explanation for the Flood is extremely close to a twelfth-century gloss published by Jeauneau (*Lectio,* p. 196) from Vatican City, Biblioteca Apostolica Vaticana, MS Arch. S. Petri H 51 (see appendix 1, no. 37).

4. (Glossing *Tim,* 23E, *utramque urbem*)

Minerva *condidit* utramque ciuitatem, id est naturalem et humanam, et *educauit* moribus et *instituit* legibus. Condidit *uestram priorem fere mille annis,* quasi fortasse parum defuit annis dico computatis, *ex indigente agro et Vulcanio semine,* idest ex illo tempore ex quo Vulcanus nimio amore succensus cum ea concubere uoluit, sed semen ea fugiente in terram cecidit, unde Erictonius fuit natus. *Hanc* uero *nostram post octo mille annorum,* sicut *continetur in sacris apicibus.* Indigentes uocantur indigene, quasi inde geniti, et uocat indigere agrum, quasi matrem Vulcanii seminis, et potest intelligi duobus modis quod dicitur de annis: uel quod illa ciuitas fere mille annis facta fuisse antequam Vulcanus in terram semen fundaret, uel quod post effusionem illius seminis fere mille <annis> facta fuit.

This gloss is dependent both on Bernard (ed. Dutton, *Bernard,* pp. 152, ll. 194 ff., 153, ll. 211 ff.) and the Vatican glosses published by Jeauneau (*Lectio,* p. 196).

5. (Glossing *Tim.* 24E)

Atlas mons est in occidente inter quem et Calpen oceanus terram intrat et uocatur mediterraneum. Et quia posset queri quomodo illa gens poterat ex Athlantico uenire quod non est commeabile, respondit quod *Tunc* erat *commeabile.* Nam insulam in ore sinus, id est recuruationis, habebat, et sic per illam et per alias insulas que erant in eodem mari illic tunc iter agentibus patebat comeatum usque ad defectum illarum insularum, id est ad initium continentis terre, id est continue, que uicina Athlantico, quia fretum diuiditur angusto littore in quo apparent uestigia ueteris portus.

The first part of this gloss is closely related, again, to the Mont-Saint-Michel glosses edited by Jeauneau (*Lectio,* p. 220). The rest is a back formation from the text of the *Timaeus* itself, which also postulates an archipelago of islands linking Atlantis with the Pillars of Hercules.

D.   Marsilio Ficino

A *placita philosophorum* on Matter and the Elements

Source: Milan, Biblioteca Ambrosiana MS S 14 sup., fols. 90v–91r. See appendix 1, no. 17. The *placita* is written into the margins of Ficino's text of Calcidius (= Waszink, p. 287, 2).

The *placita* is compiled mostly from Diogenes Laertius and from Calcidius himself. It is noteworthy that Ficino here follows Diogenes Laertius in assimilating Plato's views on matter to Aristotle's.

fol. 90v/ Plato et Filolaus habuer<unt> eandem opinionem; quere Laertium.* Opiniones de materia. Ignis: Eraclitus. Aer et infinitus: Anax<i>menes et Anaximander. Aqua: Tales, Homerus. Medium inter prima, primo. Medium inter 2a. Medium inter 3a. Quattuor elementa: Empedocles. Totum quod est et finitum: Parmenides. Totum quod est et infinitum: Xenophanes. Totum quod est sed non unum, immo in atomas divisum: Leucippus, Democritus, Epicurus. Totum quod est sed confusum omne in omnibus: Anaxagoras. Informis capacitas: Pictagoras, Plato, Aristoteles, Stoici.

Anaxagoras Clazomenius Anaximenis auditor. Tales Milesius. Anaximenes Milesius Anaximandri auditor. Anaximander Milesius Taletis auditor. Archelaus Atheniensis sive Milesius Anaxagore auditor; materiam dixit esse frigidum agens calidum. Eraclitus Ephesius nullius discipulus.** Xenophanes Colophonius Archelai auditor. Parmenides auditor Xenophanis; duo putant principia, ignem et humum: illum agens, hanc materiam.*** Mellissus Samius Parmenidis auditor: hoc omne dixit esse immutabile, immobile et infinitum unum, a quo Plato multa sumpsit, maxime hoc, quod mundus esset animal unum.# Zeno Eleates, Melissi auditor, idem dixit.

/fol. 91r/ Leucippus Eleates vel Abderites Zenonis huius auditor. Democritus Abderites Leucippi auditor. Diogenes Appolloniates, Anaximenis auditor, elementum dixit aerem, mundos infinitos, et inane infinitum. Ana<xa>goras posuit materiam esse infinita corpora omoiegenia que semper sunt, simul permixta, et numquam sit perfecta discretio, sed talis ut hoc magis apparent unum quam aliud. Democritus infinita corpora individua. Trismegistus, Pittagoras, Philolaus, Numenius, Plato et Aristoteles, Stoici de materia idem dixerunt.

Notes

*cf. Diogenes Laertius 2. 2–3.

**Ibid. 9. 5.

***Ibid. 9. 21–22.

#cf. Calcidius, ed. Waszink, p. 285, 10–12.

List of Manuscripts Not in Appendix 1

Avranches, Bibliothèque Municipale
    MS 226

Bologna, Biblioteca Universitaria
    Aldovrandi 56, vol. II

Fermo, Biblioteca Comunale
    4 C A 2.80

Florence, Biblioteca Nazionale Centrale
    Magl. XII 11
    Nazionale II I 105
    Palatino 1024, vol. II

London, British Library
>   Add. 11274
>   Add. 19968

Milan, Biblioteca Ambrosiana
>   F 19 sup.

Paris, Bibliothèque Nationale
>   lat. 12948
>   lat. 14716

Vatican City, Biblioteca Apostolica Vaticana
>   Arch. s. Petri H 51
>   Chigi E V 152
>   Reg. lat. 1308
>   Urb. lat. 1389
>   Vat. lat. 5223

Venice, Biblioteca Marciana
>   Zan. lat. 225 (1870)

Vienna, Österreichische Nationalbibliothek
>   278

NOTES

For abbreviations used in the notes, see appendix 1.

1. The *accessus* to the 1363 commentary was published in Jeauneau, *Lectio,* pp. 200–202; the same text was later republished in Klibansky, supplement, pp. 66–67.

2. On the Platonic revival of the fifteenth century, see my *Plato.*

3. Dutton, "Material Remains," pp. 205–206. I am grateful to Prof. Dutton for allowing me to consult his article in proof. The present essay is intended to supplement Dutton's study, which deals primarily with the thirteenth and fourteenth centuries.

4. For detailed information on the manuscripts discussed, see appendix 1.

5. On Forzetta's important collection, see Luciano Gargan, "Il Preumanesimo a Vicenza, Treviso e Venezia," in *Storia della cultura veneta* (Vicenza: Neri Pozza, 1976), 2:168–170. That Forzetta owned two copies of the *Timaeus* is known from a surviving inventory, but the codices have not yet been identified with any manuscripts in modern collections.

6. Pierleone's few, but interesting, glosses on the *Timaeus* discuss Aristotle's claim in *De anima* 1.3 that Plato had attributed "magnitude" to the soul, and defend Calcidius' understanding of the substance of soul as a *tertium quid* between First and Second Substance (*Tim.* 34E–35A) against Themistius and other Greek commentators. See James Hankins, "Pierleone da Spoleto on Plato's Psychogony (Glosses on the *Timaeus* in Barb. lat. 21)," in *Roma, magistra mundi: Itineraria culturae medievalis. Mélanges offerts au Père L. E. Boyle à l'occasion de son 75e anniversaire* (Louvain-la-Neuve: F.I.D.E.M., 1998), 3:337–348.

7. See Waszink, pp. clxix–clxx. The two other candidates for the main manuscript used for the *editio princeps* are also fifteenth-century Italian codices: Naples, Biblioteca Nazionale, MS VIII E 30 and Oxford, Bodleian Library, MS Canon. class. lat. 175. See appendix 1 for these manuscripts. The editor, Agostino Giustiniani (Genoese, 1470–1536), also collated an eleventh-century codex now in the British Library, MS Add. 19968. Giustiniani's preface to the 1520 *editio princeps* is published in Waszink, pp. clxvii–clxix. On Nicolaus Modrussiensis, see George McClure, "A Little-Known Renaissance Manual of Consolation: Nicolaus Modrussiensis' *De consolatione,*" in *Supplementum Festivum: Studies in Honor of Paul Oskar Kristeller,* ed. James Hankins, John Monfasani, and Frederick Purnell, Jr. (Binghamton, N.Y.: Medieval and Renaissance Texts and Studies, 1987), pp. 247–277.

8. I use the term "Chartrian tradition" rather narrowly to refer to study materials descended from, or similar in character to, the two important twelfth-century commentaries on the *Timaeus* thus far identified, i.e., those of Bernard of Chartres and William of Conches. For the controversy surrounding the "School of Chartres," see R. W. Southern, *Scholastic Humanism and the Unification of Europe,* vol. 1, *Foundations* (Oxford: Blackwell, 1995), pp. 58–101, with references to the earlier literature. While agreeing with much of Southern's argument, I believe that he does not succeed in his attempt to disassociate the twelfth-century study of the *Timaeus* from Chartres. Nor can his attempt (p. 81n) to dismiss Paul Dutton's attribution of a set of *Glosae super Platonem* to Bernard of Chartres (see below, note 13) be accepted in the absence of a serious review of the evidence assembled by Dutton.

9. Klibansky, pp. 35–36, 43. The *Policraticus* (7.5) includes a section arguing for parallels between Genesis and the *Timaeus*. A manuscript of the *Policraticus* that may have been known to Ficino is Florence, Biblioteca Nazionale Centrale MS Naz. II I 105, a thirteenth-century manuscript with fourteenth-century Italian annotations. See Mazzatinti 8:40. Fols. 21–22v of this manuscript, containing the "Epistola Johannis Anglici episcopi Carnotensis ad librum suum," has been recopied in Florentine round humanistic hand of the mid-fifteenth century.

10. See Marsilio Ficino, *Lettere, I: Epistolarum familiarium liber I,* ed. Sebastiano Gentile (Florence: Olschki, 1990), pp. 46, 82. The "Indice delle fonti" also identifies seven quotations from Calcidius and one from Cicero's translation of the *Timaeus*. For the influence of William's trinitarian interpretation of the *Timaeus* on the *Di Dio et anima,* see S. Gentile, "In margine all'epistola *De divino furore* di Marsilio Ficino," *Rinascimento,* ser. 2, 23 (1983): 40–50.

11. The library of the convent of the Santissima Annunziata contained a copy of William's *Glosae Super Platonem,* now Florence, Biblioteca Nazionale Centrale, Conv. soppr. E 8, 1398, described in Gentile, *Ritorno,* pp. 8–9, no. 7. Niccoli's library also contained a copy, later at the library of San Marco, now Florence, Biblioteca Nazionale Centrale, MS Conv. soppr. J 2, 50 (see appendix 1, no. 12). London, British Library, MS Add. 22815, of the twelfth century, contains several sets of glosses, including glosses drawn from William of Conches and Bernard of Chartres. A page added to this MS by the well-known Florentine scribe Piero Strozzi, a professional calligrapher associated with Vespasiano da Bisticci's

bookshop, makes it likely that this manuscript, too, circulated in Florence (see appendix 1, no. 15).

12. See Vatican City, Biblioteca Apostolica Vaticana, MS Urb. lat. 1389 (Jeauneau, *Glosae,* pp. 32–33), with the colophon: "Iste liber gloxarum super Timeo Platonis constitit mihi, Leonardo M. M. de mense Augusti 1434, L. 6, s. 14, d. 6." The "M. M." may indicate a member of the Malatesta family, many of whose codices passed to the collection of Federico of Urbino and thus to the Fondo Urbinate. The manuscript has many short glosses and *notabilia* in Leonardo's hand and in another fifteenth-century hand. Bessarion's manuscript of William of Conches is now Venice, Biblioteca Marciana, MS Zan. lat. 225 (1870); see Jeauneau, *Glosae,* pp. 40–41 ("Rien, dans le manuscrit, ne permet de dire que ce texte a été lu par le Cardinal"). The manuscript has Florentine decoration and is written in a round humanistic script of the mid-fifteenth century.

13. See Dutton, "The Uncovering," and idem, *Bernard,* pp. 8–21.

14. See appendix 1 and above, note 11.

15. Milan, Biblioteca Ambrosiana, MS S 14 sup., fol. 35v (= Waszink p. 148, 9): "Sed hec melius dividuntur in commenta post Calcidium." Ibid., fol. 76v (= Waszink, p. 250, 14): "Peripathetici. Geometre. De his lege in <commenta> post Calcidium." (The word or words after "lege in" were lost when the binder trimmed the pages; I restore <commenta> conjecturally on the basis of the parallel gloss at fol. 35v.) For the manuscript, see appendix 1, no. 17.

16. See appendix 1, no. 42.

17. See appendix 1, no. 23. The glosses mostly consist of parallel passages from Apuleius' *De dogmate Platonis,* Lucretius, and Bede's *De temporibus.* On Andrea Matteo III d'Acquaviva, see *Gli Acquaviva d'Aragona: Atti del VI Convegno del Centro Abruzzese di ricerche storiche* (Teramo: Centro Abruzzese di ricerche storiche, 1985).

18. S. Gentile, "Le postille del Petrarca al *Timeo* latino," in *Il Petrarca latino e le origini dell'umanesimo: Atti del Convegno internazionale, Firenze 9–22 maggio 1991,* 2 vols., published as *Quaderni Petrarcheschi* 9–10 (1992–1993), 9:129–139. A selection of the glosses was published by Pierre de Nolhac, *Pétrarque e l'humanisme,* new ed. (Paris: H. Champion, 1907), 2:141–47. The glosses are also discussed in Zintzen, "Il platonismo del Petrarca," in *Il Petrarca latino,* 9:97–98. For the rest of the bibliography, see appendix 1, no. 26.

19. The *De sui ipsius et multorum aliorum ignorantia* is edited in Francesco Petrarca, *Prose,* ed. G. Martellotti et al. (Milan: Ricciardi, 1955), pp. 710–767. On the Platonism of Petrarch, see now the article of Zintzen, "Il platonismo del Petrarca," pp. 93–113. But the complaint of literary men that Aristotle was being preferred to Plato despite the contrary opinion of the ancients as to their relative ranking was already a topos of late medieval literature: see Dutton, "Material Remains," pp. 217–219.

20. Klibansky, supplement, p. 66, thought that the "1363 commentator" must be a French master, citing "peculiarities of style and . . . the provenance of the manuscript [*sic*]." But the text is preserved in two manuscripts—Paris, Bibliothèque Nationale, MS lat. 14716, and Vatican Library, MS Chigi E V 152—and neither of these witnesses can

be a *codex descriptus* of the other. I have not been able to see the Paris manuscript, but the Chigi manuscript has an Italian (or possibly Avignonese) provenance and is written in Italian bastarda and Italian gothic cursive hands. Klibansky does not specify the "peculiarities of style" that would indicate a French authorship, unless he means simply that the author's Latin is not very classical. In any case, the Chigi manuscript was in Italy during the early Renaissance since it contains annotations in a late-fifteenth-century Italian hand. The codex displays the capital letters "E A" in the top margin of fol. 24r, in the middle of text; these could be the initials of the author, or, more likely, the owner of the codex.

21. Petrarch, *De sui ipsius et multorum aliorum ignorantia,* pp. 732, 750–56.

22. The text is broken down into four main "books" and numerous "tractates": book I = 17A–27D; II = 27D–39E; III= 39E–47E; IV = 47E–53C. (The commentary from 22C–27D is missing in the Chigi MS.) In the Calcidian and Chartrian tradition, the dialogue was usually broken into two books, i.e., book I = 17A–39E; II = 39E–53C. Most medieval commentators classed the *Timaeus* under ethics, logic, and physics; the 1363 commentator sees the work as belonging to ethics, politics, natural philosophy, and "scientiam chronicam et historialem."

23. Antonius de Romagno de Feltro corresponded with Guarino, Antonio Loschi, Omnebono della Scola, Nicolaus de Tarvisio, and other humanists of northeastern Italy; he served as chancellor to the humanist bishop Pietro Marcello; he was the author of various works on moral philosophy, including the unfinished *De paupertate,* a *quaestio* on whether it is always evil to lie (dedicated to his teacher, magister Baptista de Feltro), and a work on the four cardinal virtues based primarily on Aristotle's *Nicomachean Ethics.* See Remigio Sabbadini, "Antonio de Romagno e Pietro Marcello," *Nuovo Archivio Veneto,* n.s. 30 (1915): 225–235; E. Petersen, "Antonio de Romagno und die vier Kardinal Tugend," *Cahiers de l'Institut du Moyen Age grec et latin* 13 (1974): 63–76. In his correspondence, preserved in Biblioteca Apostolica Vaticana, MS Vat. lat. 5223, Antonio is addressed as "ser," indicating perhaps a notarial training, and "orator," i.e., ambassador.

24. See appendix 2. A, "Antonius de Romagno's Dependence on Bernard of Chartres," example 1.

25. Ibid., example 2.

26. The *accessus* is edited in appendix 2. A.

27. Venice, Biblioteca Marciana, MS Zan. lat. 469 (1856), fol 8r (36E): "Circulus firmamenti volvitur ab oriente in occidentem et iterum revertitur ad orientem, per quem boni designantur, qui, licet quandoque aberrant de bono ad malum, scilicet in occidentem, revertuntur tandem ad bonum, videlicet ad orientem. Per circulum planetarum qui volvitur ab occidente in orientem significantur peccatores qui quamvis videantur resipiscere, semper tamen revertuntur ad occidentem, idest ad peccatum, de quibus dictus est: canis revertitur ad vomitum."

28. See appendix 2. A, "Antonius de Romagno on Matter and the Elements." Antonius, like Pierleone da Spoleto (see above, note 6), was doubtless responding to Aristotle's criticism that Plato had attributed "magnitude" to the soul (*De anima* 1.3, 407a–b).

29. MS cit., fol. 14r: "Vere patimur vel putamus quod somniantes putant, quia putamus quicquid est esse in loco materiali, et putamus nichil esse nisi quod in celo vel in terra est, vel in aqua vel in aere. Quod falsum est, quia antequam hec essent, hyle et idea fuerunt in sua quadam mirabili natura secundum philosophos."

30. For the manuscripts, see appendix 1, nos. 3, 18, and 36. The Bergamo and Munich manuscripts were both copied from the Stuttgart MS; the latter was written in Padua in 1470 by Baptista Augustensis, "scriba oppidi imperialis Nordlingensis," probably while he was a student at the university (though his name does not appear in the Paduan *acta graduum*). While it is possible that Baptista himself compiled the glosses from Calcidius and other sources, it is more likely that they were the work of an arts master. Two possible Paduan candidates are (1) Cristoforo Rappi da Recanati (1423–1480), a professor of philosophy in Padua in the 1450s and '60s, who is known to have studied Plato; see Lucia Gualdo Rosa, "Un documento inedito sull'ambiente culturale padovano della seconda metà del sec. XV," *Quaderni per la storia dell' Università di Padova* 4 (1971): 1–38, and Maria Chiara Billanovich, "Cristoforo da Recanati, *artium et medicine doctor* (†1480): I libri, gli scritti," ibid., 22–23 (1989–1990): 95–132; and (2) Niccolò Leoniceno (b. 1428), a student of Ognibene da Lonigo, who studied at Padua in 1446 and taught there briefly, 1462–1464, before taking up his post in Ferrara; see Daniela Mugnai Carrara, "Profilo di Nicolò Leoniceno," *Interpres* 2 (1979): 169–212. For Leoniceno's codex of Calcidius, see appendix 1, no. 31.

31. In addition to the three codices listed in appendix 1 (nos. 17, 35, and 46), I may mention London, British Library, MS Add. 11274, written in an English hand, which contains Bruni's translations of the *Phaedrus, Apology, Crito,* and *Letters* (see Hankins, *Plato,* 2:693, no. 130); the codex may have been copied in Italy as the *Crito,* at least, was copied from Biblioteca Apostolica Vaticana, MS Vat. lat. 8611 (see Berti, p. 188). That the Paduan master's Plato manuscript contains the earlier version of Bruni's *Crito* translation also points to a non-Tuscan provenance, as Florentine bookshops usually published the definitive second version.

32. See appendix 2. C.

33. Bergamo MS (appendix 1, no. 3), fol. 72v, glossing 29E–30C: "Unum solum bonum est quod tantum bonum est et aliud nichil. Hoc est primum bonum quod, in eo quod est, bonum est. Est et secundum bonum quod et, in eo quod est, bonum dicitur, sed alio quodam sensu, quia scilicet hoc quod ipsum est bonum, ab eius voluntate fluxit cuius esse bonum est. Unde et omne album bonum est. Igitur album et est et est bonum. Sed est bonum in eo quod est quia fluxit ab eius voluntate qui bona est, non uero in eo quod est esse album dicitur, sed tantum esse album quia non est albus qui illud uoluit esse album. Sic itaque unius cuiusque natura beatitudinis capax est et opificis sui qualemcumque similitudinem recipit."

34. See appendix 1, no. 34, and appendix 2. B for an edition of the glosses. I am grateful to Contessa Anna Leopardi for permission to consult this manuscript.

35. See appendix 1, no. 17. For Ficino's lost *Institutiones,* composed before 1458, which Kristeller identifies with Ficino's early commentary on the *Timaeus* (also lost), see *Supplementum Ficinianum,* ed. Paul Oskar Kristeller (Florence: Olschki, 1938), 1:cxx, clxiii. Fi-

cino also copied excerpts from the *Timaeus* and other Platonic dialogues in Greek in Milan, Ambrosiana F 19 sup.; see A. Martini and D. Bassi, *Catalogus codicum graecorum Bibliothecae Ambrosianae* (Milan: Hoepli, 1906), 1:375–78, no. 329, and Paul Oskar Kristeller, "Some Original Letters and Autograph Manuscripts of Marsilio Ficino," *Studi di bibliografia e di storia in onore di Tammaro de Marinis* (Verona: Stamperia Valdonega, 1964), 3: 28–29.

36. Perhaps when he was tutoring Pico della Mirandola in the 1480s; see discussion later in the text.

37. At Waszink, pp. 181, 13–214, 16, Ficino identifies as Calcidius' source [pseudo-] Plutarch's treatise *De fato,* a source not identified in modern scholarship until the late nineteenth century. See A. Gercke, "Eine platonische Quelle des Neuplatonismus: 2. Chalcidius und Pseudoplutarch," *Rheinisches Museum,* n.s. 41 (1886): 26–279.

38. See Ficino's notes to Waszink, p. 76, 10 (Milan, Biblioteca Ambrosiana, MS S 14 sup., fol. 12r) "Anima mundi semper fuit"; ibid., p. 80, 20 (MS. cit., fol. 13v) "Anima mundi semper fuit sed non semper habuit rationem, sed tunc deus illi rationem inseruit cum genuit mundum"; and ibid., p. 212, 4 (MS cit., fol. 61v) "Anima cum ratione a deo creata. Irascibilis vero vis et concupiscibilis a ceteris diis." But at Waszink, p. 219, 4, Ficino insists that Plato did not literally believe in the transmigration of human souls into beasts: "*In bestias ire:* Plato intelligit: in homines similes bestias." At a later stage in his career as an expositor of Plato, Ficino would explain away entirely Plato's apparent belief in the (dangerously unorthodox) doctrine of transmigration; see my *Plato,* 1:358–359.

39. MS cit., fol. 96r = Waszink, p. 302, 5: "Due anime mundi secundum Platonem. Opinor in silva esse duas animas: unam eductam de potentia ipsius materie, quae vegetativa vel motiva, quae semper in ea fuit, quae infra corporea est et sine ratione omni motu inrationabili materiam agitabat, quam malam, idest temerariam dicimus. Alia est quam deus creavit quando voluit mundum exornare, que rationem habet et ideo ordine motus mundum <illustrat>."

40. MS cit., fol. 27r = Waszink, p. 122, 12: "Probat planetas non retrocedere sed ita nobis videri. Forte unum retro moveri dum aspicimus." MS cit., fol. 31r = Waszink, p. 134, 22: "Tres planete superiores fallunt oculos ut retrocedere videantur, dum perveniuntur ab inferioribus qui angustiores orbes suos citius peragunt. Ceteri planete nec retrocedunt nec videtur recedi." MS cit., fol. 31v = Waszink, p. 136, 5: "Cause cur planete videantur retrogradi."

41. Waszink, p. 151, 15. This passage is also frequently discussed in the medieval glossary tradition.

42. Ficino, *Opera* (Basel: Heinrich Petri, 1563; reprint, Turin: Bottega d'Erasmo, 1983), p. 969.

43. MS cit., fol. 80r = Waszink, p. 259, 5: "Ego vero ambigo ne forte radius visualis per umbrationem potius quam reflexionem mittitur, et ista natura solum ex speculis repercutitur usque ad visum." MS cit., fol. 82v = Waszink, p. 266, 2: "Ego vero puto solo comuni lumine fieri illic imagines, sed quod variis modis apparuerint ex lumine ut cuique contingere."

44. Edited in appendix 2. D.

45. I have identified Pico's hand in this MS (appendix 1, no. 20) on the basis of photographs published in *Pico, Poliziano, e l'Umanesimo di fine Quattrocento: Biblioteca Medicea Laurenziana, 4 novembre–31 dicembre 1994,* exhibit catalogue, ed. Paolo Viti (Florence: Olschki, 1994), figs. 27–29. See my figures 2.1–2.6. I plan to edit Ficino's and Pico's notes on Calcidius in a future publication.

46. See *Pico, Poliziano,* pp. 127–147 (by Sebastiano Gentile).

47. Some examples: (1) The note on Pico's fol. 38v corresponds exactly to that on fol. 20r of Ficino's manuscript (Waszink, p. 99, 17): "13 qui sunt inter cc 43 et cc 56 ex tertia parte 92 quae est 64 certissime manant et hoc est hemitonni vel limmatis causa." (2) Pico, fol. 92v (Waszink, p. 183, 15): "Plato quaedam vult providentia, quaedam fato, quaedam ex libero arbitrio, quaedam casu fieri" (see figure 2.2); Ficino, fol. 50v: "Plato: quaedam fiunt providentia, quedam fato, quedam voluntate nostra, alia fortuna, alia casu." (3) Pico, fol. 111r (see figure 2.1) and Ficino, fol. 65r (Waszink, p. 220, 10): "Animas primo [*om.* Pico] procreat, 2o applicat, 3o docet, 4o serat; 5o cadunt, 6o incorporantur, 7o animant corpus." Pico also copied on the flyleaf (fol. Ir) of his manuscript a list of words entitled "Nomina memorie"; this same list, in only slightly different order, is found in Ficino's MS, fol. 87r.

48. Waszink, p. 239, 16 and following.

49. For the textual history and sources of the commentary, see M. J. B. Allen, "Marsilio Ficino's Interpretation of Plato's *Timaeus* and Its Myth of the Demiurge," in Hankins, Monfasani, and Purnell, *Supplementum Festivum,* pp. 399–440. I shall be giving a more detailed analysis of Ficino's *Compendium in Timaeum* and the "scientific Platonism" of the sixteenth century in an article to appear in the proceedings of a Warburg Institute colloquium, *Humanism and Early Modern Philosophy,* ed. Jill Kraye (Routledge).

50. The chief sixteenth-century commentators on the *Timaeus* are Ambrosius Flandinus, O.E.S.A., *Annotationes in Timaeum* (from 1523, in Paris, Bibliothèque Nationale, MS lat. 12948); Lodovico Boccadiferro (attrib.), *In Timaeum Platonis* (from 1545, in Fermo, Biblioteca Comunale, MS 4 C A 2.80—see *Iter* 1:53—and Bologna, Biblioteca Universitaria, MS Aldovrandi 56, vol. 2, fols. 270r–277v; see below, note 61); Sebastian Fox Morzillo, *In Platonis Timaeum commentarius* (Basel: Oporinus, 1554); Fox's *De naturae philosophia seu de Platonis et Aristotelis consensione libri V* (Paris: Iacobus Puteanus, 1560) is in effect a second commentary on the *Timaeus;* Matthaeus Frigillanus, *In Timaeum Platonis ex mediis philosophorum et medicorum spatiis scholia* (Paris: Th. Richardus, 1560)—this is not a pseudonym of Marsilio Ficino, despite the British Museum Catalogue; Francesco II de' Vieri, *Libro della natura dell'Universo* (Florence, Biblioteca Nazionale Centrale, MS Magl. XII, 11); Paolo Beni, *In Platonis Timaeum sive in naturalem omnem atque divinam Platonis et Aristotelis philosophiam decades tres* (Rome: Gabiana, 1594); Cosimo Boscagli, *In Timaeum Platonis* (Florence, Biblioteca Nazionale Centrale, MS Palat. 1025, vol. 2, fols. 1–40). But the *Timaeus* is discussed in many other works of sixteenth-century Plato scholarship.

51. On Ficino's use of Cicero's *Timaeus* in his own translation of the *Timaeus,* see now Maria Cristina Zerbino, "Appunti per uno studio della traduzione di Marsilio Ficino dal *Timeo* platonico," *Respublica litterarum* 20 (1997): 123–165; I am grateful to dott.ssa Zerbino for allowing me to consult her article in typescript.

52. For Ficino's new Greek sources, see Allen, "Myth of the Demiurge," and Gentile, *Ritorno*. For the two Latin translations of Timaeus Locrus in the later fifteenth century, by Gregorio Tifernate and Francesco Filelfo, see my *Plato*, 2:436, 522.

53. See my *Plato*, vol. 1, part IV, section 3.

54. See my article "Marsilio Ficino as a Critic of Scholasticism," *Vivens Homo: Rivista Teologica Fiorentina* 5 (1994): 325–334.

55. Ficino, *Compendium in Timaeum*, in *Platonis opera* (Venice: Andrea Torresano, 1491), fol. 245r–v, in the chapter entitled, "Quomodo totus mundus ex quattuor componitur elementis et quomodo hec alia ratione sunt in celo, aliter infra lunam."

56. Ibid., fol. 245v, in the chapter "Circularis motus omnis spere semper mobili proprius est. Item ignis maxime proprium est lumen."

57. Ficino defends his theory of celestial elements, for example, by citing biblical passages where water and earth are described as being in or above the heavens; see his *Compendium*, fol. 245rb: "Audiant [the critics of Plato's elemental theory] denique sacras litteras ponentes saepe in celis aquas, ponentes terram quoque viventium" (cf. Jer. 10:13, 51:16; Dan. 3:60; Jth. 9:17). But the search for biblical parallels to Timaean science is a major theme of the *Compendium*, as of the traditional hexaemeral literature going back to the twelfth century and to Augustine.

58. Gregory Vlastos, *Plato's Universe* (Seattle: University of Washington Press, 1975), p. 68. On the "mathematization of the cosmos," see especially *Compendium*, fol. 251r–v.

59. See Dutton, "Material Remains," and my article "Antiplatonism in the Renaissance and the Middle Ages," *Classica et Medievalia* 47 (1996): 359–377.

60. See my *Plato*, vol. 1, part III.

61. Bologna, Biblioteca Universitaria, MS Aldovrandi 1996, fol. 270: "Nescio sane qua de causa passim Platonis scripta iaceant, nec ab ullo publice neque private legant. . . . Tres autem comperio causas cur Platonis scripta non ita floruerint ut debebant: prima est quia ipse de naturalibus rebus loquens cum illis miscet divina et mathematica et ita non ordinem servauit distinctum, sed omnia inordinata quodammodo confuse tractauit; alia est quia consuetudo invaluit ut Aristoteles legeretur; alia est per ignorantiam grecarum literarum." (I am grateful to David Lines for helping me obtain a microfilm of this MS.) In both the Bologna MS and the Fermo MS (see above, note 50) this text is anonymous, but it can be tentatively attributed to Boccadiferro on the basis of its presence among other texts of the same author in the Fermo MS. Boccadiferro's known works are all connected with his classroom teaching of Aristotle, but they contain numerous references to Plato, and it is known that he intended to write an epitome of the *Laws*. On Boccadiferro, see *Dizionario biografico degli italiani* (Rome: Istituto della Enciclopedia Italiana, 1960–), 11:3–4.

62. See Charles B. Schmitt, "L'introduction de la philosophie platonicienne dans l'enseignement des universités à la Renaissance," in *Platon et Aristote à la Renaissance: 16ᵉ Colloque international de Tours* (Paris: Vrin, 1976), pp. 93–104, and my article "Renaissance Platonism," in the *Routledge Encyclopedia of Philosophy* (London: Routledge, 1998), 7:439–447.

3

MARSILIO FICINO: DAEMONIC MATHEMATICS
AND THE HYPOTENUSE OF THE SPIRIT
Michael J. B. Allen

One of the enduring questions in medieval and Renaissance philosophy concerns the relationship between nature and art (in Greek *technē*), given that nature herself is full of Plinian art, given too that man's nature is defined by his art, his skills, and his ingeniousness, and given that the daemons and angels are by nature ingenious and intellectual beings. "In brief all things," wrote Sir Thomas Browne in the *Religio Medici* 16, "are artificial." One of the interesting thinkers in this regard is the Florentine Platonist Marsilio Ficino (1433–1499), who produced some of the age's most arresting analyses of the artfulness, and thus of the structure, of both human and daemonic nature and by implication of their capacities to be moved and to be acted on.

Of particular interest is material in the commentary, subtitled *De numero fatali,* that he compiled in the last decade of his life on Plato's notoriously enigmatic passage on the fatal number in book 8 of the *Republic*. But in order to understand Ficino's psychology—both of human beings and of daemons—and his speculative ideas concerning the soul's various faculties, we should first briefly consider some of the mathematical issues confronting him in Plato. For our story has an extraordinary ending and concerns the manner in which the triangular "powers" of the human spirit and habit can be the object of what we would now think of as scientific, and specifically as mathematical, manipulation.

While some interpreters have argued that Plato's metaphysics is fundamentally dualistic in that it postulates an intelligible real world and an illusory material world, Aristotle claimed in his *Metaphysics* 1.6 that Plato had divided all reality into three spheres: ideas or intelligibles, mathematicals, and sensibles. His source for this trichotomy may have been Plato's "Lecture on the Good," as Philip Merlan and others have suggested,[1] or it may have been some later development in Plato's thought. However, as early as the *Phaedo* 101B9–C9 Plato had postulated Forms of numbers, Ideal Numbers, at the same time implying that individual numbers participate in such Numbers while being inferior to them.[2] But Speusippus, Plato's nephew and his successor as head of the Academy from 347 to 339 B.C.E., had apparently dismissed the Ideal Numbers along with the other Forms, and derived the mathematical numbers directly from the One or from the One and the Indefinite Dyad, arguing, according to Aristotle, that such mathematicals were then followed by Soul. Xenocrates, who succeeded Speusippus, is said to have identified Forms with the mathematicals.[3] In the *Enneads* 5.4.2, Plotinus had also derived numbers and the Ideas from the two metaphysical principles of the One and the Indefinite Dyad, as Ficino well knew.[4]

What Speusippus' view does underscore, however, is the problem, given Plato's account in the *Timaeus,* of the role played by the mathematicals, and specifically by ratios, in the creation of Soul and souls. For the Demiurge creates the World-Soul as a kind of mathematical entity, or at least with an arithmetico-geometrical structure. Plutarch's *De animae procreatione* 2.1012D, which Ficino had worked through carefully, asserts indeed that Crantor had interpreted the psychogony in the *Timaeus* as an arithmogony, as the emergence of numbers as the first and principal sphere of reality. Speusippus and Xenocrates also had maintained that the World-Soul was a mathematical entity.[5] According to Iamblichus, Speusippus had described the soul (the World-Soul?) "as the form (*idea*) of what extends in all directions, this form being constituted according to mathematical ratios"; for his part, Xenocrates had defined the soul as "a self-changing number."[6] However, as Plutarch remarked in his *De animae procreatione* 1013CD, to say the soul is constructed with numerical proportions is not the same as saying it is itself a number, and Plato never called the soul a number.

Furthermore, a strange if isolated observation in the *Laws* 10.894A seems to argue that sensibles derive from mathematicals: "things are created when the first principle receives increase and attains to the second dimension, and from this arrives at the one which is neighbor to this, and after reaching the third becomes perceptible to sense."[7] Plato appears to be postulating two "creations": the first of that which is created in "the second dimension," the second of the sensibles in the "third" dimension. The *Timaeus* too is concerned not only with the ratios of the World-Soul but with the mathematics of sensibles. In a famous passage at 53C ff., Timaeus addresses the question concerning the basic constituent of the physical world, arguing that "every rectilinear surface is composed of triangles," meaning presumably that it can be divided up into triangles and that the triangle is the surface created or contained by the least number of straight lines. "Originally," he says, these triangles were only of two kinds, being made up of "one right and two acute angles." The archetypal triangle in the *Timaeus* (and thus, in Ficino's eyes, for Plato) is accordingly a right triangle: subordinate to it are all the obtuse and acute triangles, including the equilateral triangle as well as all non-right scalenes.

The first kind of right triangle is the isosceles with 45 degree angles subtending the right angle, that is, the half-square, and this exists obviously in only one form or nature (*mian physin,* 54A2). The second kind is the right scalene, which exists in a "countless" (*aperantous*) number of forms. But, Timaeus continues, of these countless right scalene forms we must select "the fairest" (*to kalliston,* 54A3); and the fairest triangle is that having "the square of the longer side equal to three times the square of the lesser side" (54B4–5).[8] What Plato must have in mind therefore—since Timaeus had already said at 54A7 that a pair of such triangles would form an equilateral triangle and now says that the squares, not the square roots, have the ratio—is a half-equilateral triangle, one with angles of 90, 60, and 30 degrees and a base of 1 (the lesser side), a perpendicular of $\sqrt{3}$ (the greater or longer side), and a hypotenuse of 2. This at least is the traditional view (espoused, for instance, by Albinus in his *Didaskalikos* 13, which Ficino translated, though he like others attributed it to Alcinous); and it is the one carefully propounded by Ficino himself in chapter 41 of his appendix for the final version of his *Timaeus* commentary.[9] It means, of course, that the perpendicular is not a whole number, the first instance of a triangle with all three sides as whole numbers being the famous 5–4–3 triangle beloved of the Pythagoreans. But Ficino recognized that this 5–4–3 triangle cannot itself be "the fairest" of scalenes, since the square of its "lesser" side is not a third that of the square of either of the two longer sides. In short, Timaeus must mean by the "fairest" the *hemitrigon,* the half-equilateral, of 1, $\sqrt{3}$, and 2.

Timaeus goes on to assert that both the isosceles and the half-equilateral triangles are the "original constituents" or "principles" (*archai*) of the four material elements at the heart of nature because they are the constituents of four of the five regular solids. The half-equilateral, with its $\sqrt{3}$ perpendicular, is the base principle of three of them, since 24 such scalenes constitute the tetrahedron or pyramid that makes up the molecules, as it were, of fire; 48 of them, the octahedron that makes up the molecules of air; and 120 of them, the icosahedron that makes up the molecules of water (each of the tetrahedron's 4, the octahedron's 8, and the icosahedron's 20 equilateral faces having 6 constituent half-equilaterals—6 being in traditional Platonic arithmology the perfect, the Jovian number).[10] The salient implication is that the Pythagorean theorem is the first mathematical tool needed by a natural philosopher in order to understand both the three superior elements and the fiery, airy, and watery bodies they compose, those animated by the souls of the daemons. The cube, however, which is the regular solid that makes up the molecules of earth, is constituted from 24 isosceles right triangles: that is, from 24 half-squares.[11] The fifth regular solid, the dodecahedron with 12

pentagonal faces, is assigned by the *Timaeus* 55C to "the whole" that is decorated with designs, namely with the 12 heavenly constellations (though the
Epinomis 981C assigns the dodecahedron to the aether). It cannot, however,
be constructed from the right triangles, although ancient commentators including Albinus and Ficino himself thought it consisted of 360 such triangles
(again 12 x 5 x 6), the number of the degrees of the zodiacal circle.[12] The
physical sublunar world consists of the perpetual dissolution and accretion of
the triangles constituting four of these five regular solids, which at 54D6 are
said to be like the "syllables" in nature's book, the elementary triangles being the letters of those syllables.[13] Fire, air, and water can be transformed into
each other, since they all consist of half-equilaterals, while earth can be broken up only into its isosceles triangles.

Given the figural nature of early arithmetic in the Pythagorean tradition that Ficino revived and adhered to, and given the primacy it accorded to
the monad and therefore to the odd numbers, it follows that the series of odd
numbers, $1 + 3 + 5 + 7 + 9 \ldots$, figuring as it does a gnomon of increasing
size in what Ficino calls the equilateral sequence summing 4, 9, 16, 25, . . . ,
privileges the isosceles right triangle. The scalene, by contrast, figures the
summing of even numbers in the unequilateral series of 6, 12, 20, 30, . . . ,
and the equilateral triangle, the summing of odd and even numbers in the
regular trigon series of 3, 6, 10, 15, ... But since two equal isosceles triangles constitute a square and their shared hypotenuse the diagonal of that
square, the hypotenuse can never be rational in the sense of being a whole
number. Thus we get a series of irrational hypotenuses on the model of $\sqrt{2}$
for a side of 1, of $\sqrt{8}$ for a side of 2, of $\sqrt{18}$ for a side of 3, and then of
$\sqrt{32}$, $\sqrt{50}$, $\sqrt{72}$, and so on.

However, Ficino supposed that Plato was already familiar with a formula that could determine what he and his contemporaries thought of as the
rational value of the diagonals of a particular series of these $\sqrt{2}$ squares, and
therefore of the hypotenuses of their constituent isosceles triangles; namely,
those having sides of 1, 2, 5, 12, 29, 70, and so on. This formula, which Ficino encountered in Theon of Smyrna's *Expositio* 1.31, demonstrates that if
we subtract and add 1 in alternation to the sum of the squares of the two equal
sides, then we end up with $2 - 1$, $8 + 1$, $50 - 1$, $288 + 1$, and so on; that is,
with square powers for which rational square roots exist. Put algebraically,
the formula of subtracting or adding one provides a series of positive, integral solutions for the equation $y^2 = 2z^2 + / - 1$ or $y^2 + / - 1 = 2z^2$, where y
is the diagonal and z is the side. Accordingly, Ficino thought of the hypotenuses of such a series of isosceles triangles (or the squares they constitute)
as having both rational and irrational square roots, the rational roots being

primary: for example, 50 has the irrational root of 7.0710678 . . . but the rational root of 7. Since the +1/−1 alternation providing us with rational roots for what would otherwise possess only irrational roots is regular (if the sides are odd, 1 has to be subtracted from the sum of their squares; if even, then added to it), the powers of the diagonals in the series can be viewed in the long run as maintaining a ratio of 2:1 to the powers of the side. To arrive at the sequence of such rational diagonals, we must add the value of the diagonal to the side, while adding twice the value of the side to the diagonal. The sequence of diagonals then runs 3, 7, 17, 41, 99, and that of the sides 1, 2, 5, 12, 29, 70, and so on.[14]

Ficino thought of these diagonal numbers as being "of the 5" because the first instance in the series, leaving aside the isosceles with a side of 1, is the isosceles with a side of 2 and a rational diagonal of 3, and 2 + 3 = 5. Aristides Quintilianus, incidentally, had declared in the case of another kind of perfect triangle—the Pythagoreans' right scalene of 5–4–3—that the sides at the right angle are in the relationship of epitritus (4:3), and that "it is the root of epitritus [meaning 7 (4 + 3)] added to 5 [the root of the 3:2 ratio] that Plato was referring to in the *Republic* [8.546A ff.]."[15] This "perfect" scalene provided Ficino with the keys to solving the great mystery of the Fatal Number, for the value of its sides sum to 12, which when cubed produces 1,728—the value, for him, of the Number. But if its "root" of 7 is added to 5, which is the "root" of the first isosceles, the sum is also 12. Ficino sees these roots as 4:3 and 3:2 ratios, and as being invested with musical connotations as the "consonances" of diatesseron, the perfect fourth, and of diapente, the perfect fifth.[16]

Hence for the two kinds of right triangle, Ficino knew of two ways to determine the value of the hypotenuse as an integer: first for the Pythagoreans' "perfect" scalenes with sides of 3 and 4, 9 and 12, 27 and 36, and so on, by way of the Pythagorean theorem; and second for those isosceles triangles (half-squares) with sides of 2, 5, 12, 29, and so on, by way of the same theorem after the application of Theon's plus or minus one rule.[17] He knew of no way, however, to determine rational perpendiculars for the half-equilateral triangles, Plato's "fairest" scalenes.

———

If the triangle dominates the subelemental world of planes and surfaces (the "second dimension" of the *Laws* 10 passage) and the solids they constitute, then man too must be subject to the triangle, at least insofar as he is subject in this life to time and space, his nature to Nature.[18] However, his nature is defined for Ficino by his possession of a *habitus,* from which indeed we and the monks get the word "habit," and which is etymologically linked to the verb *habere,* "to have." *Habitus* can be rendered in English as "character" and "condition," though it cannot be used to signify the interdependent notions of skill, talent, wit, and ingenuity for which Ficino deploys the term *ingenium.* The *habitus* can refer, his *Platonic Theology* maintains, to the natural optimum condition of the body—the goal, if you will, of medicine[19]—and as such it can be said "to pass into" or "to take over" our nature, or to become as it were a second nature that moves us while remaining immobile itself.[20] It can also "play the part of" or "do duty for" our "natural form."[21] The soul itself, even when separated from the body, has a *habitus* by which it is moved. Ficino believes of course that the soul only (re)acquires its true *habitus* when it has returned to its "head," that is, to its "intelligence" (*mens*).[22]

Indeed, the acquisition of such a *habitus* becomes man's primary goal, since it contains the soul's *formulae idearum,* which, when led forth into act, enable the soul to rise from the sensible to the intelligible, and to be joined with the Ideas.[23] For the true *habitus* contains the species or Ideas as they are present in us, the species indeed that correspond to all things that exist in the world in act.[24] What makes for the acquisition of a perfect *habitus,* whether of the body, the soul, the human mind, or the angelic mind, is both *praeparatio* and *affectio.*[25] The *habitus* is thus tied conceptually both to the notion of form—the *habitus* being the condition of ourselves or of some part of ourselves that most nearly approximates to the perfection of our form—and to the notion of power, the power that we have been born with but have nurtured by *praeparatio* and by *disciplina.* We can even think of it as the potentiality in our soul for becoming pure mind in the actuality of its perfect circular motion, the motion-in-rest of contemplation. Immobile itself, it nevertheless provides the soul with the "proclivity" for the absolute motion that is blessed, eternal life.

In arguing Platonically in his *De numero fatali* 16 that our "composed body" is a "discordant concord," Ficino turns predictably to two analogies, a musical one and a mathematical one, for what endows it with concord—namely an even *habitus.* Even *habitus* are like the harmonies of different voices in a choir, uniting in the diapason; and they also resemble the sums that are generated from the odd numbers in the equilateral addition series of $1 + 3 + 5 + 7$, and so on. Odd *habitus,* by contrast, are generated from even numbers, meaning from the even numbers in the unequilateral addition series of $2 + 4 + 6 + 8$, and so on. Thus an even-tempered *habitus* is like any equilateral sum—4, 9, 16. As a sum it is the child of the odd numbers, but as a product it is the result of equality and balance, of a number having multiplied itself, that is, having raised itself to the second power, squaring or square-rooting being the meaning of *dynamis* in Greek in a mathematical context, as both Plato's and Euclid's usage bears witness.[26]

If the *habitus,* our nature, is a kind of mathematical and specifically a geometrical power, squaring, then must we think of it as functioning like such a power, at least in particular contexts? In other words, does the *habitus* or nature of the soul (and of its spirit and its body) work like the square power of the hypotenuse of an isosceles right-angled triangle (or of the diagonal of a square constituted from two such triangles, which is the same thing); and is it therefore equal to double the square of either side (i.e., to the sum of the squares of both sides)? If so, we must entertain the possibility that the Pythagorean theorem has come to haunt the face of Ficino's faculty psychology. But what is the evidence that this is anything more than just an arresting image or a mere turn of phrase?

The traditional schema of the point progressing to the line to the plane (surface) to the solid goes back at least to the Pythagoreans and is repeated throughout antiquity and the Middle Ages.[27] Ficino turns to it on occasion to help define the serial subordination of the four hypostases in the Plotinian metaphysical system, the One, Mind, Soul, and Body;[28] and in doing so he often identifies the point with the One and the solid with Body—examples abound throughout his work. But he also identifies the line (and certainly the circular line) with Mind, and the plane with Soul.[29] While, to my knowledge, he nowhere advances all the elements of this series of analogies in one formal argument, he does introduce them here and there, and the schema obviously serves as one of his paradigms for metaphysical progression and hierarchical subordination. The implications for our understanding of the soul's internal structure and of its position on the Platonic scale of being in Ficino are, I believe, as bizarre as they are unexpected.

Ficino's governing text here is again the *Timaeus* 53C ff., duly bracketed by Timaeus himself as presenting views that are only "probable." Having introduced the two kinds of right-angled triangles, Timaeus had proceeded at 69C ff. to describe the creation by the Demiurge's sons of the irrational soul and at 73B ff. their taking up the triangles (i.e., before their combination into the regular solids) to mix them in "due proportions" in order to make the marrow, which will serve as a "universal seed" and a vehicle for the soul. Ficino clearly rejoiced in some at least of the figural extensions (with the puns this term implies) of the Pythagorean mathematics that Timaeus is propounding here. For his own *Timaeus* commentary explores the implications and arrives at an interpretation that identified the soul itself as the exemplary triangle, its triple powers corresponding to the three angles and the three sides of the archetypal geometrical figure. At the end of chapter 28, having observed that "mathematicals accord with the soul, for we judge both of them to be midway between divine and natural things," he writes:

> We use not only numbers to describe the soul but also <geometrical> figures so that we can think of the soul by way of numbers as incorporeal but by way of figures as naturally declining toward the corporeal. The triangle accords with the soul; for just as the triangle from one angle extends to two more, so the soul, which flows out from an indivisible and divine substance, sinks into the entirely divisible nature of the body. If we compare the soul as it were to things divine, then it seems divided. For what they achieve through one unchanging power and in an instant, the soul achieves through many changing powers and actions and over intervals of time. But if we compare the soul to things natural, then we judge it to be indivisible. For it has no sundry parts as they have, separated here and there in place, but it is whole even in any one part of the whole; nor, as they do, does it pursue everything in motion and in time, but it attains something in a moment and possesses it eternally. In this we can compare the soul, moreover, to the triangle, because the triangle is the first figure of those figures which consist of many lines and are led forth into extension or in a right angle (*in rectum*). Similarly, the soul is the first of all to be divided up into many powers—powers that are subjected in it to the understanding—and it seems to be led forth into extension when it sinks from divinity down into nature. In this descent it flows out from the highest understanding down into three lower powers, that is, into discursive reasoning, into sense, and into the power of quickening, just as the triangle too, having been led forth from the point (*signum*),[30] is drawn out into three angles. But I say the soul is the first in the genus of all to be mingled from many powers in a way, and to fall, so to speak, into extension or into a right angle (*in rectum*). For above the soul the angelic mind requires no inferior powers within itself at all and is pure and whole and sufficient.[31]

Given this fully worked out analogy of the soul with the triangle, preeminently the right-angled triangle, and given that the triangle is the premier figure of the planar realm of surfaces, Ficino clearly thinks of soul, or at least of soul in its fallen triplicity, as planar.[32] Indeed, given the variety of geometrical and arithmetical structures that govern our notion of a two-dimensional realm, the plane and its subdivisions are ideally suited to modeling the complex and ambivalent status of soul and its various faculties as intermediary between the three-dimensional body and the paradoxically linear or circular realm of pure mind—linear because it is both one and many, and circular because it is "one and equal" like the line that returns upon itself to constitute the figure that is not a figure but rather the principle and end of figures. Furthermore, the secret of the planar realm of the triangle for Ficino is the notion of power, *dynamis,* again meaning squaring or square-rooting.[33] For it is this alone that enables us to comprehend the complex, invisible proportionality and comparability of hypotenuse to side, to comprehend the geometrical ratios that govern reality.

If the *habitus* is, or functions like, a planar number, and specifically a square number or the root of such a number, it would serve in unexpected ways to validate the efficacy of, and to enlarge the scope of, a purely mathematical magic; in so doing it would privilege the beings preeminently gifted in Ficino's view for the subtleties of mathematics, namely the daemons. We might imagine a special mathematical dimension for the lower daemons on the one hand in supervising the diet, regimen, and exercise that ensures an even *habitus* in the body; and for the higher daemons on the other in disciplining the soul—over and beyond, that is, instructing it in ordinary mathematical procedures—so that it too attains an even *habitus.* But nowhere would their role be more arresting than in the case of the *habitus* of the *spiritus,* since the *spiritus* is for Ficino the object of manipulation by magicians using the resources of natural and of astral magic (and perhaps using, however unconsciously, the mathematical structures and powers that underlie such magics).[34] The *habitus* of the spirit, the hypotenuse if you will of the spirit, would be subject a fortiori to expressly mathematical manipulation—especially to the manipulation of human and daemonic geometers, those skilled above all others in the understanding of planes and surfaces. It would lend a dramatic but also a scientific dimension to the monitory exhortation in the vestibule to the Platonic Academy, "Let no one enter here who is not an adept in geometry," and to Plutarch's declaration, in a phrase he attributes to Plato, that "God is always working as a geometer" (*Aei theos geōmetrei*).[35]

Ficino had an abiding fascination with the branch of applied geometry having a singular role in daemonic magic, namely the science of optics.[36] Ficino seems to have thought of the magician using his own *spiritus* as a mirror to catch, focus, and reflect the streams or rays of *idola* or images that flow ceaselessly out from animate and inanimate objects.[37] The *idola,* and the *spiritus* that focuses the *idola,* are the means whereby he can work with and work upon anything, living as well as inert, from a distance. Aspects of his skill may be irrational or sophistical, and may be controlled in large part by his phantasy; but a particular magician, one skilled in mathematics, Ficino imagines as being able to draw on numbers, I believe, and notably on figured numbers, to effect a rational magic by way of his *spiritus* upon the *idola.* Such a magician might even consciously program his *spiritus* like a radar dish, tilting and rotating its planes according to geometrical formulae encoding and controlling particular magical operations; those formulae, in other words, best suited to affecting the dimensions, the angles, the powers that govern a physical world constituted from triangles and from the four (or for Ficino obviously the five) regular solids to which they give rise. After all, such a geometer-magus would be exercising his sovereignty over the powers governing the optical triangles formed by the objects and the *idola* he wished to perceive or manipulate, the reflecting surface of his *spiritus,* and the line of his intelligence. Clearly such triangles would themselves consist of laterals and diagonals and have rational and irrational powers; and double the sum of the degrees of their varying angles would equal the 360 degrees of the perfect circle, the zodiac, of the understanding.

In exercising these geometrical powers, the geometer-magus would be drawing on the computational and manipulative skills that Ficino and the later Platonic tradition he inherited had already assigned to the daemons. For daemons not only are the masters of mathematics, they also preside over the world of light and its optical effects and illusions, and consequently over the singular role that mirrors and prisms, reflections and refractions, play in our understanding of, and in our manipulation of, light. Moreover, they are the denizens preeminently of the world of surfaces, planes, and powers, and only the basest of them choose regularly to inhabit the three-dimensional cubicity of the physical world. In this they resemble other higher souls; for all souls are properly inhabitants, in Ficino's Platonic imagination, of the realm of planes and surfaces (preeminently rectilinear ones), though they may be imprisoned for a time in solids. In that they aspire to return to the intellectual realm, however, to become pure intellects and to contemplate the mathematicals and the Ideas of numbers, they aspire, mathematically speaking, to the "one and equal" line, the circling line of Nous; ultimately they wish to return to the unity at the apex of intelligible reality, to the One in its transcendence. Specifically, given the unique role of the triangle in Platonic mathematics and psychology (and of the Pythagorean theorem in computing the relationship of the power of the hypotenuse to the powers of the sides), we must think of the highest rational souls—those of the daemons, or at least of the higher ones who dwell far above the terraqueous orb—as the lords of triangularity and of the "comparability" that governs it, triangularity being the essence of the planar realm. We might even speculate about the devious ways the daemons practice on our mathematical sanity with irrational hypotenuses and surds or torment us with Theon's plus or minus one theorem!

The planar world occurs in Nature herself in the mirrors of lakes and pools and of other water and ice surfaces; one can also think of snow, salt, white sand, and even various rock surfaces, as well as of certain mist and cloud phenomena, that have planar qualities and whose surfaces reflect or refract light. Preeminently of course the planar world occurs in the natural faceting of precious stones, gems, and crystals. It is in the play of light on such surfaces that the presence of daemonic geometry and its science of powers can best be glimpsed by the geometer-magus. On occasions he is able even to use his own *spiritus* as a mirror-plane to capture and affect the *idola*, immaterial and material alike, that stream off objects, refiguring them by way of recourse to the laws of figured numbers. For physical light is the intermediary between the sensible and the purely intelligible realms, and in this regard it is spiritual: it resembles and therefore—given the ancient formula that like affects like[38]—can be influenced by the *spiritus*, the substance that mediates between the body and the soul and serves as the link, as light itself does, between the otherwise divided realms of the pure forms and of informed matter.

It was this eccentric nexus of concerns that, I believe, slowly emerged in Ficino's mind and led him to posit a problematic set of interdependent connections between magic, geometry, figural arithmetic, the daemons, and light in its various manifestations. Underlying the nexus is the notion of a mathematical power and the mysterious hold it exercises over our understanding of both planes and solids. And with this understanding, predictably, comes actual power to affect and change. In all this we can glimpse the profound impact on Ficino of Plato's Pythagorean mathematics, and specifically of the Pythagorean theorem, together with Theon's discussion of diagonal powers, on the one hand, and Plato's fanciful but influential presentation in the *Timaeus* of a triangle-based physics on the other.

––––––

The relationship in Ficino's mind between optics, and notably daemonic optics, and music—that is, between light wave theory and sound wave theory and the "harmonic" proportions that govern them—has yet to be explored.[39] What we must now realize, however, is that for him the planar numbers and especially the square numbers occupy a mysterious but all-powerful position between the prime numbers and the cubes;[40] and the mathematical functions of squaring and of square-rooting are envisaged as the powers that govern these planar numbers and therefore govern two-dimensional, and particularly triangular, space. This is the space that constitutes preeminently the realm of the daemons, or at least of the airy daemons and the daemons inferior to them, whose spiritual "bodies" or airy "envelopes" we might think of as themselves functioning like two-dimensional surfaces, governed by their *habitus,* by *dynameis,* by squares, and by square roots. Hence the manipulative power the daemons exercise over all two-dimensional surfaces, including each other's, and hence their innate attraction to such surfaces and especially to crystals, to faceted stones and gems, and to mirrors. But this entire planar world is, from a Platonic viewpoint, presided over by the Pythagorean geometry of hypotenuses and thus of triangles, half-squares and half-equilaterals, themselves vestiges of the greatest triangle of all, the Trinity. From the geometry of the triangle we ascend to the still more mysterious geometry of the circle and thence of the point, the unextended monad that is the image of the One.

In short, apart from the psychological and theological dimensions, four related scientific fields are affected by this Platonic triangularism: optics, music (specifically harmonics), astronomy-astrology with its star and therefore time triangles, and finally what we might call scientific daemonology. For Ficino's assumptions about the planar realm and its triangularity make the summoning of daemons a kind of science, however strained or paradoxical this claim might first appear. Communication or traffic with them has for him a mathematical, and specifically a geometrical, foundation. The notion of roots and powers—that is, of self-division and self-multiplication—suggests that daemonic agency is ever present in the realm of mathematics. But it suggests too that daemonic depths and angelic heights in ourselves are at heart mathematical conditions, *habitus* subject to mathematical laws, and that our ideas and the absolute Ideas can even be numbers with their powers and roots, as Xenocrates had originally supposed.

One of the curiosities of Platonism in the Renaissance is its impact on areas we would not immediately associate with it: daemonology, astrology, optics, and musical theory. But in many respects Ficino was simply recovering the enthusiasms and interests of the mathematical Platonism that had also interested Plotinus. It had emerged from the accounts in Aristotle and in the notices he and others had provided of the views of such early Platonists as Speusippus, Xenocrates, and Crantor, whose mathematical interests were later channeled into the works of Theon, Nicomachus, Aristoxenus, and Aristides. These were the thinkers who, in Ficino's view, had truly understood the *Timaeus* and were thus in a position to interpret the various enigmatic mathematical references elsewhere in Plato: in the *Laws,* in the *Epinomis,* and in book 8 of the *Republic.* An awareness of this mathematical Platonism (dominated by geometrical ratios) is surely called for if we are ever to establish with confidence the valencies governing early modern science, its artful exploration of Browne's "things artificial." Certainly, though awaiting further exploration, it will constitute an important chapter in the history of that science.

NOTES

1. Philip Merlan, "Greek Philosophy from Plato to Plotinus," *Cambridge History of Later Greek and Early Medieval Philosophy*, ed. A. H. Armstrong, repr. with corrections (Cambridge: Cambridge University Press, 1970), p. 16. For the "Lecture" see, for example, Ingemar Düring, *Aristotle in the Ancient Biographical Tradition* (Göteborg; [distr. Stockholm: Almqvist and Wiksell,] 1957), pp. 357–361; Konrad Gaiser, *Platons ungeschriebene Lehre* (Stuttgart: Klett, 1963), pp. 452–453; and Gilbert Ryle, *Plato's Progress* (Cambridge: Cambridge University Press, 1966), pp. 251–256.

2. W. K. C. Guthrie, *A History of Greek Philosophy* (Cambridge: Cambridge University Press, 1962–1981), 4:523.

3. Thomas L. Heath, *Mathematics in Aristotle* (Oxford: Clarendon, 1949), p. 220; Merlan, "Greek Philosophy," p. 31; Guthrie, *History,* 5:459–460, 473.

4. Merlan, "Greek Philosophy," p. 21n.

5. Ibid., p. 18.

6. Speusippus fr. 40 in *De Speusippi Academici scriptis,* ed. P. Lang (Bonn, 1911; reprinted Frankfurt: Minerva, 1964); Xenocrates fr. 60 in *Xenokrates: Darstellung der Lehre und Sammlung der Fragmente,* ed. R. Heinze (Leipzig: Teubner, 1892). Cf. Diogenes Laertius, *Lives of the Philosophers* 3.67. Merlan concludes that "the *Timaeus* and the doctrines of Speusippus and Xenocrates seem to point to some kind of equation between the mathematicals and the soul, whether we take soul to mean cosmic or individual soul" ("Greek Philosophy," p. 18). We might note, incidentally, that in his *De defectu oraculorum* 13 (*Moralia* 416CD), Plutarch says that Xenocrates assimilated the equilateral triangle (where all parts are equal) to the divine class, the scalene (where all parts are unequal) to the mortal, and the isosceles (where they are mixed) to the daemonic. See Richard Heinze in *Xenokrates . . . Fragmente,* pp. 166–167, fr. 24; D. H. Fowler, *The Mathematics of Plato's Academy: A New Reconstruction* (Oxford: Clarendon; New York: Oxford University Press, 1987), p. 299; and Guthrie, *History,* 5:467, 479.

7. Merlan, "Greek Philosophy," pp. 19–20; see also Francis M. Cornford, *Plato and Parmenides* (London: K. Paul, Trench, Trübner, 1939), pp. 14 ff., 198.

8. *Timaeus,* in *The Dialogues of Plato,* trans. Benjamin Jowett, 4th ed. (Oxford: Clarendon, 1953), 3:741.

9. See Ficino's *Opera Omnia* (Basel, 1576; reprinted, Turin: Bottega d'Erasmo, 1983), p. 1475.1: "Cum in triangulo aequilatero per divisionem a summo scalenum designaveris, scito latus huius scaleni longius, id est, erectum super angulum rectum esse, secundum potentiam triplo maiorem latere minore iacente eundemque rectum angulum continente. Nam si ex illo latere quadratum aequilaterum constitueris seorsumque ex hoc latere alterum, illud ad hoc erit triplum. Linea hypotinusa, id est subducta vel subtendens [*Op.* subtendus], est quae quasi e transverso ab hoc latere ad illud producitur opposita angulo recto ipsa, angulos utrinque acutos efficiens. Nam [*Op.* Nec] longitudine latus minus duplo superat. Haec enim totum aequilateri primi latus adaequat. Hoc vero dimidium lateris

illius existit. Si autem ex linea subducta aequilaterum fiat quadratum, aequale erit duobus quadratis, hinc quidem ex maiore latere, inde vero ex minore confectis. Pythagorae hoc inventum." In general, see A. E. Taylor, *A Commentary on Plato's Timaeus* (Oxford: Clarendon, 1928), p. 372. For Ficino's rendering of Albinus' *Didaskalikos,* see his *Opera,* p. 1953.1.

10. Cf. Albinus, *Didaskalikos* 13. For six's perfection, see my *Nuptial Arithmetic: Marsilio Ficino's Commentary on the Fatal Number in Book VIII of Plato's "Republic"* (Berkeley: University of California Press, 1994), pp. 50–53, 67–68, 129–132. Francis M. Cornford, *Plato's Cosmology: The Timaeus of Plato Translated with a Running Commentary* (London: K. Paul, Trench, Trübner; New York: Harcourt, Brace, 1937), p. 213, notes that the equilateral triangle, despite its being non-right, is present in the faces of the polyhedra constituting the first three solids.

11. Cf. Albinus, *Didaskalikos* 13; Ficino, *In Timaeum* 44 (misnumbered XLI) (*Opera,* pp. 1953.1, 1464v.1). Incidentally, Ficino thought Plato had used the cube for the Fatal Number in part because it presides over time, and thus over the twenty-four-hour cycle. Karl Popper, *Conjectures and Refutations,* 3rd ed. (London: Routledge and Kegan Paul, 1969), pp. 75–93, has argued that Plato chose the isosceles and the half-equilateral triangles because they pose the problem of $\sqrt{2}$ and $\sqrt{3}$ and therefore of the rational existence of irrational numbers. See also Stephen Toulmin and June Goodfield, *The Architecture of Matter* (London: Pelican Books, 1965), pp. 75–82.

12. Again see Albinus, *Didaskalikos* 13; Ficino, *In Timaeum* 44 (*Opera,* pp. 1953.1, 1464v.1). See also Cornford, *Plato's Cosmology,* p. 218.

13. Taylor, *Commentary on Timaeus,* p. 373.

14. For Plato and this Pythagorean mathematics, see, for example, Thomas L. Heath, *A History of Greek Mathematics,* 2 vols. (Oxford, 1921; reprinted, New York: Dover, 1981); Charles Mugler, *Platon et la recherche mathématique de son époque* (Strasbourg, 1948; reprinted, Naarden: A. W. van Bekhoven, 1969); Paul Henri Michel, *De Pythagore à Euclide: Contribution à l'histoire des mathématiques préeuclidiennes* (Paris: Les Belles Lettres, 1950); idem, *Les nombres figurés dans l'arithmétique pythagoricienne* (Paris: [En vente à la librairie du Palais de la decouverte], 1958).

15. Aristides Quintilianus, *De musica* 3.23, ed. R. P. Winnington-Ingram (Leipzig: Teubner, 1963), p. 124.25–26.

16. See my *Nuptial Arithmetic,* pp. 28–30, 74–80.

17. Ibid., pp. 56–57.

18. The following section is adapted from my *Nuptial Arithmetic,* pp. 89–99.

19. Marsilio Ficino, *Platonic Theology* 11.5 (ed. Raymond Marcel [Paris: Les Belles Lettres, 1964–1970], 2:129): "Complexio quoque plantarum et animalium quam mirabili medicinae artis solertia utitur in conservando habitu naturali aut recuperando" (Marcel renders *habitu* here as "équilibre").

20. Ibid. 15.18 (ed. Marcel, 3:98): "habitus, qui transit in subiecti naturam, immo subiecti naturam usurpat ipse sibi, certam praestat proclivitatem et subiectum movet immobilis."

21. Ibid. 15.19 (ed. Marcel, 3:103): "habitus non imaginarium quiddam est, sed naturalis formae gerit vicem."

22. Ibid. 18.8 (ed. Marcel, 3:200): "Ita enim animus habitu movetur et agit, sicut natura formis. . . . Remanere vero in anima separata habitus tum morum tum disciplinarum tam bonos quam malos . . . dicitur habitum ita in naturam converti"; 16.7 (ed. Marcel, 3:134–135): "motus et habitus animae, quatenus intellectualis rationalisque est, circuitus esse debeat . . . cum primum animus in caput suum, id est mentem, erectus, in habitum suum prorsus restituetur."

23. Ibid. 12.1 (ed. Marcel, 2:154): "Igitur mens per formulam suam ex habitu eductam in actum ideae divinae quadam praeparatione subnectitur."

24. Ibid. 15.16 (ed. Marcel, 3:83–84): "habitum reformandi, . . . respondebimus vim mentis eandem, quae et contrahit et servat habitum in eius naturam iam pene conversum, contrahere in intima sua speciem atque servare. Siquidem habitus fundatur in speciebus, species in habitu concluduntur. Proinde si habitus fit a mente, specie atque actu, ab aliquo istorum stabilitatem suam nanciscitur. . . . A specie igitur"; 13.2 (ed. Marcel, 2:210): "Mens autem . . . habitu quodam et, ut vult Plotinus [2.9.1; 3.9.1(?)], actu simul continet omnia."

25. Ibid. 14.6 (ed. Marcel, 2:268): "Nam praeparatio sive affectio formam habitumque respicit, atque certa quaedam affectio certum habitum"; and again 15.19 (ed. Marcel, 3:102), "praeparationes, quantum ad certos habitus conferunt atque cum illis proportione aliqua congruunt, tantum habitus diversos impediunt."

26. Cf. Plato's *Republic* 9.587D9 (square) with the *Theaetetus* 147D3 ff. (square root); and Euclid, *Elements* 10. See Heath, *History of Greek Mathematics,* 1:155; Taylor, *Commentary on Timaeus,* p. 372.

27. Ficino was probably introduced to the schema in Aristotle: see, for example, *Topics* 4, 141b5–22, and *Metaphysics* 3.5, 1001b26–1002b11.

28. On Ficino's variations on this system, see Paul O. Kristeller, *The Philosophy of Marsilio Ficino* (New York, 1943; reprinted, Gloucester, Mass.: P. Smith, 1964), pp. 106–108, 167–169, 266, 370, 384, 400–401; and my "Ficino's Theory of the Five Substances and the Neoplatonists' *Parmenides,*" *Journal of Medieval and Renaissance Studies* 12 (1982): 19–44, with further references. See also Tamara Albertini, *Marsilio Ficino: Das Problem der Vermittlung von Denken und Welt in einer Metaphysik der Einfachheit* (Munich: Wilhelm Fink Verlag, 1997), part 2.

29. See, for instance, Ficino's letter to Lotterio Neroni, the penultimate letter in the third book of his *Epistulae* (*Opera,* p. 750.3). Here he speaks of being either in "the undivided and motionless center" as in the one God, or in "the divided and mobile circumference" as in heaven and the elements, or in the "individual lines" that mediate between them, beginning at the center as undivided and motionless but gradually becoming divided and

"mutable" as they approach the circumference. In these lines are the souls and minds. Cf. Pico della Mirandola, *Conclusiones DCCCC, Conclusiones secundum mathematicam Pythagorae,* nos. 13 ("Quilibet numerus planus aequilaterus animam symbolizat") and 14 ("Quilibet numerus linearis symbolizat deos"), in his *Opera Omnia* (Basel, 1572), p. 79.

30. Cf. Ficino's *Timaeus* commentary 22 (*Opera,* p. 1449.3): "Quattuor apud mathematicum: signum, linea, planum atque profundum."

31. *Platonis Opera Omnia,* 2nd ed. (Venice, 1491), trans. M. Ficino, fol. 246v (sig. G[6]v) (i.e., *Opera,* pp. 1452–1453 [misnumbered 1450 and 1417]): "Congruunt animae mathematica, utraque enim inter divina et naturalia media iudicantur. Congruunt musici numeri animae plurimum, mobiles enim sunt; propterea que animam quae est principium motionis rite significant. Non solum vero per numeros sed etiam per figuras describitur anima ut per numeros quidem incorporea cogitetur, per figuras autem cognoscatur ad corpora naturaliter declinare. Convenit triangulus animae. Quia sicut triangulus ab uno angulo in duos protenditur, sic anima ab individua divinaque substantia profluens in naturam corporis labitur penitus divisibilem. Ac si cum divinis conferatur, divisa videtur; quae enim illa per unam et stabilem virtutem agunt atque subito, haec per plures mutabilesque vires actionesque peragit ac temporis intervallis. Sin autem conferatur cum naturalibus, indivisa censetur; non enim alibi habet partes alias loco disiunctas ut illa [*Op.* alia], sed etiam [*Op.* est] in qualibet totius parte tota; neque omnia mobiliter temporeque persequitur sicut illa, sed nonnihil etiam subito [*Op.* subiecto] consequitur aeterneque possidet. Licet in hoc insuper animam cum triangulo comparare, quod triangulus prima figura est earum quae pluribus constantes lineis producuntur in rectum; anima similiter prima omnium in plures distribuitur vires quae in ipsa intelligentiae subiguntur, ac produci videtur in rectum dum a divinitate labitur in naturam. In quo quidem descensu ab intelligentia summa in tres profluit vires inferiores, id est, in discursum quendam rationalem, in sensum, in vegetandi virtutem, quemadmodum et triangulus a signo productus in tres deducitur angulos. Dico autem animam ex omnium genere primam et ex pluribus quodammodo viribus commisceri et in rectum, ut ita dixerim, cadere. Mens enim angelica super [*Plat. Op.* semper] animam inferioribus intra se nullis indiget viribus, sed pura [*Op.* plura] mens est totaque mens atque sufficiens."

Chapter 28 is titled "De compositione animae et quod per quinarium in ea componenda opportune proceditur."

32. In his *Timaeus* commentary 34 (*Opera,* p. 1460), having discussed the lambda numbers, Ficino goes on to describe a metaphysical triangle governed by the double, sesquialteral, and sesquitertial properties: its apex is essence and its two sides consist of the infinite, difference, and motion, and of the limit, identity, and rest—the fundamental ontological categories explored in the *Sophist* and *Philebus;* see my *Icastes: Marsilio Ficino's Interpretation of Plato's Sophist* (Berkeley: University of California Press, 1989), chap. 2. He concludes by postulating a corresponding triangle for the soul: its unity is its essence, its will the infinite, its understanding the limit, its imagination difference, its reason identity, its power to procreate (*generandi vis*) motion, its power to join (*connectendi virtus*) rest. The whole topic awaits investigation.

33. For Ficino, either rational and irrational square roots or the squares of these roots can be deemed "powers."

34. In the *De amore* 7.4 (ed. Raymond Marcel [Paris: Les Belles Lettres, 1956], p. 247), Ficino had defined *spiritus* as "a vapor of the blood." For the various senses of *spiritus* in Ficino, see my *Platonism of Marsilio Ficino: A Study of His Phaedrus Commentary. Its Sources, and Genesis* (Berkeley: University of California Press, 1984), pp. 102–103n. The study by Ioan Petru Coulianu [Culianu], *Eros et magie à la Renaissance (1484)* (Paris: Flammarion, 1984; translated into English by Margaret Cook as *Eros and Magic in the Renaissance* [Chicago: University of Chicago Press, 1987]), has many speculative things to say about spirit and the raylike *idola,* but it should be used with great caution since it rides roughshod over Ficino's careful distinctions; this is a pity, since *spiritus,* spiritual magic, and Ficino form the nub of its concerns. In general, see the papers in the collection *Spiritus: IV° Colloquio Internazionale del Lessico Intellettuale Europeo (Roma, 7–9 gennaio 1983),* ed. Marta Fattori and Massimo Bianchi (Rome: Edizioni dell' Ateneo, 1984).

35. Plutarch, *Quaestiones Convivales* 8.2 (*Moralia* 718B–720C, specifically 718BC). Ficino gives a Latin rendering of the inscription on the Academy's entrance in his *Vita Platonis* (under *Discipuli Platonis praecipui; Opera,* p. 766.1) and glosses it thus: "Understand that Plato was intending this to apply not only to the proper measurement of lines but also of [our] passions (*affectuum*)."

36. See S. Otto, "Geometrie und Optik in der Philosophie des Marsilio Ficino," *Philosophisches Jahrbuch* 98 (1991): 290–313; Albertini, *Marsilio Ficino,* pp. 76–85.

37. Allen, *Icastes,* chap. 5.

38. Cf. Plato's *Gorgias* 510B; *Lysis* 214B; *Republic* 1.329A, 4.425C; etc.

39. But see D. P. Walker's *Spiritual and Demonic Magic: From Ficino to Campanella* (London, 1958; reprinted, Notre Dame, Ind.: University of Notre Dame Press, 1975), chap. 1; idem, *Music, Spirit and Language in the Renaissance,* ed. Penelope Gouk (London: Variorum Reprints, 1984); Gary Tomlinson, *Music in Renaissance Magic: Toward a Historiography of Others* (Chicago: University of Chicago Press, 1993), chaps. 3, 4; and my *Platonism of Marsilio Ficino,* pp. 51–58.

40. Cf. the *Timaeus* 31C–32B on the planar numbers as means.

4

SPACE, LIGHT, AND SOUL IN FRANCESCO PATRIZI'S
NOVA DE UNIVERSIS PHILOSOPHIA (1591)
Luc Deitz

I

"It is five philosophical systems that I submit to you in this book, Gregory, most blessed Father. They are all pious, and they all agree with the Catholic faith: my own, which is of recent invention; the Chaldean philosophy of Zoroaster; the Egyptian philosophy of Hermes; another Egyptian philosophy, which is a mystical one; and another still, which is properly Plato's own. I have taken enormous pains to rescue them from the ruins under which they had been buried for so long, to collect and explain them in one single volume, and to present them in their correct scientific order."[1] These are the opening lines of the dedicatory epistle that Francesco Patrizi addressed to the ailing Pope Gregory XIV[2] and to all his successors to the Holy See when he was about to publish his great systematic treatise, the *Nova de universis philosophia,* in the summer of 1591. The professor of Platonic philosophy at the Studio of Ferrara, who had been a soldier, an aspiring physician, a cotton merchant, and a bookseller before he finally went bankrupt and sought the patronage of Duke Alfonso II and the Este family,[3] was none too modest about his own achievements: "Human reason," so the epistle continues,

> is guided by reason alone; reason follows reason willingly, and even if she does not like it, reason is carried along by reason. It is with the help of reason, therefore, that men should be led to God. Accordingly, I have made every effort to devise this true and divine philosophy by relying on reason alone. With immense and unremitting effort I have, methinks, brought philosophy to completion. . . . You should therefore be the first to decree, most Holy Father, and all the popes following after you should similarly decree . . . that some of the books of the Platonic philosophers be always taught in every college and in every monastic school of your dominion. . . . You should also see to it that the rulers of the Christian world command that the same thing be done in their own colleges.[4]

The benefits of studying Platonic philosophy, and especially of studying Platonic philosophy in the form Patrizi had given it, are obvious: not only would the Italians, Spanish, and French embrace the Roman faith with even greater fervor, but, more important, the parlous and restive Germans would be brought back into the fold. Let the Jesuits take over responsibility for the teaching, and the seeds of Platonic philosophy would do the rest.[5]

Patrizi's high hopes were to be dashed straightaway. Not only did his main philosophical treatise never become a textbook for schools or universities, it also shared the fate of much of the best scholarly literature of the age and in 1596, five years after its publication, was put, *donec corrigatur,* on the *Index of Prohibited Books* (where it remained for more than three centuries).[6] Even to the present day, his work and the "correct scientific order" of its exposition remain little studied and the details of his philosophy barely understood.[7] This is unlikely to change until someone musters the courage and patience needed to produce a critical edition of the *NUP* with a detailed account of all its sources.[8] Until then, the best we can hope to achieve is to throw some interpretive light on narrowly circumscribed areas. It is exactly this that I propose to do by inquiring into how Patrizi's conceptions of space, light, and soul are related. For the sake of clarity of exposition, I should like to start with a short account of Patrizi's scheme of things in general, which, I believe, has never been adequately described. I shall then turn to the three concepts that are the chief concern of this paper, and conclude by giving what I think to be the most likely interpretation of their relationship.

II

Patrizi's philosophy, which he himself describes as "new, true, and complete," and proven "with the help of divine oracles, geometrical necessities, philosophical reasons, and conclusive experiments,"[9] is based on a number of axioms that are never clearly listed, but that may be summarized as follows:

The universe is created.[10]

The cause precedes its effect and is superior to it.[11]

The whole precedes the part.[12]

Unity precedes plurality.[13]

Nature makes no leaps (or, in other words, the progression of being is continuous).[14]

(perhaps the strangest of all) Whenever one of a pair of logical contraries is said to exist, its opposite number must exist as well.[15]

The first principle therefore—which may be called "God" and within which Patrizi, following the *Chaldean Oracles,* sometimes distinguishes between the "Father" and the "Paternal Depth"—is most properly described as *un'omnia*.[16] If we were to render this nonce word, apparently coined by Patrizi himself, in Greek, we would have to say that it is equivalent to πᾶν τὸ ἕν rather than ἕν τὸ πᾶν—"the one [principle] is everything" rather than "everything is one"—for Patrizi's metaphysics leads from unity to plurality rather than vice versa. It is clear, on Patrizi's assumptions, that the first principle must be "one" and "all" at the same time. For imagine it were only one, an unadulterated unity both in act and potency: then it would be unable to generate plurality. Or again, suppose it were only one in act, and many in potency: then it would be able to generate plurality, but not universality. A principle worthy of the name, however, must be a principle of everything; therefore it must be one in act and all in potency. What better label to choose for this than that of "un'omnia"?

According to the axiom that the progression of being is continuous, unity cannot immediately turn into plurality but needs a mean term to be able to do so. Proclus (whose Στοιχείωσις θεολογική Patrizi translated into Latin in 1583) had ascribed this role to the henads;[17] Patrizi follows him in this by assuming the existence of an *unitas primaria* (identical with the idea of the good) just below the first principle, which is in turn followed by all the secondary unities (or ideas) that are derived from it.[18] These, we are told, are superessential, and the concept of superessentiality quite naturally leads Patrizi on to introduce that of essence, which in turn, by a kind of chain reaction, triggers the remaining two members of the Proclean triad ὄν–ζωή–νοῦς: viz., life and mind, or, in the words of Patrizi, *vita* and *mens* (or *intellectus*).[19] The justification offered by Patrizi for their appearance is far from clear and seems to run as follows: essence is directly derived from the most perfect things; therefore, it must itself contain that summit of perfection which is compatible with it in the hierarchy of being. Since it is ultimately derived from God, and since God is alive, it must contain life, for life is superior to death.[20] Life implies movement, and the highest form of movement is spiritual and incorporeal movement: in other words, the cognitive process of the intellect.[21] Patrizi distinguishes three kinds of intellectual movement at this stage: ascending, reflexive, and descending. Although he does not explicitly say so, it would seem that the first corresponds to intuitive knowledge and the last to discursive reasoning, but it is not easy to see what the reflexive movement is supposed to stand for—self-consciousness, perhaps, but that is a mere guess.[22]

The ontological gap between being and becoming and between permanence and transience is bridged here, as everywhere else in the Platonic tradition, by soul, which is at the center of the enneadic scale extending below the first principle. The lower part of soul, which we would call "immanent," stretches into the bodily world and produces as its likeness nature, "whose name is better known than its meaning." Nature's means of shaping shapeless bodily bulk is quality; quality engenders a specific form; and form, finally, is inconceivable without body, which comes at the lower end of the ontological hierarchy.[23] In short, we get the following picture (figure 4.1).

This arrangement of things, which Patrizi names the "degrees" (*gradus*) of reality and likens to Jacob's ladder,[24] is new only insofar as it brings the number of hypostases (including the principle) to ten, that is, the number of the Pythagorean tetractys, to which Patrizi explicitly refers.[25] Yet anyone who has even a passing acquaintance with the writings of Proclus and Ficino will readily admit that so far, Patrizi has had little original to say in substance.

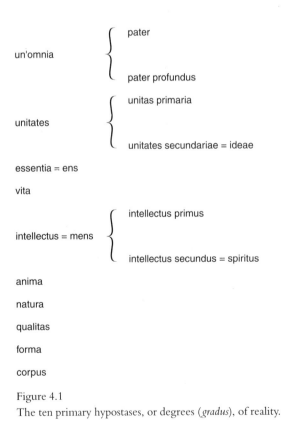

Figure 4.1
The ten primary hypostases, or degrees (*gradus*), of reality.

We must therefore try to complete the picture by briefly adumbrating the remaining elements of his ontology.

The nine degrees of reality that follow upon the One, descending vertically from top to bottom, also have nine lateral or horizontal ramifications called "series" (*series*).[26] The first series, starting from the *unitas primaria,* contains the *unitas essentiarum,* the *unitas vitarum,* the *unitas intellectuum,* and so on, until the *unitas corporum* is reached. The second series, starting from the *essentia primaria,* contains, analogously, the *essentia vitarum,* the *essentia intellectuum,* and so on, until we reach, in descending order, the last, reflexive series, which is the *corpus primarium.* Again, a diagram may help to illustrate this (figure 4.2). It is easy to see that every single constituent part of the universe (except the henads) is directly determined by two other elements. The first element immediately precedes it in the series of higher order and ultimately links it to its henad (or idea), which guarantees the identity of its generic character throughout the universe. The second of these elements immediately precedes it within the same series and determines its specific difference, which depends on the hypostasis as such. At the end of the scale, where Body dwells, we must assume absolute identity between genus and species.[27]

But this is not all. Besides degrees and series, Patrizi's universe also contains "chains" (*catenae*) of light, which extend in depth.[28] His account of these proceeds more by intimation than by explanation, but we may assume that the *unitas primaria* is at the same time the *unitas lucis,* from which proceeds the *lux unitatis;* that the *essentia primaria* is at the same time the *essentia lucis,* from which proceeds the *lux essentiae;* and so on, until once again we reach the *corpus primum,* from which there does not seem to be any light at all gushing forth.[29] We are not told what the relationship between, say, the *lux vitae* and the *anima qualitatis* is; but such questions are likely to reflect only idle curiosity, for what matters is something else. If we put together the three orders of degrees, series, and chains, they create something not formerly contained in them—viz., a third dimension, and, as a result of this, space.[30]

This allows us to draw one further inference. Patrizi's degrees, series, and chains not only beget space, they also beget space of a very definite shape. We have already seen that the interaction of the *gradus* with the *series* can be represented in the form of a triangle. If we were to illustrate the interaction of the *gradus* with the *catenae,* the picture would similarly be that of a triangle; and the same is true of the interaction of the *catenae* with the *series.* Thus, Patrizi's space can be said to be a three-dimensional figure bounded by three adjacent triangles, that is, a tetrahedron (or pyramid) whose fourth surface, by virtue of geometrical necessity, must have a triangular shape as well.

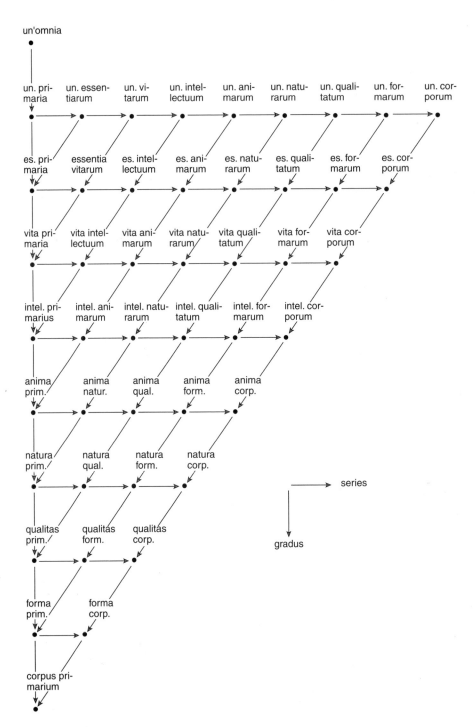

Figure 4.2
The series (*series*) emanating from each of the ten primary hypostases.

The conclusion that Patrizi's universe has the shape of a pyramid is not very meaningful in itself and might remain a mere doxographical curiosity, were it not for the fact that we can go one step further still. Among the infinite number of possible tetrahedral pyramids, only two are remarkable for their regularity: (a) the most regular irregular one, which is contained by three right-angled isosceles triangles, and (b) the only completely regular one, which is contained by four equilateral triangles. In the absence of any clear statement by Patrizi himself, it is impossible to tell which pyramid he had in mind when he devised the structure of his universe, but it is not unlikely that it was one of those just mentioned. Indeed, I would argue that he was thinking of none other than the regular pyramid, and I wish to adduce three reasons in support of this claim.

First, there is the consideration of geometrical beauty that goes with perfect symmetry: it is (for once) safe to argue that psychologically it is far more plausible that a Platonist like Patrizi would prefer symmetry and regularity to chaos and disorder (a similar feeling is expressed in *Timaeus* 30A).

Second, and perhaps more important, in Plato's *Timaeus* the regular tetrahedron is the solid associated with the element of fire.[31] If we bear in mind the fact (to which I shall be returning shortly) that in Patrizi's metaphysics light—which, needless to say, is traditionally associated with fire—plays a role second only to that of space, then it makes perfect sense that he should have depicted his universe in the shape of a regular pyramid.

Third, if we ask ourselves whether Patrizi could have been relying on ancient authorities for his counterintuitive speculations, then we can give a surprisingly unequivocal answer. The idea that the pyramid is the foundation of the noetic as well as of the natural world is explicitly stated in two Greek texts (and in two Greek texts only), the *Contra Iulianum* of Cyril of Alexandria and the *Theologumena arithmeticae* ascribed to Iamblichus.[32] Only one of these, namely the *Theologumena,* further specifies that this pyramid cannot be any other but the regular one.[33] That Patrizi knew the doctrine transmitted by the *Contra Iulianum* is easy to prove, for he included the sentence on the pyramidal shape of the universe in his edition of the Hermetic fragments.[34] He seems also to have been familiar with the more precise statement contained in the *Theologumena arithmeticae,* for among the nine surviving manuscripts of this text, one belonged to Patrizi himself and formed part of the collection that he sold to King Philip II of Spain in 1575.[35] Thus, a material witness comes to corroborate the evidence based on considerations of psychological and philosophical plausibility, and one may therefore safely judge that even if none of the three arguments listed above is conclusive per se their cumulative strength is such that we cannot avoid concluding that Patrizi's universe does indeed have the shape of a regular tetrahedron standing on one of its corners. I shall try to explain the meaning of this below; for now, another drawing may help to visualize Patrizi's speculations (figure 4.3).[36]

Patrizi rounds off his discussion of the orders of reality by stating that "reasoned argument shows that the universe contains a greater degree of firmness, stability, and solidity than bodies do."[37] These attributes plainly come about as a result of its simple pyramidal structure, and it is to the properties of this spatial structure that I now turn.[38]

III

Just as metaphysics precedes physics both in the *ordo essendi* and in the *ordo cognoscendi* (i.e., the πρῶτον καθ' ἑαυτό is the same as the πρῶτον πρὸς ἡμᾶς), so physics precedes mathematics. The opening chapter of the last and longest section of the *NUP*, the *Pancosmia*, is titled *De spacio physico;* it is followed by a chapter called *De spacio mathematico* and another called *De physici ac mathematici spacii, affectionibus.*[39] The central importance of these chapters for Patrizi's thought is proven (among other things) by their separate publication four years before being incorporated into the great systematic treatise: the chapters on physical and mathematical space under the telling Lucretian title *Philosophiae de rerum natura libri II priores,*[40] the chapter on the "affections" of space under the no less telling *Della nuova geometria libri XV.*[41]

The ontological priority of space is based, sensibly enough, on the assumption that nothing can be without space, whereas space can exist as empty space without anything to fill it;[42] its gnoseological priority is based on an anti-Aristotelian maxim often repeated by Patrizi, according to which understanding of the principles of things is not arrived at analytically at the end of the cognitive process but is the very foundation of that process. Reason and sense perception move along the same way, and one of the distinctive features of Patrizi's "new" natural philosophy is its simultaneous consistency, in his view, both with the laws of thought and with the evidence of the senses.[43]

What, then, can we know about space?[44] It would be foolish to maintain that it does not exist, for it is common knowledge shared by everybody both that space is and that it is *something.* This is proven not only by the mere existence of words like *dimensio* (dimension), *distantia* (distance), *intervallum* (interval), *intercapedo* (interval), *spacium* (space), *diastasis* (extension), or *diastema* (interval), which are concepts used by Greeks and Romans alike—a weighty argument in itself. It is proven also by our perception of distance, for who can fail to see that the heavens are above the earth or that our heads are not resting on our feet? These observations are not figments of our imagina-

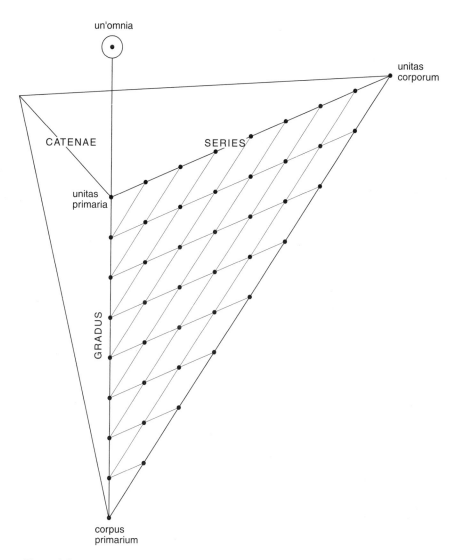

Figure 4.3
The shape of Patrizi's universe: a regular tetrahedron, made of degrees, series, and chains
(*catenae*).

tion; so the theoretical possibility of the nonexistence of space can be dismissed straight away. (George Berkeley, more radical in this respect, would later argue that we cannot validly infer the existence of space from our perception of distance.)[45]

We can similarly discard the hypothesis that space *is*, but that it *is nothing*. Patrizi's argument in this case is presented in the form of a reductio ad absurdum based on the various significations of the verb "to be," which he uses indistinctly as an existential predicate and as a copula.[46] The reasoning runs as follows:

(a) Whatever is, is something.
(b) Whatever is not, is nothing.
(c) A thing cannot be something and nothing at the same time.
(d) Space is.
∴ (e₁) Therefore space cannot be nothing.
(e₂) Therefore space must be something.

Having thus established to his own satisfaction, if perhaps not to ours, that space is to be numbered among the *entia*, Patrizi asks what its constituent parts are. He vehemently rejects Aristotle's influential theory that space is two-dimensional and identical with the inner surface surrounding any given object; in fact, he sees clearly that what Aristotle is really talking about is "place" rather than "space."[47] (The Greek word τόπος used by Aristotle can mean both, as is well-known.) Patrizi instead endorses the far more familiar view that space is three-dimensional and is made up of length (*longitudo*), width (*latitudo*), and depth (*profunditas*). This space is not the same thing as body, however; for besides being extended, bodies also have to be solid; that is, besides having a specific size and shape they also have to have a specific bulk or mass in order to offer resistance.[48] Patrizi further specifies that when applied to a natural body, the mathematical concept of length variously denotes the longest distance between its extremities or its height from top to bottom; the concept of width variously denotes the second-longest distance between two other of the body's extremities or its breadth from right to left; and the concept of depth variously denotes the smallest distance between the remaining two extremities of the body or its deepness. Yet, he adds, this is no more than a conventional way of expressing things, for when a body is turned in space its dimensions remain the same, whatever name we give to them. The terms we use are thus only expedient means of description, entirely dependent on the observer's point of view (*respectu nostri; NUP* 4, 1, 61 d)—a formulation that, it seems to me, is the closest we get in the sixteenth century to what would later become known as the isotropy of space.

Not content with rejecting Aristotle, Patrizi also takes issue with the Stoic doctrine that bodies can intermingle and interpenetrate.[49] Consider what happens when a body moves in space, he says. We must assume either that space offers resistance or that it does not. If it does, then it somehow flows through body just as body flows through it; if it does not, then it somehow recedes in front of the advancing body and again closes up behind as soon as that body changes place. Patrizi is adamant that the former hypothesis must be rejected, for the definition of natural body is resistance (of which extension is only an accidental and adventitious quality), whereas the very definition of space is extension (which has nothing to do with resistance). Body and space are thus in two different and mutually exclusive ontological categories, and to assume any kind of intersection between them would be a violation of the fifth axiom listed above, that "nature makes no leaps."[50]

Space thus offers no resistance and is penetrable by body. That part of space which is occupied by a body is what we call "place"; and when a body moves from *a* to *b,* then we say that it changes places, although it remains in the same space. If two (or more) bodies did interpenetrate, then they would have to be in the same place at the same time; that is to say, there would have to be two places where there is room for only one. But this is clearly absurd, for there can only be one body in one place at any one given time: if there were two, they would be indistinguishable (and therefore one).

Though bodies cannot interpenetrate, they can still be condensed;[51] and in order to explain the phenomenon of physical condensation, we must either suppose that some of the body's matter is lost in the process or that there are little empty holes in it that are gradually filled while the body is being compressed and gains ever greater density. Patrizi thinks that it is possible to prove empirically the existence of these vacua (which he calls *spaciola*) and of void in general; but as Charles B. Schmitt showed long ago, the only certain knowledge we can derive from Patrizi's description of his alleged empirical proofs is that he never tried to put them to the test.[52] His recourse to a number of "experiments" must be considered topical rather than factual and is a perfect example of how the so-called *experimentum crucis* was more often than not unable to fulfill the role with which it is sometimes credited by modern epistemologists;[53] it certainly had less importance in practice than in theory.

That, however, is another story. Once the existence of empty space is acknowledged, be it for the right reasons or not, the question naturally arises whether a void or voids can be found only inside the (visible) heavens or also outside it. For Patrizi, this is equivalent to asking whether the universe is finite or infinite, and his answer to this vexed question is remarkably clear and

unequivocal.[54] At the boundary of the visible universe we see the twelve constellations corresponding to the twelve signs of the zodiac and the other fixed
stars. Since they are bodies, they occupy a place, and since they occupy a
place, there must be one point that is at the farthest remove from the observer's eye. This point is necessarily at the reverse side of the body, and this
reverse side cannot be anywhere else but in space. Thus, there must be space
"behind" the stars, as it were, which is tantamount to saying that there is
space outside the (visible) universe. Patrizi rejects the hypothesis of a crys-
talline sphere at the border of the heavens and maintains, as the Stoics had,
that extracosmic space is both empty and infinite. Indeed, if it were finite, it
would have a boundary, which in turn would be in a place, which in turn
would be in space, and so on. This argument is again directed against Aristotle, who had maintained that "there is neither place nor void beyond the
heaven."[55] Lest one should think that having an infinite empty space is something of an unnecessary luxury, Patrizi adds—taking up an argument already
put forward by Cleomedes—that on the day when the visible universe is de-
stroyed by the final conflagration and goes up in smoke, it will fill a place that
is at least one hundred thousand times greater than that which it fills now,
and probably much larger; and for such eventualities infinite empty space obviously has a lot to recommend it.[56]

How, then, should space be defined? It cannot be pure "potency" (*ap-
titudo; NUP* 4, 1, 65 a) to receive body, for it is independent of body and can
exist without it (whereas the converse is not true). Nor can it be an accident,
for accidents, like categories, are predicated of substances, and substances are
those things that precede everything else on a logical and an ontological level.
Since space precedes everything else, it must be a substance, but a substance
of a very peculiar kind: Patrizi calls it a "self-subsisting hypostatical extension
that does not inhere in anything else."[57] As it is not a compound of matter
and form, or predicated of species or of individuals, it is a substance that is
different from the "substance" that heads the list of categories; it is a substance
outside the category of substance, which is the basis of everything else and
without which nothing can exist.[58] Since it is three-dimensional, it must be
said to be corporeal; but since it does not offer resistance to touch or sight, it
must equally be said to be incorporeal. Therefore its most complete definition is that of a *corpus incorporeum* or a *noncorpus corporeum:* that is to say, a "substantial extension subsisting by itself" that is homogeneous, unchangeable,
and unmovable as a whole and in its parts.[59]

Patrizi's theory of space has a number of interesting implications for his
outline of geometry. The most notable reversal of the traditional Euclidean
position is his claim that a point is not a principle (as Euclid had held) but a

principiatum, by virtue of its being a minimum of space rather than "that which has no parts."[60] To discuss Patrizi's geometry in detail would, however, lead us too far astray,[61] and so I shall turn instead to the next division of my triptych and briefly consider what he had to say about the second most important component part of the universe: light.

IV

Just like space, light is created; in fact, it is the first entity created after it.[62] Why light? Patrizi gives two reasons, one theological and the other physical. The theological argument runs as follows: "God inhabits an inaccessible light—that is to say, a light that is inaccessible in itself and did not become so only after he had said, 'Let there be light.' He has been dwelling thus inaccessibly from all eternity and still continues to do so."[63] Now it is materially and conceptually impossible for light not to shine forth;[64] therefore the concept of God analytically implies that he should manifest himself. For reasons best known to himself, he decided to do this at the very moment when he said "Fiat lux," thereby filling empty space with his visible presence.

The powers of God Almighty, who transcends all categories,[65] are thus unexpectedly restricted by the analytical dissection of concepts. What matters in the present context, however, is not this theological puzzle but the physical reason advanced by Patrizi to justify the introduction of light immediately after the creation of space. God chose to fill space with light because it is most like space and could "most easily be poured into it."[66] Like space, it is "most simple"; like space, it can have "infinite extension"; like space, it can "penetrate everything" and "fill everything"; it "cannot resist anything and yields to everything," and can therefore be penetrated and permeated by everything; "to put it in a nutshell: it is, like space, a body and bodiless." The argument for the latter claim is the same as the one previously used in connection with space: light is a body insofar as it fills the whole universe by virtue of its having three dimensions, but at the same time it is incorporeal inasmuch as it has no resistance (i.e., it is immaterial).

Patrizi waxes lyrical about the other properties of light, praising its beauty, sweetness, desirability, wisdom, goodness, power, happiness, and so forth,[67] but this is not the place to analyze these attributes in detail. He also distinguishes between different kinds of light (aerial, celestial, ethereal, and empyrean), as well as between different kinds of transparency and opacity; I cannot dwell on these matters either.[68] What is important for my purpose is the fact that Patrizi ultimately gives the same definition of light as of space. This must clearly mean something; but before we can ask what the meaning of this identity might be, we must have a quick glance at the third part of the *NUP*: the *Pampsychia,* which is devoted to soul—and in particular to the soul of the universe.

V

Besides being a philosopher, Patrizi was also a historian of philosophy; and, as such, he was clearly aware that the question of the existence of a world soul and the debate about its essence were not entirely new and that he was writing within a long tradition. For him, this tradition was that of the *prisca theologia* handed down through successive generations of thinkers, culminating in the works of Plato, and revived by Ficino.[69] Thus he informs us that Zoroaster, Hermes, Orpheus, Pythagoras, and Plato had all been of the opinion that the world was animated, as were Thales, Heraclitus, Anaxagoras, Archelaus, Parmenides, Zeno, Empedocles, all the Stoics, and—"if Plutarch is to be believed"—even Democritus.[70] Only Leucippus and Epicurus, two ridiculous philosophers, had maintained the contrary, whereas others, like Aristotle, had held the strange view that the universe was a sort of monster, partly animated and partly inanimate. Among Christian theologians and philosophers, some were to side with Aristotle (Patrizi names Origen, Jerome, Augustine, Petrus Aureoli, Duns Scotus, Thomas Aquinas, and Gaetano da Thiene); others would not accept that the heavens, and consequently the universe, are animated (these include Lactantius, Basil, Ambrose, Cyril of Alexandria, and John of Damascus); whereas a third group (here Jerome and Augustine are invoked again) would maintain at an early stage in their lives that the world is ensouled and retract their opinions later on.[71]

What is surprising, therefore, is not so much that Patrizi should have devoted a large portion of the *NUP* to discussing the time-honored question of the world soul as—in view of the great number of authorities mentioned, if not actually quoted—the quite original way in which he treats the subject.[72] A mixture of argument by authority and argument by reason, it deserves to be looked at in some detail.

To the cold eye of the analytical philosopher, the argument by authority is a species of the logical fallacy known under the name of *e consensu gentium.* Patrizi, we may assume, would turn the tables on the analytical philosopher by claiming that the very reasons that the latter might invoke *against* the validity of *prisca theologia's* claims can equally well be invoked *in favor of* them: the older a doctrine, and the greater the number of (Platonic) philosophers who have put it forward, the less likely it is to be wrong. Ignoring the fanciful chronologies—which might cause greater qualms to the historian than to the philosopher—a modern thinker might also wish to argue that Patrizi's reasoning is circular: the antiquity of a truth guarantees its authority, and the great number of (Platonic) philosophers falling in with this authority is a proof of its truth (and therefore of its antiquity). Again, Patrizi himself would probably disagree and maintain that far from being circular, his argument is consistent with the laws of thought: for truth, once revealed, is "norma sui et falsi,"[73] timeless, and independent of the contingencies of chronology.

We may safely assume that the argument by authority alone, based as it was on the belief in the perduration of the *prisca theologia,* would have seemed sufficient to Patrizi to establish the existence of the world soul. It is therefore all to his credit that he also gives a second proof of its existence, which relies to a greater extent on reason and focuses more sharply on the justification of the world soul's place and role in the hierarchy of being. Patrizi here invokes the last of the six axioms mentioned above and sets out to give, first, a list of properties that he considers to be specifically characteristic of being and second, in a rather mechanical way, a corresponding negative list detailing the properties of non-being according to the axiom of contraries. Here is a summary list of the most relevant attributes:[74]

ens	**non-ens**
sibi semper simile	semper non-ens
semper id quod erat	semper in non-ens abit
semper idem	numquam idem
non mutatur	mutatur
non alteratur	alteratur
tantum agit	semper patitur
non dividitur	dividitur
partes nullas habet	partes habet
trine non est dimensum	trine dimensum est
moles non est	moles est
mole non indiget	mole indiget
spacio non indiget	in spacio habitat
per se substat	non per se substat
αὐθυπόστατον est	ἑτερόστατον est
incorporeum est	corpus est

According to Patrizi, the real existence of the properties listed in the left-hand column entails the equally real existence of those in the right-hand column; but how are we to imagine the relation between absolute being (the *un'omnia*) and absolute non-being (the hypostasis *corpus*)? Being cannot change, for change involves alteration, and alteration is patently irreconcilable with the supreme perfection of the first principle. Furthermore, if the *un'omnia* directly imparted its power and actuality to body, it would not only change itself but would also gradually force body, which is only acted on and cannot subsist by itself, into nonentity (and thus somehow mechanically preempt the will of him "who created the universe").[75] What we need is something to mediate between being and non-being, something to bridge the ontological gap, something that is at the same time immaterial enough to come into close contact with unchanging being without running the risk of progressive annihilation, and material enough to guarantee the continued existence of the bodily world by joining it together and continually preserving it from progressive annihilation.[76]

Conceptually, such an entity is most easily described as partaking of the distinctive features of both extremities. Like Proclus and many others before him, Patrizi now "proceeds from analysis to hypostatization":[77] if it is logically (or formally) necessary that there be a *tertium quid* between *a* and non-*a,* then it is also ontologically (or materially) necessary that this entity exist. It is, of course, the world soul, whose properties can best be described by joining the contradictory descriptions of being and non-being given in the two columns above with the (in)famous pet words of later Neoplatonists, ἅμα καί: it is unchangeable and changeable at the same time, but in a different respect; it is indivisible and divisible at the same time, but in a different respect; it is identical with itself and different from itself at the same time, but in a different respect; and so on. Patrizi does not make a great effort to explain how these contrary qualities and predicates mix and mingle,[78] but he takes great pains to stress over and over again that the single most conspicuous characteristic of the world soul is its double nature of simultaneous corporeality and incorporeality.[79] Thus, we have a third entity whose distinctive property it is to be a body without body. What does all this mean? Let me conclude by suggesting two sources for Patrizi's contentions, and one interpretation.

VI

By maintaining that space, light, and soul are bodies, albeit nonbodily ones, Patrizi has made three counterintuitive claims that go against the grain of much of the philosophical tradition on which he drew.[80] Are these claims blatant superstitions, held by a vainglorious man with a deranged mind, as Francis Bacon thought?[81] Or can we make sense of them, both historically and in themselves?

It is possible to pin down precisely the provenance of Patrizi's idea that space is body. Only one ancient philosopher held it: Proclus. His opinions on this problem are cited by Simplicius in his famous *Corollarium de loco,* which is part of his *Commentary on Aristotle's "Physics"* and which Patrizi repeatedly refers to.[82] Proclus here distinguishes implicitly between space and matter on the one hand and extension and bulk on the other; once one has understood that space is essentially extension and offers no resistance, and that matter is essentially bulk and offers resistance, the initially puzzling claim that space is body by virtue of being three-dimensional makes perfect sense.[83] In fact, what Proclus actually says is that space is "immaterial body" rather than "incorporeal body,"[84] and this is, I suspect, what Patrizi really meant by his own favorite oxymoron.

Patrizi got his notion of corporeal light again from the same source. Here are Proclus' own words, as quoted by Patrizi: "Let us imagine in our mind two spheres, the first filled with the one light and the second with many bodies, but both equal in size. Place the one in a fixed position, with the center being equally fixed, and put the other inside it: you will see the universe in place, moving within the immobile light. It is itself immobile as a whole, in imitation of place, but it moves in some of its parts, in order to be in that way inferior to place."[85] This seems to me to be the key text for our understanding of that strange consequence explicitly drawn by Proclus: namely, that if space is body by being corporeal, immobile, indivisible, and immaterial, and if light is body for exactly the same reasons, then it necessarily follows that "space must be light." This identity holds for Patrizi as well.[86] To those who might wish to object that there could still be empty space (i.e., dark space not filled with light), he would reply that darkness is just a privation of light,[87] or, in other words, that darkness presupposes light (in accordance with the sixth axiom above). In fact, "dark space" is a *contradictio in adjecto,* a logical *Unding* which it is impossible to conceive of.

If space is light, then it also makes perfect sense to represent it as a regular tetrahedron standing on one of its corners, as suggested above. Although the analogy should not be pressed too far, it is clear that the volume of space, and therefore the volume of divine light in any given slice (or hypostasis), diminishes according to the distance from the Principle. The movement of descent is thus not only from the spiritual realm to the bodily realm, but also—literally, and through geometrical necessity—from infinite space to minimal space, from light to darkness, and from life to death, all of which is well in accordance with a universe designed according to Neoplatonic principles.

I am alive to the fact that the tetragonal shape, like every finite representation of the infinite, is not without difficulties, especially when it has to
be reconciled with Patrizi's other contention that the infinite universe begins
just outside the visible heavens. It is, however, heartening to see that Patrizi
himself did not think that a universe in the shape of a four-sided pyramid was
an absurd idea or a physical impossibility, as we may be inclined to do: "The
astronomers and philosophers of all nations agree in saying that the heavens
are spherical. Nonetheless, there is no reason why a pyramid, a cube, an octahedron, a dodecahedron, or an icosahedron could not also revolve when
they are fixed on an axis."[88] If pressed, I would locate the visible heavens
within the hypostasis of Body, with which it is, however, not congruent. As
Patrizi specifies, the heavens are wrapped in "lumen supercoeleste," so that
"vniuersus corporeus mundus, et intus in lumine iacet [scil. in the *lumen
coeleste*], et extra [scil. in the *lumen supercoeleste*]." According to this interpretation, the heavens would bathe in the supercelestial light within the hypostasis of Body, whereas the remaining hypostases could be considered as
the *coelum empyreum*. But I cannot help feeling that all this is vain speculation
and frivolous subtlety.[89]

It remains for us to explain the relationship of the world soul to space
and light. Once again I think it can most easily be determined with reference
to Proclus. It is well-known that many Neoplatonic philosophers, taking up
the image of the "chariots" (ὀχήματα) in Plato's *Phaedrus* (247B), devised a
subtle spiritual entity (πνεῦμα) that would serve as a kind of "vehicle" for the
soul when it traveled to a body, and as a kind of "carriage" while their union
lasted.[90] Just like individual souls, so the world soul has its vehicle. Proclus
tells us what it is in his *Commentaries on Plato's "Republic"*: "Porphyry, that
most excellent philosopher, suspected what we are writing now, and assumed
that light was the first vehicle of the cosmic soul."[91] Even if I have so far been
unable to prove that Patrizi was familiar with this particular passage, he was
certainly familiar with the doctrine expressed there, for it is also referred to,
albeit somewhat less clearly, in the *Corollarium de tempore*.[92] In fact, Patrizi's
words are but an echo of Proclus' own: "Light is the vehicle of the celestial
forces, and the bond between the upper and the lower part of the universe."[93]
If the world soul travels with light, and if they both are bonds between the
upper and the lower part of the universe, then it is only natural for soul to
share light's first and foremost properties—viz., extension without resistance—for otherwise it would be sitting very uncomfortably indeed in its
carriage. It is therefore only consistent that in addition to space and light, Patrizi should have described soul as a third entity that is "bodiless body," for no

other attributes could have summed up its role more truly, succinctly, and accurately.

Patrizi's speculations on space, light, and soul, vague and implausible as they may seem to us, offer a picture of remarkable coherence and consistency. His physics and cosmology, built around the key notion of "bodiless body," hold a very definite, if rather solitary, place in the history of science. They can be traced back with philological accuracy to the fifth century C.E., and they resurface in the work of Patrizi just as if no progress had been made in the study of nature during the intervening millennium. Patrizi was, I think, the greatest *Proclian* physicist that ever lived after Proclus—and this is certainly no small claim to fame.

NOTES

This article was written during my tenure of a British Academy Postdoctoral Research Fellowship at the Warburg Institute, University of London. I wish to thank Andrew Wong for preparing figure 4.3, Antonio Clericuzio for bibliographical and Paul Nelles for stylistic advice, and Peter Kingsley and Jill Kraye for both. Above all my thanks go to Anthony Grafton; without his enthusiasm and continuing support, I would not have tried my hand at Patrizi at all. I alone am responsible for the remaining inconsistencies and incongruities.

Throughout, the abbreviation *NUP* stands for Francesco Patrizi, *Nova de universis philosophia, in qua Aristotelica methodo, non per motum, sed per lucem, et lumina, ad primam causam ascenditur* (Ferrara: Benedictus Mammarellus, 1591). The *NUP* consists of four parts called respectively *Panaugia, Panarchia, Pampsychia,* and *Pancosmia;* references will be made to part, book, folio, and column (thus "4, 32, 153 d" means "Pancosmia, book 32, fol. 153, col. d"). The text of the so-called second edition of the *NUP* (Venice: Roberto Meietti, 1593) is identical with that of the 1591 edition: see O. Guerrini, "Di Francesco Patrizi e della rarissima edizione della sua 'Nova Philosophia,'" *Il Propugnatore* 12 (1879): 172–230, and A. Antonacci, *Ricerche sul neoplatonismo del Rinascimento: Francesco Patrizi da Cherso,* vol. 1, *La redazione delle opere filosofiche: Analisi del primo tomo delle "Discussiones"* (Bari: Editrice Salentina, 1984), pp. 108–109 n. 22.

1. *NUP,* sig. a 2v: "Quinque hoc volumine, pias omnes, omnes Catholicae fidei consonas, Gregori Pater Beatiss. tibi afferimus philosophias. Nostram recens conditam [i.e., the *NUP*], Chaldaicam Zoroastri [i.e., the *Chaldean Oracles* as collected by Plethon, plus a few fragments culled from later Neoplatonists by Patrizi himself], Hermetis Trismegisti Aegyptiam [i.e., the Greek text with a new translation of the fourteen tractates contained in Ficino's *Pimander;* a number of extracts from Cyril and Stobaeus, including the *Korē Kosmou;* the Latin *Asclepius;* and the *Definitiones Asclepii*], Aegyptiam aliam Mysticam [i.e., the *Theologia Aristotelis* in Pier Nicola Castellani's translation published in Rome in 1519, or Plato's 'mystic' teaching], et aliam Platonis propriam [i.e., Plato's 'exoteric' teaching as found in his dialogues, of which Patrizi suggests a reading order based on Neoplatonic authorities]. A nobis sane non minimo labore, e ruinis vix erutas, in unum collatas, atque illustratas, et in ordines suos scientificos distinctas."

2. I.e., Niccolò Sfondrato (1535–1591), whom Patrizi had met when they were both students at Padua.

3. The accounts of Patrizi's life rely mainly upon remarks scattered throughout his works and one autobiographical letter, dated 12 January 1587 and addressed to Baccio Valori. The text of this letter, which is preserved in the Fondo Rinuccini (Filza 27, Scatola II) of the Biblioteca Nazionale Centrale in Florence, was first published by A. Solerti, "Lettere autobiografiche di F. Patrizi di Cherso, erudito del secolo XVI," *Archivio storico per Trieste, l'Istria e il Trentino* 3 (1886): 275–281 (I have not seen this publication). A facsimile edition of the text published by Solerti can be found in V. Premeč, *Franciskus Patricijus* (Belgrade: Institut Drustvenih Nauka, 1968), pp. [100–104]. It is also printed in the not very useful volume edited by S. Cella, *Francesco Patrizi da Cherso: Pagine scelte* (Padua: Liviana Editrice, 1965), pp. 37–42, and in D. Aguzzi Barbagli, ed., *Francesco Patrizi da Cherso: Lettere ed opuscoli inediti* (Florence: Olschki, 1975), pp. 45–51. Among the modern accounts of Patrizi's life, the best are E. Jacobs, "Francesco Patricio und seine Sammlung griechischer Handschriften in der Bibliothek des Escorial," *Zentralblatt für Bibliothekswesen* 25 (1908): 19–47, esp. pp. 20–28; P. Donazzolo, "Francesco Patrizi da Cherso erudito del secolo decimosesto (1529–1597)," *Atti e memorie della Società istriana di archeologia e storia patria* 28 (1912): 1–147, esp. 7–47; and B. Brickman, "An Introduction to Francesco Patrizi's *Nova de universis philosophia*" (Ph.D. diss., Columbia University, 1941), pp. 10–20. On his time in Ferrara (1577/1578–1592) and his relationship with the Este family, see M. J. Wilmott, "Francesco Patrizi da Cherso's Humanist Critique of Aristotle" (Ph.D. diss., University of London, 1984), pp. 8–13, and C. Vasoli, "Un filosofo tra lo studio e la corte: Patrizi a Ferrara," in his *Francesco Patrizi da Cherso,* Humanistica 5 (Rome: Bulzoni, 1989), pp. 205–228.

4. *NUP,* sig. a 2v: "Ratione sola, ratio humana ducitur, Rationem ratio, libens sequitur. A ratione, ratio volens nolens etiam trahitur. Ratione igitur, sunt homines ad Deum ducendi. In hanc ergo ueram ac diuinam philosophiam, ratione sola philosophando, totis uiribus incubui. Et ingenti, sed obstinatissimo labore, ad finem, eam mihi uideor perduxisse." Sig. a 3r: "Iube ergo pater sanctissime tu primus, iubeant futuri Pontifices omnes . . . per omnia tuae ditionis gymnasia, per omnes Coenobiorum, Scholas, librorum [scil. a Platonicis philosophis scriptorum], aliquos continue exponi. . . . Cura ut christiani orbis principes, idem in suis iubeant gymnasiis."

5. See *NUP,* sig. a 3v.

6. This story has repeatedly been told; for a full survey of earlier literature, see A. L. Puliafito Bleuel, *Francesco Patrizi da Cherso, Nova de universis philosophia: Materiali per un' edizione emendata,* Quaderni di "Rinascimento" 16 (Florence: Olschki, 1993), pp. xix–xxvi, and for the objections raised against the *NUP* by the Congregation of the Index the text edited ibid., pp. xxx–xxxviii. The first *Index librorum prohibitorum* no longer to mention the *NUP* was the one edited in 1900. As the preface specifies, however, this was due not to a reappraisal of Patrizi's merits but to a mere stroke of chronological luck, for "omnes libri ante annum MDC prohibiti, abhinc ex Indice expuncti declarantur, quamvis etiamnum eodem modo damnati habendi sint, quo olim damnati fuerunt" (p. xiii).

7. It is likely that Patrizi understood the adjective *scientificus* in its original meaning of "producing knowledge" (on which see E. Benveniste, "Genèse du terme 'scientifique,'" in his *Problèmes de linguistique générale* [Paris: Gallimard, 1974], 2:247–253).

The best general descriptions of Patrizi's philosophy (taking into account both the *Discussionum peripateticarum libri IV* [Basel: Pernea Lecythus, 1581] and the *NUP*) are those by P. O. Kristeller, *Eight Philosophers of the Italian Renaissance* (Stanford: Stanford University Press, 1964), pp. 110–126, and by B. P. Copenhaver and C. B. Schmitt, *Renaissance Philosophy,* vol. 3 of *A History of Western Philosophy* (Oxford: Oxford University Press, 1992), pp. 187–195. By far the best analysis of the *NUP* is the one given by Brickman, "Introduction," pp. 21–75, but T. A. Rixner and T. Siber, *Leben und Lehrmeinungen berühmter Physiker am Ende des XVI. und am Anfange des XVII. Jahrhunderts, als Beyträge zur Geschichte der Physiologie in engerer und weiterer Bedeutung,* vol. 4, *Franciscus Patritius* (Sulzbach: J. E. von Seidel Kunst- und Buchhandlung, 1823), pp. 21–132, is still worth reading. For a brief account of Patrizi's debts in the *NUP* to the writings of Ficino, see M. Muccillo, "Marsilio Ficino e Francesco Patrizi da Cherso," in *Marsilio Ficino e il ritorno di Platone: Studi e documenti,* ed. G. C. Garfagnini, Studi e testi 15 (Florence: Olschki, 1986), 2:615–679, esp. 664–679. Patrizi's reliance on Damascius is well illustrated by T. Leinkauf, *Il neoplatonismo di Francesco Patrizi come presupposto della sua critica ad Aristotele,* Symbolon 9 (Florence: La nuova Italia editrice, 1990).

8. For the difficulties of such an undertaking, see the *Emendationes* edited by Puliafito Bleuel, pp. 7–97.

9. *NUP* 1, 1, 1 a: "Franciscus Patricius, Nouam, Veram, Integram, de vniuersis conditurus Philosophiam, sequentia, vti verissima, pronunciare est ausus. Pronunciata, ordine persecutus, Diuinis oraculis, Geometricis necessitatibus, Philosophicis rationibus, clarissimisque experimentis comprobauit."

10. *NUP* 4, 1, 61 a: "Quid autem illud fuit, quod summus opifex primum omnium extra se produxit? Quid aut debuit, aut expedijt prius produci, quam id quo omnia alia, vt essent eguerunt, et sine quo esse non poterunt?" See also 65 d.

11. *NUP* 2, 11, 22 c: "Effectus enim, a causa quidem semper uenit," and 1, 3, 5 c: "Omne enim producens, praestantius est producto" (this translates Proclus, *Inst. theol.* 7).

12. *NUP* 4, 2, 68 b: "Patuit quoque, continuum sui natura, omni diuisione antiquius ac prius esse: cuius diuisio, ac desectio, humanae cogitationis vi facta, numerum procreasse."

13. *NUP,* 2, 11, 22 d: "Ergo fas nullo modo est, ut unum, idem sit cum multitudine, quae illius proles est, et effectus" (this translates Proclus, *Inst. theol.* 5).

14. *NUP* 2, 11, 22 d–23 a: "Oportet autem omnem entium progressionem esse continuatam. Id autem fit per hyparxeon coniunctionem: Quae fit per inferioris participationem a superiore"; see also 3, 1, 49 b: "Natura enim in suis operibus non saltat, sed ordine, a proxima causa, proximum producit effectum." This is what A. O. Lovejoy, *The Great Chain of Being* (1936; reprint, Cambridge, Mass.: Harvard University Press, 1982), p. 52, called "the principle of plenitude"; for a witty, and utterly destructive, criticism of this so-called principle, see L. Vax, "Splendeur et déclin du merveilleux philosophique," in *Du banal au merveilleux: Mélanges offerts à Lucien Jerphagnon,* Les Cahiers de Fontenay 55–57 (Fontenay/St. Cloud: Ecole Normale Supérieure, 1989), pp. 275–314, esp. pp. 309–310.

15. *NUP* 3, 2, 51 b: "Contrario namque vno in rebus posito, poni necesse est et alterum"; 3, 5, 58 d: "Vno contrario in natura posito, poni est necesse, et alterum." This axiom does,

however, admit of exceptions when convenient: "Verum vt plurimum esse, duo contraria in rerum vniuersitatis generibus reperiri. Non tamen in omni" (ibid.). It should be noted that logical contraries are not the same thing as correlative modalities, which can indeed only exist as pairs (see on this R. Sorabji, *Matter, Space, and Motion: Theories in Antiquity and Their Sequel* [London: Duckworthy, 1988], p. 134).

16. *NUP* 2, 11, 22 c: "Vere sciamus, entia antequam fierent, in Deo fuisse omnia, et ex ipso, et per ipsum omnia esse facta"; 2, 11, 23 c: "Sub patre ergo Deo, et a patre, Paternum est profundum. In quo vnitas primaria, et in ea omnes vnitates, quas nomine alio ideas appellamus"; cf. with this *Chaldean Oracles*, frag. 18 des Places (*Oracles chaldaïques*, ed. and trans. E. des Places [Paris: Les Belles Lettres, 1971]). It is likely that Patrizi derived his concept of πατρικὸς βυθός in the first place from Proclus apud Simpl., *In Phys.* 614, 6 Diels (Simplicius, *In Aristotelis Physicorum libros quattuor priores commentaria*, ed. H. Diels, Commentaria in Aristotelem Graeca 9 [Berlin: Reimer, 1882]) (on which see below). See also *NUP* 2, 11, 22 c: "Primissimum illud, un'omnia est appellatum."

17. See Proclus, *Inst. theol.* 116; Patrizi's translation is *Procli Lycii Diadochi, Platonici philosophi eminentissimi: Elementa Theologica, et physica. Opus omni admiratione prosequendum. Quae Franciscus Patricius de Graecis, fecit Latina* (Ferrara: Dominicus Mammarellus, 1583).

18. *NUP* 2, 11, 23 a–b: "Et quoniam in vn'omnia, vnum primas tenebat, omnia, secundas: necessario in hac productione, vni ipsi, vnitas aliqua primaria debuit respondere. Vnitates vero reliquae responderent ipsi vn'omnia. Vnitas ergo quaedam primaria, ab vno est genita; Reliquae vnitates hanc, vt secundariae sequuntur, et sunt ab ea proximae"; and 23 c: "Et ipsa vnitas primaria Idea boni."

19. On the superessential, see *NUP* 2, 11, 23 b: "Vel etiam sunt ipsae vnitates, super entia omnia, et superessentiales." Patrizi uses the words *mens* and *intellectus* interchangeably: see *NUP* 2, 11, 23 c: "sapientes ueteres communi consensu [m]entem, et intellectum appellauere," and 3, 1, 49 c: "omnes intellectus, siue dixeris Mentes." The triad ὄν–ζωή–νοῦς, so often referred to in the writings of Proclus, goes back to Plato, *Sophist* 248A–249E, where the παντελῶς ὄν is said also to possess ζωή and νοῦς. P. Merlan (who did not think very highly of Proclus) maintained that it was merely "eine zufällig aufgeraffte Dreiheit von Begriffen . . . , deren innere Zugehörigkeit nie gezeigt wird" (see his review of *Proklos: Grundzüge seiner Metaphysik*, by W. Beierwaltes, *Philosophische Rundschau* 15 [1968]: 94–97, esp. 96). Much the same can be said of its presence in the *NUP*; see the remarks following in the text.

20. *NUP* 2, 11, 23 c: "Quoniam vero, a perfectissima venit, ipsa quoque suo gradu erit perfectissima [scil. essentia]"; "Ergo diuina illa essentia, uiuens est, et uita fruitur, optima, et sufficientissima."

21. *NUP* 2, 11, 23 c: "Per motionem autem hanc suam, cognitionem in se produxit." See on this S. Gersh, *Kinesis akinetos: A Study of Spiritual Motion in the Philosophy of Proclus*, Philosophia antiqua 26 (Leiden: Brill, 1973). (For Proclus, the "cognitive" activity of the noetic world is at the same time a "creative" activity; the same does not seem to be true for Patrizi.)

22. *NUP* 2, 11, 23 c: "Vel enim sursum, vel in se, vel deorsum motum hunc tendere necesse est." Patrizi says that the reflexive movement "procreates a second intellect" (ibid.:

"Dum uero se intelligit, et in superas suas causas reuoluitur, secundum procreat intellectum"), but this statement is muddled by his apparent confusion between *intellectus secundus, spiritus,* and *mens secunda,* which follows.

23. *NUP* 2, 11, 23, d: "Simile quiddam sibi anima, quae iam in corpus decidit, et ei [scil. intellectui] est coniuncta, in corpus profundit. Quam nos, nomine satis noto, sed significatu satis ignoto, naturam appellamus. . . . A qua [scil. natura] in corpus itidem producitur qualitas, quae naturae veluti instrumentum seruit, in corpore, et eius partibus, alterandis, et disponendis. Et dispositis formam inducit. Per quam corpus in aliam, aut aliam speciem conformetur. Sed et qualitas a natura fert similitudinem, et ei est dissimilis. Et formas easdem fert a qualitate. Et corpus a forma, ijs nimirum modis, qui proprijs horum graduum tractatibus explicabuntur."

24. *NUP* 2, 11, 24 b: "[I]lla omnia . . . aptata sunt, et ordine disposita. . . . Ut . . . omnia essent, quasi Iacobi scala, a coelis ad terram protensa, nouem gradibus disposita, per quos, Angeli, et nuncij Dei, descendendo, sapientiam, et gloriam Dei, ad nos deferrent, et ascendendo, piorum merita, et impiorum demerita ad conspectum Dei referrent." The reference is to Genesis 28:12, a text often adduced to intimate the coherence of the universe: see, e.g., A. Altmann, "The Ladder of Ascension," in *Studies in Mysticism and Religion Presented to G. G. Scholem on His 70th Birthday* (Jerusalem: Magnes Press, 1967), pp. 1–32 (reprinted in idem, *Studies in Religious Philosophy and Mysticism* [Ithaca, N.Y.: Cornell University Press, 1969], pp. 41–72); M. Idel, "The Ladder of Ascension—The Reverberations of a Medieval Motif in the Renaissance," in *Studies in Medieval Jewish History and Literature,* ed. I. Twersky, vol. 2 (Cambridge, Mass.: Harvard University Press, 1984), pp. 83–93. On "degrees," see *NUP* 2, 11, 23 d: "His autem post vnum primum, nouem gradibus, rerum tota constat vniuersitas. Qui quidem gradus ordine sunt dispositi a summo ad imum ita, vt nullum inter eos vacuum sit relictum"; and 24 b: "Gradus hi nouem, sunt primus rerum atque entium ordo, in profundum, a summo ad imum ductus."

25. *NUP* 2, 11, 24 a: "Pythagoras, quando tetractym nominabat, et per eam, sacrosancto ac summo omnium iurabat iuramento."

26. *NUP* 2, 11, 24 b: "Est alius ordo in latitudinem actus, in singulo quoque gradu. . . . Quem ordinem sicut primum illum, gradus nominauimus, sic seriem proprio nomine appellabimus. In qua serie vnitatum in latitudinem, graduum singulorum latitudinem ratio persuadet esse locatam." Strictly speaking, there are only eight series, for the last, starting from body, is reflexive and therefore not a "ramification" *sensu stricto.*

27. For a similar, though not identical structure, see Proclus, *The Elements of Theology,* ed. E. R. Dodds, 2nd ed. (Oxford: Oxford University Press, 1963), p. 255, commentary on props. 108 and 109.

28. *NUP* 2, 11, 24 c: "Tertius itidem in longitudinem est ordo, quem cathenam libuit nuncupare. Qui quidem est eiusdem generis, seu speciei, per gradus omnes transitus."

29. *NUP* 2, 11, 24 c: "Est lucis vnitas, et vnitatis lux. Est, et essentia lucis, et lux essentiae. Et vita lucis, et lux vitae, et Mens lucis, et lux mentis," etc. For possible interpretations of the *lux corporis,* see *NUP* 1, 4 and 1, 5.

30. *NUP* 2, 11, 24 c: "Ex tribus enim illis longitudine, latitudine, et profunditate, sicuti corporum nascitur soliditas, et stabilitas, et firmitudo." Puliafito Bleuel, *Nova de universis*

philosophia, p. lvi n. 205, mistakenly calls this "il modello delle quattro gerarchie sovvraposte."

31. Plato, *Timaeus* 56A. On the "fiery" (πυροειδές) and "nimble" (κινητικόν) nature of the pyramid, see also Plutarch, *De defectu oraculorum* 34 (428D).

32. Cyril of Alexandria, *Contra Iulianum* i.46; [Iamblichus], *Theologumena arithmeticae,* p. 20, 9–10 de Falco (ed. V. de Falco, rev. U. Klein [Stuttgart: B. G. Teubner, 1975]). See on this A.-J. Festugière, "La pyramide hermétique," *Museum Helveticum* 6 (1949): 211–215 (reprinted in idem, *Hermétisme et mystique païenne* [Paris: Aubier-Montaigne, 1967], pp. 131–137), and the additional remarks by P. Merlan, "Die Hermetische Pyramide und Sextus," *Museum Helveticum* 8 (1951): 100–105 (reprinted in idem, *Kleine philosophische Schriften,* ed. F. Merlan, Collectanea 20 [Hildesheim: Olms, 1976], pp. 346–351).

33. [Iamblichus], *Theologumena arithmeticae,* p. 22, 10–13 de Falco.

34. For the Greek text, see *Corpus Hermeticum,* ed. A. D. Nock, trans. A.-J. Festugière (Paris: Les Belles Lettres, 1954–1960), 4:133, fr. 28; Patrizi's translation of the sentence in question reads "Pyramis ergo subiecta naturae, et intellectuali mundo" (fol. 50v).

35. See Jacobs, "Francesco Patricio," p. 36 (on the Scorialensis Σ III. 1).

36. Given Patrizi's contentions for the isotropy of space (discussed later), the orientation of the pyramid is immaterial. For his rejection of the sphericity of the heavens see *NUP* 4, 10, esp. fol. 87 d: "Caelum igitur nostris demonstrationibus nullo modo aut est, aut dici potest sphaericum," and later discussion in text.

37. *NUP* 2, 11, 24 c–d: "Sic ratio . . . suadet, in rerum omnium vniuersitate, esse firmitatem, et stabilitatem, et soliditatem longe quam sit in corporibus, ualidiorem."

38. The foregoing attempt at reconstructing the structure of Patrizi's universe has not taken into account the slightly different description of it given in *NUP* 2, 1, which seems to belong to a different period in the redaction of the work; see Brickman, "Introduction," pp. 25, 33, with n. 14.

39. Fols. 61 a–65 d, 66 a–68 d, and 69 a–73 b, respectively.

40. *Franc. Patricii Philosophiae, De rerum natura, Libri II. priores. Alter de Spacio Physico, Alter de Spacio Mathematico* (Ferrara: Vittorio Baldini, 1587). Textual identity starts on fol. 2v, line 4 (= *NUP* 4, 1, 61 c, line 19: "Communis quaedam omnium hominum notitia . . .") and extends to the end of fol. 26v (= *NUP* 4, 2, 68 d), where the last sentence of the 1587 ed. reads: "Hosce autem libros, sequantur ij, quos Italice, de Nova Geometria edidimus"; *NUP* simply has "quos de Noua Geometria adiungemus."

41. *Della nuova geometria di Franc. Patrici Libri XV. Ne' quali con mirabile ordine, e con dimostrazioni à marauiglia più facili, e più forti delle usate si vede che le Matematiche per via Regia, e più piana che da gli antichi fatto non si è, si possono trattare* (Ferrara: Vittorio Baldini, 1587). The Latin translation of this in *NUP* 4, 3 differs slightly in the way in which it presents the material, but there is no difference in substance. For the history of doctrines on space in general, see M. Jammer, *Das Problem des Raumes: Die Entwicklung der Raumtheorien*

(Darmstadt: Wissenschaftliche Buchgesellschaft, 1960), esp. pp. 92–93. On Patrizi in particular, see J. Henry, "Francesco Patrizi da Cherso's Concept of Space and Its Later Influence," *Annals of Science* 36 (1979): 549–573, and E. Grant, *Much Ado about Nothing: Theories of Space and Vacuum from the Middle Ages to the Scientific Revolution* (Cambridge: Cambridge University Press, 1981), pp. 199–206.

42. This statement is qualified later in the text, with note 56.

43. See *De rerum natura,* fol. 2r: "Methodo, placitisque novam, rebus veterrimam, sensui, rationi, sibique consonam condere studemus, naturae philosophiam." See also *NUP* 1, 1, 1 b: "A cognitis . . . initium sumendum. Cognitio omnis, a mente primam originem: a sensibus exordium habet primum"; and 4, 1, 61 b: "sensuumque testimoniis, rationumque probationibus vtamur."

44. The following account is based on *NUP* 4, 1, which was translated (with a few lines of 4, 2) into English by B. Brickman, "On Physical Space," *Journal of the History of Ideas* 4 (1943): 224–245. There is a German translation of a few paragraphs taken from *NUP* 1, 1; 4, 1; and 2, 20, in M. Fierz, "Über den Ursprung und die Bedeutung der Lehre Isaac Newtons vom absoluten Raum," *Gesnerus* 11 (1954): 62–120, esp. 106–113 (on Patrizi, see above all pp. 79–83); an Italian translation, also of a few paragraphs only, can be found in Cella, *Francesco Patrizi,* pp. 116–129.

45. G. Berkeley, *An Essay towards a New Theory of Vision* (1709), in *Philosophical Works: Including the Works on Vision,* ed. M. R. Ayers, rev. ed. (London: Dent; Totowa, N.J.: Rowman and Littlefield, 1975), §126, p. 45.

46. Contemporary logicians generally distinguish between four meanings of the verb "to be": it can be equivalent to "∃" in "God is"; to "=" in "Elizabeth II is the queen of England"; to "∈" in "Elizabeth II is a woman"; and to "⊂" in "The woman is a human being."

47. *NUP* 4, 1, 61 a: "Quid enim illi, aliud est locus, quam spacium, longum, latumque?" See Aristotle, *Physics* 4.4, 210b34–211a1: ἀξιοῦμεν δὴ τὸν τόπον εἶναι πρῶτον μὲν περιέχον ἐκεῖνο οὗ τόπος ἐστί; 212a6: τὸ πέρας τοῦ περιέχοντος σώματος. On the various other definitions of space put forward by Aristotle, see Sorabji, *Matter, Space, and Motion,* pp. 186–201.

48. Patrizi uses four different words to express this notion: *anteresis, antitypia, renitentia,* and *resistentia* (*NUP* 4, 1, 62 d).

49. See *Stoicorum veterum fragmenta,* ed. H. Von Arnim, vol. 2, *Chrysippi fragmenta logica et physica* (Leipzig: Teubner, 1923), nos. 463–481.

50. See Aristotle, *De caelo* 1.1, 268b1: οὐκ ἔστιν εἰς ἄλλο γένος μετάβασις. This is not contradicted by the statement on fol. 65 d that finite space ("finitum quod in mundo est," on which see discussion later) has the power "corpora omnia penetrandi," for Patrizi specifies that it only penetrates the bodies' empty parts ("propriis illorum spaciis"), which are nonbodily *per definitionem.* I therefore disagree with the account of the problem given by E. Grant, "The Principle of the Impenetrability of Bodies in the History of Concepts of Separate Space from the Middle Ages to the Seventeenth Century," *Isis*

69 (1978): 551–571, esp. 569, who fails to see that space penetrates not body but the *vacua,* and strangely ignores his own correct hint put forward a few years earlier in his "Place and Space in Medieval Physical Thought," in *Motion and Time, Space and Matter: Interrelations in the History of Philosophy and Science,* ed. P. K. Machamer and R. G. Turnbull (Columbus: Ohio State University Press, 1976), pp. 137–167, esp. 159–161 and 166 n. 105: "Patrizi's 'yielding' (*cessio*) is simply an aspect of 'penetration' (*penetratio*)." In modern terminology we might perhaps say that material bodies are semipermeable to space.

51. Patrizi was convinced that any given quantity of water can be compressed to half its size (*NUP* 4, 1, 63 a), which shows that despite his invoking *experientia,* he made little effort to gain any empirical knowledge of the world. The question whether water could be compressed was, however, hotly debated in Patrizi's days; it was empirically refuted only in the middle of the seventeenth century when members of the Florentine Accademia del Cimento, "the first organization founded for the sole purpose of making scientific experiments," concluded that "a power not only thirty but a hundred and perhaps a thousand times that needed to reduce a volume of air into a space thirty times less than it first occupied, does not compress a volume of water even by as much as a hair or other lesser observable space, below what its natural extension requires": see W. E. K. Middleton, *The Experimenters: A Study of the Accademia del Cimento* (Baltimore: Johns Hopkins University Press, 1971); quotations on pp. 1, 217.

52. C. B. Schmitt, "Experimental Evidence for and against a Void: The Sixteenth-Century Arguments," *Isis* 58 (1967): 352–366, esp. 356, 363 (reprinted in idem, *Studies in Renaissance Philosophy and Science* [London; Variorum Reprints, 1981], no. VII). On these *vacua,* see *NUP* 4, 1, 63 a: "Necessario ergo . . . relinquitur, vt in spaciola vacua sibi interspersa sese receperit [scil. aqua], atque ita sit densior effecta."

53. See, e.g., C. G. Hempel, *Philosophy of Natural Science* (Englewood Cliffs, N.J.: Prentice-Hall, 1966), pp. 25–28.

54. It is surprising that A. Koyré, in his standard work *From the Closed World to the Infinite Universe* (Baltimore: Johns Hopkins University Press, 1957), completely ignores Patrizi's role in the history of this problem.

55. Aristotle, *De caelo* 1.9, 279a12–13. The theory of "a finite world immersed in an infinite space" was later rejected by Kepler on grounds very similar to those advanced by Aristotle: see Koyré, *Closed World to Infinite Universe,* pp. 86–87.

56. Compare *NUP* 4, 1, 63 d: "Si in vapores, vel in fumum resoluatur, tum centies millies, et forte amplius, maiorem occupabit locum [scil. mundus]," with Cleomedes, *Caelestia* (formerly called *De motu circulari*) 1.1, p. 2, 43–45 Todd (ed. R. B. Todd [Leipzig: B. G. Teubner, 1990]): εἰ δὲ καὶ εἰς πῦρ ἀναλύεται ἡ πᾶσα οὐσία [scil. ἡ τοῦ κόσμου] . . . , ἀνάγκη πλέον ἢ μυριοπλασίονα τόπον αὐτὴν καταλαμβάνειν; and see R. B. Todd, "A Note on Francesco Patrizi's Use of Cleomedes," *Annals of Science* 39 (1982): 311–314. Cleomedes, slightly more modest in his calculation, thought that the universe would only occupy a place at least ten thousand times greater than it does now. For Patrizi's estimate of the current size of the universe, see *NUP* 1, 8, 18 b–d.

57. *NUP* 4, 1, 65 b: "Spacium ergo extensio est hypostatica, per se substans, nulli inhaerens."

58. See on this Sorabji, *Matter, Space, and Motion,* pp. 28–29 (improving upon his "John Philoponus," in *Philoponus and the Rejection of Aristotelian Science,* ed. Sorabji [London: Duckworth, 1987], pp. 1–40, esp. 23–24).

59. *NUP* 4, 1, 65 b: "Itaque corpus incorporeum est [scil. spacium], et noncorpus corporeum. Atque vtrumque per se substans, per se existens, adeo vt etiam per se stet semper, atque in se stet: neque vnquam, neque vsquam moueatur, neque essentiam; neque locum mutet, nec partibus, nec toto." See also (e.g.) 4, 2, 68 b; 4, 23, 122 a.

60. *NUP* 4, 3, 69 ab: "[O]stendimus [.] punctum id esse, quod in spacio sit minimum. . . . Atque ita quod Euclides vti principium supposuerat, punctum partes nullas esse [= Euclid, *Elements* 1, def. 1], vti principiatum est demonstratum."

61. For some preliminary observations on Patrizi's geometry, see E. Cassirer, *Das Erkenntnisproblem in der Philosophie und Wissenschaft der neueren Zeit,* 2nd ed. (Berlin: Bruno Cassirer, 1911), 1:260–267. The discussion by H. Védrine, "L'obstacle réaliste en mathématiques chez deux philosophes du XVIe siècle: Bruno et Patrizi," *Platon et Aristote à la Renaissance: 16e Colloque international de Tours,* De Pétrarque à Descartes 32 (Paris: Vrin, 1976), pp. 239–248, is superficial and inadequate.

62. See *NUP* 1, 1; 1, 10; and 4, 4 (on which this account is mainly based). It is beyond the scope of this essay to give a general account of the history of "light metaphysics"; in any case, such comprehensive studies as I have been able to consult contained nothing illuminating about Patrizi. For the general history of the doctrines of light as a *physical* phenomenon (as opposed to a *metaphor*), see V. Ronchi, *The Nature of Light: An Historical Survey* (London: Heinemann, 1970). On Patrizi in particular, see E. E. Maechling, "Light Metaphysics in the Natural Philosophy of Francesco Patrizi da Cherso (1529–1597)" (M.Phil. thesis, University of London, 1977). I have not seen A. A. Spedicati, "Sulla teoria della luce in F. Patrizi," *Bollettino di storia della filosofia dell'Università degli Studi di Lecce* 5 (1977): 244–263 (mentioned by Muccillo, "Ficino e Patrizi," p. 668 n. 155). For a preliminary survey of the sources of the *Panaugia,* see Muccillo, pp. 665–670.

Patrizi distinguishes (as others had before him) between *lux* ("the source of light"; see 1, 1, 1 c: "Dei ipsius eiusque bonitatis imago"), *lumen* ("luminosity"; ibid.: "primaria eius [scil. lucis] proles"), and *radii* ("rays," which are also said to be he first offspring of *lux,* on which account *lumen* would be the second; 1, 3, 5 b: "Radius . . . quasi lux secunda. . . . Lumen autem est lux tertia, tum a prima, tum a secunda emanans"). For the purposes of this article, the places of *lumen* and *radii* in respect to *lux* are of no importance; "light" is indiscriminately used to translate *lux* and *lumen.*

63. *NUP* 4, 4, 73 d: "Verissimeque dictum; Deum, lucem inhabitare inaccessabilem [*sic*!], idest in seipso qui est inaccessibilis non solum, postquam eam iussit in mundo fieri, sed perpetuis antea seculis, et tota eius sempiternitate."

64. *NUP* 4, 4, 73 c: "Lux autem e se lumen non emittere, non potest"; see also 74 b.

65. Patrizi himself explicitly states in the same chapter "cum Dionysio" that God is "ὑπερούσιος, superessentialis" (74 a): see, e.g., ps.-Dionysius, *De divinis nominibus* 1.1, ed.

B. R. Suchla, Patristische Texte und Studien 33 (Berlin: De Gruyter, 1990), p. 108, l. 7; p. 109, ll. 11.13, etc.

66. *NUP* 4, 4, 73 c: "Quid autem facilius fundi potuit per spacium, quam lumen?" All the quotations in the remainder of this paragraph are from chapter 4, 4. For light as bodiless body, see also 1, 1, 2 d and 3 a; 1, 4, 10 a; 1, 8, 12 a; 4, 4, 74 b.

67. *NUP* 1, 1, 1 c; 4, 4, 75 c.

68. See above all *NUP* 1, 5–9. The distinction between different kinds of light raises the question of its homogeneity, but it seems that Patrizi introduced the distinction in the opening chapters of the *NUP* only to forget it later. Since it appears to play no role in the further development of the work, it does not require an extensive treatment here.

69. For a comprehensive genealogy of the succession of philosophers, see *NUP* 2, 9, 19 a–20 b, and for the general context the rather too superficial remarks by C. Vasoli, "L'idea della *prisca Sapientia* in Francesco Patrizi," in *Roma e l'antico nell'arte e nella cultura del Cinquecento,* ed. M. Fagiolo, Biblioteca Internazionale di Cultura (Rome: Istituto della Enciclopedia Italiana, 1985), pp. 41–56, esp. 49–51, with the literature quoted pp. 53–54 n. 2.

70. Patrizi's cautionary words (*NUP* 3, 4, 54 a: "si vere Plutarchus scripsit") show what an unusually meticulous historical scholar he was for his days. In fact, the view he ascribes to Democritus on the strength of a doxographical account by "Plutarch" is the result of a garbled reading of a few words in the Ps.-Plutarch, *De placitis* 1.7, p. 67, 7–8 Mau (*Placita philosophorum,* ed. J. Mau, in *Plutarchi Moralia,* ed. K. Ziegler et al., vol. 5, fasc. 2.1 [Leipzig: B. G. Teubner, 1971]) = p. 87 Lachenaud (*Opinions des philosophes,* ed. G. Lachenaud, in Plutarque, *Oeuvres morales,* tome 12, 2e partie [Paris: Les Belles Lettres, 1993]): see the apparatus *ad loc.* and *Die Fragmente der Vorsokratiker,* ed. H. Diels and W. Kranz, 13th ed. (Berlin: Weidmannsche Buchhandlung, 1968), vol. 2, no. 68 [55], A 74, p. 102, 14.

71. *NUP* 3, 4, 54 a and 55 b.

72. The first departure from tradition is when Patrizi says (*NUP* 3, 1, 49 b) that he intends to use the word *anima* to signify the human soul only, and the word *animus* to signify all other kinds of soul (e.g., "mundi, coeli, sphaerarum, siderum, elementorum, brutorum, stirpium, et si qui sunt alii": 50 b). Such a distinction is, however, not adhered to outside the *Pampsychia,* where Patrizi favors the far more common word *anima* for "soul" throughout. This may be another hint that the work lacks the author's *ultima manus.*

73. B. de Spinoza, *Ethica ordine geometrico demonstrata* (1677), in vol. 2 of *Opera,* ed. C. Gebhardt (Heidelberg: Winter, 1925), pars 2, prop. 43 (scholium).

74. For a fuller but even more repetitive list, see *NUP* 3, 2, 51 c.

75. *NUP* 3, 2, 51 c: "[M]undus conseruetur, vniuersitas rerum constet, quousque placitum illi fuerit, qui vniuersa condidit." The exact relationship between God and the *un'omnia* is nowhere made explicit.

76. *NUP* 3, 4, 56 d.

77. The expression is Dodds's, in Proclus, *Elements of Theology,* p. 247.

78. All we are told is that they do so "mirando quodam modo" (*NUP* 3, 2, 51 d), which is how philosophers usually express the feeling that there is "something, . . . they know not what" (J. Locke, *An Essay concerning Human Understanding* [1690], ed. P. H. Nidditch, 2nd ed. [Oxford: Clarendon Press, 1979], p. 296 [= book 2, chap. 23, §2]).

79. *NUP* 3, 2, 51 d: "[T]ertia quaedam in vniuersitate erit natura, non corporea, non incorporea. sed vtrumque et incorporea, et corporea, ita vt media quaedam sit inter vtramque. Incorporeo suo, ab incorporea pendens. Corporeo uero ad corpus vergens"; see also 4, 2, 52 a; 4, 4, 56 c; and many other passages. The logic of graduated continuity would require a sequence *incorporeum–incorporeum corporeum–corporeum incorporeum–corporeum,* which is in fact mentioned by Patrizi at 3, 2, 51 a. In practice, however, he conflates the two middle terms and does not distinguish between *incorporeum corporeum* and *corporeum incorporeum.*

80. Plotinus, for example, unambiguously stated that light and space are not bodies: *Ennead* 1, 6 (1), 3, 18; 2, 1 (40), 7, 27–28; 4, 5 (29), 7, 41–42; 6, 3 (44), 5, 29–30.

81. F. Bacon, *Descriptio globi intellectualis,* ed. J. Spedding, in vol. 3 of *The Works of Francis Bacon,* ed. J. Spedding, R. L. Ellis, and D. D. Heath (London: Longman, 1857; 2nd reprint, Stuttgart-Bad Cannstatt: Frommann-Holzboog, 1986–1994), pp. 713–768, esp. 747: "Quae enim a Platonicis et nuper a Patritio (ut diviniores scilicet habeantur in Philosophia) dicuntur [scil. de coelis et spatiis immateriatis], non sine superstitione manifesta, et jactantia, et quasi mente turbata, denique ausu nimio, fructu nullo, similia Valentini iconibus et somniis"; cf. idem, *De dignitate et augmentis scientiarum,* ed. J. Spedding, in vol. 1 of *The Works,* pp. 423–837, book 3, chap. 4, esp. p. 564: "Dogmata . . . Patritii Veneti, qui Platonicorum fumos sublimavit."

82. The Greek text of the *Corollarium* can be found in Simplicius, *In Aristotelis Physicorum libros quattuor priores commentaria,* pp. 601, 1–645, 19 Diels. The latest English translation is by J. O. Urmson, *Corollaries on Place and Time* (London: Duckworth, 1992); the passages referring to Proclus (i.e., pp. 611, 10–618, 25) can also be found (with facing Greek text) in S. Sambursky, *The Concept of Place in Late Neoplatonism* (Jerusalem: Israel Academy of Sciences and Humanities, 1982), pp. 64–81, and a French translation of them was produced by A.-J. Festugière as Proclus, *Commentaire sur la République,* (Paris: Vrin, 1970), 3:328–348. The quotations from the *Corollarium* can be found at *NUP* 1, 8, 19 c–d and 1, 9, 20 b; they refer to Simplicius, *In Phys.,* pp. 612, 29–613, 5; 614, 1–7; 616, 25–31, 34–35. Henry's "Francesco Patrizi" already contains a general reference to the *Corollarium de loco* (p. 556 n. 50), but the author (who seems to be relying on P. Duhem, *Le système du monde,* 10 vols. [Paris: Hermann, 1954–1959], for his knowledge of Proclus) fails to see the full significance of this identification.

83. See on this claim that space is body L. P. Schrenk, "Proclus on Corporeal Space," *Archiv für Geschichte der Philosophie* 76 (1994): 151–167. Patrizi also knew an explicit formulation of the distinction between corporeal resistance and spatial extension that was made by Porphyry, *Vita Pythagorae* 47, p. 58, 12–16 des Places (*Vie de Pythagore,* ed. and trans. E. des Places [Paris: Les Belles Lettres, 1982]): μαθήμασι τοίνυν καὶ τοῖς ἐν μεταιχμίῳ σωμάτων τε καὶ ἀσωμάτων θεωρήμασι τριχῇ μὲν διαστατὰ ὡς σώματα, ἄνευ δ'ἀντιτυπίας ὡς ἀσώματα προεγύμναζε κατὰ βραχὺ πρὸς τὰ ὄντως ὄντα

(τριχῇ . . . ἀσώματα is excluded from the text as an interpolation by des Places; the *Vita Pythagorae* is referred to by Patrizi in his *Chaldean Oracles* [as in note 1], fol. 3 d).

It is perhaps not superfluous to note that Descartes's argument for the corporeality of space is very different from Patrizi's and rests on his identifications of space with extension, of extension with matter, and of matter with body: the two claims have nothing in common but the name.

84. Proclus apud Simpl., *In Phys.*, p. 612, 25: σῶμα ἄϋλον (rather than *σῶμα ἀσώματον).

85. Proclus apud Simpl., *In Phys.*, p. 612, 19–35; quoted *NUP* 1, 8, 19 c (translation by Urmson, *Corollaries*, with slight alterations). This thought experiment recalls Plotinus, *Ennead* 5, 8 (31), 9, 1–15.

86. See *NUP* 4, 4 passim. Proclus apud Simpl., *In Phys.*, p. 612, 29: φῶς ὁ τόπος ἂν εἴη. See on this L. P. Schrenk, "Proclus on Space as Light," *Ancient Philosophy* 9 (1989): 87–94.

87. *NUP* 4, 4, 73 d.

88. *NUP* 4, 11, 88 a: "Coelum confessione Astronomorum et Philosophorum omnium nationum . . . esse sphaericum. Quid item prohibet, Pyramidem, cubum, octaedrum, dodecaedrum, Icosaedrum, si axibus figantur cicumuolui?"

89. Ibid. Another minor problem concerns Patrizi's statement that the shape of light is *orbicularis* (*NUP* 1, 1, 2 d). To this I have no solution to offer except to let it stand as a contradiction within Patrizi's own system (cf. note 72). Sadly unsatisfactory as this contradiction may be, I have not come across a passage in the *NUP* (or in any other of Patrizi's works, for that matter) that would allow us to explain it away or, even better, to resolve it on a higher level.

90. See on Plato's chariot and the Neoplatonists R. Cudworth, *The True Intellectual System of the Universe*, 2nd ed. (London: J. Walthoe, 1743), 2:787–793, and J. F. Finamore, *Iamblichus and the Theory of the Vehicle of the Soul*, American Classical Studies 14 (Chico, Calif.: Scholars Press, 1985), who gives a full bibliography of earlier works on the subject.

91. Proclus, *In Rem publicam*, 2, 196, 24–26 Kroll (*In Platonis Rem publican commentarii*, ed. W. Kroll [Leipzig: B. G. Teubner, 1899–1901]). Cf. Simplicius, *In Phys.*, p. 615, 33–35 (quoted in note 92 below), where this doctrine is similarly ascribed to Porphyry.

92. Proclus apud Simpl., *In Phys.*, p. 615, 33–35: καὶ δύναιτο μὲν ἂν τὸ αὐγοειδὲς ὄχημα τῆς τοῦ παντὸς ψυχῆς ἐνδείκνυσθαι, ὡς ὁ Πορφύριος ἐξηγήσατο. The doctrine here ascribed to Porphyry can already be found in the newly discovered summary accounts of the first two books of Galen's *Commentary on the "Timaeus"*: see C. J. Larrain, *Galens Kommentar zu Platons Timaios*, Beiträge zur Altertumskunde, 29 (Stuttgart: Teubner, 1992), §9, p. 86, ll. 11–13: ἂν δ' ἀσώματόν τις τὴν ψυχὴν εἶναι λέγῃ καθάπερ ὁ Πλάτων, ἀλλ' ὄχημά τι δίδωσιν αὐτῇ αὐγοειδές. Proclus' *Commentaries on the "Republic"* were included in J. Oporinus' edition of Plato's *Opera omnia* (Basel: Ioannes Valder, 1534), which Patrizi might have known.

93. *NUP* 1, 4, 11 a: "Est enim lumen virtutum coelestium vehiculum, et vinculum vniuersi superi et inferi." In view of Patrizi's undeniable use of Proclus/Simplicius as his main source for the speculations on light and space, I feel that the question, sometimes encountered in secondary literature, of whether he was "influenced" by B. Telesio's speculations on the same (see, e.g., *De rerum natura iuxta propria principia libri IX* [1586; reprint, Hildesheim: Olms, 1971], 1.25–28, 2.16–18, 4.5–17) loses some of its pertinency unless the extent and nature of this "influence" are severely qualified.

5

THE *PROBLEMATA* AS A NATURAL
PHILOSOPHICAL GENRE
Ann Blair

The Aristotelian corpus not only established for some two millennia the definitions and standards for many branches in the natural sciences, but also founded a genre that respected neither the disciplinary boundaries nor the systematic presentations for which Aristotle is famous. Although rightly called pseudo-Aristotelian in their final form—thirty-eight books containing some nine hundred problems, accumulated over the centuries by the Peripatetic school possibly as late as the fifth or sixth century C.E.—the *Problems* of Aristotle developed from an authentic Aristotelian core and spawned a vigorous tradition of editions and imitations. The latter outlasted the active use of the rest of Aristotle's natural science, with the publication of another work entitled "Problemes of Aristotle" through the eighteenth and even the nineteenth centuries.[1] Collections of *problemata,* variously copied or imitated from Aristotle and other ancient models (notably Alexander of Aphrodisias, Plutarch, and Cassius), comprised questions and answers about the causes of natural phenomena (especially relating to medicine, natural history, and meteorology) and elicited learned and popular interest to fuel over one hundred editions in the genre during the early modern period. I would venture that only books of secrets surpassed the *problemata* among works of natural philosophy in their bulk and the range of their success.[2]

The continuities and shifts in this long-lived genre illustrate the persistence and changing purposes of an encyclopedic inquiry about the particulars of nature independent of institutional and disciplinary boundaries. *Problemata* promised authoritative philosophical, that is, causal, understanding, made pleasant through the variety and familiarity of the phenomena they explained. Following the lead of Brian Lawn's masterly survey of problem literature,[3] but substituting Aristotle for the Salernitan questions as my point of departure, I wish to identify the *problemata* as a distinct subgenre within the broader category of works composed in question-and-answer format, generated by the editions and imitations of, as well as additions to, the "Problems of Aristotle" and defined by a characteristic combination of title, form, and range of topics.[4] After a general discussion of some of the other uses to which the term "problemata" was put, I shall focus on that specific genre, to show how it had both a learned and a more popular career, ranging from folio Latin editions with commentary on the classical models to inexpensive editions of problems of more recent composition in a number of vernaculars.

More unexpectedly, these two fortunae, which I call "high" and "low" for convenience, presented different texts under the same title (notably as the "Problems of Aristotle," or "of Alexander of Aphrodisias") and spawned erudite vernacular editions as well as "popular" Latin versions. This eminently versatile genre offered its readers of all levels the satisfaction of a causal understanding of daily phenomena and the pleasure of a varied accumulation of natural philosophical tidbits.[5]

I.   The Meanings and Changing Forms of *Problemata*

Πρόβλημα, derived from the Greek προβάλλω, "to set out as an obstacle," was probably first used in an intellectual rather than its original military context by Plato when describing the tactics of the Sophists defending themselves.[6] By extension, the *problema* became something difficult to overcome or to solve, a knotty problem. In a tradition starting in antiquity, independent of the one I discuss, πρόβλημα was used in geometry to describe a given task: for example, the construction of a triangle or the quadrature of the circle. This specifically mathematical usage, which has since become dominant among the modern meanings of the term, can be identified in a steady trickle of works from antiquity to the early modern period.[7] The *Oxford Latin Dictionary,* however, acknowledges only the plural *problemata,* which it defines, in medieval usage, as "subjects proposed for academic debate." This more general meaning applies to a number of early modern works titled *problemata,* or that advertise that they proceed *problematice.*[8] Typically, these neoscholastic academic works ask questions in "an?" or "utrum?" with detailed, structured answers, often complete with objections and responses *in utramque partem;*[9] by extension, the *problema* could refer to a topic or *locus* to discuss, notably in a school setting.[10] This usage anticipates the common modern notion of a "problem," with its broad range of application. In early modern academic contexts, the term could apply to both formal and informal exercises: Jan Amos Comenius, for example, defended theses at Herborn University titled *Problemata miscellanea* (1611); Johannes Kepler composed problems as paradoxical, probably somewhat playful, exercises during his studies at Tübingen.[11]

In its premodern just as in its modern usage, the term thus had multiple connotations. In his massive commentary on Aristotle's *Problems,* in 1602,

Ludovico Settala provides a fairly standard definition of the *problema* as a "thing proposed, or a question or a proposition."[12] A contemporary author of *problemata* of his own discussed at greater length five meanings of *problema*, adding a philosophico-theological usage to those I have identified: obstacle or protection; emanation (i.e., a designation of the Holy Spirit); something to be found or constructed (in mathematics); more generally, a question proposed for explanation; or, by contrast with the (simple) question, a "conjoined question," discussing whether a phenomenon is such or rather such.[13] After these introductory remarks, Martin Ruland's *Problemata* takes the more specific form of causal questions and answers, presumably in the category of what he would call the "simple question."

What I see as a genre of natural philosophical *problemata* extending from antiquity through the early modern period is a specific use of the concept following Aristotle and his ancient imitators—notably, Alexander of Aphrodisias, Plutarch, and Cassius (a second- or third-century iatrosophist known only for his eighty-four or eighty-five problems). Although the tightly paradoxical formulation of the Aristotelian *problema* loosened over time, the genre of natural philosophical and medical *problemata* involved exclusively causal questions (in "why?" and synonyms), applied to commonly known particular phenomena. The genre was built around an Aristotelian conception of philosophical knowledge as causal knowledge, applied to the explanation of ordinary rather than specialist or abstract questions. It comprises the many editions and translations of ancient *problemata* as well as modern contributions, which were published both separately and in conjunction with the "Problems of Aristotle."[14]

The pseudo-Aristotelian *problemata*, which set the standard for the genre, involve a very peculiar kind of question and answer. The question, in διὰ τί (why?), asks not about the existence or nature of a fact, but about the cause of a fact that is presumed so well known that it is not even stated before it is explained. However bizarre the "fact" may seem to us, the *problema* never includes discussion of its veracity but only of its cause.[15] As Hellmut Flashar describes it, in its characteristic form the question poses an apparent paradox—for example, of the form: why does phenomenon *x* have effect *y*, even though the analogous or corresponding phenomenon *xa* does not have effect *ya*?[16] The answer, characteristically formulated in one or more hypothetical suggestions (ἤ . . . πότερον ἤ: is it because? . . . or is it because?), resolves the apparent contradiction by upholding the principles that had seemed to be violated in the first place and by introducing distinctions of quantity, quality, or other well-known categories.

Why is it that children, who are warm, are not fond of wine, while Scythians and courageous men, who are also warm, are fond of wine? Is it due to the fact that the latter are hot and dry (for this is a characteristic condition of the man), but children are moist as well as hot? Now wine-drinking is a desire for something liquid. Their moisture, then, prevents children from being thirsty; for desire is in a sense the want of something.[17]

The paradox is resolved by introducing into the discussion a new but equally traditional variable: one must consider not only temperature but also moisture to understand the thirst for wine. The resolution of *problemata* involves the manipulation of the common pool of Aristotelian and Hippocratic notions about nature and human physiology: humors and qualities, phenomena of antiperistasis (or opposition), concoction, sympathy, and the like. Ancient *problemata* are strictly naturalistic, as is clear from the list of acceptable factors of explanation that prefaces the problems attributed to Alexander of Aphrodisias: "the solution to each question is to be sought in the habit, appearance, or action of the thing, or from its similarity, or color, or from the deception of the senses, the equivocation [of language] or from the greater or lesser force of its nature."[18]

We do not know how or for what purpose the questions that constitute Aristotle's *Problems* were first written down. They are grouped in thirty-eight topical sections, which are themselves arranged in a traditional but largely arbitrary order.[19] Within the sections, each question and answer is numbered and forms an independent unit, at best linked to its nearest neighbors by some common theme. Given its loose accumulative structure, the work is open to additions and changes of various kinds that can easily go undetected—hence the difficulty of determining which material is authentic and which interpolated in such collections. Because of their lack of structure and their relatively frequent lapses (including some 200 repetitions and more than 50 internal contradictions),[20] it is generally assumed that Aristotle's problems were not designed for "publication" as a treatise but developed from some pedagogical process, whether from a lecture notebook containing problems for discussion or from notes on classroom discussion. Aristotle referred simply to his "*problemata*"; later authors called them *problemata phusika* and, most interestingly, *problemata enkuklia*.[21] As recent studies around the term "encyclopedia" and its fabrication in the sixteenth century have shown, ἐγκύκλιος παιδεία meant general education or common culture in late antiquity rather than the "circle of learning" as was long believed.[22] *Problemata enkuklia* might thus have designated commonly known problems, or problems about commonly known phenomena. But in the Renaissance, the newly formed encyclopedic connotations of the term predominated: thus Settala explains that the *problemata* are called ἐγκύκλια because they "embrace so to speak the circle and circular connection of all things."[23] The gloss proposed by the third-century Peripatetic commentator Aspasius, however, would explain some of the peculiarities of their form: "there are many *problemata enkuklia;* they are so called because the students sitting in front of the master must answer in a row (ἐγκυκλίως) questions posed to them."[24] The juxtaposition of different answers to the same question in that characteristic "is it . . . or is it?" format might stem from the multiple opinions voiced in this kind of school exercise.

In any case Aristotle's *problemata* are not designed to generate new "scientific" or certain knowledge—they produce neither new principles nor new observations. Instead they connect two sets of givens: they match the principles of nature elaborated and learned elsewhere with common experience and observations from everyday life. *Problemata* are one of the ways of attaching particulars to the universals of *scientia* developed in systematic treatises, through commonsensical but often sophisticated reasoning.

> Why is it colder at dawn, although the sun is nearer then? Is it because at that moment the sun has been absent for a long time, so that the earth has been more chilled? Or is it because dew, like hoarfrost, falls more toward daybreak, and these are both cold? . . . Or are the night breezes at daybreak the cause of the chilling? Or are we to think that the greater cold is due to the fact that the food is digested? We are then more liable to cold because we are empty. Evidence for this is the fact that we are coldest after vomiting.[25]

Problemata are best understood as exercises in manipulating concepts of physics and medicine, using methods of argumentation acquired earlier. The goal is perhaps less to reach a single, "true" answer than to display mastery and ingenuity in the use of fundamental principles.

Aristotle himself, in a passage in the *Topics,* explicitly describes the role of the problem as an exercise in "dialectical training." A problem, like a paradox, is a question on which "either men have no opinion either way, or most people hold an opinion contrary to that of the wise, or . . . about which members of each of these classes disagree among themselves." But only certain kinds of problems are worth solving.

> It is not necessary to examine every problem and every thesis but only one about which doubt might be felt by the kind of person who requires to be argued with and does not need castigation or lack perception. For those who feel doubt whether or not the gods ought to be honoured and

parents loved, need castigation, while those who doubt whether snow is white or not, lack perception. We ought not to discuss subjects the demonstration of which is too ready to hand or too remote; for the former raise no difficulty, while the latter involve difficulties which are outside the scope of dialectical training.[26]

The *problema* is specifically designed to exercise the rational faculties in tackling difficult yet soluble questions. What may have begun spontaneously as a teaching method is thus already codified to some extent by Aristotle, although (to my knowledge) the connection between this passage in the *Topics* and Aristotle's *Problems* was not made until this century.[27]

The problems attributed to Alexander of Aphrodisias (also considered pseudo-Alexandrian) are the first to contain a methodological preface defining the genre; it circulated in the Renaissance in all of the Latin and in one of the vernacular editions (the French translation by Mathurin Heret, 1555, which included other learned features). In addition to its importance as a canonical statement of occult qualities,[28] this preface develops at greater length the Aristotelian principles of what constitutes a proper *problema*. Some questions have answers that are so clear and certain that they should not be asked. One who asks why birds have wings, or plants leaves, is without doubt lacking in good sense. Still worse, "one who asks whether nature and providential reason govern the generation and corruption of things and display order, position, . . . and beauty commits a terrible crime and deserves the severest punishment." At the other extreme, some questions are inexplicable because "they exceed the capacity of human understanding, and are known only to the immortal God, parent and author of all things." Examples in this category: Why do we laugh when tickled under the armpits or on the soles of the feet? Why does the sound of scratching marble make some people's teeth hurt? The category also includes questions on a long list of well-known occult properties, from the attractive powers of the magnet to the curative properties of different plants and the action of torpedo fish in stopping ships in their course. "Those who want to resolve these questions act rashly and there is nothing more improbable than the answers that they give."[29]

Problemata are not designed as arguments to refute the skeptics, nor as the arrogant display of idle speculation about the causes of hidden things. Instead, *problemata* pose questions "of a middling variety, that is, neither those which are obvious of themselves, nor those which are so hidden that no one can perceive them, but those which, albeit difficult and obscure, can nevertheless be explained by human erudition and understanding."[30] The *problemata* are not open to any question that might occur to one, but, following certain rules, whether composed by the master or handed down by tradition, they guide the proper exercise of curiosity.

After the preface, which elaborates on Aristotelian principles, the Alexandrian problems themselves mark the beginning of a shift away from Aristotle's typically paradoxical questions and many-layered answers. Alexander's problems sound less often like exercises in reasoning posed by a master than like requests, made by a pupil of his master, for authoritative information.

> I. Why did Homer call man Poliocrotaphus, of the hoarenes of the temples? Because that for the most part there gray haires beginne, because the forepart of the head hath more moysture and fleume in it, then the hinder.
>
> II. Why is onely the forepart of the head bald? Because it is loose and soft. . . .
>
> III. Baldnes proceedeth of drines.
>
> IV. Why are old men ful of excrements and watchful? . . . They are full of excrements, because they are colde and weake by nature, and therefore digest badly. . . . And seeing they are drie, they are also waking and watchful. . . .
>
> V. Why have children which are moyst by nature, and full of excrements, no hoare haires? Because they are moyst and hot, and fleume is moyst and cold, and have the forepart of the head as it were fleshy and thin, whereby superfluities are voyded.[31]

Little is left of the paradoxes of the Aristotelian problems, which can be traced at best in some successions of questions on a common theme (e.g., why do children not have grey hairs like the elderly since both types are "excrementitious"?—the former are moist and the latter dry). In the pseudo-Alexandrian problems, the answers are short and simple, offering information rather than manipulating concepts and causal reasoning in multiple answers. Aristotle's range of questions is also reduced to a mostly medical set of concerns about the body, disease, and the action of remedies and poisons. In the pseudo-Alexandrian text, *problemata* still play a pedagogical role, but the pedagogy involved takes the form of a dispensation of knowledge rather than an active manipulation of principles.

The notion of the *problema* as a source of unambiguous, authoritative knowledge seems to have become established during the transmission of the genre in the Middle Ages and the Renaissance. Medieval Arabic translations of the *Problems* of Aristotle assumed that the pupil was questioning the master, and some "translations," like that attributed to Thabit ibn Qurra, went so far as to rewrite the answers completely, bringing both their content and their form into accord with what was considered most authoritative then.[32] In the Latin West, a *vetustissima translatio* containing a smattering of problems attributed to Aristotle, Alexander, and Cassius probably served as the model for the Salernitan questions (eleventh-twelfth centuries), which were read in medical schools throughout Europe and which used the *problema* form to convey medical knowledge in a concise and clear format. Some versions versified the *problemata* for easier memorization.[33]

Given this well-established didactic role of the *problema,* it is not surprising that one Renaissance editor of Aristotle's *Problems* (Juan Luis Vives) explained, in mentioning the ancient expression προβλήματα ἐγκύκλια, that it designated the questions that pupils posed successively to their masters;[34] he thus unwittingly reversed the interaction that Aspasius described in explaining the origin of the expression. Similarly, in 1539 Antonio Luiz responded to critics who complained of Aristotle's indecisiveness in answering with multiple explanations (in ἤ . . . πότερον ἤ): "ἤ has the power to assert when it is placed in the answer. . . . They never learned that in Greek ἤ also means 'certainly, assuredly.' . . . Those who say that in the famous work of the *Problems* Aristotle puts everything in doubt and asserts nothing are, in my opinion, certainly very wrong."[35] Before imitating Aristotle's characteristic form of answer in his own five books of medical and philosophical problems, Luiz interpreted it as a succession of assertions rather than hypotheses, going to considerable (and implausible) lengths to defend the authoritative voice of the Philosopher through the multiple (student?) answers in his *Problems.*

II. THE EMERGENCE OF "HIGH" AND "LOW" VERSIONS OF THE "PROBLEMS OF ARISTOTLE"

The transmission of Aristotle's *Problems* to the Latin West in the thirteenth century contributed a crucial learned and authoritative dimension to the genre, although the *Problems* never became one of the canonical Aristotelian texts taught at the university; instead it formed a manuscript tradition mostly separate from the clusters of Aristotle's major works that were copied and circulated together.[36] The translation from the Greek by Bartholomew of Messina, quite faithful to the original, was apparently not much noticed,[37] but in 1300 Jean de Jandun, professor at the Faculty of Arts of Paris, complained of this neglect:

> This book of problems is commonly found corrupt and incorrect; it is not much expounded by anyone who is well-known or famous, and therefore few study it and fewer understand it sufficiently well, although there are to

be found in it numerous most beautiful and marvelously pleasing *theore-mata,* whence doubtless scholars ought to be extremely grateful to anyone who should both correct this book well and competently expound it.[38]

Already Jandun noted the corruptness of the *Problems.* With its add-on struc-ture and lack of overarching argument, allowing an easy tolerance of contra-dictions and repetitions, the genre was indeed both highly mobile and difficult to emend. Nonetheless, Jandun considered Aristotle's *Problems* a valuable source of *theoremata,* of truths of natural philosophy to complement those available elsewhere. Jandun's remark may have prompted Pietro d'A-bano of Padua to compose the *Expositio* that remained the only major com-mentary down to the early seventeenth century and set the standard for the learned discussion of the work.[39]

Pietro apparently made his own translations of the *Problems* of Aristotle and of Alexander of Aphrodisias, though neither has come down to us; in his commentary he referred to the translation of Bartholomew of Messina.[40] He provides authorities and cross-references, points out repetitions and contra-dictions, and discusses Aristotle's arguments and positions in light of current, and his own, opinion. He valued the *Problemata* for their encyclopedic qual-ity, for containing "hidden things from nearly every art and science, together with almost all philosophy in abbreviated form. [Thus it] could be under-stood only by those who had studied philosophy in all its branches."[41] Draw-ing on his own lifetime of study (the commentary was completed in 1310, six years before his death), Pietro assumed a rather learned audience, versed in the rest of Aristotle's philosophy. A recent close study of Jandun's "redac-tion" of the commentary argues that it is for the most part a copy; it had once been thought a simplified version that might have been designed for use in university teaching. Zdzislaw Kuksewicz questions whether, after expressing enthusiasm for the project, Jandun ever actually taught from Pietro's com-mentary.[42] With the exception of England, where the statutes of Edward VI at Oxford (1549) called for the *Problems* to be taught (and apparently they were), the *Problems* seems to have been mostly ignored at the universities.[43]

This neglect was not due to doubts about the authenticity of the *Prob-lems,* although the issue was discussed as a standard topic in the "schema isa-gogicum" of learned editions.[44] The varying degrees of doubt expressed in the Middle Ages and Renaissance did not shake the confidence of Francis Ba-con or Ludovico Settala in the seventeenth century nor of Jules Barthélémy Saint-Hilaire as late as the nineteenth that the pseudo-Aristotelian problems were entirely authentic.[45] The *Problems* was considered authentic much longer than were other (more completely) spurious Aristotelian works, such as the *Secreta* or *De mundo*.[46] Like the entirely authentic *History of Animals,* the *Problems* was probably neglected in large part because of its accumulation of particulars. Indeed Gilbert Jacchaeus, a philosophy professor of the early seventeenth century, explained why natural historical topics were not taught in schools: "because they do not require a master, because they are hardly transmitted demonstratively, and finally because of lack of time."[47] Certainly university pedagogy long continued to involve the method of question and answer, but applied it to "demonstrative," abstract issues, typically following a fourfold scholastic set of questions (*an? quid? quale? cur?*) and providing answers replete with distinctions, objections, and responses.[48] There was thus little formal resemblance to the simple causal questions about particulars characteristic of the *problemata;* nonetheless, Lawn finds examples of medical problems, especially those developed by the Salernitan school, serving as the basis for the more elaborate scholastic *quaestiones et responsiones*.[49]

Precisely because of its profusion of particulars, the appeal of Aristotle's *Problems,* and what Jandun called its "most beautiful and marvelously pleasing *theoremata,*" was not lost on readers outside the universities. Two Latin manuscripts belonged to the Abbey of Saint Victor in Paris.[50] The *Problems* was among the five works of Aristotle that Charles V had translated into French—along with the *De caelo* and those specifically of use for government, the *Ethics, Economics,* and *Politics* (translated by Nicole Oresme).[51] The translation of the *Problems* was the work of Evrart de Conty, the royal physician, who finished the task after the king's death in 1380. He based his version on Pietro d'Abano's commentary, distinguishing between the "text" of Aristotle and the "gloss" of Pietro, which he variously shortened, modified, and completed according to his own lights. Conty comments on the difficulty of the text, which he attributes in part to corruptions by scribes and translators, but also in part to Aristotle himself, who "speaks in this book neither to children nor to the ignorant, but to the old and the wise who have already learned and practiced in many parts of philosophy, and for that reason he speaks briefly, and hence obscurely."[52] Following Pietro and his erudition, Conty takes a learned approach to the text. At the same time, he is conscious of making it available to a wider audience and highlights the broader appeal of its questions "on many topics and from diverse sciences," which are "marvelous to propose and delightful to expose, . . . attractive and pleasant because human understanding naturally desires to know and understand the causes of the marvels of nature and takes the greatest delight in such."[53]

The readership for this translation, presumably courtly at first, is hard to gauge; the *Problems* was not without interest to the university community: Conty bequeathed his copies of Galen, and of the "*Conciliator* [another of

Pietro's works] and *Problemata* of Aristotle" to the library of the Faculty of Medicine at his death, and seven years later the faculty was concerned to ensure the return of these books from a borrower.[54] Like Nicole Oresme's translations of other works of Aristotle for Charles V, Conty's French *Problèmes* was completely forgotten in the early modern period and never printed. But Aristotle's *Problems* was widely reprinted in Latin during the sixteenth century in a translation by Theodore Gaza, the idiosyncracies of which John Monfasani discusses elsewhere in this volume.

However, the text that circulated most widely, translated under the short title "Problems of Aristotle," was a collection of problems first composed anonymously in Latin in the thirteenth or fourteenth century, probably in the German area; it is extant in some twenty medieval manuscripts and over one hundred printed editions through the early modern period.[55] This collection bears in fact no relation to the pseudo-Aristotelian problems, beyond fitting the genre of *problemata*.[56] It uses the classic form to discuss largely medical topics, and appropriates the very same title as "real" *Problemata Aristotelis*. That identity has added considerably to the difficulty of drawing a complete bibliography of Renaissance editions of the "Problems of Aristotle."[57] It is this text, which Brian Lawn first discussed and designated by its incipit, "Omnes homines" (all men naturally desire to know . . .), that alone was translated and printed in the vernaculars as the "Problems of Aristotle" during the early modern period: in German from 1492 to 1679; in French from 1554 to 1668; and in English from 1595 down to a remarkable afterlife as part of the *Works of Aristotle in Four Parts*, published into the twentieth century.[58] Thus Barthélémy Saint-Hilaire was correct to assert that his 1891 French translation of the *Problems* of Aristotle was the first vernacular translation in print; but he was evidently unaware of Conty's French translation of some 600 years earlier, which is still today extant only in manuscript.[59]

The "Omnes homines" text in its multiple editions constitutes what I call for convenience the "low" branch in the Renaissance of the *problemata* genre, which remained quite distinct from the learned one. With its 250-odd questions it is much shorter (and hence cheaper to produce) than the 900 pseudo-Aristotelian problems. The genuine *Problems of Aristotle* circulated with erudite versions of the problems of Alexander of Aphrodisias and sometimes of the "Problems of Plutarch."[60] In a parallel tradition, the "Omnes homines" was consistently printed with a "low" version of the problems of Alexander of Aphrodisias, which reduced the two sections (containing 152 and 134 problems) typical of most learned editions to a single section of 150 or 151 problems,[61] as well as a "low" selection of four problems from Plutarch's *Roman Questions* (a small subsection of the learned "problems of

Plutarch," added in part, as some editions specify, so that the pages would not go unused).[62] Early modern editions of the "Omnes homines" text rarely have any of the learned apparatus (commentary, marginal notes, index) common in Renaissance editions of the "real" Aristotelian problems.[63] Like the learned problems, the "Omnes homines" also appeared with accretions of various kinds, but different ones: the low versions of Alexander and Plutarch, together with compilations and adaptations of problems by modern authors such as Marco Antonio Zimara, Julius Caesar Scaliger, or Jean Bodin.[64] Such modern accretions were rare in editions of the learned versions of *problemata* texts: only Mathurin Heret, in his French translation of the learned version of the problems of Alexander of Aphrodisias, appended a further sixty problems—following "the advice of some other learned doctors."[65]

The "Omnes homines" volumes are consistently small (usually 12mo or 16mo; the incunabula are larger, but never bigger than quarto); and although some of the Latin editions are nicely printed, with relatively wide margins (e.g., Lyon, 1561), and a few even include a copper engraving of "Aristoteles Stagyrites" on the title page (Leipzig, 1623 and 1633; Amsterdam, 1643 and 1650),[66] the vernacular editions especially were poorly produced and became progressively cheaper in quality over time. The "Omnes homines" alone of the two versions of the "Problems of Aristotle" was printed in the vernacular, although it was also printed in Latin down to 1686; as the converse of the learned French translations of Aristotle by Conty or of Alexander of Aphrodisias by Heret (complete with marginal references, brief commentary, and alphabetical index), the long Latin career of the "low" branch defies the tidy association of "popular" texts with the vernacular and "learned" ones with Latin. Even as early as the fifteenth century—in the case of the English medical and scientific writings studied by Linda Voigts, for example—such neat linguistic divisions between high and low were breaking down.[67] Nonetheless, despite these cases of linguistic crossover, the two textual traditions of the "Problems of Aristotle" remained almost completely separate and operated seemingly without any awareness of one another.[68]

To anyone with a minimal learned concern, the "Omnes homines" text cannot have been convincing as composed by Aristotle, as its title would seem to claim (at least by the standard of the "real" *Problemata Aristotelis,* for example): the text itself frequently invokes the authority not only of Aristotle but also of Constantine (the late-eleventh-century doctor) and Albertus Magnus, among others who postdate the Philosopher. It is indeed characteristic of the popular works in the genre that they do not address the issues of authorship, authenticity, and corruptness that are important to learned editors and commentators of Aristotle's *Problems*. Rather than considering these editions to be simply "lying" about the authorship and authority of their problems in claiming that they are "of Aristotle and other philosophers and physicians," I would suggest, without further elaboration here, that the "Omnes homines" editions operate according to a different, looser notion of authorship and authority. The popular editions especially (probably both intentionally and not) took liberties with their text, combining, omitting, or changing the order of questions, often without obvious purpose.[69] In the vernacular translations especially, the text was often modified—answers shortened, questions combined or added, emphases changed. Most dramatically, the English translations substituted a new preface to the one current in French and Latin (which never appeared in German) and added some forty-four problems to the last section "of divers matters" in the "Omnes homines" text and thirty-two problems to the associated low version of Alexander of Aphrodisias.[70] Yet the various middlemen (editors, correctors, translators) involved in the production of these editions generally remain anonymous; even in the few cases in which we have a name (e.g., Damian Siffert, who composed the *Problemata Bodini*) these kinds of cultural intermediaries did not constitute "authors" in the eyes of the contemporaries whose biographical dictionaries are one of our major sources of information; as a result, we often know little of their training and motivations.[71]

With the atomized structure typical of many kinds of popular literature, the "Omnes homines" advertises a somewhat more logical structure than that of Aristotle's or Alexander's problems: topical sections follow the order of the body, as one title page boasts.[72] The work opens in many editions with a preface justifying philosophical inquiry (much like that in Alexander's *Problems*):

> All men naturally desire to know (*Omnes homines naturaliter scire desiderant*), as Aristotle prince of the philosophers writes in the first book of the *Metaphysics*. Of which the cause can be brought back to this, that each being naturally seeks its perfection and strives to become similar to the first Being, divine and immortal, insofar as it can. But certain knowledge (*scientia*) concerns the perfection of the intellect; therefore all men naturally desire to know. And another reason is that each being naturally seeks the good so that it can preserve itself in nature. But all knowledge producing *scientia* ranks among the honorable and good things, as is clear from the first book of *De anima*. Therefore naturally every man desires to know and as a consequence every *scientia* is to be desired (insofar as it can be apprehended by the human intellect). Although therefore any science (*scientia*) is worth examining, nonetheless that one is more worthy of study which is more noble and more common than the other sciences. But the philosophical science confers the greatest pleasures, as is clear from the tenth book of the

Ethics. Therefore, above all the sciences the philosophical one must be pursued more diligently, also for other reasons. For the present science is similar to the philosophical science: because the philosophical science makes the soul illustrious and lets it rejoice in this age, as Aristotle says in the *Book on Life and Death:* indeed this science does make man illustrious and works to make him similar to the first divine immortal being. . . . It is pleasing therefore to collect problems from several technical codices on the bodies of animals and especially the human body.[73]

The preface adduces the authority of Aristotle, authentic and spurious,[74] for every one of its arguments to show that the desire for knowledge is natural, noble, and quasi-divine. Rather than stressing rules for proper questioning, it emphasizes the universality of the desire for *scientia* and the good that derives from it.

The problems that follow are neither paradoxical nor multilayered but are requests for answers based on authority rather than complex reasoning. Unlike previous *problemata,* the "Omnes homines" problems also make occasional reference to religious considerations, although they generally prefer philosophical answers. The first question, why humans alone stand upright, prompts a long answer in which religious solutions are acknowledged, then set aside:

There are many answers. First, because it is the will of the Creator himself. And although this answer is true, nonetheless it does not seem solid in the question posed: for in this way it is easy to solve everything. Second, it is answered that every workman commonly makes his first work less good and after that makes his second work better. And thus God created brute animals first, and made them face toward the earth, and, as is clear in Genesis, afterward he created men, to whom he gave an honest figure, facing toward the sky. But this answer does not seem valid either to the proposed question, for it is theological, and disparages the goodness of God, which makes all things good.

A "natural" solution is preferred:

Sixth and last, it is answered that naturally each thing and work is given a figure that is suited to its motion. Just as roundness suits the sky, the pyramidal figure suits fire, whose motion is upward. Therefore, for a thing with two feet like man, the upright and pyramidal figure is most suitable. That is why among all the animals man alone bears his head upright.[75]

In the "Omnes homines" multiple answers are proposed precisely for those few questions for which a religious answer is given, so that a naturalistic one can also be included.[76] In its models and its general practice the genre is naturalistic, although it could be adapted to providential and moral conclusions, as in the case of the *Problemata Bodini,* which followed the natural theological thrust of its learned source, Jean Bodin's *Universae naturae theatrum.*

III. THE SUCCESS AND DECLINE OF THE GENRE

Brian Lawn attributes the extreme popularity of books of *problemata* to the rediscovery of Aristotle's *Problems* around 1300 and to the desire to imitate it, which continued through the late Renaissance.[77] Indeed, the early modern enthusiasm for *problemata* is clearly continuous with the medieval contributions to the genre, from Pietro d'Abano's learned commentary to the long-lived "Omnes homines" compilation. What the Renaissance added to this tradition was, in the first instance, simply the amplification made possible by printing, which turned a preexisting manuscript tradition into a very successful commercial operation and extended it to generations of new audiences. Nonetheless, such success would not have been possible if the *problemata* did not offer rewards that were meaningful to early modern readers.

From its prefatory pronouncements to its naturalistic practice, the "Omnes homines" upheld the ideal of philosophy as the true source of delight and moral edification; for the initial compiler, *scientia* made men like unto God. The "Omnes homines" and its modern accretions offered the satisfaction of a specifically philosophical desire—to understand the causes of the regular and ordinary phenomena of nature. As a result the *problemata* are noticeably distinct from the renewal of interest in the monstrous, the singular, and the marvelous with which the sixteenth century is often associated.[78] Only one short section of the "Omnes homines" is devoted to the subject of monsters: its centerpiece is a single case of a monstrous birth (of a calf and a man joined at the back) reported by Albertus Magnus, which he attributes to an overabundance of seed.[79] The structure of the *problema* rests on the assumption that the facts to be explained are self-evident, that they do not need to be asserted before they are explained, notably because they are well-known and universally true; this is precisely not the case of the singular marvels that one describes in their specificity to elicit amazement and awe at the willful interventions of God. Although the preface substituted (as of 1595) in the English "Omnes homines" for the traditional Latin one introduces the term "wonder," it too expatiates exclusively on the wonders of the human body and of the phenomena of everyday life; these ordinary "wonders" are too often neglected, the anonymous author complains, "because man's nature delighteth in novelties."[80] The "Omnes homines" thus carried on the emphasis of Albertus Magnus, one of the authorities most favored in the compilation, and his project of developing a philosophy of particulars.[81] The *problemata* in general contrast with other forms of "popular" literature about nature current in the Renaissance, like the *histoires prodigieuses,* which dwelled on singular events (rains of blood, monstrous births, and other anomalies) and their role as signs from God and as results of the divine distribution of punishments and rewards.

Among the learned, the interest in the *Problems* of Aristotle was no doubt heightened by the humanist program of paying new attention to ancient sources. The interest in the *Problems* peaked in a brief flurry of commentaries among Italians (mostly medical doctors) working independently of one another in the early seventeenth century: Julio Guastavini's commentary on books 1–10 of Aristotle's *Problems* (1608) was overtaken by Ludovico Settala's commentary on the whole work in two massive volumes, which were published first in 1602 and 1607 and reprinted together in 1632. In the meantime, one Joannes Manelfi, professor of philosophy at Monterotondo, had published a commentary on book 1 in 1630.[82]

The huge success of the genre taken as a whole was due, as in the case of the books of secrets, to the coincidence of both learned and popular interests. Audiences of all kinds read *problemata* for the access they promised to the authority of respected philosophers, ancient and modern—Zimara or Scaliger, for example, were among the most reputed scholars of Aristotle— about the particulars of everyday life, generally neglected in other works of natural philosophy. Editors from high to low vaunted the utility and pleasure to be derived from reading these works: in 1632, Settala's publisher boasted that Aristotle's problems were "curious and pleasant by their multiple variety, rich and fruitful by their remarkable erudition." For Luiz they were a philosophical shortcut to "multa breviter scire."[83] At the other end of the spectrum, a vernacular title page hailed the utility and pleasure of the work for surgeons and others curious about nature.[84] "Omnes homines" editions included a final section titled "Varia sane utilia" or in French, with a greater emphasis on the pleasure, "questions diverses, utiles et joyeuses." The refrain from a 1488 edition of the "real" *Problems* of Aristotle to the English "Omnes homines" was the line from Virgil's *Georgics* (2,490): "Foelix qui potuit rerum cognoscere causas."[85]

Marginal annotations in a few Latin editions of the "Omnes homines" indicate the kind of utility one could derive from the genre: boilerplate Aristotle, whose triple division of mental faculties into memory, imagination, and reason one reader copies from the answer to the question of why the head is not round but oblong.[86] Or one could glean tips for healthy living, such as the rule that "it is very unhealthy to eat milk and drink wine in the same meal, because they dispose one to leprosy," which another reader copies from the conclusion to a discussion on the proper succession of foods in a meal.[87] In a Latin incunabulum a diligent reader set out to enumerate all the questions in a kind of table of contents on the back flyleaf; the annotator gave up after covering the first two sections but continued to indicate in the margins the topics covered in each question. This kind of running index, a common reading technique among the learned, indicates an intention to return to the book for later reference.[88] At the "lower" end of the usage scale, German vernacular editions were bound with farmers' almanacs and home medical guides,[89] a practice that emphasized the utility of the text, which one could presumably consult on a specific issue through the topical arrangement. The pleasures the genre afforded stemmed, as various editors explained, from the causal understanding and *scientia* it provided and from the variety and familiarity of the topics treated—but no doubt also, less explicitly, from the prominence of questions about sex. Conception and birth, menstruation and lactation, gender differences of all kinds are one of the prime emphases of the "Omnes homines." Even in the miscellaneous section of "various useful questions" one learns that birds have their genitalia inside so that they are not hindered by them while flying.[90] Nonetheless, the pleasure remains linked to a quest for knowledge and distinct from the sheer bawdiness of schoolboy jokes, such as those of a *Problemata ludicra,* published with jocose (and untraceable) publication information and devoted to mocking the school use of the form.[91]

But the *problemata* did finally decline. After the last publication of Settala's commentary in 1632, there were no more editions or commentaries of the learned *Problems* of Aristotle or Alexander until the nineteenth century.[92] Francis Bacon still thought highly of both the ancient model and the general approach in 1623; and Aristotle's *Problems* were of use to traditional natural philosophers, who exploited it not only as a rich source for contradictions within Aristotle's thought, but also for its authoritative contents.[93] But the premises of the genre were gradually undermined in learned circles by skeptics like Montaigne in the 1580s and, more relentlessly, by the new kinds of questions and methods associated with the "Scientific Revolution." In attacking his contemporaries' facile belief in witchcraft in his famous essay "Of Cripples," Montaigne mocked the approach typical of the *problemata,* which sought only to explain phenomena without questioning their existence:

> [H]ow free and vague an instrument human reason is. I see ordinarily that men, when facts are put before them, are more ready to amuse themselves

by inquiring into their reasons than by inquiring into their truth. They leave aside the cases and amuse themselves treating the causes. . . . They ordinarily begin thus: "How does this happen?" What they should say is: "But does it happen?" Our reason is capable of filling out a hundred other worlds. . . . Following this custom, we know the foundations and causes of a thousand things that never were.[94]

Montaigne could also have had Aristotle's *Problems* in mind when he complained of the great authors who heaped up many causes for a phenomenon, less convinced of the truth than impressed by the ingenuity of the explanations they proposed. Why do we say "bless you" after someone sneezes? because sneezing is the windy emission produced by the most sacred part of the body, the head; therefore we give it a "civil reception." "Do not mock this subtlety; it is (they say) from Aristotle," Montaigne concludes.[95]

After learned publications in the genre had ceased, the "Omnes homines" continued to appear for yet another fifty years. The last textual additions to the genre of the "Problemata Aristotelis" were Scaliger's *Problemata gelliana* (first published in 1621; first published with the *Problemata Aristotelis* in 1643) and the *Problemata Bodini* (first published in 1602; first published with the *Problemata Aristotelis* in 1622). The last editions containing these two modern accretions marked the end of the tradition in Latin and German respectively, in 1686 and 1679. The French line had died out after 1668. The English *Problemes* appeared in freestanding editions down to at least 1666; their long afterlife as the second (and oldest) of the spurious works included in the *Works of Aristotle,* published from 1684 to the twentieth century, warrants separate study in the context of popular medicine and the chapbook industry in England, and English America, in the eighteenth and nineteenth centuries.[96] Setting aside this peculiar afterlife, I would suggest that by the late seventeenth century what had once been a respectable medieval compilation had moved increasingly down the market, most noticeably in its German and English versions; it continued to tout the authority of "Aristotle" (even in Latin as late as 1686) after specialist circles in natural philosophy had generally rejected as irrelevant its bookish knowledge and its ancient and medieval sources.

Starting in the seventeenth century, modern collections of questions in the vernacular offered themselves instead as useful and pleasant promptuaries for witty worldly conversation, or as accessible discussions of fashionable issues in the new science, politics, and ethics.[97] Releasing themselves from the questions and texts traditional in the double career of the "Problems of Aristotle," these authors advertised the novelty and curiosity of their material and their freedom from authority.[98] Although they continued to address many of the standard topics (aging, sex, digestion, etc.), they often framed questions around proverbs rather than unquestioned "facts" ("why does one say that . . . ?") and proposed answers based on psychological factors—including the state of the observer, as Brian Lawn has pointed out.[99] They adapted the question-and-answer form in general, and more specifically in "why?," to suit a more cultivated vernacular audience (notably by acknowledging a sense of distance from received "proverbial" opinion). Over the intervening centuries, the "why?" question has continued as a common formulation for works of popular science, still thriving, for example, in children's encyclopedias. It is perhaps a constant that *Omnes homines naturaliter scire desiderant,* but what passes for knowledgeable questions and answers is constantly changing.

BIBLIOGRAPHICAL NOTE

Rather than a thorough bibliography of "Aristotelian" problems, which would present many difficulties, I propose a brief outline of known editions (not including editions in *Opera omnia*). I include those listed by Brian Lawn (*The Salernitan Questions: An Introduction to the History of Medieval and Renaissance Problem Literature* [Oxford: Clarendon, 1963], pp. 99–100) and mark with an asterisk those I have consulted. Cf. also John Monfasani's article below.

I. "Genuine" *Problems* of Aristotle

In Latin, ca. 20 editions: *Problemata Aristotelis,* trans. Gaza (Rome, 1475). With *Problemata Alexandri,* trans. Giorgio Valla (Venice, *1488, 1495; Paris, 1500; Mantua, n.d.; and five other incunables without place or date). *Problemata Aristotelis,* trans. Gaza, with commentary by Pietro d'Abano, and *Problemata Alexandri,* trans. Valla (Venice, *1501, 1505, 1518, *1519; Paris, *1520). *Problemata Aristotelis,* trans. Gaza, with *Problemata Alexandri,* trans. Gaza (Paris, 1524, *1534, 1539; Lyon, *1551; Valencia, *1554; another without place or date *[1540]).

In Greek: edition of Fridericus Sylburg, with *Problemata* of Alexander and Cassius (Frankfurt, *1585).

A unique edition combining the *Problems* of Aristotle and the "Omnes homines": Venice, *1554.

II. "Omnes homines"

In Latin, under a variety of titles (see note 55) and consistently after 1541 as *Problemata Aristotelis ac philosophorum medicorumque complurium:* 20 incunables (see Lawn, p. 99; including Magdeburg, *1488; Leipzig, *1490; Cologne, *1493), and at least 45 editions until 1686 (Paris, 1501; Cologne, 1506, *[1510]; Antwerp, [1510]; Paris, 1514, *1515, *[1518], 1530; Venice, 1532, 1537). With *Problemata* of Zimara (Basel, *[1541]; [Frankfurt?], 1544). With *Problemata* of Zimara and Alexander, trans. Politian, and four problems of Plutarch (Frankfurt, *1548, *1549, 1551; Paris, *1550, 1552, 1553; n.p., *1554; Lyon, *1557; Paris, *1558; Lyon, *1561, *1569; Frankfurt, *1568; Venice, 1568; Lyon, *1569, *1570; Cologne, 1571; Lyon, *1573, *1579; Venice, 1580; Frankfurt, *1580; London, 1583; Cologne, *1601; Cologne, *1605; Frankfurt, *1609; Leipzig, *1623; Venice, 1626; Douai, 1631; Leipzig, 1633, 1671; Amsterdam, 1680); with *Problemata gelliana* of Scaliger (Amsterdam, *1643, *1650, *1686).

In German (3 successive translations, described in my "Authorship in the Popular 'Problemata Aristotelis,'" *Early Science and Medicine* 4 (1999): 1–39: 6 incunables (including n.p. [Augsburg, *1492]) and at least 25 editions to 1679 (Lawn, p. 100): including Augsburg, *1512, 1514, 1531; Strasbourg, *1515, *1520, *1540, 1543, *1546; Frankfurt, *1551, *1553, 1557, 1566, *1568, *1580, *1593, 1598; *Hamburg, 1604. With *Problemata Bodini* (Basel, 1622, *1666, *1679).

In French (anon. trans.), *Problèmes d'Aristote et autres filozofes et médecins,* at least 8 editions to 1668 (Lyon, *1554, *1570; Paris, *n.d., 1617; Rouen, 1618, *1620, *1633, *1668). With problems of Iacchino (Lyon, *1587, *1613).

In English (anon. trans.), *The Problemes of Aristotle with other philosophers and physitions,* at least 13 freestanding editions: Edinburgh, *1595; London, *1597, 1607, *1614, *1634, *1666; from Lawn, p. 100: 1649, 1670, 1679, 1680, 1684, 1696, 1704. Published as part of the *Works of Aristotle* (with *Aristotle's Masterpiece* and *Aristotle's Last Legacy* and often *Aristotle's Complete and Experienced Midwife*): probably first published in 1684, then in numerous editions including London, 1694, available in *facsimile (New York: Garland, 1986); a *"25th edition," undated; a "26th edition," London, *1764; a "30th edition," London, *1776, 1792; Philadelphia, 1792, 1798; London, 1802; New England, *1806; London, 1806, 1812; New England, 1813, *1828; London, 1829, 1830; New England, 1831; London, 1840; New York, 1849. D'Arcy Power concludes his list with editions of ca. 1922 and 1930 (*The Foundations of Medical History* [Baltimore: Williams and Wilkins, 1931], p. 176). The renewed popularity of the work also spawned a new imitation: *Aristotle's New Book of Problems,* first published London, 1725, with editions in at least *1741, 1760, 1776.

NOTES

I am grateful for comments from an anonymous reader and from those who heard versions of this paper in the Classics Department at the University of California, Irvine; the Groupe d'Etudes du Seizième Siècle at the University of Geneva; the Centre Alexandre Koyré in Paris, and the Dibner workshop on Natural Philosophy and the Disciplines.

1. Estimates as to when the *Problems* of Aristotle reached its final form range from the third century B.C.E. to the fifth or sixth century C.E. See Hellmut Flashar, trans. and intro., *Aristoteles: Problemata physica* (Berlin: Akademie Verlag, 1962), pp. 356–358—hereafter cited as "Flashar"; E. S. Forster, "The Pseudo-Aristotelian Problems: Their Nature and Composition," *Classical Quarterly* 22 (1928): 163–165; W. Prantl, "Über die Problemata Aristotelis," *Abhandlungen der Philosophisch-Philologischen Klasse der königlich Bayerischen Akademie der Wissenschaften* 6 (1850): 339–377; Pierre Louis, trans. and intro., *Aristote: Problèmes* (Paris: Les Belles Lettres, 1991), 1:xi ff. For an overview of the complex fortuna of works titled "Problems of Aristotle," see the bibliographical sketch.

2. On books of secrets, see William Eamon, *Science and the Secrets of Nature: Books of Secrets in Medieval and Early Modern Culture* (Princeton: Princeton University Press, 1994).

3. Brian Lawn, *The Salernitan Questions: An Introduction to the History of Medieval and Renaissance Problem Literature* (Oxford: Clarendon, 1963).

4. On the broader use of question and answer in pedagogical contexts, see, among other works, Brian Lawn, *The Rise and Decline of the Scholastic "Quaestio Disputata": With Special Emphasis on Its Use in the Teaching of Medicine and Science* (Leiden: Brill, 1993); Lloyd William Daly and Walther Suchier, *The Altercatio Hadriani Augusti et Epicteti Philosophi* (Urbana: University of Illinois Press, 1939)—I am grateful to Vivian Nutton for these references. For the Renaissance context, see *Le dialogue au temps de la Renaissance,* ed. Marie-Thérèse Jones-Davies (Paris: Jean Touzot, 1984), esp. Peter Mack, "The Dialogue in English Education in the Sixteenth Century," pp. 189–212; or Louis Massebieau, *Les colloques scolaires du 16e siècle et leurs auteurs (1480–1570)* (Paris, 1878; reprint, Geneva: Slatkine, 1978).

5. In this article I start from the learned tradition and make comparisons with the "popular" one. I pursue in more detail the complex popular fortuna in "Authorship in the Popular 'Problemata Aristotelis.'"

6. Flashar, pp. 297 ff.

7. As in Thomas Hobbes, *Principia et problemata aliquot geometrica* (London, 1674), or Alfonso Antonio de Sarasa, *Solutio problematis a R. P. Marino Mersenno propositi* (Antwerp, 1649).

8. See, e.g., Joseph de Ruffec, *Universa philosophia* ἀμφότερον *disputata,* dedicated to Armand Bourbon, Prince de Conty (n.d.), which announces that "Caetera disceptantur problematice."

9. See, e.g., Ianus Matthaeus Durastantes, *Problemata duo . . . an daemones sint* (Venice, 1567); Martin Weinrich, *Problematum partim physicorum, partim medicorum liber unus* (Wittenberg, 1590); Hippius Fabianus, *Problemata physica et logica peripatetica* (Wittenberg, 1604); Bartholomäus Keckermann, *Problemata nautica,* published in Johannes Isaacus Pontanus, *Discussionum historicarum libri duo* (Harderwijk, 1637); in a lighter tone, Ludovicus Rouzaeus, *Problematum miscellaneorum, antaristotelicorum centuria . . . ad studiosos* (Leiden, 1616)—I am grateful to Michel Magnien for this reference. For a Catholic example, see Louis du Pille, *Reconditioris philosophiae theoremata, problemata, paradoxa,* defended at the Jesuit Collège de Clermont, Paris, 24 August 1649.

10. As in Benedictus Aretius, S.S. *Theologiae problemata hoc est loci communes Christianae religionis methodice explicati* (Bern: Ioannes le Preux, 1603).

11. Marcelle Denis, *Comenius: une pédagogie à l'échelle de l'Europe* (Bern: Peter Lang, 1992), p. 84; Alain Segonds, introduction to Johannes Kepler, *Le Secret du monde,* trans. and ed. Segonds (Paris: Les Belles Lettres, 1984), p. ix.

12. Ludovico Settala, *In Aristotelis Problemata commentaria* (Frankfurt: haeredes Wecheli, Marnius et Aubry, 1602), p. v: "Problema quid: propositum, aut quaestio, aut propositio." Cf. "Problema est quod proponitur ad solvendum, seu, illud cuius caussa proponitur

investiganda, unde et multi Latine converterunt, quaestionem"; Iulius Guastavini, *Commentarii in priores decem Aristotelis Problematum sectiones* (Lyon: sumptibus Horatij Cardoni, 1608), fol. [4]r.

13. Martin Ruland, *Problematum medico-physicorum liber primus* (Frankfurt: e collegio musarum Paltheniano, 1608), pp. 12–15: "1. Πρόβλημα, -τος, το, Graecis est, quod obiectum atque oppositum, obex, vallum, agger . . . ; ut πρόβλημα, quidquid defensionis gratia obiicitur, sonet. 2. Deinde idem est, quod γένημα, id est, progenies, fructus, editum, enatum. . . . A Graecis Theologis πρόβλημα aliquando Spiritus Sanctus intelligitur, respectu Patris, qui sit προβολεύς, a quo emittatur et procedat. Hinc πρόβλημα translate dicitur de propugnatore atque defensore. . . . 3. Tertio, significatio huius vocabuli est mathematica. . . . ut Problema sit, in quo proponitur aliquid inveniendum, constituendum, ut invenire maximam mensuram, secare lineam, quadratum describere. 4. Quarto γενικῶς quandoque sumitur, pro quaestione seu proposito, hoc est eo, quod ad explicandum proponitur. . . . Magis tamen proprie pro quaestione coniuncta seu propositione usurpatur: ut, an opium sit frigidum? 5. Quinto, problema interdum opponitur quaestioni simplici seu absolutae, id est, quaestioni coniunctae, seu propositioni simpliciter sine comparatione factae et propositae. Et tum fere est quaestio coniuncta talis, in qua plures sententiae conferantur: ut simplex quaestio est (quinta tamen) An cometa sit stella? Problema vero est seu comparata quaestio, utrum cometa sit stella, an potius halitus in aere accensus?"

14. Occasionally the title "natural questions" is used interchangeably with "problemata," notably in Angelo Poliziano's translation of the problems of Alexander: *Super quaestionibus nonnullis physicis solutionum liber,* published in numerous editions, from Frankfurt, 1548, to Amsterdam, 1643. Conversely, what the Renaissance editions advertise as the "Problems of Plutarch" are selected from works by Plutarch known in antiquity and to classicists today as his "questions" (αἰτία)—Roman, Greek, and physical. The most famous *Natural Questions,* by Seneca and Adelard of Bath, are never called *problemata* and take a somewhat different form, with longer discursive answers within a dialogue.

15. This principle is made explicit, e.g., in Antonius Luiz, *Problematum libri quinque: opus absolutum et facundum et varium, multijugaque eruditione refertissimum* (Lisbon, 1539), fol. 3r: "Enimvero hoc quoque sciendum est, quod problema ut hic assumimus, tunc constat, quando rei manifestae causa quaeritur ignota. Ut si quaeramus propter quid dentium congelationem ab acidis fructibus factam portulaca persanat? vel propter quid ventres hieme calidiores? vel cur Aegyptus calida regio, frigida vina producat? et alia id genus. Priora ita quidem sese habere significamus, sed causam perquirimus. . . . Cum hic tantum causa dubitetur, non res ipsa."

16. Flashar, p. 299.

17. Aristotle, *Problems,* 3.7, 872a, trans. W. S. Hett, Loeb Classical Library (Cambridge, Mass.: Harvard University Press, 1936–1937), 1:81.

18. *Alexandri Aphrodisei Problemata,* trans. Ioannes Davio (Paris, 1541), fols. 6r–v: "Solutionem autem quaestionis cuiusque petendam ex rei vel ex habitu, vel specie, vel actione, vel ex rei simili consensu, vel ex colore, vel ex deceptione sensus, vel ex aequivocatione, vel ex viribus, quae magis minusve efficere possint. vel ex eo quod res durior, aut solutior,

aut maior, vel minor, vel ex tempore, aut aetate, aut more, vel parte essentiali, aut eventitia, vel ex rerum similium conditione. quorum exempla ex quaestionibus tum a nobis, tum ab aliis explanatis colligi possunt."

19. Some examples of topics by book number: 1. medicine; 2. perspiration; 3. the drinking of wine and drunkenness; 4. sexual intercourse; 5. fatigue; . . . 16. inanimate things; 18. literary study; 19. music; 20. shrubs and plants;. . . . 26. the winds; 27. fear and courage; . . . 36. the face; 37. the whole body; 38. complexion.

20. Flashar, pp. 323, 325.

21. Gellius 20.4 as mentioned in Flashar, p. 314.

22. Jean Céard, "L'encyclopédisme à la Renaissance," in *L'encyclopédisme: Actes du colloque de Caen,* ed. Annie Becq (Paris: Klincksieck, 1991), pp. 57–67, and Robert L. Fowler, "Encyclopedias: Definitions and Theoretical Problems," in *Pre-Modern Encyclopedic Texts: Proceedings of the Second COMERS Congress, Groningen, 1–4 July 1996,* ed. Peter Binkley (Leiden: Brill, 1997), pp. 3–29. For further references, see Luc Deitz, "Johannes Wower of Hamburg, Philologist and Polymath: A Preliminary Sketch of His Life and Works," *Journal of the Warburg and Courtauld Institutes* 58 (1995): 136–154, esp. n. 101–104.

23. Settala, *Commentaria* (1602), pp. v–vi: "Viri doctissimi crediderunt ea esse, in quibus agitur de rebus encycliis, quae scilicet versantur in consuetudine vitae communis, quotidiano usu et civili societate. . . . At non incommode universum hoc volumen problematum . . . posse encyclia problemata appellari, cum viro eruditissimo Theodoro Gaza, contendo: quae sententia etiam Athenaei et aliorum esse videtur, qui plurimas res naturales, medicas, mathematicas, caelestes, morales, et alias quae in consuetudine hominum versantur, cum contineat, quasi rerum omnium orbem et circularem connexionem comprehendat."

24. Aspasius, commentary on Aristotle, *Ethica Nicomachaea* 10, 30–32 (Heylbut), as quoted in Flashar, pp. 310n, 341n. Paul Moraux considers Aspasius' explanation fanciful in *Les listes anciennes des ouvrages d'Aristote* (Louvain: Editions universitaires de Louvain, 1951), p. 119.

25. Aristotle, *Problems* 8.17, 888b2, trans. Hett, 1:189.

26. Aristotle, *Topica,* trans. E. S. Forster, in *Posterior Analytics. Topica,* Loeb Classical Library (Cambridge, Mass.: Harvard University Press, 1939), 1.11, 104b2 ff, p. 299.

27. See, e.g., Jaap Mansfeld, "Physikai doxai e problemata physica da Aristotele ad Aezio (e oltre)," in *Dimostrazione, argomentazione dialettica e argomentazione retorica nel pensiero antico,* ed. A. M. Battegazzore, Atti del Convegno di filosofia Bocca di Magra, 18–22 March 1990 (Genoa: Sagep Editrice, 1993), pp. 311–382.

28. Brian Copenhaver, personal communication.

29. *Alexandri Aphrodisei problemata,* trans. Davio, fols. 4v–5v: "Haec [quam ob causam natura pennas volucrum generi largita est . . .] tam dilucida, tam certa qui discutienda in medium affert, mente proculdubio vacat. . . . qui quaerit utrum natura et ratio provida rebus oriendis, occidendisque consultet, decernatque ordinem, motum, positum,

speciem, colorem, caetera generis eiusdem: hic porro criminis teterrimi reus est, gravissimisque poenis obnoxius. Quaestionum igitur ratio, quod proposui, partim ex se certa manifestaque est: partim inexplicabilis, modumque penitus humani ingenii excedens: deo duntaxat cognita immortali, qui rerum omnium parens, atque author est. . . . Quaestiones, quas inexplicabiles dixi, huiuscemodi sunt: quam ob causam ridere hi soleant, quos in alis et costis aut plantis titillamus. cur nonnulli, cum motuum marmorum attritum, sectionemve, aut ferri stridorem, limarumve sentiunt, dentibus obtorpescunt. . . . Quocirca temere sane agunt, qui genus id quaestionum solutionibus mandare contendunt. nec dici improbabilius quicquam potest, quam rationes, quas illi adducunt." On the career of the occult properties of the torpedo fish, see Brian Copenhaver, "A Tale of Two Fishes: Magical Objects in Natural History from Antiquity through the Scientific Revolution," *Journal of the History of Ideas* 52 (1991): 373–398.

30. *Alexandri Aphrodisaei problemata,* trans. Davio, fol. 6r: "Quaerenda igitur illa sunt, quae locum obtinent medium, hoc est, non quae perspicua ex se sunt, vel tam occulta, ut percipi a nemine possint, sed quae quamvis difficilia et obscura, explicari tamen, lucemque recipere rationum, et doctrina hominis, atque ingenio possint."

31. *The Problemes of Aristotle with other Philosophers and Physitians* (London: S.G. for W.K., 1666), "Alexander Aphrodiseus his problemes," problems 1–5. In the Latin version of Davio these constitute problems 1–4; indeed in this English version no. 3 does not comprise a question, as would be usual, but is a subdivision of no. 2.

32. Remke Kruk, "Pseudo-Aristotle: An Arabic Version of *Problemata Physica X,*" *Isis* 67 (1976): 251–256. On medieval Arabic translations more generally, see Flashar, pp. 370–372.

33. Brian Lawn, ed., *The Prose Salernitan Questions* (London: British Academy, 1979), and Lawn, *Salernitan Questions,* chaps. 1–6.

34. *Problemata Aristotelis,* ed. Juan Luis Vives (Valencia: Borbonius bibliopola, 1554), preface: "Disputationes libri huius Aristoteles encyclias appellat, hoc est circulares, sive quod in circulis sint habitae et consessu auditorum rogantium doctorem et annotantium, quae ab illo responderentur."

35. Luiz, *Problematum libri quinque,* fols. 3r–v: "Nam in hoc quod ab Aristotele et a plerisque viris alijs disertissimis factitatum est, sequi voluimus, qui talem dictionem in responsionibus perpetuo adhibent. Aristoteles enim hac voce .e. cui haec latina an respondet, in solutionibus utitur, quae ambo asserendi utique vim habent, cum in respondendo ponuntur. ut si quaeram, propter quid nonnulli carpunt bonas literas et dicam. An quoniam nunquam didicerunt, vel grece .e. oti oude pote emathon [in Latin letters] per inde est ac si dicatur, utique vel certe, vel profecto hoc ideo fieri, quia nunquam didicisse eos contigit. Proinde qui aiunt ab Aristotele in Problematum praeclaro opere, omnia dubitata, nihil assertum, hij mea sententia longe sane aberrant. Haec te lector admonuisse volui."

36. Louis, in *Aristote: Problèmes,* 1:xxxvi. He counts almost sixty manuscripts ranging from the tenth to eighteenth centuries.

37. See G. Marenghi, "Un capitolo dell'Aristotele medievale: Bartholomaeo da Messina traduttore dei *Problemata physica*," *Aevum* 36 (1962): 268–285; and R. Seligsohn, *Die Übersetzung der pseudo-aristotelischen Problemata durch Bartholomaeus von Messina. Text und textkritische Untersuchungen zum ersten Buch,* Inaugural-Dissertation (Berlin: Dr. Emil Ebering, 1934).

38. Jean of Jandun, *Questiones super De physico Aristotelis* (Venice, 1560), sig. 8b (as quoted in Lawn, *Salernitan Questions,* p. 94; my translation): "Et scias quod liber ille de problematibus communiter invenitur corruptus et incorrectus, et non est multum expositus ab aliquo noto aut famoso, et ideo pauci student in eo, et pauciores intelligunt eum sufficienter: quanquam multa et pulcherrima theoremata mirabilis delectationis sunt in eo congregata. unde indubitanter ei qui illum librum bene corrigeret et exponeret competenter multas et magnas gratias deberent reddere studiosi."

39. For the existence of another medieval commentary that was never printed, see Lynn Thorndike, "Peter of Abano and Another Commentary of the Problems of Aristotle," *Bulletin of the History of Medicine* 29 (1955): 517–523.

40. On Pietro's translation, see the prologue to his commentary and an early passage in his *Conciliator:* "In quibusdam problematibus Aristoteli attributis per me translatis," as quoted in Lawn, *Salernitan Questions,* p. 92. For a detailed discussion of this commentary, see Nancy G. Siraisi, "The *Expositio Problematum Aristotelis* of Peter of Abano," *Isis* 61 (1970): 321–339; on this specific point, p. 324.

41. Pietro's commentary as paraphrased by Siraisi, "*Expositio problematum Aristotelis,*" p. 324.

42. Zdzislaw Kuksewicz, "Les *Problemata* de Pietro d'Abano et leur 'rédaction' par Jean de Jandun," *Medioevo: Saggi e rassegne* 11 (1985): 113–138. For the view that Jandun's redaction, extant in four manuscript copies, was used in teaching, see Thorndike, "Peter of Abano and Another Commentary," p. 517, as well as discussions in Lawn, *Salernitan Questions,* pp. 93, 95, and Siraisi, "*Expositio problematum Aristotelis,*" pp. 323–324. I am grateful to an anonymous reader for signaling the presence, in a fourteenth-century manuscript of the Bartholomew translation (Oxford, Merton 279), of dividers added as if to prepare a commentary. Such evidence as we have of how manuscripts (and printed books) were actually used is an invaluable complement to the textual analysis on which most arguments on these topics rely; in this case, the evidence suggests that that manuscript was used in a pedagogical setting.

43. On Oxford, see Charles B. Schmitt, *John Case and Aristotelianism in Renaissance England* (Kingston, Ont.: McGill-Queen's University Press, 1983), p. 52. Schmitt expresses surprise at the inclusion of the *problemata* in the statutes. They do not figure, for example, in the lists of works assigned at the University of Paris from 1255; see the *Chartularium Universitatis Parisiensis,* ed. Heinrich Denifle, O.P. (Paris: Frères Delalain, 1897), 1:278.

44. Authenticity was one of the seven topics to be treated in the introduction to a commentary already in late antiquity. See Jaap Mansfeld, *Prolegomena: Questions to be Settled Before the Study of an Author or Text* (Leiden: Brill, 1994), chap. 1.

45. See Louis, *Aristote: Problèmes*, 1:xxiii–xiv. In his careful edition of the *Problems* (Frankfurt, 1585), Fridericus Sylburg seriously doubted their Aristotelian origins. For another example of Sylburg's critical attention to authorship, in the case of Greek verses attributed to Phocylides, see Anthony Grafton, *Joseph Scaliger: A Study in the History of Classical Scholarship* (Oxford: Clarendon, 1993), 2:692. Less drastically, Juan Luis Vives attributed the problems to Aristotle's students or auditors: "Apparet autem opus hoc non esse ab Aristotele conscriptum, sed ex disputationibus illius, ab auditoribus collectum et congestum." (*Aristotelis problemata* [Valencia: Borbonius, 1554], preface).

46. For example, Jean Bodin, who cites the *Problems* as Aristotle's, considers *De mundo* spurious; compare the citations in his *Universae naturae theatrum* (Frankfurt: Wechel, 1597), pp. 387 (*De mundo*) and 198 (*Problems*). In the sixteenth century, the *Secreta* were generally regarded as spurious and were not generally mentioned in the lists of works of Aristotle; see Eamon, *Science and the Secrets of Nature,* p. 134. In general, see Charles B. Schmitt and Dilwyn Knox, *Pseudo-Aristoteles Latinus: A Guide to Latin Works Falsely Attributed to Aristotle before 1500* (London: Warburg Institute, University of London, 1985).

47. Gilbertus Iacchaeus, *Institutiones physicae* (Amsterdam: Ludovicus Elzevier, 1644), book 1, chap. 4, p. 13: "Hoc tamen vide: Libri de historia animalium, de partibus animalium, de natura plantarum, et caeteri hujusmodi, in scholis adolescentibus non solent praelegi. Tum quia non egent Magistro, tum quia demonstrative vix traduntur, tum denique brevitatis causa: quod adolescentibus certum annorum curriculum ad Philosophiae studia absolvenda sit veteri consilio propositum." I owe this reference to Mary Richard Reif, "Natural Philosophy in Some Early Seventeenth-Century Scholastic Textbooks" (Ph.D. diss., Saint Louis University, 1962), p. 66.

48. The case of Aquinas is discussed in James Weisheipl, "The Evolution of the Scientific Method," in his *Nature and Motion in the Middle Ages* (Washington, D.C.: Catholic University of America Press, 1985), pp. 248–249.

49. Lawn, *Salernitan Questions,* chap. 6.

50. "Hic liber est Sancti Victoris Parisiensis. Inveniens quis ei reddat amore dei." Paris, Bibliothèque Nationale (BN), MS Latin 14728, fol. 277; see also MS Latin 15120. Manuscripts outside Paris include three manuscripts containing the *Problems* that belonged to the Cathedral of Cambrai.

51. See Françoise Guichard-Tesson, "Le métier de traducteur et de commentateur au XIVe siècle d'après Evrart de Conty," *Le Moyen Français,* nos. 24–25 (1990): 131–167. I am grateful to Isabelle Pantin for this reference, on which I base my discussion of the French translation. For copies of the translation, see BN MS Fr. 24281-2 (autograph), Fr. 210, NAL 3371 (fragments).

52. BN MS Fr. 210, fol. 1va (problem 1.1), as quoted in Guichard-Tesson, "Le métier de traducteur," p. 137: "Nous devons savoir que Aristote ne parle pas en ce livre aus enfans ne aus ignorans, ains adresce sa parole aus anciens et aus sages qui ja sont apris et exercité en plusieurs parties de philozophie, et pour ce parle il briefment, et par consequant obscurement, car parole briefve et de grant sentence est communement obscure."

53. BN Fr. 210, fol. 1ra–b, as quoted in Guichard-Tesson, "Le métier de traducteur," p. 143: "Ce present livre est aussi comme une assemblee de pluseurs questions et de pluseurs demandes que le prince des philozophes, Aristote, voult estraire et eslire de pluseurs matieres et de diverses sciences, comme esmerveillables et delitables entre les questions qui se pevent faire des choses que on voit en nature. Et au voir dire, elles sont esmerveillables au proposer et delitables a exposer . . . [Elles sont] appetables et amables pour ce que l'entendement humain naturelement desire a savoir et a cognoistre les causes des merveilles de nature et en ce se delite souverainement. Et a la verité ceste cognoissance des causes fait moult a la perfection et a la félicité humaine."

54. *Chartularium Universitatis Parisiensis,* 4:53: "Evrardus de Conty, mag. medic. suum jubilaeum solvit et quosdam libros facultati medic. legat. 1403, Junio, Parisiis. Sciendum quod magister Evrardus de Conty solvit suum jubileum, scilicet dando prandium notabile in domo magistri Johannis de Montenantolio, ubi fuerunt xxxj magistri et fuit dictum ex parte sui, quia erat absens, per magistrum Guil. Carnificis et Johan. de Montenantolio, qualiter dictus magister legabat facultati in suo testamento libros, scil. textus Galeni, Consiliatorem et Probleumata Aristotelis etc." I am grateful to Jean Dupèbe for help in finding this reference. See Alfred Franklin, *Les anciennes bibliothèques de Paris* (Paris: Imprimerie générale, 1870), 2:20 n. 4: "Die 15 martii 1410 . . . Facultas diligenter inquisivit de recuperandis quibusdam libris, quos suo testamento sibi legaverat magister evrardus de Conti."

55. See the excellent discussion in Lawn, *Salernitan Questions,* pp. 99–103. The appellation "Aristotelis Problemata" is present already in medieval manuscripts of this text, e.g., Bibliothèque Mazarine, Paris, MS 991, fols. 116–149. Other examples of early titles include *Probleumata Arestotilis* (Leipzig, 1490); *Probleumata Arestotelis determinantia multas questiones de varijs corporum humanorum dispositionibus valde audientibus suaves* (Cologne, 1493); *Probleumata Aristotelis varias quaestiones cognosci admodum dignas et ad naturalem philosophiam potissimum spectantes discutientia* (Paris, 1515). The title becomes standard after the addition of the problems of Zimara as *Aristotelis ac philosophorum medicorumque complurium problemata . . . Zimarae problemata his addita* (Basel: Robert Winter, 1541).

56. Lawn concludes after a careful examination that the author of this text makes neither reference to nor tacit use of the *Problems* of Aristotle; see *Salernitan Questions,* p. 103.

57. The distinction between the two texts circulating under the same title is made neither in F. Edward Cranz, *A Bibliography of Aristotle Editions, 1501–1600* (Baden-Baden: Valentin Koerner, 1971), p. 161, nor in Louis, *Aristote: Problèmes,* 1:liii.

58. Parts of the "Omnes homines" (especially the questions) were also used in Italian and Spanish works by Ortensio Landi, Bartolomeo Paschetti, and Hieronymus Campos; see Lawn, *Salernitan Questions,* pp. 99–101. But the "Omnes homines" was never published entire as the *Problemata Aristotelis* in Italy or Spain.

59. Jules Barthélémy Saint-Hilaire, *Problèmes d'Aristote traduits en français pour la première fois* (Paris: Hachette, 1891).

60. Despite differences in cataloguing (some entries name Calphurnius Brixiensis, others Petrus Lucensis as translator), there is only one Latin translation of "Plutarch's problems" in circulation in erudite editions: it opens with a preface by Calphurnius in which he explains that the translation is the work of Petrus Lucensis.

61. In addition to this arrangement of the pseudo-Alexandrian problems rendered standard by the Aldine edition, and followed in Latin translations by Theodore Gaza and Johannes Davio, one finds a very different arrangement in the translation by Giorgio Valla, which is divided into five sections of 35, 62, 54, 121, and 78 problems respectively. These contain problems not found in the other editions, which classicists today attribute to books 3 and 4 of Alexander's problems, or sometimes call Aristotle's *problemata inedita*. For more discussion of the thorny problems of classifying these problems, see Iulius Ludovicus Ideler, *Physici et medici Graeci minores* (Amsterdam: Hakkert, 1963), vol. 1, and Hermann Usener, *Alexandri Aphrodisiensis quae feruntur Problematorum libri III et IIII*, Jahresbericht über das Königliche Joachimsthalsche Gymnasium (Berlin: Typis Academiae regiae, 1859).

62. See, e.g., *Problemata Aristotelis* (Lyon: Paganus, 1561), p. 285: "Addidimus huc ex Plutarchi problematis selectiora quaedam, simul ut haec nostra undiquaque copiosiora essent, simul ne tot paginae nobis vacarent."

63. I have found marginal summaries in one Latin edition of the "Omnes homines" (Paris: Colaeus de La Barre, 1515). The problems of Zimara also generally include marginal references in the Latin and French versions, although not in the English translation. The "Omnes homines" editions never include commentaries or indexes.

64. The problems of Zimara are standard in the editions of the "Omnes homines" after 1541 in all languages except German (Zimara was apparently never translated into German). On the earliest editions of Zimara's problems that survive, none before 1536 (Zimara had died in 1532), see Lawn, *Salernitan Questions,* p. 129. The problems of Scaliger, which first appeared in *Iul. Caes. Scaligeri adversus Desid. Erasmum orationes duae . . . quibus de novo etiam accedunt problemata gelliana* (Toulouse: apud Dominicum Bosc et Petrum Bosc, 1621), are appended to *Problemata Aristotelis* (Amsterdam: Iodocus Janssonius, 1643 and 1650; Janssonius Waesbergius, 1686). I am grateful to Michel Magnien for the reference to the edition of 1621. The *Problemata Bodini,* adapted from Bodin's *Universae naturae theatrum* by Damian Siffert, first appeared in a freestanding edition (Magdeburg: Johann Francken, 1602), then as an addition to the German *Problemata Aristotelis: Das ist Gründliche Erörterunge* (Basel: Johann Schröter, 1622; Emmanuel König, 1666 and 1679). For more on this interesting case of a vernacular transformation of a learned original, see Ann Blair, *The Theater of Nature: Jean Bodin and Renaissance Science* (Princeton: Princeton University Press, 1997), pp. 212–224. A French edition also includes some problems by Leonard Iacchinus, a Catalan physician; see *Problèmes d'Aristote* (Lyon: de Tournes, 1587) and Lawn, *Salernitan Questions,* pp. 132–133.

65. *Les Problèmes d'Alexandre Aphrodisé,* trans. Mathurin Heret (Paris: Guillaume Guillard, 1555), p. 104: "Aux lecteurs: Messieurs, la principale occasion de ceste mienne entreprise, est que ouyant monsieur Sylvius enseigner publiquement la medecine de tele grace et methode que chacun cognoist, allegant souvent et bien à propos ces problemes, m'a semblé en faire quelque grand'estime, qui m'a invité à les lire plus diligemment. . . . Nous y avons en outre adiousté quelques autres problemes, avec le conseil et advis de quelques doctes medecins, bien à propos ce me semble en cest endroit." On Heret, see Frank Lestringant, *André Thevet, cosmographe des derniers Valois* (Geneva: Droz, 1991), pp. 100–104.

66. A few editions, in Latin and in German, use woodcuts of the stereotypical "scholar" or "astronomer" on the title page: see editions of Cologne, 1510; Strasbourg, 1515 and 1540; Hamburg, 1604. For reproductions, see *Early Science and Medicine* 4 (1999): 25–30.

67. Linda Ehrsam Voigts, "Scientific and Medical Books," in *Book Production and Publishing in Britain, 1375–1475,* ed. Jeremy Griffiths and Derek Pearsall (Cambridge: Cambridge University Press, 1989), pp. 345–402, esp. p. 383. It is all the more curious, given the vernacular activity in medicine in England, that the "Omnes homines" was translated into English last, in 1595; perhaps it is a measure of the slow diffusion of a text from the German-speaking area (where it was likely composed—certainly it was translated into German first, by 1492 at the latest) through France (in 1554) to England (forty years later).

68. I have found only one edition that juxtaposes the learned problems of Aristotle and Alexander (in two sections), each with an alphabetical index (an unmistakeably learned trapping) *with* the problems of the "popular" tradition—those of Zimara, the "Omnes homines," and the four problems of Plutarch common in popular editions (although the order is reversed from the usual; no index is provided for these texts): *Aristotelis, Alexandri Aphrodisei, Marciantonii Zimarae ac Philosophorum medicorumque complurium problemata. Quibus adiectus est copiosissimus index* (Venice: in vico Sanctae Mariae Formosae, 1554). This (to my knowledge) unique edition makes no reference to the juxtaposition of two different texts titled (on inner title pages) "Aristotelis Problemata" and "Aristotelis ac philosophorum medicorumque complurium Problemata." I discuss this case more thoroughly in my "Authorship in the Popular 'Problemata Aristotelis,'" pp. 12–13.

69. For example, a French edition omits a question of Zimara that might seem politically sensitive: why do flatterers have such success with princes? (answer: because of the excessive vainglory of the latter), but then includes a number of others: why are prelates and women greedy? why do tyrants dislike men of letters? (answer: for fear that their teachings would provoke a rebellion). Compare *Problèmes d'Aristote* (Lyon: de Tournes, 1587), problems by Zimara nos. 46 (greed), 103 (men of letters); and *Problemata Aristotelis* (Amsterdam: Jan Janssonius, 1643), nos. 46 (greed), 80 (flatterers, p. 172), 104 (men of letters). Further comparison with an English edition (London, 1634) reveals more changes in the numbers of these questions: nos. 44 (greed), 74 (flatterers), 97 (men of letters). For another case in which the vernacular version modifies the division of problems in the Latin, see above note 31.

70. The forty-four extra problems appear already in the first English edition, Edinburgh: Robert Waldgrave, 1595; the thirty-two additional problems first appear in the subsequent edition, London: Arnold Hatfield, 1597.

71. One exception is the case of Walter Ryff, a prolific author of books of popular medicine active in Strasbourg, who is also responsible for three Strasbourg editions of the German *Problemata Aristotelis* in the 1540s, as I discuss in my "Authorship in the Popular 'Problemata Aristotelis.'" On Ryff, see Miriam Usher Chrisman, *Lay Culture, Learned Culture: Books and Social Change in Strasbourg, 1480–1599* (New Haven: Yale University Press, 1982), pp. 179–181, and Josef Benzing, *Walther H. Ryff und sein literarisches Werk: Eine Bibliographie* (Hamburg: E. Hauswedell, 1959).

72. *Problèmes d'Aristote et autres philosophes et medecins selon la composition du corps humain* (Lyon, 1587); see also editions of Paris: veuve Jean Regnoul, 1617, and Rouen: Adam Mallassis, 1620. Medieval collections of recipes, following classical models, were often arranged by disease, starting with diseases of the head and ending with those of the feet: see L. C. MacKinney, "Medieval Medical Dictionaries and Glossaries," in *Medieval and Historiographical Essays in Honor of James Westfall*, ed. James Lea Cate and Eugene N. Anderson (Chicago: University of Chicago Press, 1938), pp. 240–268, esp. 243. On atomized structure, see Roger Chartier, *L'ordre des livres: lecteurs, auteurs, bibliothèques en Europe entre XIVe et XVIIIe siècle* (Aix-en-Provence: Alinea, 1992), pp. 24–25.

73. *Problemata Aristotelis* (Amsterdam: J. Janssonius, 1643), sigs. A2r–v: "Omnes homines naturaliter scire desiderant, ut scribit Aristoteles princeps Philosophorum, primo Metaphysicae, cuius causa potest reddi talis, quia omne ens naturaliter appetit suam perfectionem et similiter conatur se assimilare primo Enti, divino et immortali, in quantum potest. Sed scientia est de perfectione intellectus, ergo omnes homines naturaliter scire desiderant. Rursus et alia ratio est: nam quodcunque ens naturaliter appetit bonum, ut se conservare possit in rerum natura. Sed omnis notitia scientifica, est de numero bonorum honorabilium, ut patet primo de Anima. Ergo naturaliter omnis homo desiderat scire, ex consequenti omnis scientia, in quantum intellectui humano est possibilis, est appetenda. Quamvis igitur quaelibet scientia sit perscrutanda, magis tamen illa, quae est nobilior et communior aliis scientiis. Sed philosophica scientia confert maximas delectationes, ut patet decimo Ethicorum. Ergo prae caeteris scientiis philosophica diligentius est inquirenda, etiam propter alias causas. Nam praesens scientia, est similis scientiae philosophicae: quia ipsa clarificat animam et ipsa facit delectare in hoc seculo, ut dicit Aristoteles in libro de vita et morte: quia quae etiam in tantum clarificat hominem et ipsum primo enti divino immortali, assimilari laboret, teste Seneca in epistola in talia prorumpens verba, hoc mihi philosophia promittit summopere, ut me Deo parem reddat. Libet igitur de animalibus corporalibus, praesertim de corpore humano, ex pluribus artificialibus codicibus Problemata colligere." The preface varies slightly among the different families of editions.

74. On the (spurious) *Book of Life and Death* or *Book of the Apple and of Death* (*De pomo et morte,* cited in other editions of the "Omnes homines"), see D. S. Margoliouth, "The Book of the Apple, Ascribed to Aristotle," *Journal of the Royal Asiatic Society* 24 (1892): 187–252.

75. *Problemata Aristotelis ac philosophorum medicorumque complurium* (Lyon: Theobaldus Paganus, 1569), pp. 5–6: "Quaeritur, quare inter omnia animalia homo habet faciem versus coelum elevatam? Respondetur multipliciter: Primo, quod est ex voluntate ipsius Creatoris. Et quamvis illa responsio sit vera, non tamen videtur firma in proposito: quia sic facile est omnis dissolvere. Secundo respondetur, quod omnis artifex opus suum primum communiter facit deterius et post hoc opus suum, secundum facit melius. Et sic Deus creavit bruta animalia primo, quibus dedit faciem depressam ad terram inclinatam: et secundo creavit homines, ut patet in Genesi, quibus dedit figuram honestam et ad coelum elevatam. Sed illa responsio non videtur iterum valere ad propositum, ex quo est Theologica, et primo derogat bonitati Dei, quae sua opera quantum est ex sua natura, semper facit bona. . . . Sexto respondetur, et ultimo, quod naturaliter cuilibet rei et operi talis figura est computanda, quae suo motui fiet apta. Ut Coelo competit rotunditas, igni autem competit figura pyramidis, quae motui sursum est apta. Ergo rei bipedali, ut est

homo figura diametrica, et figura pyramidis est aptissima. Ergo inter omnia animalia so-lus homo capite est elevatus." For a comparison of the different vernacular translations of this passage, see my "Authorship in the Popular 'Problemata Aristotelis,'" pp. 15–16.

76. For example: "Question. Why are the Iewes subiect unto this disease very much [piles or hemorrhoids]? Answer. The Divines do say, because they cried at the death of Christ, 'Let his blood fall upon us and our children.' . . . Another reason is, because the Iewes do eate much fleugmatike and cold meats, which doth breed melancholy bloud, which is purged by this fluxe of bloud. Another reason, is because moving doth cause heat and heat digestion, as 4. Meteor., but the Iewes doe not move nor labour, nor converse with men. Also they live in great feare, lest we should revenge the death of our Saviour, which doth also breede a coldnes in them, which dooth hinder digestion, which dooth breede much melancholy bloud in them, which is by this means purged." *The Problemes of Aristotle with other Philosophers and Phisitions* (Edinburgh: Robert Waldgrave, 1595), sigs. [D6]r–v.

77. Lawn, *Salernitan Questions,* p. 92.

78. See Jean Céard, *La nature et les prodiges: L'insolite au XVIe siècle, en France* (Geneva: Droz, 1977), and Lorraine Daston and Katharine Park, *Wonders and the Order of Nature, 1150–1750* (New York: Zone Books, 1998).

79. The section contains only three questions: "Doth nature make any monsters?"; then, following up on the case of the half-man, half-calf reported by Albertus, "whether be they one or two?"; and finally: "Why is a man borne sometime with a great head, or sixe fin-gers in one hand, or with foure? . . . When there is too much matter, then he is borne with a great head, or sixe fingers: but if there be want of matter, then there is some part too little, or lesser then there ought to be." *Problemes of Aristotle . . .* (Edinburgh: Robert Waldgrave, 1595), sig. [E8] r–Fr.

80. *Problemes of Aristotle* (Edinburgh: Robert Waldgrave, 1595), to the reader: "Every man dooth wonder (gentle Reader) at an Eclipse of the Sunne, or of the Moone, and gazeth at a Blazing Starre, and beholdeth with admiration an exquisit picture, drawne with the Pensill of a skilfull hand; yea all novelties doe please, be they never so small. But if they be once common, be they never so great wonders, no man vouchsafeth to give them the looking on. But then as *Seneca* doth say: *Non est Aethiopis inter suos insignitus color.* The selfe same doth happen in man, and in the wonderfull workmanship of his bodie, and un-speakeable excellencie of his soule: for if we regarde his excellencie, he doth surpasse all creatures under Heaven. . . . And therefore the body of man is made of a complexion most pure and delicate, and in shape comely and beautifull; and yet notwithstanding all these perfections which man hath in himselfe, few or none take delight in the studie of himselfe, or is carefull to know the substance, state, condition, quality and use of the parts of his owne bodie, although he be the honor of nature, and more to be admired than the strangest and rarest wonder that ever happened. The cause of this is no other, but because mans nature delighteth in novelties, and neglecteth to search out the causes of those things which are common. I have therefore thought good, to give thee in a knowne tongue, this little booke, written by the deepest of all Philosophers, who teacheth the use of all the parts of mans body, their nature, qualitie, propertie and use which may bring thee in read-ing of it, if reade it thou wilt, no lesse delight than profit, nor no lesse profit than delight."

81. I am relying here on the categories of analysis developed in Daston and Park, *Wonders and the Order of Nature, 1150–1750,* esp. chap. 3 on Albertus Magnus. I am grateful to the authors for circulating their work to me before publication.

82. Joannes Manelfi, *Urbanae disputationes in primam problematum Aristotelis sectionem* (Rome: apud Gulielmum Facciottum, 1630).

83. Ludovico Settala, *In Aristotelis Problemata Commentaria* (Lyon: Claude Landry, 1632), dedication by Claude Landry to Humbert de Champonay, fol. 3v: "Opus sane, velut, ab multiplici varietate, curiosum et amoenum, sic ab insigni eruditione, locuples, fructuosumque." Luiz, *Problematum libri quinque,* fol. 3r.

84. "Oeuvre fort agréable et utile aux chirurgiens et à tous ceux qui veulent apprendre les admirables secrets de la nature." *Problèmes d'Aristote* (Rouen: Behourt, 1668).

85. See *Problemata Aristotelis* (Venice: Antonius de Strate, 1488), preface of Nicolas Gupalatinus to Pope Sixtus IV; and *Problemes of Aristotle* (Edinburgh: Robert Waldgrave, 1595), to the reader.

86. "Quaeritur quare caput non est directe rotundum, sed oblongum? Respondetur, ut in ipso tres cellullae aptius possint distingui, scilicet phantasia, in fronte: logistica sive rationalis in media et memoria in posteriori cellulla." Marginal annotations read: "3. cellulae: phantasia, logistica, memoria." *Problemata Aristotelis* (Lyon: Paganus, 1561), p. 17 (section de capite). Bibliothèque Publique et Universitaire, Geneva (hereafter BPU), Md 179.

87. "Quare sit malum differre in mensa et paulatim diversa cibaria comedere? Respondetur, quia quando multum apponitur, tunc primum incipit digeri et sic partes in digerendo non aequantur et cibus decoctus corrumpitur. . . . Et per consequens est valde nocivum, in eadem mensa comedere lac et bibere vinum quia disponunt ad lepram." Marginal annotations repeat the conclusion: "et per consequens est valde nocivum, in eadem mensa comedere lac, et bibere vinum, quia disponunt ad lepram." *Problemata Aristotelis* (n.p., 1554), p. 87 (section de mamillis). BPU, Md 178.

88. *Probleumata Arestotilis* [Leipzig: Conrad Kachelofen, 1490], Harvard University Library, Inc. 2913. For other examples of this practice, see my *Theater of Nature,* pp. 195–201.

89. A German *Aristotelis Problemata* (Basel: König, 1666) is bound (in seventeenth-century vellum) with a manual on guns, *Buchsenmeisteren. Geschoss/Buchsen/Pulver . . .* (Frankfurt, 1597); a German translation of More's *Utopia, De optimo Reipublica Statu, libellus vere aureus. Ordentliche und Aussführliche Beschreibung Der . . . Insul Utopia* (Leipzig, 1612); and a *Bauwren Practica oder Wetterbuchle* (Frankfurt, 1570), Universitätsbibliothek Basel. Another copy of the same edition is bound with *Die gestriegelte RockenPhilosophia, oder Auffrichtige Untersuchung derer von vielen super-klugen Weibern hochgehaltenen Aberglauben* (Chemnitz, 1705) and with *Frauendienst darinne eigentlich beschrieben sind viel herrlicher und bewahrter Hülffs-Mittel* (Freiburg, 1675), Herzog August Bibliothek Wolfenbüttel QuN519.1.

90. *Problèmes d'Aristote* (Lyon: de Tournes, 1587), p. 163.

91. "Quid est scholasticus? R. Est filius patris sui, frater sororis, nepos avunculi, doctor omnium, praesertim in naturalibus, utpote qui se libenter adjungit puellis. . . . " *Problemata ludicra et histeriole ridiculae Animi relaxandi causa excogitata* ("Apud neminem," [1642]), p. 2.

92. Although the last selection of the "problems" of Plutarch dates from 1579, there is a later scholarly edition of the text from which the "problems of Plutarch" were drawn: *Quaestiones Romanae,* ed. M. Zuerius Boxhornius (Leiden, 1657).

93. "Of Problems there is a noble example in the books of Aristotle; a kind of work which certainly deserved not only to be honoured with the praises of posterity, but to be continued by their labours; seeing that new doubts are daily arising." *De augmentis scientiarum* (1623), in the *Philosophical works of Francis Bacon,* ed. J. M. Robertson (London: G. Routledge and Sons, 1905), p. 467, as quoted in Lawn, *Salernitan Questions,* p. 141. For traditional natural philosophers, see, e.g., Bodin, who in the *Universae naturae theatrum* delights in criticizing Aristotle for his inconsistencies, notably by citing the *Problemata* (e.g., p. 198), but still uses him as an authoritative source when convenient (e.g., p. 156). Thomas Browne (1605–1682) owned a copy of Aristotle's *Problemata* (Paris, 1562); see *A Catalogue of the Libraries of Sir Thomas Browne and Dr. Edward Browne,* ed. Jeremiah S. Finch (Leiden: Brill, 1986), section on "medici, philosophici," p. 27, no. 61.

94. Montaigne, *Essais,* III, 11; Villey ed. (Paris: PUF, 1988), pp. 1026–1027; translation by Donald Frame, *The Complete Essays of Montaigne* (Stanford: Stanford University Press, 1965), p. 785.

95. Montaigne, *Essais,* III, 6 ("Of Coaches"), opening paragraph (Villey ed., p. 899; Frame translation, p. 685); I am grateful to Francis Higman for this reference. Aristotle does indeed discuss the three types of wind produced by the body and notes that sneezing is considered sacred, although he does not mention forms of politeness associated with it; see *Problems* 33.9.

96. For editions of the English *Problemes,* which are particularly poorly preserved in library collections, see Lawn, *Salernitan Questions,* p. 100; it remains unclear which of the editions he mentions are freestanding and which part of the *Works of Aristotle.* The latest freestanding edition I have seen is London: S.G. for W.K., 1666.

Treatments of *The Works of Aristotle* have focused on the lead piece, *Aristotle's Masterpiece,* which Mary Fissell calls, after an extensive bibliographical survey, "the most widely-read book about sex and reproduction in 18th-century England and America." I am grateful to Mary Fissell for sharing with me her unpublished manuscript titled "Making Aristotle's Masterpiece: Popular Medical Books as Cultural Bricolage" (September 1997). See also Roy Porter, "Medical Folklore in High and Low Culture: *Aristotle's Masterpiece,*" in Roy Porter and Lesley Hall, *The Facts of Life: The Creation of Sexual Knowledge in Britain, 1650–1950* (New Haven: Yale University Press, 1995), pp. 33–64; Otho T. Beall, "*Aristotle's Master Piece:* A Landmark in the Folklore of Medicine," *William and Mary Quarterly* 20 (1963): 207–222; Vern Bullough, "An Early American Sex Manual; or Aristotle Who?" *Early American Literature* 8 (1973), 236–246; D'Arcy Power, *The Foundations of Medical History* (Baltimore: Williams and Wilkins, 1931), pp. 147–178. The continued circulation of this work is evident from the inclusion of *Aristotle's Masterpiece* in a list drawn up in 1954 of works exempted from seizure under the British Obscene Publications Act because they are "recognised throughout the civilised world as established classics." See

Jill Kraye, "The Printing History of Aristotle in the Fifteenth Century: A Bibliographical Approach to Renaissance Philosophy" *Renaissance Studies* 9 (1995): 189–211; quotation, 211; I am grateful to the author for this remarkable reference. I discuss this English afterlife in my "Authorship in the Popular 'Problemata Aristotelis.'"

97. For a promptuary, see Scipion Dupleix, *La curiosité naturelle en questions* (Paris, 1606). Many of the discussions at the Bureau d'Adresse of Théophraste Renaudot take the form of causal questions. For a recent treatment of Renaudot's Conférences, see Simone Mazauric, "Savoirs et Philosophie à Paris dans la première moitié du XVIIe siècle: Les Conférences du Bureau d'Adresse de Théophraste Renaudot (1633–42)" (Ph.D. diss., Université de Paris I, 1994). I am grateful to Françoise Waquet for giving me this reference and access to the work.

98. For example, see Pierre Bailly, *Questions naturelles et curieuses* (Paris: Bilaine, 1628): "je confesse que ceste entreprise est trop relevee pour moy de vouloir examiner les triviales et communes opinions de l'antiquite qui ont[eu] cours il y a si longtemps parmy nos Francois. . . . Ie me doute qu'on trouvera estrange que ie n'allegue aucun Autheur qui me serve de garand pour asseurer mon opinion, comme me voulant attribuer toutes les raisons que j'allegue. Mais je desire que l'on sçache que i'ay faict treve avec mes livres et toute langue estrangere durant que ie me suis amusé à cecy: si par occasion l'on y trouve chose qui responde au dire des anciens, il le faut attribuer a quelque confuse cognoissance que i'ay peu tirer de ma lecture passee." Au lecteur, fols. [1]r, [3]v.

99. See Lawn, *Salernitan Questions,* pp. 154–155. Pierre Bailly, for example, notes repeatedly that people notice the bad things that happen to them more than the good; hence, for example, the saying that misfortunes never come alone (*Questions naturelles et curieuses,* p. 631).

6

THE PSEUDO-ARISTOTELIAN *PROBLEMATA* AND ARISTOTLE'S *DE ANIMALIBUS* IN THE RENAISSANCE

John Monfasani

In a way, the *Problemata* and the *De animalibus* have no history in the Renaissance as translated—or, at least, very little history. Theodore Gaza's translation of the *Problemata* was printed for the first time in about 1473,[1] and for the rest of the Renaissance no humanist rival ever made it into print to challenge it.[2] The medieval version of Bartolomeo of Messina[3] was printed seven times,[4] but only because it formed an integral part of the important commentary of Pietro d'Abano (d. 1316);[5] even then, in its last five printings it was accompanied by the "new" translation of Theodore Gaza.[6] Gaza's version of the *Historia animalium, De partibus animalium,* and *De generatione animalium* (what I shall call collectively Aristotle's *De animalibus*) was first printed in 1476.[7] No other version, either medieval[8] or Renaissance,[9] appeared in print in the fifteenth or sixteenth centuries.[10] Julius Caesar Scaliger's translation of the ten books of the *Historia animalium* appeared in 1619,[11] but it was never reprinted. Consequently, Gaza reigned supreme over the *De animalibus* until the nineteenth century.

Gaza's virtual monopoly of the *Problemata* and *De animalibus* in the Renaissance is unique among the fortune of Aristotelian writings. Renaissance readers had a choice of printed Latin translations of every major Aristotelian work save for the *Problemata* and the *De animalibus*. Leonardo Bruni caused a big stir with his translation of the *Nicomachean Ethics* early in the fifteenth century, and his version continued to be printed in the sixteenth. But John Argyropoulos' translation was printed about five times more often than Bruni's in the sixteenth century, and both had plenty of competition. At least eleven other Latin translations of the *Nicomachean Ethics* became available in print in the sixteenth century. Even the medieval version of William of Moerbeke was printed at least eleven times in the sixteenth century.[12] The *Physics, Metaphysics, De anima,* all the logical works, the *Rhetorica, Politics, Economics, Magna moralia, De generatione et corruptione, De coelo, De mundo, Mechanics, Parva naturalia, Meteorologica, De gressu animalium, De motu animalium,* and the tenth book of the *Historia animalium,* not translated by Gaza, were available in at least three Latin translations in the Renaissance, some with ten or more different versions—all except, of course, the *Problemata* and the *De animalibus*, to which Theodore Gaza seems to have had exclusive rights.[13]

Gaza's dominance began in the later fifteenth century. Aldo Manuzio tells us in one of the prefaces to his *editio princeps* of Aristotle's works in Greek that he and the other bright lights of his generation—Angelo Poliziano, Giovanni Pico della Mirandola, Ermolao Barbaro, and Girolamo Donato—all learned Greek by studying the translations of Theodore Gaza.[14] Indeed, if we were to believe Manuzio, Gaza had produced a virtually perfect correspondence between his fluent Latin and Aristotle's Greek.

Because he believed that the classical editor of Aristotle's works had misordered the text, Gaza transferred book 9 of the Greek manuscripts of the *Historia animalium* to a position after book 6, thereby making it book 7. He also rearranged the sequence of text in the last few pages of book 8 (book 9 in his translation).[15] Manuzio legitimized these transpositions by incorporating them into the *editio princeps* of the Greek text. Subsequently, the major Renaissance editors of the Greek text of Aristotle's *Opera Omnia* (Erasmus, Sylburg, Camotius, and Casaubon) all resorted to Gaza's Latin translation to correct the Greek text of the *Problemata* and *De animalibus*, in essence accommodating the Greek text of Aristotle to Gaza's Latin version. Even modern editors and translators of these works cite Gaza's translation in their *apparatus criticus*.[16] Some editors, such as Josef Klek in the Teubner series and Pierre Louis in the Budé series, have condemned Gaza's penchant for periphrasis and conjecture.[17] Nonetheless they justified their reliance on him by supposing that he must have had access to some lost Greek manuscript(s) since so frequently they themselves could not find his readings in the extant Greek manuscripts.

What modern students of the *Problemata* do not know—indeed, what most sixteenth-century Renaissance scholars did not know—was that the Latin text of Gaza which they read was not the text Gaza initially published. Gaza, in fact, published his translation of the *Problemata* twice. The first time, in 1454, he dedicated it to Pope Nicholas V, two years after George Trapezuntius had finished a translation of the same work and had then been driven out of Rome in disgrace.[18] The two Greek émigrés became rivals from the moment they found each other in Rome in 1449, but their quarrel broke out into the open only with the publication in 1456 of Trapezuntius' *In Perversionem Problematum Aristotelis a Quodam Theodoro Cage Editam et Problematice Aristotelis Philosophie Protectio (The Protection of Aristotle's Problematic Form of Philosophy against the Perversion of Aristotle's "Problemata" Published by a Certain Theodore Cages*).[19] One of the main charges Trapezuntius hurled at Gaza in the *Protectio* was that Gaza had totally perverted the sequence and numbering of Aristotle's *Problems*. In my book on Trapezuntius I noted that Gaza did in fact radically reorganize Aristotle's *Problemata,* and suggested that this had gone long unnoticed because modern scholars have looked only at the second edition published twenty-one years later, wherein Gaza took into account Trapezuntius' criticism.[20] What I did not then realize was that even in the second edition, Gaza maintained his radical reordering of the *Problemata*.

Strange to say, the version of the *Problemata* that first made it into print was Gaza's original translation, which had been dedicated to Pope Nicholas V in the 1450s. It was printed at Mantua in 1473. Gaza's second edition appeared posthumously at Rome two years later, in 1475, edited by the Venetian physician Nicolaus Gupalatinus, who wrote a preface to Pope Sixtus IV.[21] Gaza is the first person to translate all thirty-eight books of the Aristotelian *Problemata* into Latin.[22] The medieval version stopped with problem 3 of book 37, and Trapezuntius went only as far as the end of book 33.[23] However, it was impossible to recognize this accomplishment, for Gaza reduced the thirty-eight books of the Greek text of the *Problemata* to twenty.[24] Some problems he simply eliminated because they were duplicated elsewhere or because they made no sense to him. Some problems he split into two because they seemed to him to involve two questions rather than one. But most of all he rearranged the problems. He folded book 6 of the Greek text into book 5 of his translation. He transferred book 9 of the Greek to the end of book 1 of his version. He inserted book 17 of the Greek into the middle of book 13 of his Latin text. He repositioned the Greek book 18 as book 12 in his Latin version. He combined books 23 and 24 of the Greek to form his book 16 and similarly made the Greek books 25 and 26 into his book 17. And he merged together books 31 through 38 of the Greek into one book, book 20, which closes his Latin translation.

Gaza made major, though not as radical, changes when translating the *De animalibus*. He decided that book 10 of the *Historia animalium* was nothing more than a misplaced Aristotelian fragment whose material better suited it to be an appendage to the *De generatione animalium*. Some Greek manuscripts in fact lack book 10. So he dropped the book altogether from his translation. He further decided that book 9 of the *Historia animalium* was misplaced in the Greek manuscripts, moving it in his Latin rendering to book 7 and making what had been books 7 and 8 in the Greek text his books 8 and 9. Finally, as noted above, he rearranged the sequence of passages at the end of book 8 (his book 9).

But whatever one may think of Gaza's reordering of the *Problemata* and of the *De animalibus*,[25] even more extraordinary was his sovereign manipulation of the words of the text. Contrary to conventional opinion, one can find

Renaissance translators who produced painfully literal translations. The six-teenth-century patristic scholar Ianus Cornarius immediately comes to mind.[26] But it still remains true that many humanists tended to paraphrase the texts they were translating for the sake of clarity and elegance. Gaza, how-ever, did more than paraphrase.[27] He significantly altered the text. For in-stance, the first problem of book 2 of the *Problemata* reads, in Hett's Loeb translation, "Why does sweat occur neither when men are straining nor when holding their breath, but rather when they let it go? Is it because if the breath is restrained, it fills the veins, so that it prevents it from escaping, *like the water from water-clocks if anyone turns them off when full?*" (ὥσπερ τὸ ὕδωρ τὸ ἐκ τῶν κλεψυδρῶν, ὅταν πλήρεις οὔσας ἐπιλάβῃ τις).[28] Gaza ren-dered the last clause: "as happens to receptacles in water, namely, that a re-ceptacle lowered into the sea collects sweet humidity within itself, and that is why it is called a clepsydra. For the humidity cannot flow when it is full and someone blocks the other side of it." In the revised version: *ut in aqua fit vasis, quod in mare demissum dulcem intra se colligit humorem, clepysdraque ob id no-minatur* [original version: *nominatum est*]. *Nequit enim effluere cum quis pleno iam vase partem alteram obturavit.*[29] Gaza inserted into the Aristotelian text words and information for which he had no textual warrant and which altered the text virtually beyond recognition. He did something similar in translating the *De animalibus,* substituting phrases and words from the *Naturalis historia* of the elder Pliny on occasion when he had trouble with Aristotle's Greek text. An early-nineteenth-century editor of the *Historia animalium,* Johann Gottlob Schneider, was annoyed enough by this practice to charge that "stulto con-silio Plinium nobis in vertendo Aristotele repraesentare quam fidem inter-pretis praestare maluit" (In translating Aristotle, [Gaza] foolishly chose to give us Pliny rather than stay true to the task of the translator).[30]

To take another example, problem 4 of book 1 of the *Problemata* asks, according to the Greek manuscripts, why one ought to use emetics (i.e., things that produce vomiting) at the change of the seasons, and then suggests, "Is it in order that there may be no disturbance at the time when the excretions are varying because of these changes?" (ἢ ἵνα μὴ γένηται συντάραξις, διαφόρων γινομένων τῶν περιττωμάτων διὰ τὰς μεταβολάς). Gaza believed, not unreasonably, that the Greek text had dropped an initial negative.[31] So the first edition of his Latin version—and all modern editions of the Greek, I might add—asks at the start why emetics ought to be avoided at the change of the seasons.[32] The emendation makes sense because the question that immediately follows seems to presume the negative: "Is it in order that there may be *no* disturbance . . . ?" Gaza had made a good conjecture; but not able to leave well enough alone, he appended to the problem a sentence that exists in no Greek manuscript. It runs as follows: "Whence it happens that some people ill digest their food and some have little desire for food" (*Unde fit ut cibum etiam alii parum concoquant, alii parum cupiant*).[33] The sentence does not clarify the problem in any significant way, and it is not clear what might have been Gaza's source or inspiration. Nonetheless, in the second edition of his translation of the *Problemata,* while dropping the negative at the start of the problem (i.e., his good conjecture),[34] he retained at the end the sentence that he had concocted and that modern editors ignore because they cannot figure out what it is doing there.

Since Gaza made similar textual changes throughout the *Problemata,* one begins to understand why Trapezuntius accused Gaza of perverting Aristotle's text—even quite apart from his reordering of the problems. However, Trapezuntius did not understand what Gaza was actually doing.

In Gaza's mind, to translate the received text of Aristotle was senseless. First of all, all the manuscripts were corrupt. "Depravati erant certe codices omnes," said Nicolaus Gupalatinus in describing Gaza's method to Pope Sixtus IV in the preface to the revised translation of the *Problemata,* which he put through the press after Gaza's death.[35] In his preface to the same pope for his translation of the *De animalibus,* Gaza had condemned the *exemplaria Graeca* as being defective (*mendosa*).[36] More important and even worse, the Greek manuscripts were all tainted by a sort of original sin, namely, the confused and faulty edition of the Aristotelian corpus published by Apellicon at Rome in the days of Sulla.[37] Gaza knew of Apellicon's activities from Strabo's *Geography* (13, 1.54), whose narrative he cited. Scholars today attribute the Roman edition of the Aristotelian corpus to Andronicus of Rhodes rather than to Apellicon, especially on the basis of a passage in Porphyry's *Life of Plotinus* (chap. 24) that escaped Gaza's notice.[38] But what Gaza did know sufficed for him to draw the conclusion that the received structure and even the very words of any Aristotelian text were a historical accident produced by an incompetent editor, and that he, Theodore Gaza, was no more obliged to follow the organization and the exact phrasing of the Greek than he was to repeat the scribal errors of the manuscripts. Consequently, to translate the authentic Aristotle he had to reedit the Greek text.[39] His goal was to recover the text Aristotle wrote, and *not* the text Apellicon edited.

As far as modern textual critics can judge—apart from those editors who suffer from the illusion that Gaza had at hand some extraordinary manuscript or manuscripts now lost to us—Gaza tended more often than not to depend on a mixture of manuscripts, which were not always the best.[40] In picking the correct reading from the various manuscripts at his disposal, and by frequently divining the correct reading when the manuscripts proved unsatisfactory, Gaza in effect created a new base text, which he considered superior to any extant Greek manuscript.[41] This is true for his translation of the *De animalibus* as well as of the *Problemata*. The result was that Gaza's translations usually make sense in precisely those passages where the Greek manuscripts and the previous Latin versions fail us. This is the reason why his readings are frequently cited in modern *apparatus critici,* even when they are rejected. Very often he was the first to identify a corruption or a lacuna and then to offer a solution.

In this respect, Gaza, who died in 1475, was literally before his time. Commissioned to make a translation, he produced instead a new edition of the Greek text in the guise of a Latin translation. Trapezuntius was absolutely right to attack him for foisting his translation on the Latin public without any real indication of how significantly he had departed from the Greek manuscripts. If Gaza had been active at the end of the century, he probably would have had a role in the *editio princeps* of the Greek text of Aristotle. But these Greek *editiones principes* did not report variants.[42] So he still would have been guilty of a daring reconstruction of the text without an *apparatus criticus* informing the reader in any detail of exactly what he was doing. Perhaps not until the nineteenth century would Gaza have come into his own. But even then he would have had trouble among his peers, since he was interested not in the *recensio* of the manuscripts but only in the *emendatio* of the text by dint of his erudition and his overly confident knack for conjecture.

In one respect, however, Gaza did fully announce to his readers what changes they should expect from him. In the preface to the first version of his translation of the *Problemata*, Gaza bragged that he was giving to readers of Latin a whole new scientific vocabulary based on proper classical usage. He violently attacked the medieval translators as ignoramuses whose barbarous rendering of Greek terms perverted Aristotle. As is well known, medieval translators often simply transliterated Greek technical terms. In the *Problemata*, Gaza proposed to replace the medieval *diaphragma* (diaphragm) with *transversum saeptum*, the medieval *colon* with *laxum intestinum*, the medieval *phlegma* (phlegm) with *pituita*, and so on.[43] In the preface to his translation of the *De animalibus*, he tempered his condemnation of traditional scientific Latin vocabulary but remained staunchly scornful of the attempts of his medieval predecessors to Latinize Greek science and philosophy.[44]

Such a transformation of vocabulary removed Gaza's translations and other similar humanist translations to a realm of discourse outside the medieval scientific tradition.[45] Gaza fully recognized this fact and expressly welcomed the destruction of the medieval interpretations and the "Aristotelian" vocabulary they had engendered.[46] Unlike his patron, Cardinal Bessarion, and his rival, George Trapezuntius, Gaza despised the medieval scholastic tradition. He saw his work as ending it and starting a revolutionary new chapter in Aristotelian scholarship. Yet in terms of technical vocabulary at least, the medieval scientific tradition successfully withstood Gaza's assault, as is obvious from the modern vernaculars.

But otherwise Gaza had his way in the Renaissance. As I have mentioned, Gaza finished the first version of his translation of the *Problemata* in 1454. This version enjoyed a notable manuscript diffusion (at the moment I know of thirteen) before being superseded by a printed revised version in 1475. After finishing the *Problemata* in 1454, Gaza immediately started on the *De animalibus,* but he did not finish this translation until the 1470s, early in the pontificate of Pope Sixtus IV. The Venetian medical doctor Nicolaus Gupalatinus saw the revised version of the *Problemata* translation through the press at Rome in 1475. In 1476 the future cardinal Ludovico Podocataro saw to the printing at Venice of Gaza's translation of the *De animalibus* from what Podocataro called the *architypus ipsius Theodori.*[47] Before they made it into print, these translations (i.e., the *De animalibus* translation and the second version of the *Problemata*) had had, as far as I can tell, a minimal manuscript diffusion. The competing coeval translations of the two texts by George Trapezuntius were as easily accessible in manuscript.[48] But Gaza had already won the image war. The patronage of Cardinal Bessarion and of Bessarion's former client Sixtus IV was critical here. No less so was the admiration Gaza enjoyed from the next generation of humanists. We have already seen Aldo Manuzio proclaim Gaza the hero of late quattrocento Italian humanists. Most important, Gaza's supporters got his translations into print, thereby relegating Trapezuntius' versions to inevitable obscurity.

Angelo Poliziano eventually shucked off his early enthusiasm for Gaza. In a well-known chapter of the first *Miscellaneorum Centuria,* he attacked Gaza for his arbitrary mistranslation of Aristotle's discussion of Hercules' melancholia in book 30 of the *Problemata.* He went on to say that Gaza was wrong to brag of having no help in translating the *De animalibus* when it is clear to anyone who compares the translations that he, Gaza, made silent use of the translation of Trapezuntius.[49] Elsewhere in the first *Centuria,* though not in the second *Centuria* (which has only become available in very recent times), Poliziano took Gaza to task again for his translation of the *De animalibus.*[50] Ermolao Barbaro (d. 1492) also came to temper his early admiration of Gaza as he worked on his emendations of Pliny and Dioscorides in the late 1480s and early 1490s. Eventually, the large majority of his references to Gaza in these emendations were to point out error rather than to express agreement.[51]

Nevertheless, after the mid-1470s Gaza's translations were the canonized versions. In the next century, when Julius Caesar Scaliger and Johannes Sambucus independently sought out a copy of Trapezuntius' translation of the *De animalibus*, they both failed.[52] By the sixteenth century, Gaza had, in effect, become the only game in town.

But a serious problem remained. Humanists may have admired Gaza's translations, but his version of the *Problemata* was essentially unusable by the medical doctors, who were in fact the largest group of professional readers of this text. Because Gaza had reordered the *Problemata,* they had a hard time finding their way in the work, and they certainly could not use his translation in conjunction with the standard medieval commentary of Pietro d'Abano. The person who first tried to find a solution to this difficulty was a medical doctor at Venice, Iacobus Surianus. In 1478 Surianus completed a manuscript (presently Additional MS 21978 of the British Library) in which for every problem he copied in the sequence of the Greek manuscripts first the medieval translation of Bartolomeo of Messina, then the translation of George Trapezuntius, and last the translation of Theodore Gaza.[53] Surianus' triplex edition of the *Problemata* is significant in that it shows that into the 1470s, before the printing of Gaza's version, Trapezuntius' translation was viewed as a viable alternative to Gaza's and also that to make Gaza usable he totally undid Gaza's reordering of the problems.

However, the key date in the transformation of Gaza's translation into a viable tool of medical learning was 1501, when another student of medicine, Dominico Massaria of Vicenza, published at Venice an edition of Pietro d'Abano's commentary on the *Problemata* together with the translations of Bartolomeo of Messina (now called the *translatio antiqua*) and of Theodore Gaza (now called the *translatio nova*).[54] We know very little about Massaria.[55] In 1511, he published at Venice *De Ponderibus et Mensuris Medicinalibus Libri Tres,* which Conrad Gesner thought highly enough of to revise; it appeared in 1584.[56] Massaria reordered Gaza's *Problemata* to conform with Pietro d'Abano's commentary and therefore, unintentionally, with the sequence of the Greek manuscripts. Massaria's edition, which was reprinted four times between 1505 and 1520,[57] also served as the model for Aldo Manuzio's 1504 edition of Gaza's scientific translations.[58]

The 1504 Aldine edition canonized Gaza as a translator of Aristotle. It contained Gaza's translations of Aristotle's *De animalibus* and *Problemata,* of Theophrastus' *De plantis,* and of pseudo-Alexander of Aphrodisias' *Problemata,* with no fewer than three sets of indices at the end. The title page carried a lengthy quotation from a preface of Ermolao Barbaro praising Gaza to Pope Sixtus IV as the translator without peer.[59] Manuzio improved on Massaria's edition of Gaza's translation of pseudo-Aristotle's *Problemata* by insert-ing into their proper place the last three problems of book 37 and all the problems of book 38 translated by Gaza but missing from Bartolomeo of Messina's translation and therefore from Pietro d'Abano's commentary.[60] The Aldine edition was reprinted three times, and became in turn the basis for most, if not all, the later editions of Gaza's translations.[61]

Most of these later editions were Latin *Opera Omnia* of Aristotle. The first of them, edited by Simon Grynaeus and published by Ioannes Opori-nus at Basel in 1538, is significant not only for its priority but also for its *De Aristotelis Operibus Censura* by Juan Luis Vives, who seems to have been the first to have asserted unequivocally that the pseudo-Aristotelian *Problemata,* though of Aristotelian origin, were in fact pseudonymous.[62] He was follow-ing in the traces of Erasmus, who in the preface to his 1531 Basel edition of the Greek text of Aristotle viewed the text as *contaminatum*.[63]

F. Edward Cranz and Charles B. Schmitt identified forty-one printings of Gaza's translation in the sixteenth century.[64] Actually, there were at least forty-four in addition to the four incunable editions.[65] We can also add twelve post-sixteenth-century printings, all but one *Opera Omnia* of Aris-totle, for a grand total of sixty printings.[66] Along the way, in 1590 Isaac Casaubon filled in some of the problems that Gaza had omitted.[67] It is this doctored version of Gaza's translation of the *Problemata* that continued to be reprinted after the sixteenth century and eventually found its way into vol-ume 3 of the 1831 Berlin Academy edition of Aristotle, from which modern scholars quote.

Gaza's success was almost total as far as the *Problemata* was concerned. After 1520, even the medieval version disappeared from "books in print." His only real competitor was the late medieval Latin pseudonymous collection of problems attributed to Aristotle, which Brian Lawn has called the "Omnes homines" problems and which had, according to Cranz and Schmitt, at least forty-six sixteenth-century printings in its Latin version, four more than Gaza's version of the Aristotelian *Problemata*.[68] Cranz and Schmitt's list of the "Omnes homines" omits the many vernacular versions and is, in fact, not complete even for the sixteenth century.[69] Moreover, the "Omnes homines" problems continued to be printed in Latin through the seventeenth century.[70] In its English version, combined with a book on midwifery and two sex man-uals, the "Omnes homines" problems continued to find readers into the twentieth century;[71] its printings thus far outnumbered those of Gaza's ver-sion of pseudo-Aristotle's *Problemata*.

Since the printings of Gaza's translation increasingly consisted of its in-clusion in *Opera Omnia* of Aristotle, where it appeared in parallel columns alongside the Greek text, and since the "Omnes homines" problems were constantly reprinted in collections of problems separate from other Aristotelian texts, one may distinguish the emergence by the late Renaissance of two distinct traditions concerning "Aristotelian" problems: a philological tradition and a semi-vulgar (in both the linguistic and colloquial senses of the word) medical tradition. The former treated a text of antique Greek origin of some significance for classical philosophy and science. This tradition became increasingly the exclusive province of professional humanists and humanistically inclined medical scholars. The other tradition favored a text of medieval Latin coinage that contained a good deal of what one might call "low" science and even religion and that interested general readers as well as medical practitioners.

For the *De animalibus,* the story of its dominance is somewhat the same. As we have seen, Ludovico Podocataro put Gaza's translation through the press at Venice in 1476. It was reprinted three more times in the fifteenth century, always at Venice.[72] These printings ensured Gaza's complete victory over Trapezuntius' competing version. The medieval translations suffered a similar fate. Neither Michael Scot's translation from the Arabic nor William of Moerbeke's translation from the Greek ever made it into print in the Renaissance. Since the major medieval authority on the *De animalibus,* Albertus Magnus' *De animalibus,* was a paraphrase with generous digressions, it could, and did, exist independently of Gaza's translation in the Renaissance. Even so, it soon faded from sight. After three incunable editions, it was printed in 1519, but not again until its inclusion in Albertus' *Opera Omnia* of 1651.[73]

Indicative of Gaza's success among Renaissance scholastics is the attitude of the sixteenth-century Italian scholastic and medical doctor Agostino Nifo. Having no other translation at his disposal, Nifo took Gaza's translation as his base text in his commentary on the *Historia animalium, De partibus animalium,* and *De generatione animalium.* He completed the commentary in 1534, but it was published only posthumously in 1546 with a dedication to Pope Paul III.[74] Under Gaza's influence, Nifo did not comment on book 10 of the *Historia animalium,* which Gaza had omitted. Nifo, who could read Greek, did report, however, that the Greek manuscripts had a different order for the books of the *Historia animalium* than that found in Gaza's translation.[75] But he preserved Gaza's scheme nonetheless, noting that Aldo Manuzio's Greek text had also followed that sequence.

Aldo Manuzio published the *editio princeps* of the Greek Aristotle in five volumes between 1495 and 1498.[76] It was in the preface to the volume containing the *De animalibus* that Manuzio eulogized Gaza as the hero of his own generation of Hellenists. Most important, though, he accepted Gaza's transposition of books 7 and 9 of the *Historia animalium*. Every Renaissance edition, Greek as well as Latin, maintained this transposition thereafter. Indeed, virtually every translation and Greek edition of the *Historia animalium* up to this day has maintained it, including the 1831 Berlin Academy edition of Immanuel Bekker, which has given us our system of numbering the pages and lines of Aristotle's writings. The only major exception to this pattern is the recent Loeb edition of *Historia animalium,* books 7–10, edited by D. M. Balme, who restored the order of the Greek manuscripts.[77]

Oddly enough, apart from the repositioning of book 9 of the *Historia animalium* as book 7 and the rearranging of the end of its book 7/8, Aldo Manuzio stuck rather closely to his Greek manuscripts. That meant that Manuzio made no further attempt to accommodate the readings of his Greek text to the Latin of Gaza's translation. He also printed book 10 of the *Historia animalium* in a separate fascicle at the end of the whole volume.[78] As already noted, Gaza had dropped this book from his translation because he judged it to contain mere *fragmenta*.[79] Manuzio accepted Gaza's view that the book consisted of *fragmenta,* as his introductory note to the *lector carissimus* makes clear.[80] But having come upon the book in one or more manuscripts after he had already printed books 1–9, Manuzio appended it to the end of the volume and thus reintroduced a book that Gaza had banished from the *Historia animalium*.

Manuzio's naïveté proved salutary for the text of the *De animalibus,* since it meant that from the start scholars who could read Greek had not only the full text of Aristotle but also a text relatively untainted by accommodation to Gaza's Latin translation.[81] The accommodation to Gaza began, however, with the next major Greek *Opera Omnia* of Aristotle, that published by Erasmus at Basel in 1531 with the help of Simon Grynaeus.[82] In his preface, Erasmus paid what seems to have become the almost obligatory homage to the translating acumen of Gaza. He also rightly acknowledged his debt to Manuzio for establishing the base Greek text. But then he went on to say specifically that "in the books on the animals not a few places have been cleaned up with the help of Theodore's translation."[83] The great Renaissance revising of the Greek text of Aristotle to conform to Gaza's translation had now begun. This process had already commenced with post-Gaza Greek manuscripts,[84] and then became even more noticeable in the second half of the sixteenth century. The important Greek *Opera Omnia* edited by Frederick Sylburg between 1584 and 1596 is symptomatic.[85] In the preface to the *De animalibus* volume he asserted that he had diligently studied the Greek texts of the earlier editions; but, he explained, "anywhere [these earlier Greek editions] were not satisfactory, I consulted the translation of Theodore

Gaza."[86] Even a cursory look at Sylburg's text and annotations will show that he constantly cited Gaza as evidence for the correct reading in the Greek. Gaza's Latin translation had become an independent, authoritative witness to the Greek text.

Until the middle of the sixteenth century, Latin readers did not have access to book 10 of the *Historia animalium:* Gaza had not translated it and no other translation had been printed. Finally, the great 1550–1552 Latin *Opera Omnia* of Aristotle produced by the Giunta Press in Venice included a translation of book 10 made expressly for this edition by the longtime teacher of Greek at Venice Giovanni Battista Feliciano.[87] Most subsequent Latin *Opera Omnia* included Feliciano's translation, and one used a rendering by Niccolò de Conti; but starting with Isaac Casaubon's Greco-Latin *Opera* of 1590, the rendering by Julius Caesar Scaliger became the standard version of book 10.[88]

Scaliger, however, had no personal connection with Casaubon's edition of 1590. He had died much earlier, in 1558. Casaubon appropriated the edition of Scaliger's translation published in 1584 by Scaliger's son Sylvius.[89] Julius Caesar Scaliger had, in fact, translated not merely book 10. As early as 1538, working mainly from the Aldine Greek *Opera Omnia,* he had completed a first draft of a translation of and annotations on the first nine books of the *Historia animalium.*[90] In the process he had become the greatest critic of Gaza since Trapezuntius. In his commentaries on Theophrastus' *Historia plantarum* and *De causis plantarum,* he consistently criticized Gaza's translation of Theophratus.[91] His *annotationes* to the *Historia animalium* were at many points a word-by-word refutation of Gaza, his favorite phrase being *Theodorus male.* Indeed, he started off the commentary with a discursus, in his first comment, on Gaza as an inadequate translator.[92] Unfortunately for Scaliger, his translation and accompanying annotations did not appear in print until 1619, edited by Philippe Maussac (who also edited other works of Scaliger). The 1619 edition was the only time Scaliger's translation was printed in the Renaissance. And no wonder. Not only had it come terribly late and long after it could have had any major influence, but it also suffered from the burden of Maussac's own *Animadversiones,* where Maussac, Scaliger's editor, corrected and criticized Scaliger.[93] Scaliger's translation was reprinted twice again, first in 1811, when Johann Gottlob Schneider adopted it in his Greco-Latin edition of the *Historia animalium.* Schneider did not think much of Scaliger's annotations, but he considered Scaliger's translation a decided improvement on Gaza's and good enough to make a new translation otiose except for some revision.[94] The Berlin Academy agreed with Schneider and adopted Scaliger's translation, as revised by Schneider, in volume 3 of the

Academy's *Opera Omnia* of Aristotle (though reproducing Gaza's translation for the *De partibus animalium* and *De generatione animalium*).

But this late vindication matters little in the face of Theodore Gaza's nearly total dominance over the Renaissance reception of the *De animalibus*.[95] The most startling aspect of this dominance is that it extended even to how the editors printed the Greek text of the *De animalibus*. The same, as we have seen, was pretty much the case for the *Problemata*.

Theodore Gaza was a serious and sophisticated student of Aristotle, and his translations are important today not for any purported lost Greek manuscript readings they preserve but rather for the understanding of the Aristotelian text they offer. He was also exceptionally fortunate in his own time. In the rough-and-tumble of Renaissance controversy, where few reputations survived unscathed, the translator whom Trapezuntius had condemned as the perverter of Aristotle, the translator whom an obscure editor had unintentionally rescued from inevitable suspicion (if not opprobrium) by making his translation of the *Problemata* conform to the needs of a medieval commentary, the translator whom Poliziano had harshly criticized,[96] the translator whom Aldo Manuzio praised while unintentionally contradicting, and the translator whom J. C. Scaliger showed in detail to have been a bold confector rather than a *fidus interpres* came in the end to control the Aristotelian texts he translated more than did any other student of Aristotle in the Renaissance.

APPENDIX 1   THE LETTER OF DOMINICUS MASSARIA TO THE READER IN THE VENICE, 1501 EDITION OF PIETRO D'ABANO'S COMMENTARY ON PSEUDO-ARISTOTLE'S *PROBLEMATA*

For an analysis of the contents of the edition and bibliography on Massaria, see notes 54 and 55 below. I have modernized the punctuation of the text but preserved its orthography, save for distinguishing u/v in accordance with modern standards. The letter is found on sig. cc4v of the edition.

Dominicus Massaria Vincentinus lectori salutem plurimum.

Quisquis es ad cuius manus divina hec commentaria in Aristotelis, philosophorum primi, Problemata pervenere, si placet priusquam preclari operis lectionem auspiceris, paucissima velim attendas diligenter.

Due tibi sese offerent Problematum e Greco in Latinum edite translationes, quorum prior, que vetustissima quidem est, tanta barbarie tantaque obscuritate deformata est ut plurimos clarissimosque philosophos sepissime defatigarit nostrumque ipsum Petrum Aponensem, maximi atque acutissimi ingenii virum et in omni scientiarum genere exercitatissimum, altera vero recentissima superioribus annis edita, interprete Theodoro Gaze, viro utriusque lingue ac dicendi artis peritissimo, que adeo aperta adeoque clara est ut a quovis vel mediocris ingenii viro intelligi facillime possit. Ego hanc, a priore non particularum solum sectione, verum problematum quoque dispositione non parum differentientem et absonam, ad illius redegi ordinem quem Petrus Aponensis commentatione non habenda neglectui presignivit quo utriusque et particularum simul et problematum ordo concordet. Sic enim utraque editio Petro exponente facile innotescet; re quidem una, solo dicendi caractere atque ordine variata non eadem videri potest. Labor itaque lectori minuetur; et, quod spernendum haud quaquam est, que sola, Petro etiam commentante, vetus editio locis non paucis caliginis ac tenebrarum plena, invia, atque inaccessibilis plerumque reddebatur, eandem disertissimi Theodori traductione, veluti maximo quodam ardentissimi solis fulgore, irradiatam, dissipata caligine porroque fugatis tenebris, planam apertam atque facile perviam studiosus quisque lector agnoscet. Quare si non quo ordine a Gaze conscripta sunt Problemata inveneris, nulla te tenebit admiratio. De industria namque id ipsum effecimus. Atque in hoc quod vides corpus digessimus, ut in hunc redacta ordinem instar membrorum coherentia convenientius quadrent, que indistincta atque promiscua in secunda conversione pernotata cognovimus. Hec de Problematibus.

Quod autem ad Petri spectat expositionem, cum illa vel impressorum negligentia vel correctorum forte nulla diligentia tot erroribus viciata esset, nunc deficiens, nunc superfluens ut locis innumeris aut problemati contrarius aut nullus interdum sensus elici posset, pro parte virili nostra, adhibitis exemplaribus vetustissimis priscisque, iuvantibus auctoribus a quibus Petrus noster excerpsit, ego ex pessima ac nullo pacto nonnumquam intelligibili, si non optimam, at mediocrem saltem et apertam satis expositionem deduxi. Siquid tamen adhuc emaculandum in Petri nostri commentariis, benigne lector, offenderis, quod non plurimum, ut arbitror, erit, nisi impressorum incuria vigiliarum nostrarum labor viciatus extiterit, mitem te mihi iudicem prebebis nec omnino damnabis quem non glorie cupiditas, non popularis aure consequende gratia ad hoc suscipiendum onus, verum magis communis litteratorum commodi futuri studium illexit. Illud etiam meminisse non pigebit ut quoties tibi inter legendum quid quod castigatione dignum prima fronte videatur occurrerit, priusquam super illo diiudices, id sepius legas relegasque ne aut rei ipsius difficultate aut Petri proprio et inusitato quodam dicendi genere impeditus aliter forte quam res sit sensisse comproberis. Vale.

APPENDIX 2   CONCORDANCES BETWEEN THE GREEK TEXT OF THE *PROBLEMATA* AND THEODORE GAZA'S LATIN TRANSLATION

This appendix includes three parts:

A. Concordance keyed to the sequence of Gaza's translation.

B. Concordance keyed to the sequence of the Greek text.

C. Table of problems omitted by Gaza and their restoration by later editors.

Here α refers to the original, 1454 version of Gaza's translation, and β refers to the revised, 1475 version of Gaza's translation.

These concordances do not take notice of Gaza's additions to and subtractions from individual problems. For the Greek text I have followed Bekker's numbers. In the case of Gaza's translation, the numbering in α and β agree except in those few instances where β includes a problem omitted by α (see table at C). In these instances, I have followed the numbering of β. Neither the *editio princeps* of α nor that of β number the problems. The dedication copy of α, Vatican City, Biblioteca Apostolica Vaticana, Vat. lat. 2111 (= G) does number the problems, but not always correctly. Eleven times the scribe miscounted a problem (at 2:12, 2:33, 5:22, 7:8, 9:6, 9:34, 17:9, 17:77, 17:81, 20:50, and 20:67), which led to a mistaken sequence. As a check on G, I consulted Milan Biblioteca Ambrosiana, MS, A 249 inf., transcribed by Niccolò Perotti. However, Perotti also nodded sometimes when numbering the problems. For instance, after 20:30 he numbered the next problem 20:22 instead of 20:31, which means that in his numbering the final problem in the original translation is given as no. 92 instead of, correctly, as no. 101.

A. Concordance Keyed to the Sequences of Gaza's Translation
(★ = a problem omitted in α but included in β)

Gaza	Greek	Gaza	Greek	Gaza	Greek	Gaza	Greek
1:1	1:1	1:35	1:36	1:69	9:11	2:30	2:30
1:2	1:2	1:36	1:37	1:70	9:12	2:31	2:31
1:3	1:3	1:37	1:38a	1:71	9:13	2:32	2:32
1:4	1:4	1:38	1:38b	1:72	9:14	2:33	2:33
1:5	1:5	1:39	1:39			2:34	2:34
1:6	1:6	1:40	1:40	2:1	2:1	2:35	2:35
1:7	1:7	1:41	1:41	2:2	2:2	2:36	2:36
1:8	1:8	1:42	1:42	2:3	2:3	2:37	2:37
1:9	1:9	1:43	1:43	2:4	2:4	2:38	2:38
1:10	1:10	1:44	1:44	2:5	2:5	2:39	2:39
1:11	1:11	1:45	1:45	2:6	2:6	2:40	2:40
1:12	1:12	1:46	1:46	2:7	2:7	2:41	2:41
1:13	1:13	1:47	1:47	2:8	2:8	2:42	2:42
1:14	1:14–15	1:48	1:48	2:9	2:9		
1:15	1:16	1:49	1:49	2:10	2:10	3:1	3:1
1:16	1:17	1:50	1:50a	2:11	2:11	3:2	3:2
1:17	1:18	1:51	1:50b	2:12	2:12	3:3	3:3
1:18	1:19	1:52	1:51	2:13	2:13	3:4	3:4
1:19	1:20	1:53	1:52	2:14	2:14	3:5	3:5
1:20	1:21	1:54	1:53	2:15	2:15	3:6	3:6
1:21	1:22	1:55	1:54	2:16	2:16	3:7	3:7
1:22	1:23	1:56	1:55	2:17	2:17	3:8	3:8
1:23	1:24	1:57	1:56	2:18	2:18	3:9	3:9
1:24	1:25	1:58	1:57	2:19	2:19	3:10	3:10
1:25	1:26	1:59	9:1	2:20	2:20	3:11	3:11
1:26	1:27	1:60	9:2	2:21	2:21	3:12	3:12
1:27	1:28	1:61	9:3	2:22	2:22	3:13	3:13
1:28	1:29	1:62	9:4	2:23	2:23	3:14	3:14
1:29	1:30	1:63	9:5	2:24	2:24	3:15	3:15
1:30	1:31	1:64	9:6	2:25	2:25	3:16	3:16
1:31	1:32	1:65	9:7	2:26	2:26	3:17	3:17
1:32	1:33	1:66	9:8	2:27	2:27	3:18	3:18
1:33	1:34	1:67	9:9	2:28	2:28	3:19	3:19
1:34	1:35	1:68	9:10	2:29	2:29	3:20	3:20

Concordance A (continued)

Gaza	Greek	Gaza	Greek	Gaza	Greek	Gaza	Greek
3:21	3:21	4:28	4:27	5:37	5:40	8:7	10:7
3:22	3:23	4:29	4:28	5:38	5:41	8:8	10:8
3:23	3:24	4:30	4:29	5:39	5:42	8:9	10:9
3:24	3:25a	4:31	4:30	5:40	6:1	8:10	10:10
3:25	3:25b	4:32	4:31	5:41	6:2	8:11	10:11
3:26	3:26	4:33	4:32	5:42	6:3	8:12	10:12
3:27	3:27			5:43	6:4	8:13	10:13
3:28	3:28	5:1	5:1	5:44	6:5	8:14	10:14
3:29	3:29	5:2	5:2	5:45	6:6	8:15	10:15
3:30	3:30	5:3	5:3	5:46	6:7	8:16	10:16
3:31	3:31	5:4	5:4			8:17	10:17
3:32	3:32	5:5	5:5	6:1	7:4	8:18	10:18
3:33	3:33	5:6	5:6	6:2	7:6	8:19	10:19
3:34	3:34	5:7	5:7	6:3	7:7	8:20	10:20
3:35	3:35	5:8	5:8	6:4	7:8	8:21	10:21
		5:9	5:9	6:5	7:5	8:22	10:22
4:1	4:1	5:10	5:10			8:23	10:23
4:2	4:2	5:11	5:11	7:1	8:1	8:24	10:24
4:3	4:3	5:12	5:12	7:2	8:5[1]	8:25	10:25
4:4	4:4	5:13	5:13	7:3	8:6	8:26	10:26
4:5	4:5	5:14	5:14	7:4	8:7	8:27	10:27
4:6	4:6a	5:15	5:15	7:5	8:9	8:28	10:28
4:7	4:6b	5:16	5:16	7:6	8:10	8:29	10:29
4:8	4:7	5:17	5:17	7:7	8:11	8:30	10:30
4:9	4:8	5:18	5:18	7:8	8:12	8:31	10:31
4:10	4:9	5:19	5:19	7:9	8:13	8:32	10:32
4:11	4:10	5:20	5:20	7:10	8:14	8:33	10:33
4:12	4:11	5:21	5:21	7:11	8:15	8:34	10:34
4:13	4:14	5:22	5:22	7:12	8:16	8:35	10:35
4:14	4:13	5:23	5:23	7:13	8:17	8:36	10:36
4:15	4:14	5:24	5:24	7:14	7:8	8:37	10:37
4:16	4:15	5:25	5:25	7:15	8:18	8:38	10:41
4:17	4:16	5:26	5:26	7:16	8:19	8:39	10:42
4:18	4:17	5:27	5:27	7:17	8:20	8:40	10:43
4:19	4:18	5:28	5:28	7:18	8:21	8:41	10:44
4:20	4:19	5:29	5:29	7:19	8:22	8:42	10:45
4:21	4:20	5:30	5:30			8:43	10:46
4:22	4:21	5:31	5:31	8:1	10:1	8:44	10:47
4:23	4:22	5:32	5:32	8:2	10:2	8:45	10:48
4:24	4:23	5:33	5:35	8:3	10:3	8:46	10:49
4:25	4:24	5:34	5:36	8:4	10:4	8:47	10:50
4:26	4:25	5:35	5:37	8:5	10:5	8:48	10:51
4:27	4:26	5:36	5:39	8:6	10:6	8:49	10:52

Concordance A (continued)

Gaza	Greek	Gaza	Greek	Gaza	Greek	Gaza	Greek
8:50	10:53	9:28	11:28	10:5	12:5	12:4	18:4
8:51	10:54	9:29	11:29	10:6	12:6	12:5	18:5
8:52	10:55	9:30	11:30	10:7	12:7	12:6	18:6★
8:53	10:56	9:31	11:31	10:8	12:8	12:7	18:7★
8:54	10:57	9:32	11:32	10:9	12:9	12:8	18:8★
8:55	10:58	9:33	11:33	10:10	12:10	12:9	18:9[2]
8:56	10:59	9:34	11:34	10:11	12:11a	12:10	18:10★
8:57	10:60	9:35	11:35	10:12	12:11b		
8:58	10:61	9:36	11:36	10:13	12:12	13:1	15:1
8:59	10:62	9:37	11:37	10:14	12:13	13:2	15:2
8:60	10:63	9:38	11:38	10:15	13:1	13:3	15:3
8:61	10:64	9:39	11:39	10:16	13:2	13:4	15:4★
8:62	10:65	9:40	11:40	10:17	13:3	13:5	15:5[3]
8:63	10:66	9:41	11:41	10:18	13:4	13:6	15:7
8:64	10:67	9:42	11:42	10:19	13:5	13:7	15:8
		9:43	11:43	10:20	13:6	13:8	15:9
9:1	11:1	9:44	11:44	10:21	13:7	13:9	15:10
9:2	11:2	9:45	11:45	10:22	13:8	13:10	15:11
9:3	11:3	9:46	11:46	10:23	13:9	13:11	15:12
9:4	11:4	9:47	11:47	10:24	13:10	13:12	15:13
9:5	11:5	9:48	11:48	10:25	13:11	13:13	16:1
9:6	11:6	9:49	11:49	10:26	13:12★	13:14	16:2
9:7	11:7	9:50	11:50			13:15	16:3
9:8	11:8	9:51	11:51	11:1	14:1	13:16	16:4
9:9	11:9	9:52	11:52	11:2	14:2	13:17	16:5
9:10	11:10	9:53	11:53	11:3	14:3	13:18	16:6
9:11	11:11	9:54	11:54	11:4	14:4	13:19	16:7
9:12	11:12	9:55	11:55	11:5	14:5	13:20	16:8
9:13	11:13	9:56	11:56	11:6	14:6	13:21	16:9
9:14	11:14	9:57	10:40	11:7	14:7	13:22	16:10
9:15	11:15	9:58	11:57	11:8	14:8	13:23	16:11
9:16	11:16	9:59	10:38	11:9	14:9	13:24	16:12
9:17	11:17	9:60	10:39	11:10	14:10	13:25	16:13
9:18	11:18	9:61	11:58	11:11	14:11	13:26	17:1
9:19	11:19	9:62	11:59	11:12	14:12	13:27	17:2
9:20	11:20	9:63	11:60	11:13	14:13	13:28	17:3
9:21	11:21	9:64	11:61	11:14	14:14	13:29	19:1
9:22	11:22	9:65	11:62	11:15	14:15	13:30	19:2
9:23	11:23			11:16	14:16	13:31	19:3
9:24	11:24	10:1	12:1			13:32	19:4★
9:25	11:25	10:2	12:2	12:1	18:1	13:33	19:5[4]
9:26	11:26	10:3	12:3	12:2	18:2	13:34	19:6
9:27	11:27	10:4	12:4	12:3	18:3	13:35	19:7

Concordance A (continued)

Gaza	Greek	Gaza	Greek	Gaza	Greek	Gaza	Greek
13:36	19:8	13:79	19:50	14:42	21:6	16:7	23:7
13:37	19:9			14:43	21:7	16:8	23:10
13:38	19:10	14:1	20:1	14:44	21:8	16:9	23:9
13:39	19:11	14:2	20:2	14:45	21:9	16:10	23:11
13:40	19:12	14:3	20:3	14:46	21:10	16:11	23:12
13:41	19:13	14:4	20:4	14:47	21:11	16:12	23:13
13:42	19:14	14:5	20:5	14:48	21:12	16:13	23:15
13:43	19:15	14:6	20:6	14:49	21:13	16:14	23:14
13:44	19:16	14:7	20:7	14:50	21:14	16:15	23:8
13:45	19:17	14:8	20:8	14:51	21:15	16:16	23:16
13:46	19:18	14:9	20:9	14:52	21:16	16:17	23:17
13:47	19:19	14:10	20:10	14:53	21:17	16:18	23:18
13:48	19:20	14:11	20:11	14:54	21:18	16:19	23:19
13:49	19:21	14:12	20:12	14:55	21:19	16:20	23:20
13:50	19:22	14:13	20:13	14:56	21:20	16:21	23:21
13:51	19:23	14:14	20:14	14:57	21:21	16:22	23:22
13:52	19:24	14:15	20:15	14:58	21:22	16:23	23:23
13:53	19:25	14:16	20:16	14:59	21:23	16:24	23:24
13:54	19:26	14:17	20:17	14:60	21:24	16:25	23:25
13:55	19:27	14:18	20:18	14:61	21:25	16:26	23:26
13:56	19:28	14:19	20:19	14:62	21:26	16:27	23:27
13:57	19:29	14:20	20:20			16:28	23:28
13:58	19:30	14:21	20:21	15:1	22:1	16:29	23:29
13:59	19:31	14:22	20:22	15:2	22:2	16:30	23:30
13:60	19:32	14:23	20:23	15:3	22:3	16:31	23:31
13:61	19:33	14:24	20:24	15:4	22:4	16:32	23:32
13:62	19:34	14:25	20:25	15:5	22:5	16:33	23:33
13:63	19:35	14:26	20:26	15:6	22:6	16:34	23:34a
13:64	19:36	14:27	20:27	15:7	22:7	16:35	23:34b
13:65	19:37	14:28	20:28★	15:8	22:8	16:36	23:35
13:66	19:38	14:29	20:29[5]	15:9	22:9	16:37	23:36
13:67	19:39a	14:30	20:30	15:10	22:10	16:38	23:37
13:68	19:39b	14:31	20:31	15:11	22:11	16:39	23:38
13:69	19:40	14:32	20:32	15:12	22:12	16:40	23:39
13:70	19:41	14:33	20:33	15:13	22:13	16:41	23:40
13:71	19:42	14:34	20:34	15:14	22:14	16:42	24:41
13:72	19:43	14:35	20:35			16:43	24:1
13:73	19:44	14:36	20:36	16:1	23:1	16:44	24:2
13:74	19:45	14:37	21:1	16:2	23:2	16:45	24:3
13:75	19:46	14:38	21:2	16:3	23:3	16:46	24:4
13:76	19:47	14:39	21:3	16:4	23:4	16:47	24:5
13:77	19:48	14:40	21:4	16:5	23:5	16:48	24:6
13:78	19:49	14:41	21:5	16:6	23:6	16:49	24:7

Concordance A (continued)

Gaza	Greek	Gaza	Greek	Gaza	Greek	Gaza	Greek
16:50	24:8	17:31	26:9	17:74	26:51	18:31	29:12
16:51	24:9	17:32	26:10	17:75	26:52	18:32	29:13
16:52	24:10	17:33	26:11	17:76	26:53	18:33	29:14
16:53	24:11	17:34	26:12	17:77	26:54	18:34	29:15
16:54	24:12	17:35	26:13	17:78	26:55	18:35	29:16
16:55	24:13	17:36	26:14	17:79	26:56		
16:56	24:14	17:37	26:15	17:80	26:57	19:1	30:1
16:57	24:15	17:38	26:16	17:81	26:58	19:2	30:2
16:58	24:16	17:39	26:17a	17:82	26:59	19:3	30:3
16:59	24:17	17:40	26:17b	17:83	26:60	19:4	30:5
16:60	24:18	17:41	26:18	17:84	26:61	19:5	30:6
16:61	24:19	17:42	26:19	17:85	26:62	19:6	30:7
		17:43	26:20			19:7	30:8
17:1	25:1	17:44	26:21	18:1	27:1	19:8	30:9
17:2	25:2	17:45	26:22	18:2	27:2	19:9	30:10
17:3	25:3	17:46	26:23	18:3	27:3	19:10	30:11
17:4	25:4	17:47	26:24	18:4	27:4	19:11	30:12
17:5	25:5	17:48	26:25	18:5	27:5	19:12	30:13
17:6	25:6	17:49	26:26	18:6	27:6	19:13	30:14
17:7	25:7	17:50	26:27	18:7	27:7		
17:8	25:8	17:51	26:28a	18:8	27:8	20:1	31:1
17:9	25:9	17:52	26:28b	18:9	27:9	20:2	31:2
17:10	25:10	17:53	26:29	18:10	27:10	20:3	31:3
17:11	25:11	17:54	26:31	18:11	27:11	20:4	31:4
17:12	25:12	17:55	26:33	18:12	28:1	20:5	31:5
17:13	25:13	17:56	26:34	18:13	28:2	20:6	31:6
17:14	25:14	17:57	26:35a	18:14	28:3	20:7	31:7
17:15	25:15	17:58	26:35b	18:15	28:4	20:8	31:8
17:16	25:16	17:59	26:36	18:16	28:5	20:9	31:9
17:17	25:17	17:60	26:37	18:17	28:6	20:10	31:10★
17:18	25:18	17:61	26:38	18:18	28:7	20:11	31:11[6]
17:19	25:19	17:62	26:39	18:19	28:8	20:12	31:12
17:20	25:20	17:63	26:40	18:20	29:1	20:13	31:13
17:21	25:21	17:64	26:41	18:21	29:2	20:14	31:14
17:22	25:22	17:65	26:42	18:22	29:3	20:15	31:15
17:23	26:1	17:66	26:43	18:23	29:4	20:16	31:16a
17:24	26:2	17:67	26:44	18:24	29:5	20:17	31:16b
17:25	26:3	17:68	26:45	18:25	29:6	20:18	31:17
17:26	26:6	17:69	26:46	18:26	29:7	20:19	31:18
17:27	26:7	17:70	26:47	18:27	29:8	20:20	31:19
17:28	26:8	17:71	26:48	18:28	29:9	20:21	31:20
17:29	26:4	17:72	26:49	18:29	29:10	20:22	31:21
17:30	26:5	17:73	26:50	18:30	29:11	20:23	31:22

Concordance A (continued)

Gaza	Greek	Gaza	Greek	Gaza	Greek	Gaza	Greek
20:24	31:23	20:44	33:1	20:64	34:3	20:84	36:1
20:25	31:24	20:45	33:2	20:65	34:4	20:85	36:2
20:26	31:25	20:46	33:3	20:66	34:5	20:86	36:3
20:27	31:26	20:47	33:4	20:67	34:6	20:87	37:1
20:28	31:27	20:48	33:5	20:68	34:7	20:88	37:2
20:29	31:28	20:49	33:6	20:69	34:8	20:89	37:3
20:30	31:29	20:50	33:7	20:70	34:9	20:90	37:4
20:31	32:1	20:51	33:8	20:71	34:10	20:91	37:5
20:32	32:2	20:52	33:9	20:72	34:11	20:92	37:6
20:33	32:3	20:53	33:10	20:73	34:12	20:93	38:1
20:34	32:4	20:54	33:11	20:74	35:1	20:94	38:2
20:35	32:5	20:55	33:12	20:75	35:2	20:95	38:3
20:36	32:6	20:56	33:13	20:76	35:3	20:96	38:4
20:37	32:7	20:57	33:14	20:77	35:4	20:97	38:5
20:38	32:8	20:58	33:15	20:78	35:5	20:98	38:6
20:39	32:9	20:59	33:16	20:79	35:6	20:99	38:7
20:40	32:10	20:60	33:17	20:80	35:7	20:100	38:8
20:41	32:11	20:61	33:18	20:81	35:8	20:101	38:9
20:42	32:12	20:62	34:1	20:82	35:9	20:102	38:11[7]
20:43	33:13	20:63	34:2	20:83	35:10		

1. α translates 8:2–4.

2. In α this is 12:6, the last problem in book 12.

3. In α from this point on the corresponding problem in book 13 is one digit lower.

4. In α from this point on the problems of book 13 are numbered two digits lower; see number 2 above.

5. In α from this point on the problems of book 14 are numbered one digit lower.

6. In α from this point on the problems of book 20 are numbered one digit lower.

7. The corresponding last problem in α is 20:101.

B. Concordance Keyed to the Sequence of the Greek Text

(★ = a problem omitted in α but included in β)

Greek	Gaza	Greek	Gaza	Greek	Gaza	Greek	Gaza
1:1	1:1	1:9	1:9	1:17	1:16	1:25	1:24
1:2	1:2	1:1	1:10	1:18	1:17	1:26	1:25
1:3	1:3	1:11	1:11	1:19	1:18	1:27	1:26
1:4	1:4	1:12	1:12	1:20	1:19	1:28	1:27
1:5	1:5	1:13	1:13	1:21	1:20	1:29	1:28
1:6	1:6	1:14	1:14a	1:22	1:21	1:30	1:29
1:7	1:7	1:15	1:14b	1:23	1:22	1:31	1:30
1:8	1:8	1:16	1:15	1:24	1:23	1:32	1:31

Concordance B (continued)

Greek	Gaza	Greek	Gaza	Greek	Gaza	Greek	Gaza
1:33	1:32	2:16	2:16	3:16	3:16	4:21	4:22
1:34	1:33	2:17	2:17	3:17	3:17	4:22	4:23
1:35	1:34	2:18	2:18	3:18	3:18	4:23	4:24
1:36	1:35	2:19	2:19	3:19	3:19	4:24	4:25
1:37	1:36	2:20	2:20	3:20	3:20	4:25	4:26
1:38	1:37	2:21	2:21	3:21	3:21	4:26	4:27
1:38b	1:38	2:22	2:22	3:22	Missing	4:27	4:28
1:39	1:39	2:23	2:23	3:23	3:22	4:28	4:29
1:40	1:40	2:24	2:24	3:24	3:23	4:29	4:30
1:41	1:41	2:25	2:25	3:25a	3:24	4:30	4:31
1:42	1:42	2:26	2:26	3:25b	3:25	4:31	4:32
1:43	1:43	2:27	2:27	3:26	3:26	4:32	4:33
1:44	1:44	2:28	2:28	3:27	3:27		
1:45	1:45	2:29	2:29	3:28	3:28	5:1	5:1
1:46	1:46	2:30	2:30	3:29	3:29	5:2	5:2
1:47	1:47	2:31	2:31	3:30	3:30	5:3	5:3
1:48	1:48	2:32	2:32	3:31	3:31	5:4	5:4
1:49	1:49	2:33	2:33	3:32	3:32	5:5	5:5
1:50a	1:50	2:34	2:34	3:33	3:33	5:6	5:6
1:50b	1:51	2:35	2:35	3:34	3:34	5:7	5:7
1:51	1:52	2:36	2:36	3:35	3:35	5:8	5:8
1:52	1:53	2:37	2:37			5:9	5:9
1:53	1:54	2:38	2:38	4:1	4:1	5:10	5:10
1:54	1:55	2:39	2:39	4:2	4:2	5:11	5:11
1:55	1:56	2:40	2:40	4:3	4:3	5:12	5:12
1:56	1:57	2:41	2:41	4:4	4:4	5:13	5:13
1:57	1:58	2:42	2:42	4:5	4:5	5:14	5:14
				4:6a	4:6	5:15	5:15
2:1	2:1	3:1	3:1	4:6b	4:7	5:16	5:16
2:2	2:2	3:2	3:2	4:7	4:8	5:17	5:17
2:3	2:3	3:3	3:3	4:8	4:9	5:18	5:18
2:4	2:4	3:4	3:4	4:9	4:10	5:19	5:19
2:5	2:5	3:5	3:5	4:10	4:11	5:20	5:20
2:6	2:6	3:6	3:6	4:11	4:12	5:21	5:21
2:7	2:7	3:7	3:7	4:12	4:13	5:22	5:22
2:8	2:8	3:8	3:8	4:13	4:14	5:23	5:23
2:9	2:9	3:9	3:9	4:14	4:15	5:24	5:24
2:10	2:10	3:10	3:10	4:15	4:16	5:25	5:25
2:11	2:11	3:11	3:11	4:16	4:17	5:26	5:26
2:12	2:12	3:12	3:12	4:17	4:18	5:27	5:27
2:13	2:13	3:13	3:13	4:18	4:19	5:28	5:28
2:14	2:14	3:14	3:14	4:19	4:20	5:29	5:29
2:15	2:15	3:15	3:15	4:20	4:21	5:30	5:30

Concordance B (continued)

Greek	Gaza	Greek	Gaza	Greek	Gaza	Greek	Gaza
5:31	5:31	8:16	7:15	10:22	8:22	10:66	8:63
5:32	5:32	8:17	7:16	10:23	8:23	10:67	8:64
5:33	Missing	8:18	7:18	10:24	8:24		
5:34	Missing	8:19	7:19	10:25	8:25	11:1	9:1
5:35	5:33	8:20	7:20	10:26	8:26	11:2	9:2
5:36	5:34	8:21	7:21	10:27	8:27	11:3	9:3
5:37	5:35	8:22	7:22	10:28	8:28	11:4	9:4
5:38	Missing			10:29	8:29	11:5	9:5
5:39	5:36	9:1	1:59	10:30	8:30	11:6	9:6
5:40	5:37	9:2	1:60	10:31	8:31	11:7	9:7
5:41	5:38	9:3	1:61	10:32	8:32	11:8	9:8
5:42	5:39	9:4	1:62	10:33	8:33	11:9	9:9
		9:5	1:63	10:34	8:34	11:10	9:10
6:1	5:40	9:6	1:64	10:35	8:35	11:11	9:11
6:2	5:41	9:7	1:65	10:36	8:36	11:12	9:12
6:3	5:42	9:8	1:66	10:37	8:37	11:13	9:13
6:4	5:43	9:9	1:67	10:38	9:59	11:14	9:14
6:5	5:44	9:10	1:68	10:39	9:60	11:15	9:15
6:7	5:45	9:11	1:69	10:40	9:57	11:16	9:16
7:1	Missing	9:12	1:70	10:41	8:38	11:17	9:17
7:2	Missing	9:13	1:71	10:42	8:39	11:18	9:18
7:3	Missing	9:14	1:72	10:43	8:40	11:19	9:19
7:4	6:1			10:44	8:41	11:20	9:20
7:5	6:5	10:1	8:1	10:45	8:42	11:21	9:21
7:6	6:2	10:2	8:2	10:46	8:43	11:22	9:22
7:7	6:3	10:3	8:3	10:47	8:44	11:23	9:23
7:8	6:4	10:4	8:4	10:48	8:45	11:24	9:24
7:9	Missing	10:5	8:5	10:49	8:46	11:25	9:25
		10:6	8:6	10:50	8:47	11:26	9:26
8:1	7:1	10:7	8:7	10:51	8:48	11:27	9:27
8:2	Missing[1]	10:8	8:8	10:52	8:49	11:28	9:28
8:3	Missing[2]	10:9	8:9	10:53	8:50	11:29	9:29
8:4	Missing[3]	10:10	8:10	10:54	8:51	11:30	9:30
8:5	7:5	10:11	8:11	10:55	8:52	11:31	9:31
8:6	7:6	10:12	8:12	10:56	8:53	11:32	9:32
8:7	7:8	10:13	8:13	10:57	8:54	11:33	9:33
8:8	7:17	10:14	8:14	10:58	8:55	11:34	9:34
8:9	7:7	10:15	8:15	10:59	8:56	11:35	9:35
8:10	7:9	10:16	8:16	10:60	8:57	11:36	9:36
8:11	7:10	10:17	8:17	10:61	8:58	11:37	9:37
8:12	7:11	10:18	8:18	10:62	8:59	11:38	9:38
8:13	7:12	10:19	8:19	10:63	8:60	11:39	9:39
8:14	7:13	10:20	8:20	10:64	8:61	11:40	9:40
8:15	7:14	10:21	8:21	10:65	8:62	11:41	9:41

Concordance B (continued)

Greek	Gaza	Greek	Gaza	Greek	Gaza	Greek	Gaza
11:42	9:42	13:8	10:22	16:8	13:21	19:24	13:51
11:43	9:43	13:9	10:23	16:9	13:22	19:25	13:52
11:44	9:44	13:10	10:24	16:10	13:23	19:26	13:53
11:45	9:45	13:11	10:25	16:11	13:24	19:27	13:54
11:46	9:46	13:12	10:26★	16:12	13:25	19:28	13:55
11:47	9:47			16:13	13:26	19:29	13:56
11:48	9:48	14:1	11:1	17:1	13:27	19:30	13:57
11:49	9:49	14:2	11:2	17:2	13:28	19:31	13:58
11:50	9:50	14:3	11:3	17:3	13:29	19:32	13:59
11:51	9:51	14:4	11:4			19:33	13:60
11:52	9:52	14:5	11:5	18:1	12:1	19:34	13:61
11:53	9:53	14:6	11:6	18:2	12:2	19:35	13:62
11:54	9:54	14:7	11:7	18:3	12:3	19:36	13:63
11:55	9:55	14:8	11:8	18:4	12:4	19:37	13:64
11:56	9:56	14:9	11:9	18:5	12:5	19:38	13:65
11:57	9:58	14:10	11:10	18:6	12:6★	19:39a	13:66
11:58	9:61	14:11	11:11	18:7	12:7★	19:39b	13:67
11:59	9:62	14:12	11:12	18:8	12:8★	19:40	13:68
11:60	9:63	14:13	11:13	18:9	12:9[5]	19:41	13:69
11:61	9:64	14:11	11:14	18:10	12:10★	19:42	13:70
11:62	9:65	14:15	11:15			19:43	13:71
		14:16	11:16	19:1	13:29	19:44	13:72
12:1	10:1			19:2	13:30	19:45	13:73
12:2	10:2	15:1	13:1	19:3	13:31	19:46	13:74
12:3	10:3	15:2	13:2	19:4	13:32★	19:47	13:75
12:4	10:4	15:3	13:3	19:5	13:33[6]	19:48	13:76
12:5	10:5	15:4	13:4★	19:6	13:33	19:49	13:77
12:6	10:6	15:5	13:5[4]	19:7	13:34	19:50	13:78
12:7	10:7	15:6	13:6	19:8	13:35		
12:8	10:8	15:7	13:7	19:9	13:36	20:1	14:1
12:9	10:9	15:8	13:8	19:10	13:37	20:2	14:2
12:10	10:10	15:9	13:9	19:11	13:38	20:3	14:3
12:11a	10:11	15:10	13:10	19:12	13:39	20:4	14:4
12:11b	10:12	15:11	13:11	19:13	13:40	20:5	14:5
12:12	10:13	15:12	13:12	19:14	13:41	20:6	14:6
12:13	10:14	15:13	13:13	19:15	13:42	20:7	14:7
				19:16	13:43	20:8	14:8
13:1	10:15	16:1	13:14	19:17	13:44	20:9	14:9
13:2	10:16	16:2	13:15	19:18	13:45	20:10	14:10
13:3	10:17	16:3	13:16	19:19	13:46	20:11	14:11
13:4	10:18	16:4	13:17	19:20	13:47	20:12	14:12
13:5	10:19	16:5	13:18	19:21	13:48	20:13	14:13
13:6	10:20	16:6	13:19	19:22	13:49	20:14	14:14
13:7	10:21	16:7	13:20	19:23	13:50	20:15	14:15

Concordance B (continued)

Greek	Gaza	Greek	Gaza	Greek	Gaza	Greek	Gaza
20:16	14:16	21:23	14:59	23:25	16:25	25:7	17:7
20:17	14:17	21:24	14:60	23:26	16:26	25:8	17:8
20:18	14:18	21:25	14:61	23:27	16:27	25:9	17:9
20:19	14:19	21:26	14:62	23:28	16:28	25:10	17:10
20:20	14:20			23:29	16:29	25:11	17:11
20:21	14:21	22:1	15:1	23:30	16:30	25:12	17:12
20:22	14:22	22:2	15:2	23:31	16:31	25:13	17:13
20:23	14:23	22:3	15:3	23:32	16:32	25:14	17:14
20:24	14:24	22:4	15:4	23:33	16:33	25:15	17:15
20:25	14:25	22:5	15:5	23:34a	16:34	25:16	17:16
20:26	14:26	22:6	15:6	23:34b	16:35	25:17	17:17
20:27	14:27	22:7	15:7	23:35	16:36	25:18	17:18
20:28	14:28★	22:8	15:8	23:36	16:37	25:19	17:19
20:29	14:29[7]	22:9	15:9	23:37	16:38	25:20	17:20
20:30	14:30	22:10	15:10	23:38	16:39	25:21	17:21
20:31	14:31	22:11	15:11	23:39	16:40	25:22	17:22
20:32	14:32	22:12	15:12	23:40	16:41		
20:33	14:33	22:13	15:13	23:41	16:42	26:1	17:23
20:34	14:34	22:14	15:14			26:2	17:24
20:35	14:35			24:1	16:43	26:3	17:25
20:36	14:36	23:1	16:1	24:2	16:44	26:4	17:29
		23:2	16:2	24:3	16:45	26:5	17:30
21:1	14:37	23:3	16:3	24:4	16:46	26:6	17:26
21:2	14:38	23:4	16:4	24:5	16:47	26:7	17:27
21:3	14:39	23:5	16:5	24:6	16:48	26:8	17:28
21:4	14:40	23:6	16:6	24:7	16:49	26:9	17:31
21:5	14:41	23:7	16:7	24:8	16:50	26:10	17:32
21:6	14:42	23:8	16:15	24:9	16:51	26:11	17:33
21:7	14:43	23:9	16:9	24:10	16:52	26:12	17:34
21:8	14:44	23:10	16:8	24:11	16:53	26:13	17:35
21:9	14:45	23:11	16:10	24:12	16:54	26:14	17:36
21:10	14:46	23:12	16:11	24:13	16:55	26:15	17:37
21:11	14:47	23:13	16:12	24:14	16:56	26:16	17:38
21:12	14:48	23:14	16:14	24:15	16:57	26:17a	17:39
21:13	14:49	23:15	16:13	24:16	16:58	26:17b	17:40
21:14	14:50	23:16	16:16	24:17	16:59	26:18	17:41
21:15	14:51	23:17	16:17	24:18	16:60	26:19	17:42
21:16	14:52	23:18	16:18	24:19	16:61	26:20	17:43
21:17	14:53	23:19	16:19	25:1	17:1	26:21	17:44
21:18	14:54	23:20	16:20	25:2	17:2	26:22	17:45
21:19	14:55	23:21	16:21	25:3	17:3	26:23	17:46
21:20	14:56	23:22	16:22	25:4	17:4	26:24	17:47
21:21	14:57	23:23	16:23	25:5	17:5	26:25	17:48
21:22	14:58	23:24	16:24	25:6	17:6	26:26	17:49

Concordance B (continued)

Greek	Gaza	Greek	Gaza	Greek	Gaza	Greek	Gaza
26:27	17:50	27:6	18:6	30:12	19:11	32:10	20:40
26:28a	17:51	27:7	18:7	30:13	19:12	32:11	20:41
26:28b	17:52	27:8	18:8	30:14	19:13	32:12	20:42
26:29	17:53	27:9	18:9			32:13	20:43
26:30	Missing	27:10	18:10	31:1	20:1	33:1	20:44
26:31	17:54	27:11	18:11	31:2	20:2	33:2	20:45
26:32	Missing			31:3	20:3	33:3	20:46
26:33	17:55	28:1	18:12	31:4	20:4	33:4	20:47
26:34	17:56	28:2	18:13	31:5	20:5	33:5	20:48
26:35a	17:57	28:3	18:14	31:6	20:6	33:6	20:49
26:35b	17:58	28:4	18:15	31:7	20:7	33:7	20:50
26:36	17:59	28:5	18:16	31:8	20:8	33:8	20:51
26:37	17:60	28:6	18:17	31:9	20:9	33:9	20:52
26:38	17:61	28:7	18:18	31:10	20:10★	33:10	20:53
26:39	17:62	28:8	18:19	31:11	20:11[8]	33:11	20:54
26:40	17:63			31:12	20:12	33:12	20:55
26:41	17:64	29:1	18:20	31:13	20:13	33:13	20:56
26:42	17:65	29:2	18:21	31:14	20:14	33:14	20:57
26:43	17:66	29:3	18:22	31:15	20:15	33:15	20:58
26:44	17:67	29:4	18:23	31:16	20:16	33:16	20:59
26:45	17:68	29:5	18:24	31:17a	20:17	33:17	20:60
26:46	17:69	29:6	18:25	31:17b	20:18	33:18	20:61
26:47	17:70	29:7	18:26	31:18	20:19		
26:48	17:71	29:8	18:27	31:19	20:20	34:1	20:62
26:49	17:72	29:9	18:28	31:20	20:21	34:2	20:63
26:50	17:73	29:10	18:29	31:21	20:22	34:3	20:64
26:51	17:74	29:11	18:30	31:22	20:23	34:4	20:65
26:52	17:75	29:12	18:31	31:23	20:24	34:5	20:66
26:53	17:76	29:13	18:32	31:24	20:25	34:6	20:67
26:54	17:77	29:14	18:33	31:25	20:26	34:7	20:68
26:55	17:78	29:15	18:34	31:26	20:27	34:8	20:69
26:56	17:79	29:16	18:35	31:27	20:28	34:9	20:70
26:57	17:80			31:28	20:29	34:10	20:71
26:58	17:81	30:1	19:1	31:29	20:30	34:11	20:72
26:59	17:82	30:2	19:2			34:12	20:73
26:60	17:83	30:3	19:3	32:1	20:31		
26:61	17:84	30:4	Missing	32:2	20:32	35:1	20:74
26:62	17:85	30:5	19:4	32:3	20:33	35:2	20:75
		30:6	19:5	32:4	20:34	35:3	20:76
27:1	18:1	30:7	19:6	32:5	20:35	35:4	20:77
27:2	18:2	30:8	19:7	32:6	20:36	35:5	20:78
27:3	18:3	30:9	19:8	32:7	20:37	35:6	20:79
27:4	18:4	30:10	19:9	32:8	20:38	35:7	20:80
27:5	18:5	30:11	19:10	32:9	20:39	35:8	20:81

Concordance B (continued)

Greek	Gaza	Greek	Gaza	Greek	Gaza	Greek	Gaza
35:9	20:82	37:1	20:87	38:1	20:93	38:7	20:99
35:10	20:83	37:2	20:88	38:2	20:94	38:8	20:100
		37:3	20:89	38:3	20:95	38:9	20:101
36:1	20:84	37:4	20:90	38:4	20:96	38:10	Missing
36:2	20:85	37:5	20:91	38:5	20:97	38:11	20:102[9]
36:3	20:86	37:6	20:92	38:6	20:98		

1. α translates.

2. α translates.

3. α translates.

4. In α from this point on the corresponding problem of book 13 is numbered one digit lower.

5. This is 12:6 in α.

6. In α from this point on the corresponding problem in book 13 is numbered two digits lower; see note 4 above.

7. In α from this point on the corresponding problem of book 14 is numbered one digit lower.

8. In α from this point on the corresponding problem of book 20 is one digit lower.

9. In α 20:101 is the last problem, corresponding to 38:11 of the Greek text.

C.   Table of Problems Omitted by Gaza and Their Restoration by Later Editors

Note that only the second and third columns reflect Gaza's own choices. The other columns reflect the vagaries of editors. Note also that in the revised translation, Gaza dropped some problems that he had translated in the first version and translated other problems that he had omitted in the first version. For the original translation I used G and the ca. 1473 *editio princeps*.

Problem (as numbered by Bekker)	Translatio Originalis (1454)	Translatio Emendata (1475)	Editio Massaria (1501)	Editio Aldina (1504)	Editio Casaubon (1590)	Editio Berolina (1831)
3:22	Missing	Missing	Missing	Missing	Text in Italics[1]	Text
5:33	Missing	Missing	Missing	Missing	Text[2]	Text
5:34	Missing	Missing	Missing	Missing	Text[3]	Text
5:38	Missing	Missing	Missing	Missing	Text[4]	Text
7:1	Missing	Missing	Missing	Missing	Missing	Missing
7:2	Missing	Missing	Missing	Missing	Missing	Missing
7:3	Missing	Missing	Missing	Missing	Missing	Missing
7:5–8	6,7,8,5[5]	6,7,8,5,	6,7,8,5,	6,7,8,5	5–8	5–8
7:9	Missing	Missing	Missing	Missing	Text[6]	Text
8:2	Text	Missing	Missing[7]	Missing	Missing	Missing

Concordance C (continued)

Problem (as numbered by Bekker)	Translatio Originalis (1454)	Translatio Emendata (1475)	Editio Massaria (1501)	Editio Aldina (1504)	Editio Casaubon (1590)	Editio Berolina (1831)
8:3	Text	Missing	Missing	Missing	Missing	Missing
8:4	Text	Missing	Missing	Missing	Missing	Missing
8:22	Text	Text	Text[8]	Text	Text	Text
10:38	Text	Missing	Missing	Missing	Missing	Text
10:39	Text	Missing	Missing	Missing	Missing	Text
10:40	Text	Missing	Missing	Missing	Missing	Text
13:12	Missing	Text	Text	Text	Text	Text
15:4	Missing	Missing	Missing	Missing	Missing	Missing
18:6	Missing	Text	Text	Text	Text	Text
18:7	Missing	Text	Text	Text	Text	Text
18:8	Missing	Text	Text	Text	Text	Text
18:10	Missing	Text	Text	Text	Text	Text
19:4	Missing	Text	Text	Text	Text	Text
20:28	Missing	Text	Text	Text	Text	Text
26:12	Text	Text	Partial Text[9]	Partial Text[10]	Text	Text
26:30	Missing	Missing	Missing	Missing	Text[11]	Text
26:32	Missing	Missing	Text[12]	Missing	Missing	Missing
30:4	Missing	Missing	Missing	Missing	Text[13]	Text
31:10	Missing	Text[14]	Text	Text	Text	Text
38:10	Missing	Missing	Missing	Missing	Text[15]	Text

1. Casaubon's translation; he noted in the margin that Gaza omitted this problem.

2. Casaubon repeated here without notice Gaza's translation of problem 1:46.

3. Casaubon repeated here without notice Gaza's translation of problem 1:52.

4. Casaubon repeated here without notice Gaza's translation of problem 1:39.

5. No problem is omitted in 7:5–8. Gaza's sequence was initially preserved because at this point it conformed with that of the particular Greek manuscript tradition followed by Bartolomeo of Messina and was therefore traditional. See concordance B above at 7:5.

6. Casaubon repeated here Gaza's translation of problem 1:38, revising it to correspond to the Greek of 7:9. In a note at the start of book 7, he acknowledged the repetition but said nothing about the revision.

7. Massaria substituted 8:22 here.

8. Seemingly missing because Massaria transposed 8:22 to 8:2.

9. Massaria gave here only the opening lines of 26:12 (up to *anniversarii prohibent* in Gaza's translation). Note also that he numbered this problem 26:11 because the commentary skips the real 26:11.

10. The same partial text as in Massaria's edition.

11. Casaubon repeated here (as problem 26:33 in his numeration) without notice Gaza's translation of problem 23:16.

12. Massaria put here Gaza's translation of 26:12.

13. Casaubon repeated here without notice Gaza's translation of problem 5:25.

14. Note that this edition transposed 31:9 and 31:10.

15. Casaubon repeated here without notice Gaza's translation of problem 21:24.

Notes

The following abbreviations are used in the notes:

AL   *Aristoteles Latine Interpretibus Variis,* ed. Academia Regia Borussica (Berlin: Reimerus, 1831).

BAV   Vatican City, Biblioteca Apostolica Vaticana.

CS   F. E. Cranz, *A Bibliography of Aristotle Editions, 1501–1600,* 2 ed. with addenda and revisions by Charles B. Schmitt, Bibliotheca Bibliographica Aureliana, 38 (Baden-Baden: Koerner, 1984). Note that CS numbers without suffixes are identical to those in *Index Aureliensis: Catalogus Librorum Sedecimo Saeculo Impressorum,* vol. 2 (Baden-Baden: Koerner, 1996).

G   BAV, MS Vat. lat. 2111, dedication copy of the original version of Theodore Gaza's translation of pseudo-Aristotle's *Problemata.*

GA   *De generatione animalium.*

Goff   F. R. Goff, *Incunabula in American Libraries: A Third Census of Fifteenth-Century Books Recorded in North American Collections* (New York: Bibliographical Society of America, 1964).

GW   *Gesamtkatalog der Wiegendrucke,* 9 vols. to date (Leipzig: K. W. Hiersemann, 1925–).

H   L. Hain, *Repertorium Bibliographicum, in Quo Libri Omnes ab Arte Typographica Inventa usque ad Annum M. D. Typis Expressi Ordine Alphabetico Recensentur,* 2 vols. in 4 (Stuttgart: J. G. Cotta, 1826–1838).

HA   *Historia animalium.*

IA   *Index Aureliensis: Catalogus Librorum Sedecimo Saeculo Impressorum,* 13 vols. to date (Baden-Baden: Koerner, 1962).

PA   *De partibus animalium.*

Verzeichnis   *Verzeichnis der im deutschen Sprachbereich erschienenen Drucke des XVI. Jahrhunderts,* 24 vols. to date (Stuttgart: Anton Hiersemann, 1983–).

1. There are two *editiones principes* of Gaza's translation of the *Problemata.* The first is of Gaza's original translation, dedicated to Pope Nicholas V (d. 1455). This was printed at Mantua about 1473 without Gaza's preface by Johannes Vurster and Johannes Baumeister (H 1729; GW 2452; Goff G-1030). The second *editio princeps* is of Gaza's revised translation, published at Rome by Johannes Reinhard, who finished printing it on 19 May 1475 (H 1730; GW 2453; Goff G-1031). Gaza had died earlier in 1475. This edition carries a preface by Nicholas Gupalatinus, a *medicus Venetus* and disciple of Gaza, addressed to Pope Sixtus IV. All subsequent editions contain the revised translation. The remaining two incunable editions are composite editions containing Giorgio Valla's translation of pseudo-Alexander of Aphrodisias' *Problemata* and Giovan Pietro d'Avenza's translation of Plutarch's *Problemata,* first printed at Venice by Antonius de Strata, 1488–1489 (GW 860; Goff A-387), and then reprinted, probably at Venice, in 1489 or later (GW 861, which knows only the copy in the Bibliothèque Nationale in Paris). The edition of 1499 reported by Flashar is a ghost (Aristotle, *Problemata Physica,* trans. H. Flashar [Berlin:

Akademie Verlag, 1962], p. 376). For Valla's translation in the composite editions, see F. E. Cranz, "Alexander Aphrodisiensis," in *Catalogus Translationum et Commentariorum: Mediaeval and Renaissance Latin Translations and Commentaries,* ed. P. O. Kristeller, E. F. Cranz, and V. Brown, 7 vols. to date (Washington, D.C.: Catholic University of America, 1960–), 1:130–32, 2:419.

2. CS, p. 219, reports no competing Renaissance Latin translation in print up to 1600. Nicasius Ellebodius made a translation of books 1–11 that survives in Milan, Biblioteca Ambrosiana, MS D 291 inf., fols. 1r–45v (annotations: fols. 47r–72r; his annotations are also extant in the autograph MS O 246 sup., fols. 27r–39v); see D. Wagner, "Zur Biographie des Nicasius Ellebodius (d. 1577) und zu seinen 'Notae' zu den aristotelischen Magna Moralia," *Sitzungsberichte der Heidelberger Akademie der Wissenschaften: Philosophisch-historische Klasse* 5 (1973): 15, 28. The only other fifteenth-century translation was by George Trapezuntius, Gaza's fellow Greek émigré and archrival. For the history of his translation, see my *George of Trebizond: A Biography and a Study of His Rhetoric and Logic* (Leiden: Brill, 1976), pp. 74–75, 150–151, 176–177; and *Collectanea Trapezuntiana: Texts, Documents, and Bibliographies of George of Trebizond* (Binghamton, N.Y.: Medieval and Renaissance Texts and Studies, 1984), pp. 707–709, where I list ten Renaissance manuscripts.

3. *Aristoteles Latinus,* ed. G. Lacombe, A. Birkenmajer, M. Dulong, E. Franceschini, and L. Minio-Paluello, 2 vols. (vol. 1: Rome: Libreria dello Stato, 1939; vol. 2: Cambridge: Cambridge University Press, 1955), 2:1345, lists fifty-two manuscripts (not counting the items in parentheses). To this list B. Lawn, *The Salernitan Questions: An Introduction to the History of Medieval and Renaissance Problem Literature* (Oxford: Clarendon, 1963), p. 93 n. 3, adds four more manuscripts. An edition of book 1 in Bartolomeo's translation is to be had in R. Seligsohn, *Die Übersetzung der pseudo-aristotelischen Problemata durch Bartholomaeus von Messina, Text und textkritische Untersuchungen zum erstem Buch,* Inaugural-Dissertation (Berlin: Dr. Emil Ebering, 1934); but see Marenghi in note 18 below for an edition of this and other books of Bartolomeo's translation. The early-fourteenth-century scholastic Pietro d'Abano claimed to have made an independent translation; but if he did, the text does not survive; see Lawn, pp. 92–93; for Pietro as a translator, see M.-T. d'Alverny, "Pietro d'Abano, traducteur de Galien," *Medioevo* 11 (1985): 19–64 (reprinted in her *La transmission des textes philosophiques et scientifiques au Moyen Age,* ed. C. Burnett [Aldershot, Hants.: Variorum, 1994], art. XIII).

4. There are two incunable editions (Mantua: Paulus de Butzbach, 1475 [H 16; Goff P-436]; and Venice: Johannes Herbort, 1482 [H 17; Goff P-437]) and five sixteenth-century editions (Venice: Bonetus Locatellus, 1501 [CS 107.694]; Venice: Gregorius de Gregoriis, 1505 [CS 107.726]; Venice: Luca-Antonius de Giunta, 1518 [CS 107.852]; Venice: Octavianus Scotus, 1519 [CS 107.863]; and Paris: A. Boucard, impensis J. Petit, 1520 [CS 107.868]). We must note here that CS mistakenly treat as editions of the "Vulgate translation (Bartholomaeus de Messina)" what were in fact editions of the purely medieval Latin confection known from its incipit as the "Omnes homines" collection of problems. This mistake carried over to C. B. Schmitt and D. Knox, *Pseudo-Aristoteles Latinus: A Guide to Latin Works Falsely Attributed to Aristotle before 1500* (London: Warburg Institute, 1985), where this medieval forgery is not listed; see the remarks of J. Kraye, "The Printing History of Aristotle in the Fifteenth Century: A Bibliographical Approach to Renaissance Philosophy," *Renaissance Studies* 9 (1995): 209. Concerning "Omnes homines,"

see Lawn, *Salernitan Questions,* pp. 99–103. The editions listed by CS that are of the "Omnes homines" problems rather than of Bartolomeo's translation are in CS's 107 series nos. 687, 692A, 720 (identified in CS as Gaza's translation), 727, 735, 756, 759, 795, 816, 820, 829, 901, 926A, 934A, 955A; and in the 108 series, nos. 002A, 002B, 087A, 140, 159, 180, 207, 240B, 273I, 292, 292A, 294, 340D, 359B, 363A, 433, 459A, 496A, 506, 541A, 550A, 553 (the index mistakenly lists 533, which in fact does not exist: the numbering in the *Index Aureliensis* skips from 108.527 to 108.538), 554A, 562, 570B, 584, 637, 641, 650B, 656, and 662. For incunabula of the Latin version of the "Omnes homines" problems, see *GW* 2454–2461.

5. Pietro's was the standard medieval commentary; see N. G. Siraisi, "The *Expositio Problematum* of Peter of Abano," *Isis* 61 (1970): 321–339; and Lawn, *Salernitan Questions,* pp. 92–95. For another, unprinted medieval commentary (usually condensing Peter of Abano) probably by William Burley, see Lawn, p. 95, and L. Thorndike, "Peter of Abano and Another Commentary on the Problems of Aristotle," *Bulletin of the History of Medicine* 29 (1955): 517–523, who did not recognize the possible attribution to Burley. Finally, Lawn, pp. 96, 208, calls attention to two other anonymous commentaries and four epitomes.

6. As we shall see shortly, the first of these composite editions of Bartolomeo's and Gaza's translations, that of Venice, 1501, is quite important.

7. Venice: Ioannes de Colonia and Ioannes Manthen, 1476 (*GW* 2350, H 1699, Goff A-973), with a preface of Theodore Gaza to Pope Sixtus IV, edited by "Ludovicus Podocatharus Cyrius ex archetypo ipsius Theodori," according to the colophon on sig. ff6r. Gaza did not translate the *De motu animalium* or the *De incessu animalium.*

8. There were two main medieval versions: the *vetus* in 19 books (the *HA* in 10 books, the *PA* in 4 books, and the *GA* in 5 books), which Michael Scot produced in the first half of the thirteenth century from the Arabic and which formed the basis of the commentaries of Peter of Spain and Albertus Magnus; and the *nova* in 21 books (because of the inclusion of the one book of the *De incessu animalium* and of the one book of the *De motu animalium,* both of which were missing from the Arabic tradition) by William of Moerbeke from the Greek in the second half of the thirteenth century. An anonymous version of the *PA* exists in a *codex unicus,* Padua, Bibl. Antoniana, MS Scaff. XVII, 370, fols. 88r–117v (= *Aristoteles Latinus,* 2:1031–1032, no. 1503). There is also evidence of other anonymous translations of different parts of the *De animalibus;* see *Aristoteles Latinus,* 2:788, 1289 (note that the "translation" of the *HA* by Petrus Gallegus in BAV MS, Vat. lat. 1288, fols. 131r–155r [ibid., 2:1258–1259, no. 1937], is really an epitome; see A. Pelzer, "Un traducteur inconnu: Pierre Gallego, franciscan et premier évêque de Carthagène (1250–1267)," in *Miscellanea Francesco Ehrle,* Studi e testi 37 [Vatican City: Biblioteca Apostolica Vaticana, 1923], 1:414–422, 435–447). Concerning the translations of Michael Scot and William of Moerbeke, see *Aristoteles Latinus,* 1:174–180 and 2:1288–1289, where, if one does not count items with question marks or in parentheses, there are listed 234 manuscripts containing all or part of William's translation and 60 manuscripts of Michael's.

9. In the fifteenth century, George Trapezuntius translated the same three parts of the *De animalibus* as Gaza; see my *George of Trebizond,* pp. 72–73, 77–78; and *Collectanea Trapezun-*

tiana, pp. 705–707, where I list nine manuscripts. In the sixteenth century, the only meager competition was the Latin translation of and commentary on book 1 of the *PA* by the Paduan professor Nicolaus Thomaeus Leonicus (d. 1531), which appeared posthumously at Venice (Ioannes de Farris et fratres) in 1540 (CS 107.998), edited by his nephew Magnus Leonicus, and then again at Paris (Iacobus Bogardus) in 1542 (CS 108.044). Gaza never translated book 10 of the *HA.* In the course of the sixteenth century, three different translations appeared to make good this omission, the first by the teacher of Greek at Venice Giovanni Battista Feliciano (Ioannes Bernardus Felicianus) in the famous Giunta *Opera Omnia* of Aristotle at Venice, 1550–1552 (CS 108.193); for a while this version became the standard version in the *Opera Omnia* (cf. CS 108.400 [Lyons: haeredes Iacobi Iuntae, 1556], 108.430D [ibid., 1561], 108.579 [Venice: ad signum seminantis, 1572], 108.599 [Venice: Iunta, 1579], 108.610 [Venice: Gaspar Bindoni, 1576], 108.644 [Lyons: Iacobus Berjon, 1580], 108.646 [Lyons: Iunta, 1580?], and 108.669 [Venice: Ioachim Bruniolus and Nicolaus Moretti, 1584–1585]). Julius Caesar Scaliger (d. 1558) produced a Latin version of book 10, which his son Sylvius published in 1584 (Lyons [*recte* Geneva]: Jacques Stoer for Antoine de Harsy; CS 108.665) and which thereafter replaced Felicianus' version in the *Opera Omnia* of Aristotle (CS 108.708 [ed. Isaac Casaubon; Lyons: Iacobus Bubonius and Guillelmus Laemarius, 1590), 108.722 [Frankfurt a. M.: Marnius et Abrius Wechel, 1593], and 108.755, [ed. Iulius Pacius; Geneva: Guillelmus Laemarius, 1597]).

10. A possible exception is the anonymous translation of the *HA* reported by CS 108.645 as Lyons: Stephanus Brignol, 1580, known only from the copy in Olomouc, Universitní Knihovna, which I have not seen. However, as CS themselves acknowledge (108.636), Le Président Baudrier and J. Baudrier, *Bibliographie Lyonnaise: Recherches sur les imprimeurs, libraires, relieurs et fondeurs de lettres de Lyon au XVIe siècle* (Lyons, 1895–1921; facsimile reprint, Paris: F. de Noble, 1964–1965), 6:372, 10:289, report that Brignol printed Gaza's translation of the *HA* in 1580 as a volume in the Latin *Opera Omnia* of Aristotle published by Jeanne Giunta. It is very difficult to imagine that Brignol would that same year print on his own a new, anonymous version to compete with his printing of Gaza's translation. So my guess is that CS 108.645 is either a ghost or at most nothing more than a variant printing of CS 108.636, i.e., Gaza's translation.

11. Scaliger, trans., *Historia de Animalibus,* ed. Jacques Maussac (Toulouse: Raymundus Comerius, 1619). See K. Jensen, "The MS-Tradition of J. C. Scaliger's *Historia de Animalibus,*" in *Acta Scaligeriana: Actes du Colloque International organisé pour le cinquième centenaire de la naissance de Jules-César Scaliger (Agen, 14–16 septembre 1984),* ed. J. Cubelier de Beynac and M. Magnien, Recueil des travaux de la Société académique d'Agen, ser. 3, 6 (Agen: Société académique d'Agen, 1986), pp. 257–286. I thank Luc Deitz for calling this article to my attention.

12. CS, pp. 172–73, list 77 sixteenth-century editions of John Argyropoulos' Latin translation (his translation became the standard version of the *Opera Omnia;* there were 7 incunabula editions: *GW* 2359, 2361–2366), 38 of Ioannes Perionius' (including the editions of the revision by Nicolaus Grouchius), 18 of Ioannes Bernardus Felicianus', 14 of Leonardo Bruni's (plus 9 incunabula editions: *GW* 2359–2360, 2367–2374), 12 of Dionysius Lambinus' (including the two editions of the version revised by Matthias Bergius), 11 of William of Moerbeke's (in addition to 6 incunabula editions: *GW*

2359–2360, 2375–2378), 7 of Nicolaus Grouchius', 7 of Adrianus Turnebus', 5 of Antonius Riccobonus', and 1 each of Marcus Antonius Muretus' (only book 5), Jacobus Strebaeus', Victorinus Strigelius', Iacobus Camerarius', and Laelius Peregrinus'.

13. See CS, pp. 160–224 ("Systematic Index of Aristotelian Works"). See also S. Perfetti, "'Cultius atque integrius': Teodoro Gaza, traduttore umanistico del *De Partibus Animalium*," *Rinascimento,* 2nd ser., 35 (1995): 253–286, whose first section (pp. 253–254) has the rubric, "Il successo editoriale delle versioni gaziane." Dr. Perfetti's article came to my attention after my article was already with the editors; so it was satisfying to discover that on virtually every issue we had in common he had independently come to the same conclusion as I.

14. *GW* 2334; Goff A-959. Writing in vol. 3 (January 1497) to the dedicatee of the edition, Alberto Pio, prince of Carpi, Manuzio explained: "Quod si hos de animalibus libros cum iis conferes quos miro successu Theodorus Gaza, licet Graecus homo, tamen et Latinae et Graecae eruditorum omnium aetatis suae facile princeps, fecit Latinos, brevi, quantum profeceris, non poenitebit. Ibi enim utriusque linguae proprietatem licet cognoscere, quod et nobis et Graecis est apprime necessarium. Nullus est (mihi crede) Graecus liber in quo facilius disci Graeca lingua possit ab hominibus nostris propter Theodorum. Sic Graece didicit Hermolaus Barbarus, sic Picus Mirandulus, avunculus tuus, sic Hieronymus Donatus, sic Angelus Politianus, summo viri iudicio, summo ingenio, ac undecumque doctissimi, sic denique quicunque Graecas literas callet temporibus nostris. Idem et tibi, mi Alberte, censeo faciendum ut, cum et tu non mediocri sis ingenio iis quos praedixi, et eloquio par fias et scientia rerum." The preface is available in G. Orlandi, *Aldo Manuzio editore: Dediche, prefazioni, note ai testi* (Milan: Edizioni il Polifilo, 1976), 1:13–14, and B. Botfield, *Prefaces to the First Editions of the Greek and Roman Classics and of the Sacred Scriptures* (London: H. G. Bohm, 1861), pp. 202–203; this passage is also quoted by Perfetti, "'Cultius atque integrius,'" p. 257.

15. On his rearrangement of the end of book 8, see Aristotle, *History of Animals, Books VII–X,* ed. and trans. D. M. Balme, prepared for publication by A. Gotthelf, Loeb Classical Library (Cambridge, Mass.: Harvard University Press, 1991), p. 542.

16. In the case of the *Problemata*, their task was made easier by the fact that Gaza's translation was available on pp. 415–474 of *AL,* which was vol. 3 of the Berlin Academy edition of Aristotle.

17. *Aristotelis Quae Feruntur Problemata Physica,* ed. C. E. Ruelle, H. Knoellinger, and (*post utriusque mortem*) J. Klek (Leipzig: Teubner, 1922), p. x; *Aristote; Problèmes,* ed. P. Louis (Paris: Les Belles Lettres, 1991–1994), 1:xlvii.

18. See notes 1 and 2 above. Trapezuntius' translation was never printed in the Renaissance. G. Marenghi has edited his and Bartolomeo of Messina's versions of book 11 in Aristotle, *Problemi di fonazione e di acustica,* ed. and trans. G. Marenghi (Naples: Libreria scientifica editrice, 1962), pp. 124–135; and of books 1, 6–9, 14, 27–28, and 31–33, in Aristotle, *Problemi di medicina,* ed. and trans. G. Marenghi (Milan: Istituto editoriale italiano, [1966]), pp. 275–336.

19. Monfasani, *George of Trebizond,* pp. 151–167; idem, *Collectanea Trapezuntiana,* pp. 411–421.

20. Monfasani, *George of Trebizond,* p. 153.

21. On Gupalatinus, see Mary A. Rouse and Richard H. Rouse, "Nicolaus Gupalatinus and the Arrival of Print in Italy," *La Bibliofilia* 88 (1986): 221–251.

22. For the sake of convenience, I term a "book" what the medieval and Renaissance translators called a *particula.* In the Renaissance one also finds the term *sectio.* A given *particula* or *sectio* of the Aristotelian *Problemata* could have as few as three or as many as sixty problems.

23. As Trapezuntius explained in his colophon, his Greek exemplar stopped at this point; see my *Collectanea Trapezuntiana,* p. 707. The medieval manuscripts and editions mask their omission by numbering the last problem as problem 1 of book 38, when in fact it is problem 3 of book 37. Bartolomeo of Messina thus omitted problems 4–6 of book 37 and problems 1–11 of book 38.

24. See appendix 2 above. Trapezuntius attempted a concordance in BAV, MS Urb. lat. 1322, fols. 138v–139r; see Monfasani, *Collectanea,* p. 59. The Pierpont Morgan Library in New York has a copy of the 1473 *editio princeps* (shelf mark: f 1264) in which a fifteenth-century owner noted in the margin the start of each book of the *antiqua translatio* (he sometimes called it the *altera translatio*) when the latter's numbering differed from Gaza's. Two sixteenth-century attempts at a concordance can be found in manuscripts of the Biblioteca Ambrosiana in Milan. One is that in the Pinelli miscellany, D 195 inf., fols. 20r–v, based on the 1475 printed edition and the medieval translation, in three columns; column 1 lists the *particula* (i.e., book) number in Gaza's translation and the folio number (*charta*); column 2 gives the title; column 3 reports the corresponding number of the *particula(e)* in the medieval version. The rubric for and introduction to the concordance are found on fol. 19r (fol. 19v is blank). The other concordance is in the massive Pinelli miscellany R 109 sup., fols. 319r–320v. The concordance proper is on fol. 319–v, where next to the title of each of the twenty books of Gaza's translation are listed the number of problems it contains, the folio number on which it starts (e.g., "ch[arta] 23ᵃ"), and the number or numbers of the corresponding book or books of the Greek manuscripts; fol. 320r is blank; fol. 320v has the title "Come si respondano i problemi d'Aristotile divisi in vinti particole (nel modo che furno [sic] dati la prima volta fuori [?] dal Gaza) alla divisione ordinaria delle 38 particole."

25. In the early seventeenth century, Daniel Heinsius made a comparable attempt to re-order Aristotle's *Poetics.* One major difference, however, was that he explained what he had done and gave the reasons why. See P. S. Sellin, "From *Res* to *Pathos:* The Leiden 'Ordo Aristotelis' and the Origins of Seventeenth-Century Recovery of the Pathetic in Interpreting Aristotle's *Poetics,*" in *Ten Studies in Anglo-Dutch Relations,* ed. J. Van Dorsten (Leiden: For the Sir Thomas Browne Institute, at the University Press, 1974), pp. 72–93.

26. For a survey of the problems and issues involved in the translating of Greek scientific works in the Renaissance, see C. B. Schmitt, *Problemi dell'aristotelismo rinascimentale,* trans. A. Gargano (Naples: Bibliopolis, 1985), pp. 99–132; and B. Copenhaver, "Translation, Terminology, and Style in Philosophical Discourse," in *The Cambridge History of Renaissance Philosophy,* ed. C. B. Schmitt et al. (Cambridge: Cambridge University Press, 1988), pp. 77–110.

On Cornarius (d. 1558), see I. Backus, *Lectures humanistes de Basile de Césarée: Traductions Latines (1439–1618)* (Paris: Institut d'études augustiniennes, 1990), pp. 42–54 (p. 42: "Sa méthode de traduction, qui marque un retour à la méthode médiévale *de verbo ad verbum*. . .").

27. For a study of Gaza's pattern of paraphrasing in the *PA*, see Perfetti, "'Cultius atque integrius,'" pp. 274–280.

28. Aristotle, *Problems*, trans. W. S. Hett, rev. ed., Loeb Classical Library (Cambridge, Mass.: Harvard University Press, 1953), 1:47.

29. I cite the revised version from Aristotle, *AL*, p. 419, and the original version from G, fol. 19r.

30. Aristotle, *De Animalibus Historiae Libri X*, ed. J. G. Schneider (Leipzig: Hahn, 1811), 1:vi; cf. also Balme's comment in the introduction to the Loeb edition of books 7–10: (p. 49) "Many of his [Gaza's] Latin equivalents seem to originate in Pliny." See also Perfetti, "'Cultius atque integrius,'" pp. 278–280, for the *PA*.

31. I.e., one should write οὐ δεῖ instead of δεῖ at 1.4, 859a25 to correct the Greek manuscripts.

32. G, fols. 5v–6r: "Cur vomitus, quom tempora immutantur, citari minime debeant?"

33. *AL*, p. 415; the original translation adds *etiam* between the first *alii* and *parum* (G, fol. 6r).

34. Where he had conjectured *minime* (see note 32 above), he now wrote *maxime:* "citari maxime debeant" (G, fol. 5v).

35. *Editio princeps,* Rome, 1475 (see note 1 above), fol. 2r: "Testis ego sum, qui eo dictante scribebam [scrihebam *editio*], quantum laboris insumpserit senex doctissimus annum continuum in emendandis plurimis librariorum erroribus. Depravati erant certe Greci codices omnes. Ipse tamen exactissime iudicio, ut optimum interpretem decet, tum ob lingue Grece sibi vernacule atque Latine elegantie peritiam summam, tum quia Paripatetice [*sic*] secte studiosissimus semper extitit, id in Problematis fecit, quod in aliis quoque rebus fieri solet, ut ex multis corruptis ac perversis quoddam integrum atque optimum factum sit."

36. *Editio princeps,* Venice, 1476 (see note 7 above), sig. R a3r: "Accedit ad hec altera causa laboris, quod exemplaria Graeca (libros hos De animalibus dico) mendosa admodum habemus vel librariorum culpa vel eo casu quem apud Strabonem geographum legimus. In his enim emendandis elaborare interpres sine dubio debet ne ipse errasse in convertendo videatur."

37. Ibid., sig. R a4r: "Sed Appellicon Theius, de quo Strabo, plura, ut alia permulta, in exscribendis Aristotelicis libris depravavit."

38. See I. Düring, *Aristotle in the Ancient Biographical Tradition* (Göteborg, [distr. Stockholm: Almqvist and Wiksell,] 1957), pp. 414–425.

39. Gaza likened his work to that of preparing an edition. Speaking of Giovanni Andrea Bussi's edition of Pliny's *Natural History,* he asked (preface to translation of the *De anima-*

libus, editio princeps, sig. R a3v): "Quid autem cum vel ipsa Latina exemplaria, quorum ope iuvari nitimur, non minus sint emendanda quam Greca?"

40. For the *Problemata,* G. Marenghi has identified Gaza's base exemplars as the late-fourteenth-century manuscript Venice, Biblioteca Marciana, Zan. Gr. 259 (coll. 892) (= x, in the sigla of the editors) and especially the fifteenth-century manuscript Milan, Biblioteca Ambrosiana, A 174 sup. (= A^m); see his "Un capitolo dell'Aristotele medioevale: Bartolomeo da Messina traduttore dei *Problemata physica,*" *Aevum* 36 (1962): 278, 281 n. 59; his edition of Aristotle, *Problemi di fonazione,* p. 29; and his "Per un'edizione critica dell' Ἀριστοτέλους Προβλημάτων Ἐπιτομὴ Φυσικῶν," *Edizione nazionale dei classici greci e latini: Bollettino del comitato,* n.s., 19 (1971): 102 n. 3. For the *HA* Balme has concluded that "Gaza mostly agrees with the α family, but quite often with β, only rarely with γ. Sometimes he agrees with Guil. [William of Moerbeke] alone and may be making conjectures based on Guil. There is no positive evidence that he used MSS now unknown." See Balme's Loeb edition and translation of books 7–10 of the *HA,* p. 49.

41. After a fashion Gaza was imitating his patron, Cardinal Bessarion, who collected manuscripts of Greek texts and then made "improved" copies based on collations and conjecture. For a survey of this practice in regard to Aristotelian texts, see the chapter "Gli studi aristotelici" in E. Mioni, *Vita del Cardinale Bessarione* published as *Miscellanea Marciana* 6 (1991): 136–148. See also Balme and Gotthelf in their Aristotle, *History of Animals,* pp. 39–40.

42. Gaza's *Annotationes* in Badia, Montecassino MS 649, on his translation of the *De animalibus* are interesting in this respect. Despite his numerous conjectures, the *Annotationes* almost never speak of them. Instead they consist mainly of lexicographical and grammatical comments and paraphrases with a leaven of philosophical comments. Strangely, he does mention a conjecture he did not use at *HA* 2, 505a9 (Badia 649, p. 9): "*corpori admota est.* In Graecis testibus est σῶμα, quod significat corpus, sed videtur esse mendum quoniam braniae ori proprie applicantur et propterea scribit στόμα potius, id est, os, et in dandi casu, στόματι, hoc est, ori."

43. The whole passage is extraordinary in its sweeping condemnation of medieval scientific and philosophical vocabulary (G, fols. 2v–3r).

44. R, sig. a2v: "[Aristotle's meaning] etiam propter incultum, horridum, et ineptum sermonem interpretis vix intelligi potest." See Perfetti, "'Cultius atque integrius,'" pp. 263–271, for a detailed analysis of Gaza's neoclassical refashioning of scientific vocabulary in the *PA.* As Perfetti, p. 267, and G. Pozzi in his edition of Ermolao Barbaro's *Castigationes Plinianae et in Pomponium Melam* (Padua: Antenore, 1973), 1:cxli, point out, already in the 1470s Giorgio Merula criticized Gaza for preferring an *obscura periphrasis* to the clearer Greek transliteration of a term.

45. For literature, see Schmitt, *Problemi dell'aristotelismo rinascimentale,* and Copenhaver, "Translation, Terminology, and Style."

46. G, fol. 4r: "Hic est ille doctus, viri philosophi, cuius interpretationem publice docetis. . . . Tum errores et sordes istorum interpretum adhuc legitis atque defenditis, qui tanti mali causam dederunt? Non in ignem proiicitis? Non nulla potius quam eiusmodi habere vultis?"

47. See note 7 above.

48. See my *Collectanea Trapezuntiana,* pp. 706, 708, where I list nine manuscripts of the *De animalibus* and ten of the *Problemata.* I know of only two manuscripts of the revised translation of Gaza's *Problemata* (Florence, Biblioteca Laurenziana, Gadd. 89, sup. 59; and Florence, Biblioteca Nazionale, Magl. XII.49), and I suspect that both derived from the printed edition. There is a manuscript dedication copy of the translation of the *De animalibus* (BAV, Vat. lat. 2094), but it bears no trace of Gaza and may have in fact been completed after his death, since only the pope could have paid for such a wonderously deluxe manuscript. I am not sure whether the other two manuscripts known to me (Paris, Bibliothèque Nationale, lat. 6793; Seville, Biblioteca Universitaria, 332.155) derive from the Vatican manuscript or from the printed edition.

49. Poliziano, *Miscell. Cent.* 1:90, in *Opera Omnia* (Basel: Nicolaus Episcopius 1553; reprint, ed. I. Maïer, Turin: Bottega d'Erasmo, 1971), 1:301–303. Poliziano's charge that Gaza hypocritically plagiarized from Trapezuntius has been amply documented in the case of the *PA* by Perfetti, "'Cultius atque integrius,'" pp. 11.268–271, 277, 281–283, who also shows, however (pp. 283–286), that Gaza did not simply produce a "traduzione di una traduzione." On Poliziano's later critical attitude toward Gaza, see now also V. Fera, "Poliziano, Ermolao Barbaro e Plinio," in *Una famiglia veneziana nella storia: I Barbaro* (Venice: Istituto veneto di scienza, lettere ed arti, 1996), pp. 200–203, 210–215; and idem, "Un laboratorio filologico di fine Quattrocento: La Naturalis historia," in *Formative Stages of Classical Traditions: Latin Texts from Antiquity to the Renaissance,* ed. O. Pecere and M. D. Reeves (Spoleto: Centro italiano di studi sull'Alto medioevo, 1995), pp. 449–451.

50. See Poliziano, *Miscell. Cent.,* 1.72 in (*Opera Omnia,* 1:283). In *Miscell. Cent.* 2.40, 46, in A. Poliziano, *Miscellaneorum Centuria Secunda,* ed. V. Branca and M. Pastore Stocchi (Florence: Fratelli Alinari, 1972), 4:63.32–64.40, 82.42–44, Poliziano cited Gaza without criticism; see V. Branca, *Poliziano e l'umanesimo della parola* (Turin: Einaudi, 1983), p. 286 n. 182; I owe my appreciation of Branca's note to Perfetti, "'Cultius atque integrius,'" p. 258.

51. See Pozzi's discussion of these references in his edition of Barbaro's *Castigationes Plinianae,* 1:cxvii, cxviii, cxl–clxii. Pozzi gave as his conclusion (p. cxlii): "L'atteggiamento di sconfinta ammirazione per il maestro greco, testimoniata al tempo del Temistio, ha dunque subito una revisione, non però un rovesciamento." For Barbaro's praise, see note 59 below.

52. Scaliger mistakenly thought that book 10 of Trapezuntius' translation had been printed somewhere in Germany and desired a copy; see his letter to Charles Sevin, in *Epistolae et Orationes Nunquam Antehac Excusae* (Leiden: Christophorus Raphelengius, 1600), pp. 146–147, which is quoted among the testimonia in Aristotle, *Historia de Animalibus, Iulio Caesare Scaligero Interprete, cum Eiusdem Commentariis,* ed. J. Maussac (Toulouse: Raymundus Colomerius, 1619), sig. ★★★1v; Sambucus had been promised an "autograph" copy by the Paduan professor of medicine Hieronymus Mercurialis; see his letters of 1573, 1574, and 1582 in H. Gerstinger, *Die Briefe des Johannes Sambucus (Zsamboky), 1554–1584* (Vienna: Böhlaus, 1968), pp. 141, 156, 264, 268, 286–88, 305–306; cf. also idem, "Johannes Sambucus als Handschriftensammler," in *Festschrift der Nationalbibliothek in Wien, Herausgegeben zur Feier des 200 Jährigen Bestehens des Gebäude* (Vienna: Österreichische Staatsdruckerei, 1926), pp. 343–347.

53. See P. O. Kristeller, comp., *Iter Italicum: A Finding List of Uncatalogued or Incompletely Catalogued Humanistic Manuscripts of the Renaissance in Italian and Other Libraries* (London: Warburg Institute; Leiden: Brill, 1963–1996), 4:108; and my *Collectanea Trapezuntiana*, p. 25.

54. CS 107.687. The edition combines Pietro d'Abano's commentary with the composite edition described in note 1 above. Its title runs *Problemata Aristotelis cum duplici translatione, antiqua videlicet et nova, scilicet, Theodori Gaze, cum expositione Petri Aponi. Tabula secundum magistrum Petrum de Tussignano per alphabetum. Problemata Alexandri Aphrodisei. Problemata Plutarchi* (Venice: Bonetus Locatellus, "1501, tertio Kalendas Sextiles" [= 30 July]). The contents of the volume are as follows:

Sig. aa1v: register of quires (aa[10], bb[8], cc[4], a[8]–z[8], &[8], symbol for "-us,"[8] symbol for "=rum,"[8] A[8]–L[8]).

Sig. aa2r–cc4v: alphabetical index keyed to the *particulae* and problems rather than page numbers, prepared by Petrus de Tussignano.

Sig. cc4v: letter of Dominicus Massaria to the reader (edited in appendix 1 above).

Fols. 1r–272r: commentary of Pietro d'Abano with the translations of Bartolomeo of Messina and Theodore Gaza (for the defective state of Bartolomeo's translation and its misleading numeration at the end, see note 23 above).

Fol. 272r–v: a gathering of seventeen Aristotelian problems translated by Gaza that Massaria could not figure out where to place. They are, according to the order of the Greek manuscripts (see concordance B in appendix 2 for their numbering in Gaza), 21:21–26, 20:27–34, 38:8–9, 38:11.

Fol. 273r: Victor Pisanus Ludovico Mocenigo (letter concerning Giorgio Valla and his translation of pseudo-Alexander of Aphrodisias' *Problemata*).

Fol. 273v: Preface of Giorgio Valla to Giovanni Marliani for his translation.

Fols. 274r–288r, col. 1: Pseudo-Alexander of Aphrodisias, *Problemata,* trans. Giorgio Valla, with excerpts from various authors interspersed as commentary (Ambrosius, Quintilian, Albertus Magnus, Pliny, etc.).

Fol. 288r, cols. 1–2: Preface of Nicolaus Gupalatinus to Pope Sixtus IV for Theodore Gaza's translation of the pseudo-Aristotelian *Problemata*.

Fol. 288r, col. 2: Letter of Giovanni Calfurnio to Marco Aurelio concerning Giovan Pietro d'Avenza (da Lucca) and his translation of Plutarch's *Problemata*.

Fols. 288v–296r: Plutarch, *Problemata,* trans. Giovan Pietro d'Avenza, with an eight-line poem of Calfurnio to Dominicus Siliprandus Mantuanus at the end, followed by the printer's colophon (referring exclusively to the edition of Plutarch) and printer's mark.

55. All the authorities (T. Pesenti, "La cultura scientifica: medici, matematici, naturalisti," in *Storia di Vicenza,* ed. F. Barbieri and P. Preto [Vicenza: Neri Pozza, 1987–1993], 3.1:261; S. De Renzi, *Storia della medicina italiana* [Naples, 1845–1848; reprint, Bologna: Forni, 1966], 3:464; and F. Angiolgabriello di Santa Maria, *Biblioteca e storia di quegli scrittori così della città come del territorio di Vicenza* [Vicenza: Vendramini Mosca, 1772–1782], 4:17–19) only know him from his work on medical measurement, where he called himself Dominicus Vince<n>tinus Arcignaneus (i.e., of Arzignano, near Vicenza) and Dominicus Massarius Vincentinus. In the preface he said that he had been reading medical texts for many years, but he does not actually identify himself as a doctor. The same preface suggests that he had some knowledge of Greek and Arabic. If and to what degree he was related to the well-known contemporary Vincenzan medical doctor Alessandro Massaria, I cannot say.

56. The first edition was printed at Venice "per Ioannem Tacuinum de Tridino"; Conrad Gesner's edition came out at Zurich "apud Froschovenum." I have consulted the latter at the Yale University School of Medicine. In between these two editions there were three others in medical miscellany volumes: Pavia: Bernardinus de Garaldis, 1516 (first work: Pliny the Elder, *Opus Aureum,* which is a conflation of excerpts from the *Natural History*); Lyons: "apud insigne angeli," 1534 (first work: *opera* of Ioannes Mesue); and unknown printer, 1541 (first work: *opera* of Ioannes Mesue).

57. These four editions are Venice: Gregorius de Gregorius, 1505 (CS 107.726); Venice: Luca Antonius de Giunta, 1518 (CS 107.852); Venice: Octavianus Scotus, 1519 (CS 107.863); and Paris: A. Boucard, 1520 (CS 107.868).

58. CS 107.720. For an extensive description, see *Aldo Manuzio tipografo, 1494–1515,* ed. L. Bigliazzi, A. Dillon Bussi, G. Savino, and P. Scopecchi (Florence: Octavo, 1994), pp. 127–128, no. 83; for the preface, see Orlandi, *Aldo Manuzio editore,* 1:76.

59. The passage was taken from Barbaro's preface to his translation of Themistius' paraphrase of Aristotle's *Posterior Analytics* published in 1481; see E. Barbaro, *Epistolae, Orationes, et Carmina,* ed. V. Branca (Florence: "Bibliopolis," 1943), p. 9; the passage is also quoted by Perfetti, "'Cultius atque integrius,'" p. 257.

60. Gaza left out one of the eleven problems in book 38 (see concordance B in appendix 2 above). Aldus bragged about this feat on fol. 1r, after the listing of the contents of the edition: "Aristotelis Problemata in duas de quadraginta sectiones in quibus quatuordecim [i.e., the last three problems of book 37 and the eleven of book 38] quae circa finem deerant quaestiones in quarumque locum totidem ex iis quas alibi in eo ipso volumine habentur falso suppositae fuerant in suum locum restituendas curavimus. Qui error in omnibus est Problematum libris qui ante Venetiis excusi leguntur."

61. The first reprint was a copycat edition by Balthazard de Gabiano at Lyons in three volumes ca. 1505 (CS 107.731; see Baudrier and Baudrier, *Biographie Lyonnaise,* 7:15–17); the second was an Aldine reprint in 1513 (CS 107.809; not an exact reprint: see Bigliazzi et al., *Aldo Manuzio tipografo,* pp. 127–128, no. 83); and the last was published by Octavianus Scotus at Venice in 1525 (CS 107.893). Only a methodical collation of variants would establish the printing stemma.

62. CS 107.968; see especially *Vives: Edicions Princeps,* ed. E. González, S. Albiñana, and V. Gutiérrez (Valencia: Universitat de Valencia, 1992), pp. 51, 196–197, with two plates. Vives' *Censura* appears on sig. α3r–6v of vol. 1. See esp. sig. α5v: "Apparet autem opus hoc non esse ab Aristotele conscriptum, sed ex disputationibus illius ab auditoribus collectum et congestum. Multae sunt in eo repetitiones, quem nunquam reliquisset autor ipse si in ordinem digessisset quae disputaverat et consignasset monumentis literarum. Reliquit autem studiosus coacervator, dum maiorem diligentiam adhibet in cogendo quam iudicium in disponendo. Multae insunt in eis rationes frigidae, leves, dilutae, alienae ab Aristotelici ingenii gravitate atque acrimonia, aliae obscurae et molestae." The *Opera Omnia* of Vives edited by G. Mayans, 8 vols. (Valencia, 1782–1790), does not contain the *Censura.*

63. Erasmus' preface to John More is in Erasmus of Rotterdam, *Opus Epistolarum,* ed. P. S. Allen, H. M. Allen, and H. W. Garrod (Oxford: Clarendon, 1906–1958),

9:138.222–229: "Opus Problematum in primis eruditum apparet a studiosis contaminatum, vel hoc iudicio, quod toties eadem iterantur. Et haud scio an hoc sit opus quod Laertius hoc titulo recenset, Problematum ex Democrito libri sex. Commemorantur et alii duo libri Problematum τεθεαμένων, unus Encycliorum, rursus alii θέσεων et Protaseon, parte cerptas."

Interestingly, Gianfrancesco Pico della Mirandola was not an influence here. In 3.2–6 of his *Examen Vanitatis,* published in 1520, he cast doubt upon the authenticity and integrity of many of Aristotle's writings (*Opera Omnia* [Basel: Henricus Petri, 1573; reprint, Hildesheim: Georg Olms, 1969], pp. 1021–1042), but not on the *Problemata.* Indeed, this early Pyrrhonian skeptic had a naive faith in Gaza; cf. *Examen Vanitatis* 3.5 (*Opera Omnia,* p. 1038): "Ad Latinum Aristotelem ubi venerimus, mirum quam ipse non erit Aristoteles—libros eos semper excipio quos Theodorus Gaza convert e Graeco." On the importance of Pico in this regard and his influence on Vives, see J. Kraye, "Like Father, Like Son: Aristotle, Nicomachus, and the *Nicomachean Ethics,*" in *Aristotelica et Lulliana . . . Charles H. Lohr . . . dedicata,* ed. F. Domínguez et al. (The Hague: Nijhoff, 1995), pp. 158–162.

64. CS lists forty-two editions, but one, 107.730 (Cologne: Martin von Werden, ca. 1505), is an error; it contains the "Omnes homines" problems and not Gaza's translation: I checked the copy at Yale University's Harvey Cushing/John Hay Whitney Medical Library. There were five printings of the Massaria edition (see notes 54 and 57 above) and four of the Aldine edition (see notes 58 and 61 above). The remaining thirty-two printings are as follows:

One edition of just three books of the *Problemata:* Paris: ex officina Calvarini, 1550 (108.181).

One edition of the *Problemata* with pseudo-Alexander of Aphrodisias' and Plutarch's *Problemata:* Valencia: Ioannes Mey, 1544 (108.288B).

Eight editions accompanied by Gaza's translation of ps.-Alexander of Aphrodisias' *Problemata:* Paris: n.p., ca. 1520 (107.874); Paris: Simon Colinaeus, 1524 (107.890); Paris: Simon Colinaeus, 1534 (107.943); Basel: (Andreas Cratander), 1537 (107.956); Paris: Iacobus Kerver, 1539 (107.984); Lyons: Paulus Mirallietus, 1550 (108.178A; Cranz, "Alexander Aphrodisias," 2:419, reports the printer as N. Baccaneus; see the next edition); Lyons: Paulus Mirallietus, exc. Nicolaus Baccaneus, 1551 (108.212); Lyons: Simphorianus Beraud, 1572 (108.573).

Twenty in Latin *Opera Omnia* of Aristotle: Basel: Oporinus, 1538 (107.968; see note 62 above); Basel: Oporinus, 1542 (108.033; ed. Hieronymus Gemusaeus); Basel: Oporinus, 1548 (108.137; reprint of the 1542 edition); Lyons: Io. Frellonius, 1549 (108.160); Venice: Giunta, 1550–1552 (108.193; ed. Ioannes Baptista Bagolinus); Venice: Cominus de Tridino, 1560–1562 (108.423); Lyons: Symphorianus Barbierus, 1561 (108.429; in Gemusaeus' edition); Lyons: haeredes Iacobi Iuntae, [1561] (108.430F); Venice: Giunta, 1562–1574 (108.456); Basel: Io. Hervagius, 1563 (108.457); Lyons: Symphorianus Barbierus or Ioannes Frellonius, 1563 (108.460 and 108.460A); Venice: ad signum seminantis, 1572 (108.579); Venice: Giunta, 1575 (108.599); Venice: Gaspar Bindoni, 1576 (108.610); Lyons: A. I. Martinus, 1578 (108.629); Lyons: Io. Iunta, 1579 (108.636); Lyons/Geneva: Iacobus Berjon, 1580 (108.644); Lyons: Honoratus and Michaelis, 1581 (108.652); Venice: Ioachim Bruniolus, 1584–1585 (108.669); Frankfurt: M. and A. Wechel, 1593 (108.721–723).

Two in Greco-Latin *Opera Omnia:* Lyons: Iacobus Bubonius and Guillelmus Laemarius, 1590 (108.708; ed. Isaac Casaubon); and Geneva: Guillelmus Laemarius, 1597 (108.755; ed. Iulius Pacius; vol. 1: *De Animalibus;* vol. 2: *Problemata).*

65. For descriptions of the incunable editions, see note 1 above. The three missing printings are the Venice 1501 reprint by Albertinus Vercellensis of the composite edition of pseudo-Alexander Aphrodisias and others; see Goff A-387a; *IA* 1:339, no. 103.322; and Cranz, "Alexander Aphrodisiensis," 1:131; and two extracts, namely, book 31 (*Problemata quae ad oculos pertinent;* this is the first part of book 20 in Gaza's arrangement), with Giorgio Valla's *De Natura Oculorum:* Strasbourg: Henricus Sybold, May (1529?) (*Verzeichnis,* 1.1:533, no. A 3624), and book 1 (*Problemata de Re Medica,* which also constitutes book 1 in Gaza's arrangement) with Giorgio Valla's *De Corporis Commodis et Incommodis* and *De Differentia Pulsuum:* Strasbourg: Henricus Sybold, July (ca. 1530) (*Verzeichnis,* 1.1:534, no. A 3625). I consulted the latter edition at the National Library of Medicine, Bethesda, Maryland. Gaza's translation is on sig. D 7v–F 6r.

66. Gaza's translation of book 1 is reproduced in *Urbanae Disputationes in Primam Problematum Aristotelis Sectionem Ioannis Manelfi Eretani Sabini* (Rome: Gulielmus Facciottus, 1630) (seen at the University of Pennsylvania, Philadelphia).

Three printings in Latin *Opera:* Lyons: Horatius Cardon, 1603–1614; *AL* (1831); Frankfurt a. M.: Minerva, 1962 (facsimile reprint of the Venice 1562–1574 edition by Giunta).

Eight printings in Greco-Latin *Opera:* Geneva: Petrus de la Roviere, 1605 (ed. Isaac Casaubon); Geneva: Petrus de la Roviere, 1606 (ed. Iulius Pacius); Geneva: Petrus de la Roviere, 1607; Paris: Typis Regiis, 1619 (vol. 1: *De Animalibus;* vol. 2: *Problemata);* Paris, Typis Regiis, 1629; Paris: Aegidus Morellus, 1629 (ed. Gulielmus Du Val); Paris: Aegidius Morellius, 1639; Paris: I. Billaine, 1654.

67. Casaubon's method was to repeat Gaza's translation of similar or nearly identical problems elsewhere in the *Problemata.* See appendix 2.C above.

68. See note 4 above. The Aristotelian *Problemata* and the "Omnes homines" problems are also discussed in Ann Blair's essay in this volume.

69. Lawn, *Salernitan Questions,* pp. 99–100 n. 5, lists a number of printings not found in CS. I have come across at Yale University a 1560 printing of Paris not recorded by either.

70. Ibid.

71. See Kraye, "Printing History of Aristotle," p. 210.

72. Venice: Ioannes et Gregorius de Gregoriis, 1492 (*GW* 2351; Goff A–974); Venice: (Simon Bevilaqua), ca. 1495 (*GW* 2352; Goff A–975); Venice: Bartholomaeus de Zanis, 1498 (*GW* 2353; Goff A–976).

73. H. Stadler was incorrect to report four incunable editions in Albertus Magnus, *De Animalibus Libri XXV nach der Kölner Urschrift,* Beiträge zur Geschichte der Philosophie des Mittelalters: Texte und Untersuchungen 15–16 (Münster i. W.: Aschendorff, 1916–1921), 1:xi; he gave credence to the ghost reported by Hain (no. 548: Venice, 1498). The three incunabula were Rome: Simon Chardella, 1478, ed. Fernando of Cordova

(*GW* 587; Goff A–223); Mantua: Paulus de Butzbach, 1479 (*GW* 588; Goff A–224); and Venice: Ioannes et Gregorius de Gregoriis, 1495 (*GW* 589; Goff A–225). For the 1519 edition of Venice, heredes O. Scoti ac sociorum, see *IA*, 1:261 no. 102.543.

74. Venice: Hieronymus Scotus, 1546, with Nifo's lengthy preface to Pope Paul III at the start. Each commentary has its own numeration. The author's colophon to the *GA*, the last text in volume, reads (p. 216 in the *GA*) "Salerni, XIII Iunii, MDXXXIIII."

75. Ibid., pp. 201 (numeration for the *HA*,), col. 1 (first comment in book 7), and 259, col. 2 (first comment in book 9). Nifo did revise Gaza's translation, however; see Perfetti, "'Cultius atque integrius,'" p. 262 n. 36, and his "Metamorfosi di una traduzione: Agostino Nifo revisore dei 'De animalibus' gaziani," *Medioevo* 22 (1997): 259–301.

76. For literature and a description, see Bigliazzi et al., *Aldo Manuzio tipografo,* pp. 31 no. 4, 38 no. 11, 51–52 no. 22 and 23, 55 no. 25.

77. Dr. Allan Gotthelf informs me that these restorations will be carried out as well in Balme's *editio maior,* forthcoming from Cambridge University Press.

78. In some copies, such as that in the Pierpont Morgan Library in New York, book 10 is bound in immediately after book 9. This obviously makes sense textually, but it also renders pointless much of Manuzio's introduction (quoted in note 80 below).

79. Gaza argued as follows in the preface to his translation (R, sig. a4r–v): "Sunt etiam exemplaria tum Greca, tum vero Latina que habeant fragmentum quoddam Historiis additum. Sed id causas quasdam materiales agentesque generationis humane exponit; non historiam complectitur. Itaque non inter Historie libros hoc ego ponendum duxi. Sed si collocandum uspiam est, libris de generatione meo quidem iudicio coniungendum est." For a defense of book 10 as plausibly an authentic Aristotelian work, see D. M. Balme, "Aristotle Historia Animalium Book Ten," in *Aristoteles Werk und Wirkung Paul Moraux Gewidmet,* ed. J. Wiesner (Berlin: De Gruyter, 1987), 2:191–199.

80. Orlandi, *Aldo Manuzio editore,* p. 14; Botfield, *Prefaces,* p. 201: "En tibi, lector carissime, fragmenta ea quae Gaza in prooemio De Animalibus in nonnullis codicibus tum Graecis tum Latinis inveniri ait; quae suo fortasse loco impressa legeres si suo tempore in manus nostras venissent. Nunc vero hoc loco adiecta maluimus quam te iisdem qualibuscunque fraudari. Vale."

81. For a while there seems to have been uncertainty about what to do with book 10. In his edition of the Greek text of the *HA* in 1527 (Florence: Philippus Iunta, as part of a collection of works of Aristotle and Theophrastus in Greek; CS 107.899), Nicolaus Leonicus Thomaeus included book 10 and even numbered it as such, but a colophon he placed at the end of book 9 said that this was the end of the *HA*. Book 10 has no colophon. However, in his Greek *Opera Omnia* of Aristotle four years later, Erasmus unequivocally treated the *HA* as ending with book 10.

82. CS 107.928; Ioannes Bebelius was the printer.

83. Erasmus, *Opus Epistolarum,* 9:139.272–73: "In libris De Animalibus loca non pauca repurgata sunt, praesidio tralationis Theodoricae."

84. Cf. Balme's statement in the introduction of his Loeb edition of the *HA* (p. 47): "Gaza's influence was inordinate, and is apparent in the manus priores of nL^c and in the later hands in E^{a}mS^{c}O^c."

85. CS 108.664; Frankfurt a. M.: "apud Andreae Wecheli heredes, Claudium Marnium et Ioannem Aubrium."

86. Ibid., vol. 4, sig. † 2r–v: "Quae sicubi non satisfacerent, Theodori Gazae versionem consului. . . . Notarum libello attuli nunc meas, nunc alienas coniecturas. Non enim Gazam tantummodo in eiusmodi locis adii, sed etiam siquid ab aliis notatum invenirem, diligenter huc comportavi."

87. On Feliciano, see Cranz, "Alexander Aphrodisias," 1:90–91; and Eugene F. Rice, Jr., "Paulus Aegineta," in Kristeller, Cranz, and Brown, *Catalogus Translationum et Commentariorum,* 4 (1980): 165–166.

88. For editions containing Feliciano's and Scaliger's translations, see note 9 above; de Conti's is in CS 108.423; Venice: Cominus de Tridino, 1560.

89. Working under the spell of Gaza, Sylvius—and, later, Maussac in his 1619 edition (see note 52 above)—called book 10 *a fragmentum,* even though in his commentary Scaliger himself says no such thing. Indeed, though he worked from the Aldine edition, Scaliger argued correctly that book 10 should follow book 7—i.e., book 9 of the manuscripts— thereby undoing Gaza's reordering.

90. See especially Jensen, "MS-Tradition."

91. I consulted Theophrastus, *De Historia Plantarum Libri Decem, Graece et Latine . . . Accesserunt Iulii Caesaris Scaligeri in Eosdem Libros Animadversiones,* ed. I. Bodaeus Stopel (Amsterdam: Henricus Laurentius, 1644); K. Jensen, *Rhetorical Philosophy and Philosophical Grammar: Julius Caesar Scaliger's Theory of Language* (Munich: Wilhelm Fink, 1990), pp. 44–45, cites the 1566 Lyons-Geneva edition of Scaliger's commentary on the *De Causis Plantarum* and the 1584 Lyons edition of his *Animadversiones.*

92. Aristotle, *Historia Animalium Scaligero interprete,* pp. 2–3; Jensen, "MS-Tradition," pp. 278–283. Scaliger's son Joseph Justus Scaliger maintained the family tradition; see *Scaligerana ou Bons mots, recontres agreables et remarques judicieuses et sçavantes de J. Scaliger,* ed. Tannegui Le Fèvre and P. de Colomiès (Cologne, 1695), 242: " . . . Theodoro Gazae, magno certe viro et docto, qui tamen in librorum Aristotelis De Animal. versione lapsus est aliquibus locis."

93. Maussac's *Animadversiones* are found in Aristotle, *Historia Animalium Scaligero interprete,* pp. 1233–1248.

94. Aristotle, *De Animalibus Historiae,* ed. Schneider, 1:v–vi.

95. Except for the several sixteenth-century translations of book 10 of the *HA* and the one edition of Scaliger's translation of the whole *HA,* Gaza monopolized the printing of the *De animalibus.* I know of forty-four complete or partial editions of Gaza's translation of the *De animalibus.*

For the four incunables, see notes 7 and 72 above.

For the four editions of Gaza's collected translations first published by Aldo Manuzio in 1504, see notes 58 and 61 above.

For the twenty-two Latin and Greco-Latin *Opera Omnia*, see the end of note 64 above, to which we need to add the inclusion of Gaza's translation of the *PA* and *GA* in the 1831 Berlin Academy Aristotelian *Opera Omnia* and of the *GA* in vol. 3 of the 1854 Paris, Fermin Didot, *Opera Omnia* of Aristotle.

There were two editions outside the Aristotelian *opera* of Gaza's translation joined to Petrus Alyconius' translations of the *De motu animalium* and *De gressu animalium:* Paris: Simon de Colines, 1524 (CS 107.891); and Paris: Simon de Colines, 1533 (107.938).

There were five editions of the same five texts joined to Gaza's translation of Theophrastus' *Historia plantarum* and *De causis plantarum:* Basel: Andreas Cratander, 1534 (CS 107:939); Basel: Andreas Cratander, 1550 (108.175); Lyons: Nicolaus Bacqueonius, 1552 (108.233); Lyons: heredes Iacobi Iuntae, 1552 (108.233A); Lyons: heredes Iacobi Iuntae, 1560 (108.400; *De gressu* in Nicolaus Leonicus Thomaeus' version and *De motu* in an anonymous translation); and Lyons: heredes Iacobi Iuntae, typ. Theobaldi Pagani, 1580 (108.646; *De gressu* and *De motu* as the preceding edition).

There were two sixteenth-century editions of the *HA, PA,* and *GA* grouped together without any other texts: Paris: Prigentius Calvinus, 1542 (each text with its own title page; called to my attention by Allan Gotthelf, who owns a copy); and Venice; Hieronymus Scotus, 1545 (CS 108.110).

There was also one edition of the *PA* alone: Paris: Calvarinus, 1542 (108.046).

There was a single printing of the *GA,* accompanied by John Philoponus' commentary in the translation of Nicolaus Petreius: Venice: Io. Antonius et Stephanus ac Fratres de Sabio, 1526 (CS 108.898, but not recognized in the index as containing Gaza's translation).

Finally, there were the two commentaries that used Gaza's translation: Agostino Nifo's of 1546 on the *HA, PA,* and *GA* (see note 74 above), and Daniel Furlanus' of 1574 on the *PA* (Venice: Ioannes Baptista Somachius; omitted by CS; seen at the British Library, London).

96. Poliziano is not hostile in referring to Gaza's translation of Theophrastus in *Cent. Miscell.* 1:72 (in *Opera Omnia,* 1:283).

II

Natural Disciplines

7

Epistemological Problems in Giovanni Mainardi's Commentary on Galen's *Ars parva*
Daniela Mugnai Carrara

I

Galen's *Technē iatrikē* (*Ars medica*), generally known in the Latin Middle Ages as the *Ars parva* and subsequently, under the influence of medical humanism, by the more exact title of *Ars medicinalis,* was used for medical teaching from late antiquity and was a formal part of the curricula of university faculties of medicine from the Middle Ages until the eighteenth century.[1] The work thus had an extraordinarily long and uninterrupted life. Both its conciseness and the genuine obscurity of a number of passages—an obscurity certainly not lessened in the work's numerous translations—necessitated many interpretive expositions over the course of time. For centuries, the rich tradition of commentary that originated in this way provided material for methodological discussions that made use not only of the tools of logic but also of the theoretical positions supplied by natural philosophy and Aristotelian epistemology, the foundations of the systematization of medical culture in the West. Because of the nature of medicine as a discipline on the border of theory and practice, these commentaries provided the occasion for reflection on general concepts about the nature of scientific knowledge and, to a certain extent, the occasion for their modification. As far as medicine itself was specifically concerned, concepts of fundamental importance were developed and modified over the course of the centuries: these included notions of health, disease, and the neutral state, as well as the entire set of problems about the scope and the subject of the theoretical considerations and practical activities of the physician. In addition, the same context always gave rise to prolonged and lively discussions about the scientific status of medicine, a discipline that encompassed in its own proper sphere both theoretical considerations and practical applications.[2]

Within the curricula of the medieval faculties of medicine, the *Ars parva,* in addition to being read, explained, and commented on in the course of studies, was one of the canonical texts (along with the *Aphorisms* of Hippocrates) from which *puncta* were extracted for the *tentamen* and then for the real graduation examination.[3] Hence, the commentaries on this work of Galen constituted a genuine and distinct literary genre, inserted into the heart of the academic institution. We still lack a complete census of commentaries. Nevertheless, the manuscript sources and printed editions so far available make it clear that the major figures in medical culture were profoundly grounded in this Galenic text; they also lead one to suppose that a significant proportion of all academic teachers of medicine felt themselves duty bound to produce something, in the form of a commentary or *quaestiones,* connected with the interpretive problems raised by the *Ars parva.*[4]

In the period of medical humanism, notwithstanding the renovation of medical culture at various levels resulting from the new methods promulgated by "philologist physicians," the faculties of medicine remained in general tied to medieval teaching and institutional tradition. As a result, the use made of the *Ars parva* in this period offers a privileged vantage point from which to investigate the interaction of the new tendencies of medical humanist culture with the methods and issues traditional in university culture and teaching.

Before we begin an examination of some of Giovanni Mainardi's proposed solutions to the problems posed by the *Ars parva,* it is perhaps appropriate briefly to characterize the movement of renovation that constituted medical humanism.[5] A sketch of some of the main features will enable us better to put the approaches and proposals of the Ferrarese physician in context and to evaluate their real significance. The new culture manifested itself in two ways: on the one hand, its proponents rejected and were bitterly critical of the *auctoritates* of medical scholasticism and vigorously promulgated a return to the pure sources of Greek medicine and botany; on the other, they brought the presentation of the classical texts of medieval medical scholasticism up to date by bringing them into line with the formal requirements of the new humanist culture.[6]

Humanism began to have significant effects on medical culture in the last decades of the fifteenth century. That was the time when generations educated in humanistic schools began to become culturally productive in the learned professions, into which, once they had completed their training with the technical instruction gained in the universities, they imported the fertile seeds of humanist method. At the same time, nonspecialists, too, were beginning to feel the need for access to the scientific as well as the literary patrimony of classical antiquity; it is sufficient here simply to mention the cases of Angelo Poliziano and Ermolao Barbaro. And it was precisely in the area of scientific culture that the awareness dawned that Greek and Latin culture had had different roles, a realization that made possible a more complex and

realistic evaluation of those two worlds, which up to that time were unconsciously confused in the vague idealization of classical antiquity. We find ourselves confronted for the first time with personages, of whom Leoniceno
remains the classic example, in whom philological competence (acquired
thanks to the *paideia* of humanist teachers) is united with the traditional and
sophisticated philosophical medical culture imparted in the universities.[7]
This union bore fruit in the work of exceptionally well-prepared scholars
who could handle competently both the linguistic and the technical aspects
of ancient scientific texts and thus could renovate scientific thought.[8]

The most significant change that humanists introduced into medical
culture was not, in my opinion, the rejection of scholastic language in favor
of a formal renewal of language and style following classical models, though
that was one important aspect. Such linguistic rejection and renewal expressed a profound value, namely the recognition that the corpus of knowledge transmitted must be clearly and securely accessible to understanding.
But of much greater importance was the recovery of Greek sources and their
direct use, without mediation.

Direct contact with the "living and pure sources" of Greek scientific
culture was made possible by the intense activity of numerous philologists who made available, in the original texts and in the new translations, the
entire scientific corpus: the authors who wrote on philosophy, medicine,
botany, mathematics, and astronomy. In some cases, as in that of Galen,
which is directly relevant here, important texts were recovered that either
had remained completely unknown during the Latin Middle Ages or had circulated in abbreviated form. Where medicine was concerned, broader and
deeper knowledge of the classical authors provided a secure instrument for a
critique of the organization and procedures of university teaching, a critique
focusing on the need to free medical teaching from questions and issues that
were substantially extraneous to the subject. The effort to render medicine
independent of philosophy and thus make it an autonomous discipline—
an intention that was one of the most pronounced aspects of the work of
Mainardi but certainly not a common trait of all medical humanists—did not
involve, however, a rejection of Aristotelianism *tout court*. The salient point
was rather a sharp rejection of the scholastic systematization of medicine as
the humanists set aside texts and teaching methods strongly influenced by
philosophical issues, especially those filtered through the texts of Arab authors and their followers, the "moderns." Aristotelian natural philosophy and
logic continued to provide the fundamental concepts that made possible the
formulation of medical theories; but the Aristotelianism of humanist physicians was unquestionably an Aristotelianism in crisis. The direct comparison

of Aristotle and Galen both made ever more obvious the contrast between the two authors on some essential points and made ever more urgent the choice between loyalties.

II

Giovanni Mainardi of Ferrara (1462–1536) was certainly one of the most outstanding figures of medical humanism. Mainardi, who had been Leoniceno's pupil and was his successor in the chair of *medicina teorica* at Ferrara's Studium (1524), like his teacher brought forward a wide program of reformation of medical culture. But in his case, the proposed reformation had a stronger bias toward the practical aspects of medicine.[9] He enjoyed a rich and varied life, both personally and professionally: university teacher and successful doctor, personal physician at the court of Mirandola (1493–1502), royal physician at the Hungarian court (1513–1518), and physician of Alfonso d'Este at Ferrara (from 1518). He traveled extensively and was in contact with many personalities on the intellectual scene of his time. After the untimely death of Giovanni Pico della Mirandola in 1494, Mainardi edited (along with his pupil Gianfrancesco Pico—nephew of Giovanni Pico) the *Disputationes adversus astrologiam divinatricem,* one of the fundamental texts of the new Renaissance culture.

The wide range of his interests and the humanistic foundation of his approach to specific problems of medical culture are clearly revealed in the twenty books of his *Epistolae medicinales.* This best-selling work, whose complete edition, after several partial editions (the first in 1521), was published only after Mainardi's death (Basel, 1540), combines the traditional genre of *consilia* with humanistic and philological discussions on a variety of medical, botanical, and pharmacological themes: topics range from questions of terminology and identification of diseases and remedies to the taxonomy of skin diseases and the cure of the plague and supposedly new diseases such as syphilis, as well as treatments for gastric disorders and internal maladies. His strong interest in botany and pharmacology and his attempt to bring them back to their original purity are well represented in this work and are also behind his *Annotationes et censurae in Mesue Simplicia et Composita* (1535), a classic text of medieval medical tradition.

The specific character of his university teaching, with its scholastic approach to traditional themes of medical culture but with the novelty of the humanistic philological method, comes to the fore in his commentary on the first book of Galen's *Ars parva.* This work, first published in Rome in 1525,[10] is one of the first Renaissance expositions of Galen's text to appear after the pioneering interpretation by Mainardi's former teacher Leoniceno of the three ordered doctrines of which Galen speaks in the proem of the *Ars parva*. Moreover, it and the commentary of Giovanni Battista da Monte are among the most important and widely disseminated commentaries produced under the influence of medical humanism. Mainardi's work bears witness to his endeavor to make a distinctive personal contribution to the convincing new interpretation of Galen's three ordered doctrines as simple "ordines docendi."[11] Without substantially modifying Leoniceno's revolutionary interpretation, Mainardi proposes a whole series of notable exegeses of specific points. His commentary, much more closely tied to university teaching than was Leoniceno's work, offers a valuable opportunity to investigate the extent to which medical humanism was able to make a breach in the scholastic medical system into which Galen's text had been integrated; it also allows us to see some of the differences of opinion within medical humanism, despite a common nucleus of important positions.[12] An analysis of the whole of Mainardi's commentary would far exceed the limits of this paper. I want simply to offer some examples of his method of proceeding taken both from his own introduction and from his commentary on Galen's proem. From these points one can easily identify his positions on the much-discussed problems connected with the structure of *Ars parva* and, what is more important, with the epistemological status of medicine, since he treats these subjects almost exclusively at the beginning of the work, following the usual scheme of the *accessus ad auctores*.

In form, Mainardi follows the tradition of the medieval commentators, but he always inserts innovations, both in interpreting the position of the cited authors and texts and in presenting his own opinions. The leitmotif of the whole work is supplied by the constant presence of Galen, who appears almost as a tutelary deity: "we who follow the opinion of Galen," "we who follow Galen do not hold the opinions of anyone else," "I defend myself with the shield of Galen"—these and other similar phrases are standard formulae that recur throughout the commentary.

There are a number of other noteworthy features in Mainardi's commentary that mark it as a work of startling modernity. Above all the endeavor, made necessary by the wider and deeper knowledge of Galen's thought, was to make medicine an autonomous discipline with respect to philosophy. Mainardi seeks to give medicine its own dignity and particular excellence, which in no way depend on participation in the epistemology of Aristotelian science. The effort is to eliminate, to the extent allowed by the text itself, any aspects particularly related to logic—that is, precisely those aspects on which the medieval commentators had particularly insisted. Mainardi is extremely critical of his medieval predecessors, from Pietro d'Abano to Drusianus (Pietro de' Torrigiani, or Turisanus, the *Plusquam commentator*), from Gentile da Foligno to Jacopo da Forlì, not to mention Giovanni Sermoneta: he is even readier to recognize, at least in one case, the merits of the "Arab commentator on Galen" (Haly ibn Ridwan), though he offers many criticisms of him as well.[13]

His version of the polemic against Avicenna, the classic topos of medical humanism, is extreme: "No one should oppose to me here or elsewhere the authority of Avicenna; really I do not consider him among the medical authors but among the writers who have gathered the sayings of others." Aware of the temerity of this judgment, Mainardi adds that he discounts Avicenna's opinion only when "Galen's opinion, or invincible reason, or the evident truth of the thing itself" constrains him to dissent from the author of the Canon.[14] Generally, following Leoniceno's decisive recommendation—that Galen should be explained from Galen and not from the fantasies of commentators—Mainardi rests his own interpretation on Galen's authority, collected from statements of Galen in other works.[15] Besides Galen, the most frequently cited authors are the Greek commentators on Aristotle (Alexander of Aphrodisias, Themistius, Ammonius, Eustratius). It should be noted, however, that Leoniceno's range of authors cited is much richer than Mainardi's.

The historical interest manifested by Mainardi on many occasions is another particularly interesting feature of his work. Not only did he begin his preface with a biography of Galen, but more than once he presents the reader with a historical reconstruction of the origin and development of problems before giving his own interpretation. This concern with putting things in historical perspective perhaps resulted from knowledge of the proem of Celsus' *De medicina,* recently rediscovered by humanists.[16] In any case, it seems that inserting the problems into a historical process contributes importantly (perhaps without Mainardi's fully realizing it) to a relativistic assessment of the various interpretations given over the course of time. In a cultural context in which the authorities of the past, including the recent past, were rarely questioned, Mainardi's historical approach gave him one more legitimate reason to propose his own interpretations.

Attention is also paid to issues connected with the organization and transmission of medical knowledge. Naturally, these issues were important for the medieval commentators as well, but the interest in them among humanists was of a very different kind.[17] In Mainardi's case, attention to organization of teaching involved referring to authors different from the traditional ones as well as deliberately deciding to avoid as far as possible the numerous questions traditional in commentaries on the *Ars parva* that had more to do with dialectic than with medicine. According to Mainardi's curt judgment, such questions were a waste of time for the physician.

Mainardi's reflections about the methods and procedures of research and of what is now called scientific discovery seem open to innovation. He affirms that "someone who is discovering something in a certain way teaches himself." Nothing prevents him from subsequently teaching someone else "by the same procedures [lit., order] that he has taught himself."[18] This extremely felicitous and unusual image with which Mainardi defines the process of research not only breaks the rigid structure of the medieval relation between teaching and learning, between master and pupil, since in this case the learner is a pupil of a very particular kind; more important, it reveals an open and accepting attitude toward the possibilities offered to anyone who follows a line of "discovery" (invention) in an art or science. His remark becomes even more significant if it is linked to his negative judgment of the excessive obsequiousness toward the *auctoritates* among his predecessors. Such an attitude, according to Mainardi, had enormously damaged medicine, impeding new developments different from those recorded in the works of the past.[19]

Some solutions are proposed by Mainardi on the basis of his own translation of the Galenic text. Mainardi was convinced that many problems that were particularly difficult to solve had originated in misunderstanding of and consequent bad translations from the Greek text. He himself therefore translated afresh the pericopes of Galen's text to which he appended his commentary. This new translation was especially helpful in allowing him to handle concepts of health, sickness, and the neutral state. Mainardi in fact translated the first two of these as *saluber* and *insaluber,* suggesting also the suitability of *aegrotativus* (and *aegrotabilis*), instead of *sanus* and *aeger* (the medieval terms). By so doing, he stressed disease as a process, not an ontological entity—a conceptualization certainly more attuned with discussion of the latitude of qualities and of the passage from one qualitative state to another.[20]

III

Let us now examine in detail some points of the commentary. Mainardi proposes to abbreviate the treatment of arguments that had become classic topoi in the *accessus*. Therefore he does not follow the use of the "moderns," who write in the proem of every work a huge quantity of things, smuggling them in as Aristotelian when in reality they are entirely extraneous to Aristotle's thought and, in any case, "have more to do with dialectic than with medicine."[21] But he could not, obviously, completely free himself from the constraints imposed by the traditional genre of commentary and by the audience of students he was addressing. Thus, he limits himself to information about the author, the title, and the subject of the work, and the order that the author had followed in the exposition. Following the biographical information with which Mainardi prefaced his commentary are brief notes about the title.[22] Cutting short the disquisitions of his predecessors on this subject (which he condemns as "puerile"), he confines himself to observing that in the Greek manuscripts we find the title *Ars medicinalis* and not *Ars parva*. A more interesting inquiry about the title, Mainardi remarks, would be why Galen had used the term *ars* only for this work.

Previously, Mainardi had held that the term *ars* referred to the teaching of medicine by the method of definition here used by Galen. But after more careful reflection, he concluded that the reason for this terminological choice was that all the main points of medicine were encompassed, as in a compendium, in this work. The other works of Galen take their titles from the part of medicine they cover. By contrast, the *Ars parva* deals with the essential elements of medicine, according to the very definition of medicine, which, since it is valid, encompasses the principles—that is, the essential elements—on which all the specific aspects of medicine rest: bodies, signs, and causes. The *Ars parva,* therefore, presents statements that are the results of demonstrations carried out elsewhere.[23] As for the subject of the work, Mainardi notes that the Greeks, when speaking of a single work, are concerned to designate not the "subject" but the *scopos* and *prothesis,* that is, the "goal" and the "intention," which have a wider scope than just the subject. The intention, expressed by Galen himself, is to teach medicine by the definitive doctrine: the subject, then, is that of the whole art, namely health.[24]

Mainardi does not agree with those who consider the text an epilogue, a summary, as it were, of Galen's entire output. Nor does he agree with those who consider it a handbook for beginners. The difficulty of the work and Galen's own statement oblige us to consider it an aid for the mnemonic recapitulation of the whole discipline rather than an introductory text. In Mainardi's view, once students have mastered this work with the help of a good teacher, they will be able to tackle the other works of Galen on their own. Conversely, a good exposition of this work seems the most efficacious and appropriate way for an excellent teacher to crown his didactic efforts.[25]

This last topic does not reflect an idle classificatory whim, as at first sight one might surmise, but is inscribed in the general framework of discussions about the best way of arranging in a rational order the prescribed books of the academic curriculum. The urge to reform the medical curriculum was typical of humanists; it gave Giovanni Battista da Monte the occasion, some years later, to write his two prefatory letters "de ordine legendi Galeni opera" for the Giunta Galen of 1541 and 1550.[26]

The most interesting aspect of this introductory section, from the epistemological point of view, is Mainardi's treatment of problems concerning the subject of medicine. These problems are directly linked to the discussion about the scientific status of the discipline in the commentary on the first pericope of the book, which concerns the controversial definition of medicine that Galen places as an epigraph to the *Ars parva*. Right from this point, Mainardi anticipates the arguments that lead him to deny the status of *scientia* to medicine and proudly to claim it instead as an art—but an art of high epistemological profile to which all other liberal arts and philosophy itself must serve as propaedeutics.

As for the problem of establishing the real subject of medicine, Mainardi assails the belief of many of his predecessors that the human body was the primary subject of medicine. Such a position was unacceptable to Mainardi because it rendered medicine dangerously dependent on philosophical speculation about the elements. Following Galen, Mainardi denies that the body, the undoubted object of the operative part of medicine, is also the subject of medicine's theoretical consideration.[27] The true subject of medicine is health, and for the sake of health the physician develops his theoretical reflections, operates, and finally is acknowledged in his professional specialty with respect to other workers (*artifices*). Since medicine is a productive or, better, a restorative art, it is defined by what it restores, not by that on which its restorative action is conducted. Many restorative arts can deal with the same subject: for example, in restoring a house, different arts work on the roof, the walls, and the floor. These different arts are not distinguished from one another by theoretical consideration of the house itself as a unit, but by that which each of them repairs. Since medicine is obviously unable to produce human bodies, but can preserve or restore health, it must be distinguished from the other arts by health and not by the human body.[28]

To this discussion of the definition of the subject of medicine in Mainardi's preface we can add his comment on the definition of medicine given in the work itself. According to Galen, "Medicine is the science of things that are healthy, not healthy, and neutral. It does not change anything if someone says 'unhealthy.' What is important is to understand the term 'science' in the common sense."[29] This definition had been identified through the pseudo-Galenic *Introductio sive medicus* as that given by Herophilus, and Mainardi himself places great confidence in that attribution; indeed, in many cases he escapes apparently insoluble problems with the hypocritical assertion, "in any case this definition is not by Galen." However, Galen's use of it was still perplexing. Galen's epistemological attitude is, in fact, rather ambiguous. It encompasses two different concepts of medicine, neither of which matches the definition of a science according to Aristotelian criteria. Galen's anatomical, physiological, exhortative, and polemical works transmit an iatrosophistic concept of Alexandrian origin. This position considers medicine as a *technē theōrētikē,* strictly linked to logico-mathematical knowledge and endowed with a high epistemological profile owing to the control of causes provided by anatomy. By contrast, the clinical works transmit the Hippocratic concept of medicine as a *technē poiētikē* which produces and maintains health. It would therefore be a productive art, like painting, sculpture, architecture, and shipbuilding. But it could also be seen as a *technē epanorthōtikē*—that is, similar to the techniques through which houses, shoes, and clothes are repaired. In either of the latter two forms, medicine is an empirical technique that occupies a very low place in the Aristotelian hierarchy of scientific knowledge.[30]

The Arabs received primarily the Alexandrian iatrosophistic concept of medicine. With the reception of Arab medicine and, contemporaneously, Aristotelian philosophy in the West, that concept was inserted into the Western tradition of empirical medicine.[31] Although Averroës' *Colliget* (1.1) speaks of medicine as "ars operativa," Avicenna's opinion (*Canon* 1.1) that medicine was a science, subordinate to natural philosophy, was the primary source of inspiration for the medieval commentators on the *Ars parva*. In the university context, the stress on the learned aspects of medicine, the strengthening of its ties to natural philosophy, and emphasis on its high epistemological profile were developments guaranteed to earn for physicians the dignity and the honors of a learned profession endowed with great social prestige, on the model offered by the faculties of law.[32] As we can see from commentaries by Taddeo Alderotti, Pietro d'Abano, Torrigiano de' Torrigiani, Jacopo da Forlì, and Ugo Benzi, the question of whether medicine should be defined as science or as art was discussed interminably. While some commentators tried to deal with the unequal epistemological level of theoretical and practical medicine by claiming, as Taddeo Alderotti does, the status of science for the theory of medicine and that of art for its practice, others, such as Bartolomeo da Varignana and Dino del Garbo, declared medicine an art, stressing its practical aim and reclaiming its independence from natural philosophy. Mainardi, like the rest of his predecessors (except perhaps for Leoniceno, who was aware of the eclecticism of Galen's thought),[33] does not seem to have been aware of the flexible meanings that the terms *epistēmē* and *technē* were acquiring in Galen's day, at the very time that philosophy was losing the connotation of profound knowledge of causes and taking on the sense of a theoretical *technē*. In that environment—largely thanks to Galen's own efforts—medicine was rising to the status of a theoretical *technē*, as the ruling scientific discipline in the cultural panorama of the period.[34]

But for Mainardi, as for most previous Latin commentators, the contrast between science and art was a very sharp one, so that he felt obliged to try to solve the problem of positioning medicine as one or the other. After noting, following Leoniceno, that even though Galen uses the definition put forward by Herophilus, he does not seem entirely to approve it, Mainardi emphasizes how important it is to understand the term *scientia* not in the strict sense but in the common sense, broadening its meaning to include productive arts such as medicine.[35] Mainardi realized that the problem of whether or not medicine could be allocated the status of *scientia* was a very old one. He provided his readers with a historical reconstruction of the various solutions proposed, taking as his starting point the pseudo-Galenic *Introductio sive medicus*. The Methodists held that all of medicine should be considered a science; Erasistratus thought instead that the part of medicine that dealt with causes and matter belonged to science, whereas the curative and prognostic part was conjectural. Galen always held it to be a productive art, like those of architecture, shipbuilding, and other similar things, which no one thought should be considered sciences. Ammonius, Eustratius, and Averroës were of the same opinion. However, Avicenna followed the opinion of the Methodists and considered as science both the part of medicine that reflects on principles, which came to be called theory, and the part that teaches how to operate, which came to be called practice. Most of the moderns follow Avicenna and think that medicine can be called both science and art, believing that these two definitions are not contradictory, provided neither is understood in the strict sense.[36]

Mainardi supports his own opinion with the authority of Galen: "We who adhere to the opinion of Galen do not deny that in a certain way [medicine] can be called a science; however, we hold that in the proper and absolute sense it is an art, because it has an operative *habitus* and reaches its goal—health—not by necessary but by contingent means. Moreover, it has to do not with being but with generation and the things that can be produced by us. All these characteristics, according to Aristotle, are distinctive traits distinguishing art from science."[37] Mainardi decisively rejects the solution put forward by Pietro d'Abano, who proposed an ontological distinction between the moment of discovery and what some historians might call a period of "normal science."[38] Pietro wanted in this way to distinguish a period of the art, which would correspond to the moment of finding out and establishing

the discipline, and a period of science, when the discipline was already perfected. Mainardi thought this a ridiculous idea.[39]

———

As we have already noted, the problem of the relations between medicine and
philosophy is closely connected with that of the scientific status of medicine.
Galen, in harmony with the culture of his time, was relatively uninterested
in metaphysical problems and therefore made natural philosophy, ethics, and
logic propaedeutic to medicine in his system.[40] Avicenna, on the contrary,
subordinated medicine to natural philosophy, and the medievals for the most
part followed him.[41] Mainardi firmly denies that medicine could be referred
to any part of philosophy, not even to natural philosophy. He cites as support
for his opinion the Aristotelian topos according to which the activity of the
physician begins where that of the natural philosopher leaves off.[42] He specifies that the physician and the philosopher could both deal with disease and
health, but from very different points of view. The physician considers the
things that lead to operation, while the philosopher is concerned with speculation for own sake.[43]

Mainardi held that at this point it was useful to put the problem in historical context. As Celsus informs us, the ancients considered medicine to be
a part of "wisdom," since its first founders were philosophers. Hippocrates
subsequently separated medicine from philosophy, but because he himself
was a philosopher, he left some philosophical elements within it. Later, many
of his successors—above all those who belonged to the sect of the rational-
ists, who were more philosophers than physicians—introduced many additional philosophical and dialectical elements. These elements made medicine
more prestigious but also more distanced from its proper end; they made
physicians worthy of admiration but not actually better, since their ability to
discuss improved more than did their ability to cure. For this reason, Galen
reproached them.[44]

Mainardi underlines the cultural comprehensiveness as well as the specific character of medicine. The former idea, which was certainly not foreign
to the medieval commentators, took on a very different meaning in his work.
For him, medicine, although requiring full cultural preparation in the liberal
arts and philosophy itself, remained confined within the epistemological
framework of an art. Medicine was assigned—mistakenly, in his view—the
status of a science because anyone who wanted to learn it as it had been transmitted must be an expert in all the liberal arts and all philosophy, even though
medicine itself was an art and should not be called a science in the proper
sense.[45] Mainardi stresses that the structure of medicine is directed toward operation, a focus not characteristic of a science. If it were allowed that medi-

cine has the status of science, we would be obliged to admit that all the artisanal and vile crafts could be considered sciences too and that they too made
use of true demonstrations, since they prove many things through cause and
effect. For this reason, if the people who work in these crafts were philosophers and logicians, as are those who practice medicine, they too would have
imported into their crafts many of the same philosophical aspects found in
medicine.[46] Mainardi goes on:

> Someone might think that the fact that medicine is directed toward oper-
> ation does not mean that it cannot be a science . . . ; to this I will reply that
> it is impossible for any of the arts to be a science because, in addition to
> possessing a *habitus* that is productive and aimed toward external operation,
> the intentions that lead to their goals are reached in a contingent way.
> Moreover, if one affirms that there are many aspects in medicine that are
> not directed toward operation and can therefore be shown by demon-
> stration, I will absolutely admit that there are such real demonstrations.
> However, precisely because they are demonstrations, they no longer be-
> long to medicine—that is, to an art—but instead become part of natural
> philosophy.[47]

In this way Mainardi dismissed the endeavor of Drusianus (Turisanus) to
claim medicine as a science by restricting the definition of art to the curative
part and stating that all of medicine was speculative, but not for the sake of
speculation alone.[48]

Mainardi's deeper knowledge of Galen also allowed him to take a definitive position on the problem of the division of medicine into theory and
practice. This division, probably of late Alexandrian origin and patterned on
the division of philosophy, was firmly established by the Avicennian systematization of medicine and fitted well into the organization of university studies through the separation of chairs.[49] Though Drusianus was aware that
Galen had not mentioned the theory/practice division, he did not seem to
find the omission important.[50] But for Mainardi, Galen's silence on this issue,
and the fact that he always spoke of medicine as a productive art, was sufficient reason to condemn the division into theory and practice as artificial and
illegitimate.[51]

———

Finally, Mainardi's epistemological views led him to take a noteworthy position on the already long-standing *disputa delle arti*.[52] For him, as we have seen,
it was not possible to include medicine among the sciences in the strict Aristotelian sense of the term. To do so, in his view, would completely miscon-
strue medicine's specific character and goals. By contrast, he was fully ready

to follow the alternate route of underlining the great cultural and professional dignity of the arts:

> I would not want to be accused of doing medicine damage by including it among the company of arts in which pettifogging lawyers are accustomed to degrade us, as if it was something vile to profess an art and be called masters of it, something that they despise. The term "*art*" in fact designates something so noble that even the imperial dignity, than which there is nothing greater on earth, is defined, according to Quintilian, with the name of art. Nor does the name of art abrogate the dignity of medicine because medicine shares it with humbler arts. Indeed the name of man does not take dignity away from kings even though they share it with commoners. Furthermore art represents something noble because those who possess an art are always considered superior to those who lack it. We say not only that medicine is an art, but that it is the noblest of the arts, which Galen himself, in *De constitutione artis medicinalis,* holds as superior even to rhetoric. As far as the term "master" is concerned, it is given not only to those who practice medicine but also to those who are ready to learn it, since, as Pietro d'Abano says, those who are future physicians must be already masters of other disciplines. The excellence attached to the term master is shown by expressions such as "Roman magistrate" and "master of the knights" and "great master," used at the court of the king of France; and what is most important of all, Christ, king of kings, does not refuse the name of master. Let us leave, however, the lawyers with their quibbles. We do not blush to be called masters of the noblest of arts.[53]

Mainardi proudly claims for medicine and the profession of the physician the dignity of a special cultural and professional position, different from that of either the philosopher or the lawyer, in a period in which those two figures still enjoyed hegemony in the cultural scene. He thereby signals indisputably that his world was consciously undergoing great cultural and social changes. Mainardi recognizes that these changes are affecting the discipline of medicine, which is now in effect inserted into a cultural system in evolution. His awareness appears in his call for a continuing openness in scientific research:

> One must think that there are many more things still to be investigated than those that have so far been discovered by human ingenuity, so much so that even today the saying of Aristotle is true that the enormous number of things we know is only the least part of the things we do not know. Therefore, because many things remain to be investigated in all the sciences, for a long time our predecessors have been wrong. They based themselves on things that had already been discovered and treated what-

ever their predecessors wrote as an oracle, and therefore they added nothing to the arts. This failure, especially in the field of medicine, was a great sin and did much damage.[54]

The manifold aspects of Renaissance relationships between natural philosophy and the various disciplines that emerge in the articles in this volume give us a picture of a complex situation. Within the field of medicine, patterns of approach to new trends in culture and practice are varied and diverse, as Vivian Nutton's vivid portrayal of learned medicine in Tudor England also shows. Mainardi's critical view was the product of a very different cultural background. He practiced religiously Leoniceno's recommendation to elucidate Galen by Galen himself and not by means of alien philosophical lucubrations. Leoniceno's insistence on this point combined perfectly with Mainardi's striving to depict medicine as an independent discipline, free from heavy philosophical debts. His positions are even more significant, given the context in which they are formulated: commentary on a standard text in the university medical curriculum, that is, one of the *loci naturales* where the effects of the marriage between medicine and philosophy were most evident.

Despite the criticisms by many medical humanists of the excessive penetration of medicine by Aristotelian logic, epistemology, and natural philosophy, medical theory remained deeply embedded in a general foundation of Aristotelian philosophy. Although humanist physicians were generally inclined to side with Galen against Aristotle, the pull of a complete, well-structured, sophisticated tradition was very difficult to resist. Mainardi's position is therefore especially noteworthy. Indeed, some years after the publication of his commentary, another of Leoniceno's pupils, Giovanni Battista da Monte, in his highly successful Paduan university teaching, constantly stressed the need for medicine to maintain strong links with philosophy as the only way to attain a methodical and rational practice.

NOTES

Many thanks to Nancy Siraisi for her kindness—not least for the translation of the text read at the seminar's session, which I have not substantially altered for the printed version.

1. The text of *Ars parva* is printed in Claudius Galenus, *Opera omnia,* ed. C. G. Kühn (Leipzig: Off. Libr. C. Cnoblochii, 1821–1833; facsimile reprint, Hildesheim; Olms, 1964–1965), 1:305–412.

2. On the relationship of medicine and logic, especially in the *Ars parva,* see Jole Agrimi and Chiara Crisciani, "Medicina e logica in Maestri bolognesi tra Due e Trecento: Problemi e temi di ricerca," in *L'insegnamento della logica a Bologna nel XVI secolo,* ed. Dino

Buzzetti, Maurizio Ferriani, and Andrea Tabarroni, Studi e memorie per la storia dell'
Università di Bologna, n.s., 8 (Bologna: Istituto per la Storia dell'Università di Bologna,
1992), pp. 188–239. On the relationship of medicine and philosophy, see Paul Oskar Kristeller, "Philosophy and Medicine in Medieval and Renaissance Italy," in *Organism, Medicine, and Metaphysic,* ed. E. F. Spicker (Dordrecht: Reidel, 1978), pp. 29–40; Graziella
Federici Vescovini, "Medicina e filosofia a Padova fra XIV e XV secolo: Jacopo da Forlì
e Ugo Benzi da Siena (1380–1430)," in her *"Arti" e filosofia nel secolo XIV: Studi sulla
tradizione aristotelica e i "moderni"* (Florence: Enrico Vallecchi, 1983), pp. 231–278; and
Charles B. Schmitt, "Aristotle among the Physicians," in *The Medical Renaissance of the Sixteenth Century,* ed. Andrew Wear, Roger K. French, and Ian M. Lonie (Cambridge: Cambridge University Press, 1985), pp. 1–15.

3. On the curriculum and examination procedures at the University of Ferrara, where
Giovanni Mainardi taught, see Vincenzo Caputo and Riccardo Caputo, *L'università degli
scolari di Medicina ed Arti dello Studio Ferrarese (sec. XV–XVIII)* (Ferrara: Tipografia artigiana,
1990); for the *Ars parva,* see Statuto 57, 127–128; pp. 8, 44. See also Vincenzo Caputo, *I
collegi dottorali e l'esame di dottorato nello Studio Ferrarese: Gli Statuti del Collegio ferrarese dei
dottori di Medicina ed Arti (sec. XV–XVII)* (Ferrara: Università degli Studi di Ferrara, 1962),
pp. 51–55, 114–118.

4. There is a partial census of commentaries on *Ars parva* in Justus Niedling, *Die mittelalterlichen und frühneuzeitlichen Kommentare zur "Techne" des Galenos,* inaugural dissertation
(Paderborn: Druck der Bonifacius-Druckerei, 1924). On the medieval commentaries on
Ars parva, see Per-Gunnar Ottosson, *Scholastic Medicine and Philosophy: A Study of Commentaries on Galen's Tegni (ca. 1300–1450),* 2nd ed. (Naples: Bibliopolis, 1984).

5. On medical humanism, see Walter Pagel, "Medical Humanism—A Historical Necessity in the Era of the Renaissance," in *Essays on the Life and Work of Thomas Linacre, ca.
1460–1524,* ed. Francis Maddison, Margaret Pelling, and Charles Webster (Oxford:
Clarendon, 1977), pp. 375–386; Richard J. Durling, "Linacre and Medical Humanism,"
in ibid., pp. 77–106; Jerome J. Bylebyl, "The School of Padua. Humanistic Medicine in
the Sixteenth Century," in *Health, Medicine, and Mortality in the Sixteenth Century,* ed.
Charles Webster (Cambridge: Cambridge University Press, 1979), pp. 335–370; idem,
"Medicine, Philosophy, and Humanism in Renaissance Italy," in *Science and the Arts in the
Renaissance,* ed. John W. Shirley and F. David Hoeniger (Washington, D.C.; Folger Shake-
speare Library, 1985), pp. 27–49; *Humanismus und Medizin,* ed. Rudolf Schmitz and Gun-
dolf Keil, Mitteilung II der Kommission für Humanismusforschung (Weinheim: Acta
Humaniora, 1984); Vivian Nutton, *John Caius and the Manuscripts of Galen,* supplementary
vol. 13 ([Cambridge]: Cambridge Philological Society, 1987); idem, "Greek Science in
the Sixteenth-Century Renaissance," in *Renaissance and Revolution: Humanists, Scholars,
Craftsmen, and Natural Philosophers in Early Modern Europe,* ed. J. V. Field and Frank A. J. L.
James (Cambridge: Cambridge University Press, 1993), pp. 15–28; and idem, "The Rise
of Medical Humanism: Ferrara, 1464–1555," *Renaissance Studies* 11 (1997): 2–19.

6. On the adaptation of a medieval standard text—Avicenna's Canon—to the humanist
trend, see Nancy G. Siraisi, *Avicenna in Renaissance Italy: The "Canon" and Medical Teaching in Italian Universities after 1500* (Princeton: Princeton University Press, 1987).

7. On Nicolò Leoniceno, see Dominico Vitaliani, *Della vita e delle opere di Nicolò Leoniceno vicentino* (Verona: Tipolitografia Sordomuti, 1892); Daniela Mugnai Carrara, "Profilo

di Nicolò Leoniceno," *Interpres* 2 (1979): 169–212; and eadem, *La biblioteca di Nicolò Leoniceno Tra Aristotele e Galeno: Cultura e libri di un medico umanista,* Accademia Toscana di Scienze e Lettere "La Colombaria" 118 (Florence: Olschki, 1991).

8. For the role played by humanism in scientific thought, see Eugenio Garin, "Gli umanisti e la scienza," *Rivista di Filosofia* 3 (1961): 259–278; Marie Boas, *The Scientific Renaissance, 1450–1630* (London: Collins, 1962); and Paola Zambelli, "Rinnovamento umanistico, progresso tecnologico e teorie filosofiche alle origini della rivoluzione scientifica," *Studi Storici* 3 (1965): 507–546. See also Eugenio Garin, "Rinascimento e Rivoluzione scientifica," in his *Rinascite e rivoluzioni: Movimenti culturali dal XIV al XVIII secolo,* 2nd ed. (Bari: Laterza, 1976), pp. 297–326.

9. On Giovanni Mainardi (known also as G. Manardo and G. Manardi), see *Atti del convegno internazionale per le celebrazioni del V centenario della nascita di G. Manardo* (Ferrara: Università degli Studi di Ferrara, 1963); Paola Zambelli, "Giovanni Mainardi e la polemica sull'astrologia," in *L'opera e il pensiero di Giovanni Pico della Mirandola nella storia dell' umanesimo* (Florence: Sansoni, 1965), 2:205–279; and Vaclaw Urban, "Consulti inediti di medici italiani (Giovanni Manardo, Francesco Frigimelica) per il vescovo di Cracovia Pietro Tomicki (1515–1532)," *Quaderni per la Storia dell'Università di Padova* 21 (1988): 75–103.

10. On later editions, see J. Hill Cotton in *Dictionary of Scientific Biography,* ed. C. C. Gillespie (New York: Scribner and Sons, 1981), s.v. "Manardo, Giovanni."

11. Nicolò Leoniceno, *De tribus doctrinis ordinatis secundum Galeni sententiam* and *Antisophista medici Romani,* in his *Opuscula, per A. Lemnium adnotata* (Basel, 1532), 62A–83A, 146C–174C. On these works, see Daniela Mugnai Carrara, "Una polemica umanistico-scolastica circa l'interpretazione delle tre dottrine ordinate di Galeno," *Annali dell'Istituto e Museo di Storia della Scienza di Firenze* 8 (1983): 31–57.

12. On the different opinion of another leading medical humanist, Giovanni Battista da Monte, on the crucial issue of the independence of medicine from philosophy, see Schmitt, "Aristotle among the Physicians," p. 12.

13. See Giovanni Mainardi, *In artem Galeni medicinalem commentarius,* in Claudius Galenus, *Artis medicae liber primus a Iohanne Manardo commentariis illustratus, cui Nicolai Leoniceni Quaestio de tribus doctrinis praefixa est* (Padua, 1564), fols. 22v–24r. All subsequent citations of Mainardi's commentary are from this edition.

14. Ibid., fol. 34v: "Nemo autem neque hic, neque alibi in hac mea commentatione Avicennae autoritatem mihi opponat, eum enim in auctorum medicinae catalogo minime me habere profiteor, sed scriptorum qui aliorum dicta collegerunt, ut alias quandoque scripsi, et aliquando, deo optimo maximo aspirante, latius explicaturus sum. Quod temeritati nemo bonus mihi adscribet, maxime ubi de Galeni agitur opinione, non enim ab Avicenna secedo, nisi quando vel Galeni sententia, vel invincibilis ratio, vel ipsa rei aperta veritas me cogit dissentire."

15. Using this Alexandrian hermeneutic criterion, Leoniceno reverses the then-standard approach to the text. For the medieval commentators it was quite usual to explain difficult passages of Galen's text with the support of other philosophical and medical authorities. Cf. Leoniceno, *Antisophista medici Romani,* 151C: "Galenus siquidem ex Galeno est

intelligendus. Caetera omnia sunt nugae et falsae latinorum expositorum qui Arabes in plerisque sunt imitati imaginatione." On this point, see Daniela Mugnai Carrara, "Nicolò Leoniceno e Giovanni Mainardi: aspetti epistemologici dell'umanesimo medico," in *Alla corte degli Estensi: Filosofia, arte e cultura a Ferrara nei secoli XV e XVI,* ed. Marco Bertozzî, Atti del Convegno internazionale di Studi, Ferrara, 5–7 March 1992 (Ferrara: Università degli Studi, 1994), pp. 19–40.

16. For the use by Renaissance scholars of Celsus' proem to *De medicina,* but also of Galen's *De sectis* and the pseudo-Galenic *Introductio sive medicus* (works also used by Mainardi) on the many opinions of ancient medical schools, see Nancy G. Siraisi, "Giovanni Argenterio and Sixteenth-Century Medical Innovation: Between Princely Patronage and Academic Controversy," in *Renaissance Medical Learning: Evolution of a Tradition,* eds. Michael R. McVaugh and Nancy G. Siraisi, Osiris, 2nd ser. 6 (Philadelphia: History of Science Society, 1990), p. 173.

17. On the organization of medieval medical education, see Nancy G. Siraisi, *Arts and Sciences at Padua: The Studium of Padua before 1350* (Toronto: Pontifical Institute of Mediaeval Studies, 1973); eadem, *Taddeo Alderotti and His Pupils: Two Generations of Italian Medical Learning* (Princeton: Princeton University Press, 1981); Jole Agrimi and Chiara Crisciani, *Edocere medicos: Medicina scolastica nei secoli XIII–XV* (Naples: Guernini, 1988); and Nancy G. Siraisi, *Medieval and Early Renaissance Medicine: An Introduction to Knowledge and Practice* (Chicago: University of Chicago Press, 1990).

18. Mainardi, *In artem Galeni,* fol. 27v: "Sunt quidam . . . quibus ego minime assentiendum duco, qui enim invenit, quodammodo seipsum docet. Nihil autem vetat quo minus eo quo semet docuit ordine alium docere valeat."

19. See below, note 54.

20. Leoniceno's translation was "Medicina est scientia salubrium et insalubrium et neutrorum. Nihil vero differt et si quis loco insalubrium aegrorum dixerit." Lorenzano translated the same passage: "Medicina est sanabilium scientia, aegrotabilium et neutrorum. Nec interest si dixeris valetudinariorum." On the editions of these new humanistic translations of *Ars parva,* see Richard J. Durling, "Chronological Census of Renaissance Editions and Translations of Galen," *Journal of the Warburg and Courtauld Institutes* 24 (1961): 251. On the concepts of health, disease, and neutral state in some medieval commentaries on *Ars parva,* see Ottosson, *Scholastic Medicine and Philosophy,* pp. 126–194. For Mainardi's translation, see note 29 below.

21. Mainardi, *In artem Galeni,* fol. 3r: "De subiecto ad fastidium in cuiuslibet libri exordio scribunt recentiores, multa perperam quasi Aristotelica confingentes, ab Aristotelis mente penitus aliena. Quae cum sint alio loco a nobis declarata, ad dialecticamque potius quam ad medicinam spectent, ab eis in praesentia supersedere satius duxi."

22. Mainardi (ibid., fols. 1r–2r) draws biographical information from other works of Galen: *Methodus medendi, De anatomicis adgressionibus, De pharmacis secundum genus, De simplicibus medicamentis, De differentiis pulsum, De antidotis.*

23. Mainardi, *In artem Galeni,* fols. 2r–3r.

24. Ibid., fols. 3r–4v.

25. Ibid., fols. 4v–5r.

26. See Daniela Mugnai Carrara, "Le epistole prefatorie sull'ordine dei libri di Galeno di Giovan Battista da Monte: Esigenze di metodo e dilemmi editoriali," in *Vetustatis Indagator: Scritti offerti a Filippo Di Benedetto* (Messina: Centro Interdipartimentale di Studi Umanistici dell'Università di Messina, 1999), pp. 207–234.

27. Mainardi, *In artem Galeni*, fol. 3r: "Diximus igitur . . . iuxta Galeni sententiam, corpus humanum medicinae subiectum statui aliquo pacto non posse, subiectum dico considerationis non operis. Conveniunt enim omnes illud subiectum non esse quod per accidens et secundario, non per se et primo consideratur. Tale esse corpus humanum a Galeno capite penultimo libri de partibus artis medicinalis didicimus."

28. Ibid., fol. 3r–v: "Verum autem subiectum secundum eiusdem eodem in loco sententiam sanitas existit, ut quam medicus per se primo considerat ad quam omnem reliquam refert considerationem, et per quam potius quam per aliud quodvis ab omni alio artifice separatur. Cum enim medicina ars sit factiva, imo potius refectiva, iuxta eiusdem Galeni sententiam in libro de medicinalis artis constitutione, per id quod reficit, non per id circa quod operatur, est a caeteris artibus distinguenda, quando et per hoc caeterae refectivae artes distinguuntur et quidem merito cum nihil vetet varias circa eandem rem reficiendam artes versari. Ut exempli gratia, circa domum, alia quidem ars est, quae imbricum, alia quae parietes, alia quae pavimenta instaurat, nec inter se domus ipsius consideratione, utpote quae una est distinguuntur, sed eius potius ratione, quod in ipsa domo reparatur. Cum igitur medicina non corpus humanum facere, sed sanitatem conservare vel reficere possit, aliaeque artes circa idem humanum corpus aliud scilicet in eo vel conservando vel reficiendo versari possint, per sanitatem non per corpus humanum a caeteris est segreganda."

29. Galenus, *Opera omnia*, 1:307–308: "Ἰατρική ἐστιν ἐπιστήμη ὑγιεινῶν καὶ νοσωδῶν καὶ οὐδετέρων οὐ διαφέρει δὲ οὐδ' εἰ νοσερῶν τις ἔποι. Τοῦ μὲν οὖν ἐπιστήμης ὀνόματος κοινῶς τε καὶ οὐκ ἰδίως ἀκούειν χρή." Mainardi translates: "Medicina est scientia salubrium, insalubrium et neutrorum. Non differt autem si aegrotativorum quis dixerit. Nomen vero scientia communiter et non proprie audire oportet" (30r); see also above, note 20.

30. On Galen's epistemological thought, see Michael Frede, "On Galen's Epistemology," in *Galen: Problems and Prospects,* ed. Vivian Nutton (London: Wellcome Institute for the History of Medicine, 1981), pp. 65–86; Mario Vegetti, "Modelli di medicina in Galeno," in ibid., pp. 47–63; and Stephania Fortuna, "La definizione della medicina in Galeno," *La parola del passato* 42, no. 234 (1987): 181–196.

31. See Heinrich Schipperges, "Die arabische Medizin als Praxis und Theorie," *Sudhoffs Archiv* 43 (1959): 317–328; John M. Riddle, "Theory and Practice in Medieval Medicine," *Viator* 5 (1974): 157–184; and Ottosson, *Scholastic Medicine and Philosophy,* pp. 68–76.

32. Siraisi, *Taddeo Alderotti and His Pupils,* p. 13; eadem, "Taddeo Alderotti and Bartolomeo da Varignana on the Nature of Medical Learning," *Isis* 68 (1977): 27–39; and eadem, "Medicine, Physiology, and Anatomy in Early Sixteenth-Century Critiques of the Arts and Sciences," in *New Perspectives on Renaissance Thought: Essays in the History of*

Science, Education, and Philosophy, in Memory of Charles B. Schmitt, ed. John Henry and Sarah Hutton (London: Duckworth, 1990), pp. 214–229.

On the relationship of Italian Aristotelianism and medicine, see Antonio Poppi, *Introduzione all'Aristotelismo Padovano* (Padua: Antenore, 1970); Charles B. Schmitt, "Filosofia e scienza nelle Università italiane del XVI secolo," in *Rinascimento: Interpretazioni e Problemi* (Bari: Laterza, 1979), pp. 353–398; Eugenio Garin, *Aristotelismo veneto e scienza moderna,* Saggi e testi 16 (Padua: Antenore, 1981); and Giancarlo Movia, "Struttura logica e consapevolezza epistemologica in alcuni trattatisti padovani di medicina del sec. XV," in *Scienza e filosofia all'Università di Padova nel Quattrocento,* ed. Antonio Poppi (Trieste: Lint, 1983), pp. 375–394.

33. See Leoniceno, *De tribus doctrinis,* 80D: "Oportuit enim ipsos cum de Galeni opinione disceptarent, eundem Galenum non Aristoteli, a quo non raro dissentit, sed magis Platoni, cuius semper summus fuit imitator, ostendere consentientem." See also 73A and 81B.

34. See Margherita Isnardi, "Techne," *La parola del passato* 16, no. 79 (1961): 257–296.

35. Mainardi, *In artem Galeni,* fol. 30r: "Quanquam haec definitio Herophili fuerit, ut ex introductorio et libro salubrium sexto aperte colligitur, nec a Galeno usquequaque probata, sicuti inferius ostendetur, placuit tamen Galeno ea hoc loco uti, veluti valde nota et satis commode suo proposito servienti, in qua scientiae nomen (quemadmodum ipse dicit) communiter accipere oportet, ut factivas etiam artes qualis est medicina comprehendat. Non proprie ut videlicet ex adverso contra artem distinctam et veris scientiis tantummodo conveniens." See also fols. 37v–38r.

36. Ibid., fol. 31r.

37. Ibid., fol. 31v: "Nos Galeni haerentes sententiae posse aliquo modo dici scientiam non negamus, proprie tamen et absolute esse artem putamus, cum sit habitus recta ratione factivus et non sit eorum quae necessario fiunt, sed finem suum hoc est sanitatem contingenter nanciscatur, quae arti adversus scientiam distinctae sexto moralium adscribit Aristoteles. Atque circa generationem, hoc est, res quae fieri a nobis possunt, non circa esse, id est, res necessarias versentur, quo etiam discerniculo in fine postremorum resolutivorum artem a scientia Aristoteles separavit."

38. As in Thomas Kuhn, *The Structure of Scientific Revolutions,* 2nd ed. (Chicago: University of Chicago Press, 1970).

39. See Pietro d'Abano, *Conciliator controversiarum quae inter philosophos et medicos versantur* (Venice, 1565; facsimile reprint Padua: Antenore, 1985), Diff. 3, fol. 6r. Also, Mainardi, *In artem Galeni,* fol. 31v: "Non enim Aponensi in hac parte standum, differentia tertia sui Conciliatoris exponenti, artem esse circa generationem, id est appellari artem dum invenitur, scientiam circa esse, id est, ubi iam inventa sit, ut propterea secum fateamur medicinam ab Hippocratem eam faciente, dici potius artem potuisse, quam a nobis, confiteri enim pariter oporteret omnem scientiam esse artem dum invenitur et omnem artem scientiam dum est inventa, quod certe est valde ridiculum."

40. On Galen's philosophical thought, see Pier Luigi Donini, "Galeno e la filosofia," in *Aufstieg und Niedergang der römischen Welt,* part 2, 36.5, ed. Wolfgang Haase (Berlin: De

Gruyter, 1972–), pp. 3484–3504; R. James Hankinson, "Galen's Philosophical Eclecticism," in ibid., pp. 3505–3522.

41. Avicenna, *Liber Canonis* (Venice, 1582), 1.1.1, fol. 3v. Ottosson, *Scholastic Medicine and Philosophy,* pp. 68–88.

42. "Ubi desinit physicus ibi medicus incipit." For the use of this Aristotelian passage (*De sensu et sensata* 1, 436a18–b2), on which is based the traditional relationship of medicine and philosophy, see Schmitt, "Aristotle among the Physicians," pp. 9–10.

43. Mainardi, *In artem Galeni,* fols. 31v–32r: "Non potest etiam medicina ad aliquam philosophiae partem referri, quod enim neque ad mathematicam neque ad divinam satis per se evidens est. Sed quod neque ad naturalem, Aristotelis vulgata sententia constare potest, inde dicentis medicum exordium capere, ubi desinit philosophus naturalis. Quod non ita intelligendum est, ut negetur eisdem de rebus utrunque considerare, cum de sanitate et morbo inter ea quae parva naturalia vocant scripserit Aristoteles secundoque de partibus animalium dixerit, ad naturalem philosophum attinere aliquo modo de causis morborum pertractare. Quod primo quoque Therapeutices affirmavit Galenus et secundo libro Anatomicarum aggressionum, ait diversam utriusque esse circa dissectiones considerationem, medicumque ea tantummodo considerare quae ad opus conducunt, nudam vero speculationem ad philosophum pertinere, quod primo Colliget scripsit Averrois."

44. Ibid., fols. 32r–v: "Verum quoniam, ut scribit Celsus, primi medicinae inventores fuere philosophi, medicina ab antiquis sapientiae pars credebatur, donec eam a philosophia separavit Hippocrates. Sed quoniam et ipse philosophus fuit, nonnulla quoque philosophica suae immiscuit medicinae, licet ad ipsam professionem contracta, sicut quando in libro elementorum et de natura humana, corpus humanum ex elementis compositum probavit, quia doleret. Posteriores quoque, et hi praesertim qui rationalem sectam professi sunt, quoniam et ipsi philosophi et quandoque magis quam medici, multa philosophica interdumque dialectica immiscuerunt, quae medicinam quidem ipsam venustiorem reddunt, sed a proprio fine multum divertunt, medicosque maiori admirationi, sed non propterea meliores reddunt, cum disserendi illis potius adsit quam curandi peritia et propterea eos a Galeno omnibus in locis reprehendi videmus."

45. Ibid., fol. 32v: "Indeque natum puto ut medicina inter scientias a multis numeretur, quoniam qui eam, ut nunc scripta est discere cupiunt, liberales artes omnes et universam philosophiam callere opus sit, licet ipsa per sese medicina ars et non proprie scientia sit dicenda."

46. Ibid., fols. 32v–33r: "Quod si quis eam veram scientiam esse contendat quoniam in ea verae demonstrationes fiunt . . . quarum causa, Galenus tum in fine huius libri tum alibi saepe instructum esse in demonstrationibus eum oportere mandat qui sit ad discendum medicinam accessurus. Dicam quod sicuti ordinatio ad opus facit medicinam non esse proprie scientiam, sed solum communiter, ita demonstrationes suas non proprie sed communiter dici demonstrationes, sicuti quinto libro sui Colliget, caput 8, voluit Averrois. Si quis vero neget ordinationem istam ad opus auferre nomen verae scientiae et demonstrationis is fateri cogetur omnes sellularias vilesque artes veras dici scientias debere et veras facere demonstrationes, cum et in illis multa per causas et per effectus probentur. Quod si

hi qui eas professi sunt, sicuti hi qui medicinam dialectici et philosophi fuissent, multa quoque in idem in illas transtulissent."

47. Ibid., fol. 33r: "Et si quis adhuc resistat quoniam ordinatio eiusmodi rationem scientiae ab Aristotele primo libro postremorum resolutivorum non videtur auferre, quae est rem per causam cognoscere et quod illius est causa et quod non contingit aliter se habere. Respondebo per hanc ultimam particulam artes omnes a vera scientia excludi, quoniam cum sit habitus factivi et ad extrinsecum opus ordinati, finis intentiones non necessario consequuntur nec sunt de his quae necessario fiunt, sicut ex sexto moralium superius ostendimus et propterea id de quo ars est, aliter habere contingit. Et si adhuc non vis cedere, dicens, multa esse in medicina quae non ita secum habent illam ad opus ordinationem, quin sine illa possint demonstrari, confitebor utique veras illas esse demonstrationes, sed ita a medicina sicuti ab eius genere, id est ab artis ratione, decidere et ad naturalem philosophiam conscendere."

48. Ibid., fol. 33r–v. See Pietro Torrigiano de' Torrigiani, *Plusquam Commentum in parvam Galeni Artem Turisani Florentini medici praestantissimi* (Venice, 1557), 8B.

49. Nancy G. Siraisi, "Changing Concepts of the Organization of Medical Knowledge in the Italian Universities: Fourteenth to Sixteenth Centuries," in *La diffusione delle scienze islamiche nel Medioevo europeo,* Convegno internazionale, Rome, 2–4 October 1984 (Rome: Accademia Nazionale dei Lincei, 1987), pp. 291–321; Agrimi and Crisciani, *Edocere medicos,* pp. 21–47.

50. Torrigiano, *Plusquam Commentum,* 10B.

51. Mainardi, *In artem Galeni,* fol. 34r–v: "Nunc an [medicina] theorica vel practica vel utraque dici possit videamus. Galeno ergo quoque hic haerentes, sicuti eam etiam proprie scientiam negavimus, ita nec theoricen, nec practicen proprie loquendo affirmamus. Exigit enim ratio, ut a quocunque genus, ab eodem et species submoveatur. Nec si totam factivam esse fateamur practicen propterea dicere cogimur, practica enim a praxis deducitur, quae vox latine actionem significat, quam esse aliud a factione septimo libro primae philosophiae Aristoteles testatur. A praxi vero id est ab actione morales scientiae practicae vocantur. Aliquid tamen esse in medicina non inficiamur, quod aliorum comparatione theoricum dici possit, quod et in libri theologicarum sententiarum proemio Scotus quoque testatur, totam medicinam practicen vere esse dicens, haberi tamen in ea, ad quod dici aliquo modo theoricum possit, licet et ipse practici nomen non bene intellexerit, quae res multa eum de praxi superfluo scribere nec dicam male coegit."

52. On the *disputa delle Arti,* see *La Disputa delle Arti nel Quattrocento,* ed. Eugenio Garin, 2nd ed. (Rome: Istituto Poligrafico e Zecca dello Stato, 1982); Giulio F. Pagallo, "Nuovi testi per la 'disputa delle arti' nel Quattrocento: La 'Quaestio' di Bernardo da Firenze e la 'Disputatio' di Domenico Bianchelli," *Italia Medioevale e Umanistica* 2 (1959): 467–481; see also *Sapere e/è potere: Discipline, dispute e professioni nell'università medioevale e moderna,* 3 vols. (Bologna: Istituto per la storia di Bologna, 1990).

53. Mainardi, *In artem Galeni,* fols. 33v–34r: "Nec velim vitio mihi verti quasi de medicina pessime merito, quoniam eam in artium numero repono, quo legulei infringere nobis solent, quasi vile sit artes profiteri vocarique, quod ipsi dedignantur, magistri. Nomen enim artis adeo nobilem signat, ut imperatoria quoque dignitas, qua nulla aliquando in ter-

ris maior fuit, Quintiliano teste, artis nomine censeatur. Nec dignitatem artis nomen abrogat medicinae, quia sit vilioribus commune, sicuti nec hominis nomen regibus, quia sit illis cum plebecula commune. Alioqui de se nobile quid ars repraesentat, cum qui arte pollent, ea carentibus semper praeponantur. Non solum autem artem dicimus esse medicinam, sed artium nobilissimam, quam et rethoricam maiorem, libro De artis medicinalis constitutione facit Galenus. Magistri etiam dicuntur non solum qui iam medicinam tenent, sed et qui ad ediscendam eam accedunt quia scilicet aliarum disciplinarum magistros esse debent, ut bene scripsit Aponensis qui medicorum scholas petituri sunt. Quantae vero praestantiae magistri nomen sit et magistratus Romanorum et magistri equitum et magni magistri apud Galliarum reges adhuc custoditum nomen ostendunt et quod maius his omnibus est, quod rex regum Christus magistri nomen non recusavit. Valere igitur cum suis ambagibus leguleios sinentes, nobilissimae artis magistros dici non erubescamus."

54. Ibid., fol. 28v: "Sed illud potius tenendum esse longe plura quae nondum vestigari potuerunt, quam ea quae humano ingenio sunt adinventa, ut adhuc verum sit illud Aristotelicum maximam eorum quae scimus partem, minimam esse eorum quae ignoramus. Quare cum adhuc in omnibus scientiis plurima supersint investiganda, hoc unum longo tempore peccaverunt maiores nostri, quod inventis stantes oraculique loco habentes quaecunque a senioribus scripta erant, nihil artibus adiecerunt, quod potissimum in medicina, magna cum iactura hactenus peccatum est."

8

"A Diet for Barbarians": Introducing Renaissance Medicine to Tudor England

Vivian Nutton

Should one wish to choose any one region in which to examine in detail the introduction of Renaissance medicine and what it stood for, the example of Tudor England would surely be high on the list of preferred subjects. Its medicine and that medicine's practitioners are, when it began in 1485, obscure—few, save for the Welsh or hunters after the exotic, now remember Lewis of Caerleon, royal physician, mathematician, astrologer, and spy—yet it ends in 1603 with one of the most famous names in medical history, William Harvey, newly returned from Padua and failing, at least for the moment, to gain entry into the London College of Physicians.[1] Within little more than a century, England and its physicians had moved from northern darkness almost to center stage in European medicine. From letters, private papers, and publications—to say nothing of their grave monuments—one can gain an insight into the hopes and aspirations of those who, directly or indirectly, brought about this change and can see clearly what they themselves thought most important in the development of their medicine. Even if what they have to say touches rarely on natural philosophy in the narrow sense, as opposed to investigations of the wider world, at the very least it serves as a reminder that natural philosophy was but one key to unlock the secrets of nature.

It is important to stress, at the very outset, the low state of English learned medicine in the later Middle Ages, even as compared with its continental neighbors, let alone with Italy. In 1500 the two universities of Oxford and Cambridge between them produced at most five or six M.D.'s a decade, with Oxford somewhat more prolific than Cambridge.[2] A few foreign practitioners might come to England, usually in the train of prelates and princes. Henry VII employed a German, Jacobus Fries; a Frenchman, Jean Veyrier of Nîmes; and, most famous of them all, Giambattista Boerio of Genoa.[3] The timorous Ferdinando de Molina in 1490 was moved to make his will because "I am now in way to depart for to go to Oxford."[4] That town in 1500 saw the prosecution of an Italian, Dionisio of Nola, for practicing surgery without a license, and the town of Coventry was briefly home to a Greek,

Nicholas Rayes.[5] But the contribution of these visitors to English medicine was minimal; few stayed for long, or had eminent pupils. Nor was there much movement of physicians from England to Italy—the Hundred Years' War with France and the English civil wars saw to that. Between Thomas the Englishman in 1401 and William Hadcliffe at Padua in 1446, no Englishman went to Italy to study medicine, and between John Free in 1460 and John Chamber in 1503 I count a mere eleven medical travelers to Italy.[6] When they returned, it was far more often to political or ecclesiastical preferment, as Walter Lacey enjoyed, than to day-to-day medical practice.[7]

The great age of English medieval medicine—with John of Gaddesden, John of Arderne, and Mertonian natural philosophers like Simon Bredon, whose works were copied and circulated on the Continent—had long since departed, and the writings of English medical men were unknown abroad, even if they had been worth reading.[8] Roger Marchall's *Lanterne of fisicians* and the *loci communes* of John Argentine are poor things indeed.[9] They show how firmly fixed English medicine of the 1460s and 1470s was in the medicine of the 1300s, if not the 1200s. Signs of an acquaintance with such luminaries as Taddeo Alderotti are few; and although in the 1480s one can trace the gradual arrival of contemporary practical medical texts by Cermisone and Bartolomeo Montagnana, their apparent impact was small.[10] Institutionally, the situation was no better. England lacked any organization for the control or improvement of medicine in general—a result of its political fragmentation as much as of the weakness of its doctors. Its hospitals were numerous but usually tiny, and frequently tottering on the edge of bankruptcy;[11] there were no civic physicians or municipally paid healers; and such public health regulations as there were were poorly enforced.

The gradual establishment and consolidation of the Tudor dynasty, under Henry VII and still more under his son and successor Henry VIII, was the prerequisite for any wider medical developments, for, as David Starkey has argued, it was in the forms of politics and statecraft that Renaissance ideas came first to be felt.[12] England became more stable, more firmly governed, and wealthier, and both monarchs began to adopt openly fashions taken from France and Italy. In medicine, the new trend can been seen in the request by Henry VII around 1500 for a copy of the statutes of the hospital of S. Maria Nuova in Florence to serve as a basis for his new hospital of the Savoy in London.[13] Begun in 1508, though not completed for almost a decade, the Savoy hospital was a tangible, indeed monumental, sign of the new medical renaissance, even if the result was more English than Italian.

It is tempting to see in Henry's request the first evidence for the influence of a scholar, physician, and humanist—Thomas Linacre, newly returned home from Italy. Born in 1460 and educated at Oxford, where from 1484 he was a fellow of All Souls, Linacre was to play a decisive role in the development of English medicine, even after his death in 1528.[14] His career is highly unusual: not least because he spent eleven years or more continuously in Italy, first with Politian and Chalcondylas in Florence, then for three years in Rome, and finally for five or six years in Venice and Padua, where he took a medical degree in 1496. How much Greek Linacre knew when he left England has been vigorously disputed, but all are agreed that it was his period in Italy that turned him into one of the finest Greek scholars of his day, specializing above all in scientific and medical translation. The 1490s, the years of his Italian sojourn, were a crucial decade in the transformation of medicine and science. The clarion call from Leoniceno and his fellow hellenists in northern Italy for the replacement of traditional Latin authors by their Greek sources was loud and rousing. Linacre, a friend of Aldus, was one of those who responded by translating texts from the Greek into a more classical Latin, beginning in 1499 with a translation of Proclus, *De sphaera*. His first publications on medicine, however, did not appear for almost twenty years: Galen's *De sanitate tuenda* in 1517; the *Methodus medendi* in 1519; *De temperamentis* in 1521; *De facultatibus naturalibus* in 1523; *De usu pulsuum* in 1524; *De symptomatum differentiis* in 1524; and, posthumously, a fragment from Paul of Aegina, *De diebus criticis,* in 1528. In quantity, and even more in quality, this was a considerable achievement. Basing himself largely on his own Greek manuscripts, Linacre turned into elegant and accurate Latin the most important of Galen's works on practical medicine.

But it was as a pedagogue, not a physician, that Linacre reappears at the English court in 1500, charged with the education of the young prince Arthur, and it was not until nearly ten years later, in 1509, that he was appointed a royal physician. A friend of Colet, Erasmus, and their circle, he was actively engaged in education—he wrote three grammar books for schools—and he numbered Thomas More among those to whom he taught Greek. It was this combination of Erasmian humanism (to use a shorthand term), Greek, and medicine that was to have an enormous impact on English medicine, for one would not go far wrong in describing the practice of learned medicine in England down to the end of the sixteenth century as being in the Linacre tradition.[15]

It was an influence not only mediated through Linacre's own personality, impressive though that was, and through his friendship with other humanists such as More and, later, Juan Luis Vives. It was also expressed in more permanent ways. Linacre was a very wealthy man, amassing, out of the income provided by various canonries and rectories, a considerable fortune in books, land, and cash. At his death, he founded a lectureship in physic at St. John's College, Cambridge, and two at Oxford. St. John's appointed their first lecturer in 1525, but it was not until 1559 that the first such lecturer was admitted at Oxford. Linacre's will makes it clear what was to be taught: the new Galen, using, for the most part, Linacre's translations, with a strong bias toward practical therapy. They were specifically enjoined to deal with "literal" questions—that is to say, explication; and they were to avoid those that "Galen callyth logical," that is, more disputatious debates about natural philosophy, in part simply to save time and make it possible to cover Linacre's syllabus within two and a half to three years.[16] It was a bias later followed by Henry VIII when he in turn came to establish the new Regius Professorships of Physic at the two universities in the early 1540s. Along with the other new professorships of Hebrew, Greek, divinity, and civil law, medicine was now to participate fully in the new humanism, the learning that took texts from antiquity as the basis of sound theory and practice.[17] The impact on their respective universities of the Linacre lecturers has been well studied by Gillian Lewis and, even if one takes a less sanguine view of their achievements than she does, two things are clear: some of the holders of the post were men of distinction, even if not as well qualified in medicine as we might expect; and their books and publications display that prejudice in favor of the classics called for by their founder.[18]

Second, and even more significant, it was at the urging of Linacre, and of other Italian graduates in medicine around the court, that in 1518 the London College of Physicians was set up to govern medical practice in London and its immediate environs.[19] This was, in effect, the first time that such a governing institution had been created in London—an attempt a century earlier had failed within two years—and, at least in theory, it mandated for the first time a graduate qualification for the practice of physic in London. Its model was that of an Italian college, like that of Padua or Venice: a body of elite physicians charged with laying down and enforcing standards of practice within the locality.[20] This is not the place to recount in detail the vicissitudes of the College or to explain the difficulties faced in imposing the authority of a small committee—with never more than twenty-five members in all until the end of the century—over a burgeoning metropolis.[21] It is enough here to emphasize two points. First, like the College of Physicians at Lyons, the London physicians saw their role as superior even to that of the universities, and their standards as far outstripping even those of a Paduan M.D. And, second, the College's aim was to impose a Galenic medicine on all English medical practitioners. Exactly what was initially implied by this is unclear, since the earliest statutes have not come down to us, but Sir George

Clark has argued convincingly that the ferocious examination in the works of Galen with a little Hippocrates, as approved in 1563, must have gone back at least to 1541, if not earlier.[22] Institutionally, then, the London College of Physicians maintained, for at least a century, the preferences and prejudices of Thomas Linacre, if not of Galen of Pergamum.

It is only too easy to deride the London College for its ambitions, its outdated learning, its bookishness, and its elitism. Seen from the perspective of the 1590s, or even the 1570s, the efforts of successive councils and presidents, most notably John Caius, to impose the classical writ of the College on all throughout England who might wish to practice medicine appear ludicrously overoptimistic, and its leading spokesmen antiquarian bigots. But, as is becoming clear, in 1518 when the College was founded, and indeed into the 1550s, the new Greek-based medicine was seen as the utmost in modernity. By purifying the medicine of the Middle Ages of ignorant accretions, by using new and better translations of Galen, one could avoid many errors in practice—the program advocated by Leoniceno in Ferrara and eagerly taken up by other northern Italian Hellenists—and win new knowledge from texts whose longevity of itself guaranteed their value.[23]

Nor, until the introduction of Paracelsian medicines and ideas in the 1560s, was there any clear alternative to humoral medicine save empiricism. Even if there might be disagreement on details, the general principles of classical medicine were never challenged. Besides, Linacre's own translations, notably of Galen's *Method of Healing,* had rescued major practical Galenic texts from medieval neglect; and as the next generation of scholars was to show, they offered many apparently new ideas on therapy.[24]

It was a program that fitted perfectly with the new ideals of the utility of scholarship put forward by Erasmus, Colet, and their friends: the purification and improvement of learning by a return *ad fontes,* to the mainly Greek springs of their various disciplines. The young men of the 1520s who were to carry out this program—Thomas Lupset, Edward Wotton, and, above all, John Clement—were given royal support, financial as well as moral; they were provided with posts at the new humanist foundations in Oxford; they communicated regularly with Thomas More and his London circle; and they shared in the reforming interests characteristic of Erasmus, in theology as well as in medicine.[25] In their writings, in their libraries, and in their letters, we may glimpse their priorities—and their dislikes. The older Aristotle of the Oxford schools is replaced by Plato; the medieval scholastics by the church fathers, notably Chrysostom; logical analysis by exegesis and emendation.

This new English learning can claim, as its most enduring monument in medicine, the Aldine *editio princeps* of Galen, published in Venice in 1525.

It was seen through the press by three Englishmen, including John Clement, and by one Saxon, Georg Agricola, later to be more famous for his mineralogy than for his medicine. In the next decade, the editors of the Basel edition of Galen in 1538 gratefully acknowledged the valuable help they had received from notes sent from Britain. One can trace this tradition of medical textual scholarship in Greek through John Caius at Cambridge, and George Edrych at Oxford, down to Theodore Goulston at Oxford and London at the beginning of the seventeenth century.[26] In the quality of their Greek learning, these men compare favorably with their Continental counterparts, and, what is often forgotten, their publications often had a directly didactic purpose. Caius' editions and translations of Galenic anatomy were intended for practical use, and Edrych's commentary on Paul of Aegina's surgery was dedicated "pro iuuenum studiis ad praxim medicam."

In essence, what is being done in England amounts to little more than the continuation of the program and methods first announced by Leoniceno: the acquisition, collation, translation, and elucidation of Greek medical and scientific books and manuscripts in order to reach a better understanding of the principles on which medicine had for centuries been based. It was a program supported at the highest level by king and by court. When in the 1540s there arrived in England a Portuguese *converso,* Manuel Brudus, a member of a family that had long treated members of the Spanish nobility, he enjoyed the powerful patronage of the king's steward, Sir William Sidney, and leading English courtiers like Sir John Baker and Sir Thomas Audley. In return he dedicated to them his book *On Diet in Fever according to Hippocratic Principles,* in which he explained that the English diet of good red meat and beer was medically necessary for those who live in cold northern climates.[27] His little book is a neat exposition of modern humanist medicine, well suited to an audience already familiar with its main principles and able to appreciate the practical benefits of the new learning.

Those who were responsible for its propagation in England were also, like Leoniceno, eager explorers of the whole natural world. John Clement and George Owen were keen botanists, an interest they shared with William Turner despite their religious differences.[28] Many of the early members of the London College of Physicians were singled out for praise by William Bullen for their interest in botany or zoology, and even a diplomatic bag might contain seeds and specimens from abroad intended for a leading London physician.[29] One can detect a slight shift in emphasis over the generations. Clement, Owen, and Edward Wotton are rather more bookish than their successors: Wotton's treatise *De differentiis animalium,* printed after a long delay in Paris in 1552, contents itself largely with identification and with organization of material assembled out of classical texts. Conrad Gesner, to whom Wotton presented a copy, passed a harsh but not unjust verdict upon it: "he took a good deal from Athenaeus, but he did not take everything, nor was he as careful as I am myself."[30]

It is in the next generation, with those who came to maturity in the 1540s, such as John Caius and William Turner, that practical experience of the plants and animals themselves comes to the fore. True, their work, whether like Turner on plants, or like Caius on birds and animals (his book on *English Dogs* is still well worth reading today), is largely descriptive: concentrating on the identification and naming of the natural world, and taking Aristotle and Dioscorides as the starting points.[31] But both men impart a sense of the importance of observation and practical understanding of plants and animals. They examined them out of an Aristotelian enthusiasm for the natural world—even for such unlikely subjects as tinkers' curs, which, "with marueilous paceience beare bigge budgettes fraught with Tinckers tooles, and metall meete to mend kettels, porrige pottes, skellets, and chafers, and other such like trumpery requisite for their occupacion and loytering trade, easing him of a great burthen which otherwise he himself should carry upon his shoulders."[32] One has only to read Turner on the plants of the Rhineland, or Caius on the humble puffin or the greyhound, to be convinced that their energy and enthusiasm did not stop at the printed page or at their library door.[33] Gillian Lewis has drawn attention to a booming interest in botany and in botanical books from the 1540s onward in Oxford, and she has suggested that many Oxonians may have carried this passion for plants and herbs with them after their university days, even into the wilder reaches of North Wales.[34] All this signifies the transition from the world of Leoniceno to that of Conrad Gesner, a friend of both Caius and Turner and, like them, a practical man as well as a bibliophile. It marks, one might say, a return to Aristotle—not to Aristotle the logician but to Aristotle the naturalist—and one might indeed think of it as a contribution to natural philosophy, in the widest sense.

The same generation, and in particular John Caius, can also be credited with the introduction of the new anatomy from Italy into England. It was once thought that David Edwards, who taught medicine and Greek at Corpus Christi College, Oxford, around 1524, and who later migrated to Cambridge, had learned his anatomy at Padua around 1525.[35] Unfortunately, the Englishman abroad who was called Odoardus was Edward Wotton; although Edwards certainly did at least once dissect a corpse—whether in Oxford or in Cambridge is not clear—his learning appears to have been largely home-grown. This is not to say that some of it, as displayed in his *In anatomen introductio luculenta et brevis,* printed in London in 1532, does not derive from reading an Italian exemplar, in this case Alessandro Benedetti, or that it is not also a testimony to the introduction of the new Greek technical terms into medicine.[36] But there is no evidence that Edwards knew the newly published and newly translated texts of Galen that, for effectively the first time, revealed the anatomical discoveries of that ancient physician and the central place that they held in his thought and writings.

The same could not be said of John Caius, that doughty defender of the status quo, who, like his mentor Galen, was passionate in his advocacy of dissection.[37] He lectured on anatomy himself, and his statutes for his refounded Cambridge college demanded at least one annual anatomy for its medical students. He collated manuscripts of Galen's *Anatomical Procedures,* which he edited with a commentary, and he also edited and translated into Latin *On Bones.* According to his autobiography, it was his work on anatomy that he prized most highly, not least because he had shown up the follies of Vesalius in translating Galen without a full mastery of Greek and, still more, in proclaiming that Galen had never dissected a corpse—which, of course, depends entirely on what one means by dissection.[38] That there was a market for the new anatomy in England is also clear from the success of Thomas Geminus in his plagiarisms of the *De humani corporis fabrica* of Vesalius, as well as from the number of copies of the *Fabrica* circulating in Oxford and Cambridge within a year or two of its publication.[39] Richard Caldwell, sometime fellow of Brasenose College, Oxford, was one of those most involved in 1570 in setting up the Lumleian Lectures in surgery at the London College of Physicians, and he himself produced a translation, via an earlier Latin version, of the *Tables of Surgerie* of Jean Tagault.[40] Another anatomical publicist, John Banester, author of the highly derivative *History of Man, Sucked from the Sap of the Most Approved Anatomists* (published 1578), had a license from Oxford to practice medicine and left his tiny ivory-and-boxwood manikin, which he presumably used in his anatomical demonstrations, to Cambridge.[41] As we know from Peter Jones's work on the books of Thomas Lorkyn, the long-lived Regius Professor of Physic, anatomical study was pursued enthusiastically in Cambridge; the very latest of discoveries were eagerly debated well into the 1580s, if not beyond.[42] William Harvey, a scholar and later fellow of Caius College, also reports on seeing at least one dissection carried out while he was there before he left for Italy.[43] One can draw a similar picture of the introduction of the new anatomy into London, and of the propagation of the new humanist medicine by leading members of the London College. They were joined in this by the learned surgeons forming the elite of the Company of Barber Surgeons, who took their knowledge of Galen and of ancient surgery at secondhand, via the French of Tagault or Vidius.[44]

In short, if one looks at English medicine around 1580, one cannot fail to be impressed by the vigor, if not always by the quality, of the work being done and by the great changes that had taken place since Linacre returned from Italy. There was now little to distinguish what was taking place in the English universities from that of, say, Montpellier, though not perhaps Bologna or Padua; and while the members of the London College were un-tiring (and unsuccessful) in their attempts to control the swarms of irregular practitioners who flocked to the ever-expanding and ever-wealthier capital, the same problems afflicted most of the medical colleges of northern Europe.[45]

What part in all this was played by natural philosophy? The answer is, sadly, almost none. Aristotle still formed part of the staple of the arts course in both Oxford and Cambridge, and in 1560 a Swiss student, Johann Ulmer, reported back very favorably on the medical teaching at Oxford in which the eight books of Aristotle's *Physics* were read daily.[46] How much of them the weary student could master at 6 A.M., when the lectures were held, or whether he was any better equipped to cope with an hour of Galen *On the Affected Parts* immediately afterward, is a matter to be left to the imagination. But compared with what is going on in northern Italy or at Wittenberg, there appears to have been little interaction between natural philosophy and medicine in England.[47]

There is, however, one possible exception to this. John Caius in 1544 published at Basel a treatise, *De methodo medendi,* which he republished with a few slight changes at Louvain in 1556.[48] Its opening pages, in traditional fashion, consider the precise meaning to be given to the three types of method outlined by Galen at the beginning of the *Ars medica* (*Ars parva*). Caius is brusque in his definition of method; it is a way and rationale for teaching and learning, based on the nature of the thing to be investigated, and his preferred advice is that one should follow Galen and Plato in breaking down a larger topic into more manageable parts and proceeding from there.[49] Caius is aware of the vigorous debate on this begun by Leoniceno—given his Italian connections, it would have been very surprising if he were not—but it is difficult to determine just what influence this debate had on him, for several reasons.[50]

First, his treatise is about a specific method, that of healing. Once Caius has explained his general understanding of what a method is, the rest of the first book is taken up entirely with recommendations for medical practice, which Caius divides up into the conservation, preservation, and rectification of the body's health. Book 2 is entirely concerned with the treatment of diseases. In all this one needs both method, which deals with universal principles, and practice, which deals with individual instances; these are the two legs of medicine.[51] In other words, although the preface might suggest engagement with wider questions of natural philosophy, the bulk of this treatise pays no attention to them.

Second, Caius merely takes over the conclusions of Leoniceno that Galen's recommendations in the *Ars medica* were aimed at teaching, and that discussions on the epistemological value of Galen's three methods of approach can be subordinated to a focus on their utility in promoting a specific method of healing.[52]

Third, it is above all Galen who provides the information and model for Caius. Aristotle is mentioned only in passing, and with apparently less regard than Plato, whose methodology Galen had appreciated highly.[53]

Finally, and perhaps most crucial for my purposes, the arguments and indeed most of the wording of this book are not Caius' own. They are taken over directly from the lectures that the greatest Galenist of the sixteenth century, Giovanni Battista da Monte, had just given in Padua, on Galen's *Method of Healing for Glaucon*.[54] Caius' justification for this at the end of his life, that he was bringing to wider notice in a more elegant form the most significant conclusions for medical practice of the greatest physician and teacher of the day, rings as hollow today as it did then; and Caius' long list of predecessors, including Galen, who have taken over large chunks of others' writings and ideas in their own publications succeeds only in cloaking plagiarism with pedantry.[55] As we can see from the other published versions of da Monte's lectures, Caius, despite his protestations, was merely his master's voice.[56] Thus, even if we allow that this tract shows an awareness of wider debates in natural philosophy, it is hard to credit it all, or even mostly, to John Caius.

If we exclude this hybrid production, there is very little evidence for any of the English medical writers being influenced directly by any of the wider debates in natural philosophy taking place in Italy. Their hero was Galen, the anatomist, the therapist, and, one should not forget, the logician. The second possibility of a strong influence on medicine from natural philosophy comes with the work of William Harvey, and in particular with his *Exercitationes anatomicae de motu cordis et sanguinis in animalibus* of 1628. This has been recently emphasized by Roger French in his argument for the crucial role of Harvey and his discovery of the circulation of the blood in the transition from the medieval world of Aristotelian natural philosophy to the world of the eighteenth century.[57]

In one sense, French is saying nothing new. Thirty years previously Walter Pagel had argued strongly that Harvey's thought world was still that of the Aristotelian universe, with its Aristotelian causes and its ideas on the perfection of circular motion; and he had connected some of Harvey's own arguments with those being put forward in Italy by Aristotelians such as Cremonini and Cesalpino.[58] Nor is there any dispute that Harvey owed much to that great Aristotelian anatomist Fabricius ab Aquapendente, his teacher at Padua.[59] Thus natural philosophy, in an Aristotelian sense, clearly does have a part to play in Harvey's work—but what that part was, as French unwittingly demonstrates, and when it began to exercise its influence are far from easy to determine.

French takes the strong line that Harvey was influenced considerably throughout his life by the natural philosophy of Aristotle as expressed in his *Physics* and its related books about the natural world. It was something that he had learned as a student in Cambridge, and it was only confirmed for him in Padua, where he was exposed, perhaps for the first time, to Aristotle's writings on animals, which had not formed part of the traditional syllabus of natural philosophy. Harvey's Aristotelianism found its expression in his language of discovery and in the careful proofs he offered for it in a manner reminiscent of a university disputation in philosophy or medicine. And, of course, French is right to point out that whether one accepted or rejected Harvey's discovery frequently depended far more on one's preexisting attitude toward a wider natural philosophy than on any single or specific argument put forward by Harvey.[60]

But once one begins to look for detailed evidence of influence from natural philosophy, French's arguments either collapse at crucial points or rely more on faith than on documentation. What lectures Harvey heard on Aristotle in Cambridge are unknown; they will have included lectures on the *Organon, Physics,* and *De anima,* but how the lecturers interpreted these texts or what subsidiary guides were used, two crucial questions, cannot be answered with any degree of certainty.[61] It may, however, be relevant to note that at least in the opinion of Charles Schmitt, who knew Renaissance Aristotelianisms better than most, Harvey's use of Aristotle was very different from that of the English tradition represented by John Case.[62] The intellectual career of Gabriel Harvey, a decade or so before his more famous namesake, would appear to show that Ramism was being rejected in favor of a stricter but much more elementary Aristotelianism, such as was later visible in the summaries of Bartholomaeus Keckermann, widely read in Cambridge in the 1610s.[63]

Whatever Harvey read of Aristotle in Cambridge or in Padua, overt acknowledgment of Aristotelian physics is rare in *De motu cordis,* although, as Gweneth Whitteridge has shown, the proofs that form the second half of the book correspond exactly to the rules laid down by Aristotle.[64] Besides, if Andrew Wear's argument is correct, Harvey was following "the way of the anatomists," which was neither that of the philosophers nor that of the physicians, and which depended heavily on the precedent and the injunctions laid down by Galen.[65] Indeed, Galen is far more prominent in Harvey than is Aristotle, and one could with some force argue for a continuation in Harvey of the tradition of Galenism represented (in their different ways) by Fabricius and by John Caius. The quantitative argument about the sizes of the veins and arteries coming to and from the heart and the consequent meditation on the amount of fluid they might contain have Galenic precedents familiar to Harvey.[66] Harvey's consideration of the purpose of the elegantly and artistically contrived structure of the heart, its fibers and the veins, would have gladdened the heart of any Galenist brought up on *The Usefulness of Parts*. At least one of his experiments with ligatures was anticipated by Galen, and one might compare Harvey's careful use of logic to establish the truth of his observations with Galen's recommendations for his ideal anatomist.[67] Although chronologically much later, the notes that Harvey made around 1644 in the margin of one of his copies of Galen are of considerable significance for understanding how his mind worked. The texts Harvey was then reading were only peripherally concerned with practical medicine, but he underlined every single word that had any connection with logic and proof—"plausibility," "judgment," "demonstration," "accurate," and so on—all of which bespeaks an unusual interest in precision of argument.[68]

Even for Harvey, then, a certain skepticism is required in assessing the part played in his discoveries by natural philosophy, whether in the narrow sense of Aristotelian physics or in a larger one that goes on to encompass all aspects of science. From one perspective, Harvey unites an English intellectual tradition of medical Galenism and of studying the natural world of plants and animals with a more sophisticated anatomical tradition deriving from Italy and, through Fabricius, concentrating on comparative anatomy and physiology. In this, Harvey is not untypical of the leading figures in English medicine in the sixteenth century, which, in its passage from obscurity to a blaze of success, depended little if at all on natural philosophy, except as it was mediated by and through Galen. Instead, its main focus was practical rather than theoretical: it aimed at medical rather than intellectual benefits. It was not at all insular, for one can point to English scholars on the Continent, and to an increasing number of foreigners coming to England and even elsewhere in Britain. In the sophistication and precision of what was done, particularly to edit and interpret Galen, English medicine performed at a level that at least equaled the best that Italy could provide.

It was a tradition that began by emphasizing the advantages of Greek and of Greek medicine, and, as represented by the hierarchy of the London College, it gained institutional permanence. It was a tradition that encouraged observation and description of the natural world of plants and animals, and, certainly from the 1540s if not earlier, the importance of dissection as the foundation of medicine. It was supported at the very outset by the monarch and the court; as such, it was merely one of the ways in which England was transformed in the first half of the sixteenth century into a Renaissance monarchy. Although by 1580 orthodox Galenists were often finding their attempts to prosecute or force out Paracelsian practitioners frustrated by wealthy and eminent patrons, this was not the case earlier in the century. Besides, even in 1600, Galenists continued in control of the two universities and of the London College.

This pattern was not repeated in every other European country; France, Spain, Germany or Denmark developed in different ways and with different emphases—some political, some religious, others intellectual or more strictly medical. The clash between Aristotelian natural philosophers and Galenist physicians familiar to us from accounts of life at Bologna or Padua does not appear to have occurred in England, where Paracelsianism and Protestantism were more vigorous opponents.[69] But how to identify these differences—and, still more, how to explain them—is not at all easy. At least in some places, medicine as an academic discipline might remain relatively immune from the blandishments of natural philosophy. But whether that immunity was due to the authority of Galen, to the attitudes imparted by the new medical humanism, or to the cussedness and traditionalism of many of its English practitioners must remain an open question.

NOTES

1. Pearl Kibre, "Lewis of Caerleon, Doctor of Medicine, Astronomer and Mathematician," *Isis* 43 (1952): 100–108; Sir Geoffrey Keynes, *The Life of William Harvey* (Oxford: Clarendon, 1978), p. 39.

2. F. M. Getz, "The Faculty of Medicine before 1500," in *The History of the University of Oxford,* ed. Jeremy I. Catto and Ralph Evans, vol. 2, *Late Medieval Oxford* (Oxford: Clarendon, 1992), pp. 373–405; Damian R. Leader, *A History of the University of Cambridge, vol. 1, The University to 1546* (Cambridge: Cambridge University Press, 1988), pp. 202–210.

3. The data on most individual physicians are conveniently found in Charles H. Talbot and E. A. Hammond, *Medical Practitioners in Medieval England* (London: Wellcome Historical Medical Library, 1965), hereafter cited as TH. On Fries, see TH, pp. 96–98; Veyrier, TH, p. 192; Boerio, TH, pp. 117–119. For good overviews of medieval English

medicine, see Carole Rawcliffe, *Medicine and Society in Later Medieval England* (Stroud: Alan Sutton, 1995); Faye Getz, *Medicine in the English Middle Ages* (Princeton: Princeton University Press, 1998).

4. TH, p. 47.

5. TH, pp. 36, 228–229.

6. On Thomas, see G. B. Parks, *English Travellers in Italy* (Stanford: Stanford University Press, 1955), p. 636; Hadcliffe, TH, pp. 398–399; Free, TH, p. 147; Chamber, TH, p. 131–132. Others abroad include John Argentine, TH, pp. 114–115; Henry Bagot, TH, p. 75; William Buckingham, TH, p. 386; John Clerke, TH, pp. 133–134; Thomas Denman, TH, pp. 339–340; Donat of Ireland, Parks, p. 625 (possibly to be identified with Denys of Ireland; ibid., p. 627); Walter Lacey, TH, p. 369; Thomas Linacre, TH, pp. 348–350; John Oxney, Parks, p. 634; John Racour, TH, p. 177.

7. One would like to know more of the medical career of Robert Sherborn, later bishop of Chichester; see TH, pp. 300–302.

8. Getz, "Faculty of Medicine before 1500," pp. 389–393. Whether the elder John Caius, "doctor in medicinis" by 1495, obtained his degree in Italy is unclear; see Damian R. Leader, "Caius Auberinus: Cambridge's First Professor," in *A Distinct Voice: Medieval Studies in Honor of Leonard E. Boyle, O.P.,* ed. Jacqueline Brown and William P. Stoneman (Notre Dame, Ind.: Notre Dame University Press, 1997), pp. 322–327.

9. See TH, pp. 314–315 (Marshall), with L. E. Voigts, "A Doctor and His Books: The Manuscripts of Roger Marchall (d. 1477)," in *New Science out of Old Books: Studies in Manuscripts and Early Printed Books in Honour of A. I. Doyle,* ed. R. Beadle and A. J. Piper (Aldershot: Scholar Press, 1995), pp. 249–314; Oxford, Bodleian Library, Ashmole MS 1437 (Argentine). See also Damian Riehl Leader, "John Argentein and Learning in Medieval Cambridge," *Humanistica Lovaniensia* 33 (1984): 71–85.

10. Argentine in his commonplace book mentions a "Pilula Taddei"; copies of Montagnana (d. 1467) were owned by John Racour (TH, p. 177) and by William Goldwyn (d. 1482; TH, p. 396), who also had a work by Cermisone (d. 1441).

11. See Nicholas Orme and Margaret Webster, *The English Hospital, 1070–1570* (New Haven: Yale University Press, 1995).

12. David R. Starkey, "England," in *The Renaissance in National Context,* ed. Roy Porter and Mikuláš Teich (Cambridge: Cambridge University Press, 1992), pp. 146–163.

13. Katharine Park and John Henderson, "The First Hospital among Christians," *Medical History* 35 (1991): 164–188.

14. Basic documentation, and much else, is in Francis Maddison, Margaret Pelling, and Charles Webster, eds., *Essays on the Life and Work of Thomas Linacre, c. 1460–1524* (Oxford: Clarendon, 1977), on which this paragraph largely depends.

15. V. Nutton, "John Caius and the Linacre Tradition," *Medical History* 23 (1979): 373–391; Jonathan Woolfson, *Padua and the Tudors: English Students in Italy, 1485–1603* (Cambridge: James Clarke & Co., 1998), pp. 73–102.

16. John M. Fletcher, "Linacre's Lands and Lectureships," and R. Gillian Lewis, "The Linacre Lectureships Subsequent to Their Foundation"; both in Maddison, Pelling, and Webster, *Thomas Linacre,* pp. 107–197, 223–264.

17. F. Logan, "The Origins of the So-called Regius Professorships," in *Renaissance and Renewal in Christian History,* ed. Derek Baker, Studies in Church History 14 (Oxford: Published for the Ecclesiastical History Society by B. Blackwell, 1977), pp. 272–277.

18. Lewis, "Linacre Lectureships."

19. Charles Webster, "Thomas Linacre and the Foundation of the College of Physicians," in Maddison, Pelling, and Webster, *Thomas Linacre,* pp. 198–222.

20. Sir George Clark, *A History of the Royal College of Physicians* (Oxford: Clarendon Press for the Royal College of Physicians, 1964): 64–66; Gweneth Whitteridge, "Some Italian Precursors of the Royal College of Physicians," *Journal of the Royal College of Physicians* 72 (1977): 67–80.

21. On the fortunes of the College, see Clark, *History;* Margaret Pelling and Charles Webster, "Medical Practitioners" in *Health, Medicine, and Mortality in the Sixteenth Century,* ed. Charles Webster (Cambridge: Cambridge University Press, 1979), pp. 165–235; Harold J. Cook, *The Decline of the Old Medical Regime in Stuart London* (Ithaca, N.Y.: Cornell University Press, 1986).

22. Clark, *History,* pp. 88–105.

23. For this perspective, see Daniela Mugnai Carrara, *La Biblioteca di Nicolò Leoniceno. Tra Aristotle e Galeno: Cultura e libri di un medico umanistico,* Accademia Toscana di Scienze e Lettere "La Colombaria" 18 (Florence: Olschki, 1991); Lawrence I. Conrad, Michael Neve, Vivian Nutton, Roy Porter, and Andrew Wear, *The Western Medical Tradition, 800 BC to AD 1800* (Cambridge: Cambridge University Press, 1995), pp. 250–264; V. Nutton, "The Rise of Medical Humanism: Ferrara, 1464–1555," *Renaissance Studies* 11 (1997): 3–19.

24. Jerome J. Bylebyl, "Teaching *Methodus Medendi* in the Renaissance," in *Galen's Method of Healing,* ed. Fridolf Kudlien and Richard J. Durling (Leiden: Brill, 1991), pp. 157–189.

25. Maria Dowling, *Humanism in the Age of Henry VIII* (London: Croom Helm, 1986); Vivian Nutton, *John Caius and the Manuscripts of Galen* ([Cambridge]: Cambridge Philological Society, 1987), pp. 38–49, 58–61.

26. For Caius, see Nutton, *John Caius;* for Edrych, Gillian Lewis, "The Faculty of Medicine," in *The History of the University of Oxford,* ed. James K. McConica, vol. 3, *The Collegiate University* (Oxford: Clarendon, 1986), pp. 238, 242; for Goulston, Daniel Béguin, "L'Edition Goulston et les prétendus manuscrits perdus de Galien," *Revue d'Histoire des Textes* 19 (1989): 341–349; Vivian Nutton, "The Galenic Codices of Theodore Goulston," *Revue d'Histoire des Textes* 22 (1992): 259–268.

27. M. Brudus Lusitanus, *Liber de Ratione Victus in singulis Febribus secundum Hippocratem ad Anglos* (Venice: heirs of P. Ravanus, 1544); a second edition was published at Venice in

1559. The much-traveled Brudus ended his life as one of the doctors of the sultan of Turkey.

28. William Turner, *De re herbaria* (London, 1538), sig. A 1v; idem, *A new herbal* (London: S. Mierdman, 1551), sig. A iiv; Charles E. Raven, *English Naturalists from Neckham to Ray* ([Cambridge]: Cambridge University Press, 1947), p. 69. For Clement, note also his "three fair herberes and a great cage for birds" in his garden in Bucklersbury; see A. W. Reed, "John Clement and His Books," *The Library*, ser. 4, 6 (1925–1926): 333.

29. William Bullein, *A Dialogue between Soarness and Charity, The Bulwarke of Defence against all Sicknesse* (London: John Kingston, 1562), pp. 4–5; on the diplomatic bag, see J. Gairdner, *Letters and Papers of the Reign of Henry VIII,* vol. 13.2 (London: Stationery Office, 1891), p. 16 n. 45.

30. Vivian Nutton, "Conrad Gesner and the English Naturalists," *Medical History* 29 (1985): 930–97.

31. Whitney R. D. Jones, *William Turner: Tudor Naturalist, Physician, and Divine* (London: Routledge, 1988); for John Caius as a naturalist, the best account still remains that of Raven, *English Naturalists.*

32. John Caius, *A Treatise of English Dogges,* in *The Works of John Caius,* ed. Edwin S. Roberts (Cambridge: Cambridge University Press, 1912), p. 27.

33. For the story of Caius' pet puffin, see Conrad Gesner, *Historia animalium, I–IV* (Zurich: C. Froschover, 1551–1558), 3:768. Caius' further comment is written in the margin of his own copy of Gesner, Gonville and Caius College, Cambridge, classmark L. 19.4, and repeated with slight verbal changes in his *De rariorum animalium atque stirpium historia libellus* (London: W. Seres, 1570), fols. 21v–22r.

34. Lewis, "The Faculty of Medicine," pp. 247–249.

35. The suggestion of Edwards's Paduan studies goes back to Alfred B. Emden, *A Biographical Register of the University of Oxford, A.D. 1501–1540* (Oxford: Clarendon, 1974), p. 185, and has been followed by most recent scholars, including Lewis, "The Faculty of Medicine," p. 255. For a disproof, see Nutton, *John Caius,* p. 74.

36. Edwards's book is reprinted, with introduction and translation by Charles D. O'Malley and Kenneth F. Russell, as *David Edwardes, Introduction to Anatomy, 1532* (London: Oxford University Press, 1961). They do not note the immense debt to Alessandro Benedetti, *Historia corporis humani sive Anatomice* (Venice: B. Guerraldus, 1502, and often reprinted); see now Alessandro Benedetti, *Historia corporis humani sive Anatomice,* ed. and Italian trans. Giovanna Ferrari (Florence: Giunti Gruppo Editoriale, 1998).

37. To the old biography of Caius by John Venn in his *Works,* ed. Roberts, add Clark, *History,* pp. 106–124; Christopher Brooke, *A History of Gonville and Caius College* (Woodbridge: Boydell Press, 1985), pp. 55–88; and Nutton, *John Caius.*

38. John Caius, *De libris suis* (London: W. Seres, 1570), reprinted in *Works,* ed. Roberts, pp. 75–83.

39. I know of at least six copies in circulation before 1550.

40. Francis W. Steer, "Lord Lumley's Benefaction to the College of Physicians," *Medical History* 2 (1958): 298–305.

41. *Pace* Lewis, "The Faculty of Medicine," p. 246, the manikin is to be found in the Cambridge University Archives.

42. Peter Murray Jones, "Thomas Lorkyn's Dissections 1564\5 and 1566\7," *Transactions of the Cambridge Bibliographical Society* 9 (1988): 109–229; idem, "Reading Medicine in Tudor England," in *The History of Medical Education in Britain,* ed. Vivian Nutton and Roy Porter (Amsterdam: Rodopi, 1995), pp. 153–183.

43. Keynes, *Life of Harvey,* p. 17.

44. Vivian Nutton, "Humanist Surgery," in *The Medical Renaissance of the Sixteenth Century,* ed. Andrew Wear, Roger K. French, and Ian M. Lonie (Cambridge: Cambridge University Press, 1985), esp. pp. 96–99; Margaret Pelling, "Appearance and Reality: Barber-Surgeons, the Body, and Disease," in *London, 1500–1700: The Making of the Metropolis,* ed. A. Lee Beier and Roger Finlay (London: Routledge, 1986), pp. 82–112. For some speculative interpretation of this boom in anatomy, see Jonathan Sawday, *The Body Emblazoned: Dissection and the Human Body in Renaissance Culture* (London: Routledge, 1995).

45. Margaret Pelling, *The Strength of the Opposition: The College of Physicians and Unlicensed Medical Practitioners in Early Modern London* (London: Macmillan, 1999).

46. Lewis, "The faculty of Medicine," p. 231.

47. On Italy, see Charles B. Schmitt, "Aristotle among the Physicians," in Wear, French, and Lonie, *Medical Renaissance,* pp. 1–15. For Wittenberg, see Vivian Nutton, "Wittenberg Anatomy," in *Medicine and the Reformation,* ed. Ole Peter Grell and Andrew Cunningham (London: Routledge, 1993), pp. 11–32; Sachiko Kusukawa, *The Transformation of Natural Philosophy: The Case of Philip Melanchthon* (Cambridge: Cambridge University Press, 1995).

48. John Caius, *De methodo medendi libri duo ex Cl. Galeni Pergameni et Jo. Baptistae Montani sententia* (Basel: H. Froben and N. Episcopius, 1544); reprinted in John Caius, *Opera aliquot et versiones* (Louvain: A. M. Bergagne, 1556). The latter is reprinted in *Works,* ed. Roberts, pp. 1–56, and this is the edition cited hereafter.

49. Ibid., pp. 7–14.

50. Daniela Mugnai Carrara, "Una polemica umanistico-scolastica circa l'interpretazione delle tre dottrine ordinate di Galeno," *Annali dell'Istituto e Museo di Storia della Scienza di Firenze* 8 (1983): pp. 31–57.

51. Caius uses this (traditional) metaphor at length at *De methodo medendi,* p. 56.

52. For this debate, see, as well as Mugnai Carrara, "Una polemica," Neal W. Gilbert, *Renaissance Concepts of Method* (New York: Columbia University Press, 1960); Andrew Wear, "Galen in the Renaissance," in *Galen: Problems and Prospects,* ed. Vivian Nutton (London: Wellcome Institute for the History of Medicine, 1981), pp. 238–245; and Bylebyl, "Teaching *Methodus Medendi.*"

53. Contrast Caius' single reference to Aristotle, in passing, at *De methodo medendi,* p. 9, with the fulsome praise of Plato on pp. 7–12.

54. Although Caius hints at this in his title, he makes no reference to da Monte in his text; and the casual reader would be easily misled into thinking that da Monte, like Galen, provided merely a starting point for Caius' own cogitations. The similar production by Crato (see note 56 below) is both more honest and more independent of da Monte's words.

55. Caius, *De libris suis,* pp. 73–75.

56. Caius' wording can be compared with that in Walenty Lublin's edition of da Monte's *In libros Galeni De arte curandi ad Glauconem explanationes* (Venice: B. Constantinus, 1554) and in the (much freer) version put out by Johann Crato von Crafftheim (with da Monte's approval), *Methodus therapeutica ex sententia Galeni et Joannis Baptistae Montani* (Basel: J. Oporinus, 1555); the three are printed consecutively in Johannes Baptista Montanus, *Opuscula* (Basel: P. Perna, 1558). For da Monte's reaction to Caius' publication, see his *In primi libri Canonis Avicennae primam fen commentaria* (Venice: V. Valgrisi, B. Constantinus, 1557), p. 8.

57. Roger French, *William Harvey's Natural Philosophy* (Cambridge: Cambridge University Press, 1994).

58. Walter Pagel, *William Harvey's Biological Ideas* (Basel: Karger, 1967).

59. On Fabricius ab Aquapendente, see Andrew Cunningham, "Fabricius and the 'Aristotle Project' in Anatomical Teaching and Research at Padua," in Wear, French, and Lonie, *Medical Renaissance,* pp. 195–222.

60. See especially French, *Harvey's Natural Philosophy,* pp. 51–70.

61. Prof. M. Feingold in discussion that followed this paper suggested that while the Aristotelian texts remained in the curriculum, they were studied as part of language training, and that by 1600 little trace, if any, remained of the advanced Aristotelianism common in Italy and Germany. If he is correct, Harvey's "natural philosophy" is far more likely to have derived from Italy than from Cambridge.

62. Charles B. Schmitt, "William Harvey and Renaissance Aristotelianism," chap. 6 of *Reappraisals in Renaissance Thought* (London: Variorum, 1989).

63. Virginia F. Stern, *Gabriel Harvey: His Life, Marginalia, and Library* (Oxford: Clarendon; New York: Oxford University Press, 1979), pp. 21–22, 70. Cf. W. T. Costello, *The Scholastic Curriculum at Early Seventeenth-Century Cambridge* (Cambridge, Mass.: Harvard University Press, 1958).

64. See William Harvey, *An Anatomical Disputation Concerning the Movement of the Heart and Blood in Living Creatures,* trans. Gweneth Whitteridge (Oxford: Blackwell Scientific Publications, 1976).

65. Andrew Wear, "William Harvey and the Way of the Anatomists," *History of Science* 21 (1983): 223–249; see also his introduction to William Harvey, *The Circulation of the Blood and Other Writings* (London: Dent, 1990).

66. Owsei Temkin, *The Double Face of Janus* (Baltimore: Johns Hopkins University Press, 1977), pp. 162–166.

67. Charles R. S. Harris, *The Heart and the Vascular System in Ancient Greek Medicine* (Oxford: Clarendon, 1973), pp. 381–383.

68. Vivian Nutton, "Harvey, Goulston, and Galen," chap. 14 of *From Democedes to Harvey* (London: Variorum, 1988).

69. Some good examples of conflict are found in Schmitt, *Reappraisals in Renaissance Thought*. The clash between Aristotle and Galen (and the new learning) is famously exemplified in the debate between Corti and Vesalius, summarized in Charles D. O'Malley, *Andreas Vesalius of Brussels* (Berkeley: University of California Press, 1964), pp. 98–100.

9

From the Laboratory to the Library:
Alchemy According to Guglielmo Fabri
Chiara Crisciani

In the history of Latin alchemy, much remains to be learned about the period from John of Rupescissa to Paracelsus. In particular, fifteenth-century alchemical texts, which include both examples of alchemical research and assessments of alchemy, have been among the least studied by historians. Yet these are precisely the texts that may be expected to illuminate the process whereby the three major shifts in emphasis that characterized alchemy between the end of the Middle Ages and the early modern period were disseminated and received. These changes were the relative discredit into which transmutatory alchemy had fallen, the increasing importance of therapeutic doctrines and goals in the alchemy of the elixir and fifth essence, and the emergence in alchemical literature of linked alchemical and religious themes that do not always refer to work in the laboratory.

These three developments were interrelated in various complex ways that have yet to be fully clarified. They evidently evolved from trends already present in medieval alchemy; but they also belong to a general restructuring both of the scientific disciplines of alchemy and medicine and of forms of knowledge—empirical, rational, prophetic, and magical. The work of Guglielmo Fabri that is the subject of the present paper provides one noteworthy example of a fifteenth-century alchemical text in which continuity and innovation go hand in hand and in which previously developed topics are reworked and transformed. Fabri seems to be at a crossroads between the trends and problems of late medieval alchemy and their development in the early modern period. He provides us with a useful vantage point for evaluating continuity and innovation, the utilization of traditional sources and concepts, and the introduction of new themes and approaches destined to undergo further development in the future.

I

The *Liber de lapide philosophorum et de auro potabili,* which as far as I know is unedited, seems to have been written about 1449 and certainly before the end of the fifteenth century.[1] The author, Guglielmo Fabri de Die, does not present himself as an alchemist; indeed, he does not present himself at all.[2] From the few pieces of information that can be found about him we know that he was a doctor "of law and medicine" and that he was part of the entourage of Duke Amadeus VIII of Savoy during the years in which Amadeus was an antipope elected by the Council of Basel under the name of Felix V; Guglielmo's duties as a court functionary seem to have been chiefly connected with his legal qualifications.[3] But Fabri evidently served as a faithful adviser and a cultivated secretary; no doubt, too, the pope appreciated him as a physician because of his transalpine, or French, origin and education. Thus, it was specifically to Guglielmo Fabri that the pope turned for what at first sight looks like a consilium.

De lapide opens with the pope's description of his disease and bitter remarks about the incompetence of his attending physicians (fol. 245r–v). Felix V—an uneasy duke in search of a new form of majesty, a hermit-prince, an aged pope disturbed by the choices he had made and frustrated in his hopes[4]—now also suffered in body. In addition to the weakness of age, he was also afflicted by a kind of paralysis of one hand and one foot, which was so painful that he groaned aloud in the course of his conversation with Fabri. The attending physicians—Italians who followed the "common way of *vulgares medici*"—had asserted that no cure was possible. However, he had heard of another "secret kind of healing" (*medendi genus secretum*) described by Arnald of Villanova and other Frenchmen, which was able to provide a remedy for the "discomforts of age" (*incommoda senectutis*), and therefore perhaps also for the affliction from which he was suffering.[5] Groaning and sighing, the pope cited some learned examples of prolongevity; he wondered, angrily, why such remedies had now vanished or were known only to very few people. Eventually, after much skirting around the subject, he came to potable gold and the transmutatory art, saying to Guglielmo: "What do you have to say about that medicine of the philosophers that they call the elixir? . . . Is it possible that something of truth or power is to be found in these things?" (fol. 245 v: "post multos circuitos devenit ad aurum potabile et ad artem trasmutatoriam dicens: 'Quod ais tu de illa medicina philosophorum quam elixir dicunt? . . . Estne possibile latere in illis aliquid veritatis et virtutis?'").

This, supposedly, is the situation and the question that gave rise to Fabri's text. De lapide presents itself as a report of dialogues between the author and the pope in a series of meetings. The dialogues are interspersed with short treatises, with which Fabri found it more appropriate to respond to the pope's demands, given their relevance and the pope's imperiousness. The text is thus a mixture of genres: the report of meetings and dialogues, characterized by a loose conversational tone and by the rhetoric of patronage, frames three treatises on specific themes mostly written in an impersonal scholastic style. They deal with (1) the possibility of transmutation, which is treated by means of a *demonstratio* based on the Aristotelian four causes;[6] (2) the therapeutic efficacy of potable gold, inserted in a reasoned survey on the nature and virtue of gold; and (3) the interpretation of some particularly occult terms, namely *telchem* or *thelesim* and *yxir.*

Beyond the frame of dialogue, the three texts are also unified by a limitation that the pope imposed on Fabri, as well as by Fabri's own choice. Felix V insisted that the style, forms of argument, and authorities used must be exclusively Peripatetic. Fabri was in complete agreement and chose to proceed in all three treatises by starting from authorities, moving on to rational demonstration, and ending in *experimentum.* On this basis, he undertook to satisfy the pontiff, travelling along "the correct path of philosophers" (*semitam rectam philosophantium*).[7] And in fact Fabri's whole attitude is completely philosophical: he confidently handles doctrines, trends, and works (at least, some of the works) of Latin alchemy, but he has nothing to do with the *opus,* as at one point he declares and as his entire text reveals. Neither in the section on transmutation nor in that on potable gold does Fabri ever give any indications about specific ingredients, operations, or processes. The term *opus* in its technical sense never appears in *De lapide;* the work is a purely doctrinal treatise by a philosopher and never assumes the operative character present in other contemporary texts by physicians interested in alchemy.[8]

II

It will have become apparent that in presenting the pope's first question, Fabri reformulated it: where the pope sought concrete information about the truth and power of the elixir, Fabri transformed this request into a broader inquiry into the art of transmutation and potable gold. The question was only on one subject, the answer on two. Evidently, for Fabri the elixir is the pivotal concept of transmutatory alchemy and could also be linked to remedies—for instance, potable gold—pertaining to a medical alchemy. Already, then, we learn that in the mid-fifteenth century a question on the elixir suggested two objects and two different projects, which could be either unified or separate.

The so-called pseudo-Lullian tradition of the elixir combined these projects. According to that tradition, a preparation composed solely of mineral ingredients is the instrument both of perfecting imperfect metals and of healing the diseases of man, for whom it can also prolong life.[9] In one sense, the medical tradition of the fifth essence was also unifying: although in *De consideratione quintae essentiae* John of Rupescissa sought only pharmacological and therapeutic results, he nonetheless maintained that quintessence of wine could reinforce the therapeutic virtues of artificially prepared gold.[10] But the very fact that Fabri discussed the question of the elixir in two separate sections, one on transmutation and the other on potable gold, shows that—at least for the moment and formally—he opted for a distinction between transmutatory and medical alchemy. This attitude is consonant with the position of those fourteenth- and fifteenth-century physicians who considered from various points of view, and for the most part maintained a distinction between, the disciplines of alchemy and medicine.[11] Yet in reality Fabri oscillated between distinction and convergence of transmutatory and medical projects, as we can see from his passages on alchemical theory and especially from the sources that he used.

Let us briefly examine his alchemical theory. It joins the most traditional alchemical theory of the nature of metals with a theory of transmutation that in essentials follows Albertus' *De mineralibus,* in which transmutation is explained as a process of purification of metals. Fabri cites word for word the medical analogy with which Albertus closed his analysis of the possibility of transmutation.[12] Albertus had named the elixir, but without giving any details about its composition or role.[13] Fabri, too, is not clear about what the elixir is. In his discussion of the formal cause of transmutation, the elixir is defined as form that will give perfection, or, rather, that will give true nature to imperfect metals, which are to be understood as matter with respect to this form that they are waiting to assume. But where does this form come from and how does it work? Here again Fabri uses suggestions from Albertus and from the *Summa perfectionis* of "Geber," decontextualizing them and depriving them of all operative reference. He uses these suggestions to construct a solution that is as original in its totality as it is oversimplified.

The form, according to Fabri, burns whatever is corruptible and imperfect in the metal and saves "whatever *humiditas radicalis* there is in the metal, digesting the humidity and converting it into perfect gold" ('*quod est in metallo de humiditate radicali, illam digerens et convertens in perfectum aurum*'). In other words, the elixir eliminates "all the *superfluitates* from the metal and maintains only the parts of quicksilver existing in the metal . . . because it attracts what is similar to its nature, but repels what is contrary to it." As a result of this oversimplification, Fabri explains transmutation as a purification of corruptible superfluities followed by a *digestio* that reinforces the *humiditas radicalis*. Hence, the *humiditas radicalis* is the central structure both in the metal to be perfected and in the perfect metal (fols. 247v–248r). In the second section on potable gold Fabri notes—still basing himself on Albertus—that by its own nature, gold has a matter that is totally purified from two superfluous and dangerous kinds of humidity. Therefore, among its many virtues, it can eliminate superfluous humors from the human body and can support humidity analogous to its own, that is, the radical moisture of man, given that in gold "only radical humidity remains."[14]

The connection that Fabri proposes between metallurgical alchemy and "medical metallurgy" rests therefore on the theoretical efficacy of the concept of radical moisture:[15] as the elixir works on the *humiditates* of the metal, so potable gold works on the radical moisture. The connection is undoubtedly original: I have not found any trace of it, in this formulation, in other authors or in the texts that Fabri uses, in which indeed elixir, radical moisture, and gold are linked in various other ways.[16] However, in Fabri the connection is based on an analogy that maintains a distinction between the processes that it connects. Therefore the identity of functions between elixir and gold does not allow one to identify them; and it does not follow that a single agent "cures" metals and men, as was affirmed in the *Testamentum,* one of the main pseudo-Lullian treatises. Moreover, unlike ps.-Lull, ps.-Arnald, and Rupescissa, Fabri does not maintain that elixir transforms gold in any way and that gold, precisely because it has been treated in this way, becomes more suitable for strengthening human radical moisture. These stages in the process would make Fabri's conception into a simplified but organic whole like those proposed by those authors. But this sequence is specifically excluded by the passage he quotes from Albertus, to which I have just referred: for Albertus and for Fabri, it is natural gold, not treated gold, that is free of those dangerous humidities.

When one turns to Fabri's sources, it becomes clear that he was well-acquainted with various writings from which he could have drawn the idea of a single agent capable of curing metals and men, had he chosen to do so. In fact, throughout the second section, he makes explicit use of precisely those texts—the *Testamentum,* the *De retardatione accidentium senectutis* of "Bacon" and/or "Arnald"—in which alchemical procedures and therapeutic virtues are organically united.[17] Moreover, he lists most of the therapeutic properties of gold according to the canonical medical texts, but, in addition, he frequently follows word for word (although without explicit quotation) the emphatic descriptions of the virtue of the alchemical remedy found in the texts named. Finally, in what he calls an "aphorism," which is in fact a little hymn of praise, Fabri exalts the celestial sun and the terrestrial sun (that is, gold), which are linked both by analogy and by a relationship of influence, through which the first infuses the second with the most ample array of virtues. The comparison, the praises, and the enthusiasm are those of Rupescissa in *De consideratione* and of Arnald in *De vinis*.[18]

In other words, Fabri uses texts in which metallurgical and medical purposes are united in order to propose a theory that instead distinguishes between these purposes. He does not justify—or even perhaps perceive—this incongruity. He has decided not to analyze the various doctrines of the texts he uses at all profoundly from a theoretical point of view but rather to exploit these doctrines in, so to speak, a rhetorical way. It is as if he were drawing on a generalized store of interpenetrating medical, alchemical, and medico-alchemical knowledge, unified in his eyes more by homogeneity of language and the possibility of intertextuality than by theoretical coherence. In this textually fluid environment, potable gold, elixir, and fifth essence can be interchangeable—especially for anyone, who, like Fabri, considers only their exceptional properties and virtues, does not inquire into the theoretical or scientific presuppositions, and does not try out the practical procedures of manufacture.

Given this oscillation between distinction (in treating transmutatory and medical alchemy in two sections), analogy (founded on the theory of radical moisture), and unification (in his use of sources), it was no accident that Fabri finally postulated the fundamental unity of metallurgical and medical alchemical projects in *thelesim*. This substance, totally vague and especially secret because of its absolute *fortitudo*, lies at the heart of the *Emerald Tablet*, which, as a generic wisdom text, served as a foundation for any and all trends within alchemy. Whoever possesses the *yxir* or *thelesim* of which Hermes speaks, concludes Fabri, obtains infinite riches and overcomes the disabilities of every disease.[19]

III

The analogy drawn by Fabri between the two radical moistures—that of gold and that of human beings—is, as already noted, an original contribution that is based on considerations taken from texts of both metallurgical alchemy and medical alchemy. Fabri himself, however, does not develop this potentially fertile idea theoretically.[20] Indeed, the idea, however interesting, may not be the most significant contribution of his text taken as a whole. More striking is the complex image that *De lapide* presents of alchemy as a discipline—of its connections not only with medicine but also with other values and branches of knowledge.

It will already have become obvious that Fabri was well aware of the difference between the metallurgical and the medical approaches in Latin alchemy. Furthermore, Fabri knew perfectly well that metallurgical alchemy had aroused doubt and perplexity from the very beginning and that in his own day it was discredited—invalidated by ecclesiastical prohibitions and rendered suspect by the tricks of empirics and incompetent practitioners (fol. 205r). Nevertheless, his *demonstratio* guaranteed the possibility of the art of transmutation.

Certainly, Fabri had an up-to-date knowledge of the principal disputes and arguments of the "quaestio de alchimia" that in the thirteenth and fourteenth centuries accompanied the introduction and development of alchemical research in the West, finally silting up into the enormous *quaestio-tractatus* of Petrus Bonus of Ferrara and into the judicial accusations of the inquisitor Nicolás Eymerico.[21] Many of the arguments in Fabri's demonstration are in fact traceable in preceding *quaestiones*. They are, however, reproposed not in a *quaestio* but in a *demonstratio*—which, like a *quaestio,* certainly links Fabri to the use of syllogistic argument but, unlike a *quaestio,* allows him to avoid the listing and refutation of positions different from his own.[22] Moreover, his *demonstratio* is not "scholastic," in the sense that it does not originate from commentary on a text—unlike, for example, the *quaestiones* on alchemy by Themo the Jew and Pomponazzi, both of which are part of commentaries on the *Meteorologica.*[23] It is not even "alchemical," in the sense that it does not aim at removing specific doubts, errors, and perplexities—as "Geber" did in his *Summa,* before proceeding with his treatise on specialized alchemical theories and instructions. Fabri's *demonstratio,* although following the rules of the genre (to demonstrate by means of the Aristotelian four causes), is at once systematic and open: he is free both merely to hint at themes that were obligatory points for discussion in the *quaestiones* and to introduce digressions in which he expands on themes that he thinks significant.

Thus, for example, earlier authors who wrote on metallurgical alchemy had frequently laid great stress on demonstrating the epistemological link of *subalternatio* as a way of guaranteeing a place for their discipline in scientific naturalistic knowledge. In contrast, Fabri simply assumes a hierarchical structure in which alchemy is a *scientia* or, rather, a *pars philosophiae,* which is subalternate to natural philosophy and more specifically to *De mineralibus.* This science elaborates theories that should guide the operations of practical alchemists, who are called *mechanici* and *subalternati:* they are to be ruled and controlled by the true philosopher (*De lapide,* fols. 246r–v, 250r). Similarly, the relationship between alchemy as an *ars* and nature, which had been central to the *quaestio de alchimia,* is no longer a problem—or, at any rate, not for Fabri—and therefore is not discussed.[24] On this subject, Fabri repeatedly declares himself in agreement with Albertus' position and simply mentions a relationship of *imitatio-ministratio,* like that which medicine and agriculture have with nature (fols. 246v, 252v).

However, Fabri introduces on his own account a long paragraph on the nature and virtues of fire into his discussion of the efficient instrumental cause of performing the *opus.* Here, along with physical properties and scientific authorities, appear ps.-Dionysius and his *Angelic Hierarchy,* Pythagoras and Plato, the Chaldeans, and Scripture; and an analogy is proposed between the Holy Trinity and the fiery trinity of light, live coals, and flame. Fabri was not the first author to divide fire in this way; moreover, since he (following the ps.-Baconian *Speculum secretorum*) endowed the fire of alchemists with almost the same transmutatory efficacy as the *lapis,* and since the *lapis* could be (according to Bonus) a perfect analogy of the Holy Trinity, it is not surprising that this fiery trinity could be compared to the divine Trinity. Nevertheless, the explicit analogy may be Fabri's own. At any rate, despite having checked a number of likely texts, I have not yet found any earlier sources for this trinitarian analogy, which Fabri asserts was set forth by great philosophers.[25]

Again, at the end and as a complement to the *demonstratio,* Fabri inserts a whole passage on the role of ethics in transmutation (fols. 249r–250v). The philosopher who is an inquirer into this art (*philosophus huius artis inquisitor*), who is content with inquiry after truth alone, "must be not only a natural philosopher but also a moral philosopher." So, in addition to a rational *subalternatio* of alchemy to natural philosophy, which makes it a true science, we find a sort of *subalternatio* to ethical values, which makes it a virtuous science. In the literature of metallurgical alchemy there were certainly admonitions and exhortations about moral qualities necessary for the alchemist.[26] Fabri knows these moral qualities, links them to the Aristotelian *Ethics,* and amplifies them into a reflection that unites scientific knowledge with prophetic-religious knowledge, philosophical virtue, and political power.

Let us see how. Fabri here is commenting on the prologue of the *Tractatus aureus* of Hermes. From it, he infers that Hermes was a prophet because he declares that he has been divinely inspired and because he "tells in advance" (before revelation) his faith in one God, in free will, in resurrection, and in the Last Judgment.[27] Thus alchemy appears to be *partim divina*[28]—that is, it derives also from inspiration and prophetically expresses religious truth; therefore, alchemical truths are also concealed in the images and analogies used by the prophets of old. For this reason, the alchemist prays God to unveil meanings to him and to help him in operating.[29] Such knowledge, which is simultaneously rational and holy, necessarily required those high moral qualities that in fact the alchemists recommend. But—and here Aristotle intervenes—virtue is also the ability to choose with free will and operate *delec-*

tabiliter. How then can the alchemist—at once philosopher, illuminated by inspiration, and virtuous by reason of his freedom—allow himself to be constrained by the unjust violence of princes who too often oppress seekers after truth? Fabri has no doubts: the *valens philosophus* would prefer to die for the alchemical art rather than reveal that which is the fruit both of divine inspiration and of his free will as a researcher. Three examples of "scientific suicide" to defend knowledge from tyrannical oppression confirm the philosophical virtue of alchemist philosophers, which is inseparable from their true knowledge.[30]

Warnings about difficult relations between alchemists and the powerful are topoi in the literature of metallurgical alchemy.[31] Rupescissa too warns, with prophetic zeal, against those tyrants who, at the advent of Antichrist, will oppress evangelical men.[32] Fabri transforms the warnings by weaving them together with classical examples and Aristotelian philosophy. Thus transformed, they become the basis on which Fabri constructs his concept of the correct and ideal relation between alchemists and princes,[33] to which I shall return. Finally, at the end of this passage—the themes of which are obviously very dear to his heart—Fabri stresses that theoretical knowledge is inseparable from philosophical virtue and from the proper exercise of power. Alchemy and its goals are legitimate and acquirable without those crimes that usually go along with seizing power; indeed, the acquisition of power without infamous crimes could in a certain way be favored by alchemy (fols. 249v–250r). Its legitimacy resides above all in the fact that it is a just method of searching for truth because it proceeds "first of all by theory with the true doctrine of philosophy."[34] Certainly, tyrants are not virtuous, because they are ignorant and aggressive; equally lacking in virtue are the deceitful practices of fraudulent and wandering alchemists, because they proceed "casually like empirics," far from the light of theory and of true philosophy. Tyrants and empirics are thus joined at opposite but symmetrical extremes with respect to an ideal center where the powerful and virtuous knowledge of the alchemists and the wisdom of their protectors is to be found. Tyrants have power but not knowledge and wisdom. Therefore their power is crude and fragile. Empirics have only apparent knowledge; therefore it is inefficacious and deceitful. Both are in any case ethically deviant and distorted, far from philosophical virtue.

IV

This analysis, which has brought out some values—scientific, ethical, and religious—of the image of alchemy held by Fabri, will, I believe, have made clear the sense in which I spoke of Fabri as standing at a crossroads. That will become clearer still if we consider what Fabri has to say about the other, medical, variety of alchemy. We have already seen how much Fabri uses texts from this branch, even though he did not completely accept their basic concept. More generally, Fabri is one of the earliest and most noteworthy compilers and constructors of the legend of Lull the alchemist, his relations with Arnald, his stay in England, and his relations with kings.[35]

This legend is important because it provided a supporting structure that, from the fifteenth century onward, accompanied the accumulation of the Lullian corpus and supplied an explanation for the resemblance between the positions held in the most ancient core of the corpus and those found in texts attributed to Arnald and John Dastin. Fabri knew these texts and was aware of the common orientation of their authors.[36] He named them many times and described them as operating together for—or rather, with—King Edward of England. About Edward himself, he said: "How many labors King Edward of England undertook! In the dress of a hermit, he went around the whole world for the sake of this art."[37]

Several points are worth stressing here. It was around these personages that Fabri constructed his own model of ideal alchemical patronage, which we are thus able to read. There is a king, directly interested in alchemy to the point of personally undertaking tiring tasks, journeys, and the hermit's solitary life; three wise alchemists come to him from afar, and he treats them with every honor; they instruct him; they work with him and for him; they perform the *opus;* and the king divides its results with them. Besides sharing knowledge and wealth, Edward would like to share his power, namely the kingdom, with them too; but they "say that to reign and to philosophize are two incompatible things" (fol. 253r), and leave their books to the king and to posterity. We can recognize elements here—the eremitical life, the refusal of power, the bequest of writings; the mastery and collaboration between the wise man and the king—which go back to two archetypes of the relation between the wise man/alchemist and king: the *Liber* of Morienus and the *Secretum secretorum*. As usual, Fabri reworks these elements in new forms; in this case perhaps he uses them also to characterize his own relationship with the pope. Indeed, just as the three alchemists put philosophy in first place, so when Felix V compliments Fabri on his excellent demonstration and wants him next to prove alchemy through experiment, Fabri replies that the philosopher must rest content solely with speculative inquiry into truth.[38]

Moreover, Fabri, like modern historians, perceives agreement in doctrine among the three alchemists and traces a genealogy that has its origin in Roger Bacon.[39] In fact, Fabri made much use of "Bacon," both *De retarda-*

tione (which, although much used without attribution, is in one instance explicitly quoted as by Arnald) and the *Speculum secretorum*—in the latter case specifying that Bacon "was a Peripatetic." This is yet further evidence of Fabri's sensitivity to the specific characteristics of medical alchemy. Also like modern historians, however, he was not able to trace explicit derivation from Bacon and had to content himself with simply pointing out a chronological relation: "Roger came before them."

V

Fabri was very familiar with and sympathetic to the authors of medical alchemy, but his sympathy fell short of complete accord. He contributed to the construction of a legend that highlights a doctrinal trend synthesizing alchemy and medicine; but he also readily broke up the synthesis or at least did not adopt it consistently. Moreover, Fabri was also theoretically competent in metallurgical alchemy. He provided a logical demonstration of the possibility of metallurgical transmutation and confirmed its legitimacy. But he did not touch on the operative doctrines and the *opus*, nor did he appear to value them particularly highly. In fact, he warns the pope that he has written the demonstration for him "not in order that you should spend money in trying out such secrets," but only to show "the possibility of the thing."[40] As for *experimentum,* which according to Fabri's own declaration at the beginning was indispensable to a Peripatetic proof, appeal to the evidence of sense experience is not lacking.[41] But when at the end of the demonstration the pope asks for *experimenta* in the sense of operative proofs, Fabri, as we have seen, reminds him of the primary value of philosophical truth alone. After that, a long silence follows. Again, at the end of the section on gold, at the moment of "descent to experiments," Fabri affirms that these would be too incredible; and in any case he prefers to replace them with the praise of gold in the aphoristic hymn about the harmony between the celestial and the terrestrial sun.[42]

How, therefore, did he conceive of alchemy, of trends within alchemy, of the alchemical *opus,* the elixir, and potable gold? They were not subjects to which he intended to give a profound theoretical and critical analysis in order to reach a definitive solution. Nor were they goals that would demand the operative intervention that gives rise to actual processes and tangible results. Instead, they were for him essentially, primarily, and above all exceptionally interesting textual objects. Fabri analyzed them by means of doctrinal and exegetical exercises and applied to them the techniques of erudition, namely collecting and amplifying.

This interpretation of his text explains its apparent theoretical uncertainty by highlighting Fabri's relative indifference to specific comparisons and definitive solutions. It may also explain why Fabri chose to deal with the subject not within the constraints of the *quaestio* format but in a more elastic *demonstratio*. Again, this interpretation reinforces the idea that the most convincing and substantial relation between alchemy and medicine on which Fabri focused is that which emerges from his learned and erudite gathering together and common use of the disciplines' respective texts. So it is hardly surprising that the pleasures of erudition are judged by Fabri to be among the principal aims and results of his research. Indeed, Fabri recognizes that his rational inquiry about potable gold and also his demonstration on transmutation are undertaken chiefly "as a recreation and exercise" (fol. 251r: "causa solaci et me exercitandi volo experiri ad quod potero pervenire"). What are involved are exercises in doctrine and logical ability, which are interesting, pleasurable, certainly also rigorous, but not radical. They are pleasurable for Fabri and also for the pope who, while he groans from pain at the beginning of the text and during the first meeting, in the end gets up laughing. Perhaps he has not solved his physical problems, but it has evidently done him good to devote himself "with so much pleasure to these difficult arguments" (fol. 253v).

At the same time, erudition can be a good interpretive tool. In fact, the objects (alchemy, potable gold, elixir, etc.) presented by the texts gathered and used by Fabri all involve language that is to a certain extent metaphorical and allusive, as Fabri and the pope recognized at the very beginning of their meetings. These objects, accordingly, require two kinds of exegesis. One is the reduction to univocal Peripatetic semantics undertaken by Fabri in the first section. The other, different but no less valid, is that of erudition and allegorical interpretation. I think that the excursus on fire and the story of the deeds of the three alchemists in part have this function. From this point of view, moreover, we can consider the entire third (and last) section of *De lapide,* which is devoted precisely to the analysis of mysterious terms—*telchem aut thelesim* and *yxir*: "these are two very secret words among the ancient philosophers and prophets" about which "all seers prophesied and wrote an infinite number of allegories" (fols. 251r–253v).

Here also, as always, Fabri starts from the texts and follows the analysis of his favorite Albertus, but in this case he uses the *Speculum astronomiae.* He distinguishes two arts to which these terms pertain. One is licit transmutatory alchemy (*yxir*); the other (linked with *thelesim-telkem*), although placed by "great philosophers" side by side with philosophy and medicine, is an ill-defined astrological-magical science.[43] The goals of this science, as listed by

Thebith Bencorat, are prohibited by divine and human law.[44] Arnald devoted some space to it in his *De sigillis,* but readily—Fabri points out—came down from the heavens of this difficult and fallible science "to the bowels of the earth," that is, to research on minerals, and found out the truth, which is alchemy.[45] In fact the alchemical science of the elixir fulfills in a legitimate and natural way, without danger to either body or soul, all the goals that the forbidden science of the *thelesim* promises (fols. 252v–253r). As we can see, in this section Fabri seems to distinguish two sciences—magic and alchemy—and two meanings of *thelesim.* In one sense, *thelesim* is equivalent to a magical link, pertains to magic, and is forbidden. In the other sense, *thelesim* is equivalent to *yxir,* as the permissible, efficacious, and true *thelesim* of Hermes—that is, the highly generic elixir-*thelesim-fortitudo* of the *Tablet.* Up to this point, Fabri is presenting an interesting—and unusual—reduction of magic to alchemy. He transforms the illicitness and vagueness of magic into the legitimate rationality of alchemy.

But this treatment of alchemy also makes it assume the broad ends, the power, and the sacred aura of the magic art. As a result, in this section the elixir has become highly generic and polyvalent; it therefore can be interpreted in many different ways. In fact, it becomes a nucleus of aggregation and amplification that produces a paratactic list of scriptural quotations, mythical fables, and references to poets (fol. 252r). This exuberant series of definitions, so interesting in its richness and potentially capable of amplifying the meaning of "alchemy," is not further developed in any way, and its innovative character remains implicit. It surely confirms that for Fabri, alchemy had to do with religiosity and mythology. But it also shows that at the time when he wrote, these links were not yet the basis for a concept of alchemy aimed at perfection of the soul and spirituality of the adept. Nor were they yet the basis for the elaboration of a fully developed mythological hermeneutics tracing the art back to ancient poet-philosophers and thus capable of interpreting elusive alchemical instructions by means of mythological *fabule.*

As is well known, these developments became obvious later, after Fabri, during the sixteenth century.[46] They had already made an embryonic but very significant appearance in the *Pretiosa margarita* of Petrus Bonus.[47] However, even a brief comparison of Fabri's work with this text reveals how different the two are in style and intention. When Bonus speaks of the ancient philosopher-alchemists, as he is considering their alchemical theories and *dicta* regarding the *lapis* (a union of body, spirit, and soul, which is finally revealed in all its glory at the end of the *opus,* arising out of depurated matter, etc.), he remarks that for the very reason they were alchemists, they were prophets too. In other words, Bonus starts out from his own theory on the *lapis*; he considers some particular technical phases of the *opus* and the concrete features of the *lapis* mentioned in the alchemical texts. He thus points out that the ancients, who knew the truth of alchemical phenomena because they had seen them during their operative working, must necessarily have glimpsed some facts and truths of the Christian religion, which had the same characteristics. Fabri expresses the same opinion, but he is simply commenting on the generic statements of the prologue of Hermes' *Tractatus aureus.* Bonus maintains that alchemy, "partim divina," is an all-pervasive science that permeates all other forms of knowledge and draws them to itself. This is why the poets' verses and *fabule* can be interpreted as having reference to the *opus,* of which they show specific stages and features.[48]

Unlike Petrus Bonus, Fabri rests his list of scriptural and poetic allusions neither on a fully articulated, explicit conception of alchemy nor on a philosophical analysis of alchemical language. Moreover, he has no interest or competence in the operations and techniques of alchemists, on which Bonus had based his reflection on alchemist-prophets and poet-alchemists. Therefore, Fabri's references to Scripture and the poets do not serve to explain. Rather, he uses wide erudition and symbolic amplification to stress and exalt just one concept: the power of a *thelesim*-elixir, the mysteriousness and elusiveness of which are more and more emphasized and not explained.[49]

VI

Other physicians more or less contemporary with Fabri also took an interest in alchemy and in the relation of alchemy and medicine. In 1456 a petition was presented to King Henry VI of England that was very different from many others in which a license to practice transmutatory alchemy was requested. The signatories, among whom were various medical practitioners and court physicians, proposed researches aimed at utilizing the philosopher's stone as a medicine that would be much more powerful than those handed down by the ancients. It is the "mother of medicines," the agent of perfection for all bodies in the pseudo-Lullian tradition. The petition was accepted; but we are better informed about the manuscript of the *Testamentum* prepared by Kirkeby, one of the signatories, than about the outcome of the research.[50]

Moving on to other cases closer to Fabri, either culturally or professionally, we can better describe the position of Michele Savonarola, university professor and physician at the Estense court from 1440; he wrote a *Libellus de aqua ardente* for Leonello d'Este.[51] In this work, following the tendency of fourteenth-century physicians, Savonarola distinguished between alchemy and medicine. He doubted the possibility of manufacturing the quintessence, praising instead the therapeutic value of *aqua ardens,* which can prolong life. In his treatise, gold maintains its traditional properties and as such is an ingredient in numerous medicinal recipes. However, he held it to be indigestible; he therefore defined potable gold as "something to laugh at," while *aqua ardens* was a "precious treasure," especially useful in times of plague.[52] Savonarola's position on the relation of medicine and alchemy is characterized by distinction of fields, attention to the possibility of the exchange of ingredients and techniques, and knowledge of alchemical theories and operations.

Antonio Guaineri, professor at the University of Pavia and court physician to Amadeus VIII, appears more confused about the subject. Perhaps he was one of those physicians incapable of curing Amadeus because they were following "the way of vulgar medical practitioners," whose ignorance the pope lamented with Fabri at the beginning of our text. Certainly, Guaineri's texts, which collectively amount to a *practica,* are dedicated to operative medicine. In them he competently describes tools and techniques of distillation, sublimation, and fermentation; and he is very attentive—with a mixture of contempt and professional interest—to the practices and remedies of the *vulgares* and to the skills of pharmacists, goldsmiths, and *gemmarii.* In Guaineri's work, texts are cited that are unusual to find in a *practica,* such as Albertus' *De mineralibus* and the *Secretum secretorum.* He reports some alchemical remedies and products of techniques that are common to alchemy and pharmacology. He describes a recipe for an excellent ointment for paralysis (evidently not very efficacious on his august patient), which he got from a hermit who, having become expert in compounding medicines during years of vain alchemical research, subsequently became a physician. As for potable gold, it appears "absurd" to him for the same reasons as it did to Savonarola, but he adds: "I have, however, heard from two alchemists worthy of faith that they can undoubtedly manufacture it."[53] Finally, one member of Piedmont's Albini di Moncalieri family of physicians, all of whom were linked to the court of Savoy, left a splendid recipe for potable gold; in his dryly technical instructions, Albini stressed that with a little quantity of this gold, "in three days wonders will occur."[54]

In their diversity, these positions bear witness to the widespread interest in alchemy among physicians and in particular to the physicians' lively attention to potable gold. Moreover, all these authors and their ideas are in various ways linked to courtly needs and culture. Fabri has only these two very general characteristics in common with this group; in many specific features, his text is quite different.

Fabri's distance from the *opus* and from practice allows him to isolate and define specific themes and comparisons, as well as to devote his analysis exclusively to aspects of his own choice, avoiding the standard contexts in which alchemical themes are developed in medical texts. Moreover, if we agree that Fabri's way of writing is to be understood primarily as a form of erudition, we can see how this erudition enriches and enlarges the themes under discussion, when it is set against the narrow technicality of Albini's recipe for potable gold, Guaineri's casual anecdotes about alchemists, and Savonarola's sober considerations on *aqua ardens*. In other words, having defined his subject and removed it from practice, Fabri made it into a new object—specifically, a textual object—around which he could collect anthropological notes, poetic and mythological references, and ethical evaluations of a kind that we cannot find in most medical texts. Indeed, the rules governing the writing of such texts would make these subjects inappropriate and out of place. Thus, Fabri proposed to the pope neither efficacious remedies nor *experimenta;* instead of the "secret medicine" (*medendi genus secretum*) which the pope had requested, he offers a self-sufficient collection of quotations, symbols, interesting stories, unforeseen textual interweavings, and themes developed in unusual ways. That is, he offers not a cure for the pope's physical disease but satisfaction for his intellectual curiosity in a context not of therapy but of erudition. This has to be intended as an edifying pedagogy as well as a development giving a new connotation to the ancient motto continually repeated by the alchemists: "Liber aperit librum."

VII

After having described the formation of an ideal alchemical patronage in which king and alchemists work together with practical success, Fabri embarks on a different kind of relation, which he presents as based principally on pleasurable study. This transformation does not escape the pope, nor does he seem happy with a textual object that he can possess only by erudition. Therefore, eventually, he exclaims "with a certain vehement outcry": "Who can say, in these days, 'I have the true elixir?'" (fol. 253v).

To this last cry—a lament more than a question—not even historians can give a precise response, because of the many different directions in which alchemy moved during the fifteenth century, in which the elixir could have so many different meanings. Certainly Fabri had no response. But as others had dedicated texts to the pontiff in order to produce the *opus* or had produced the *opus* for kings, so too Fabri offers an *opus* to his august interlocutor. What Fabri produces is a literary opus, manufactured by working on textual ingredients that he manipulates and assembles, thus transforming them. This opus is evidently produced in the library, not in the laboratory. When Fabri leaves the pope—as he informs us—he retires to his house (perhaps into his *studiolo*?), entrusts himself to his *ingeniolum,* peruses a great many books—and writes.

The lack of operative referent and the distance between the *opus* and manual work, typical of many alchemical texts written in the Middle Ages and the early modern period, have justly been viewed as a sign and result of a process of "spiritualization" of alchemy. In this view, the art was de-emphasizing the program of transmutation and turning either into medical-pharmacological research or into a search for spiritual perfection of the adept. But much remains to be learned about fifteenth-century alchemy. Fabri's is only one of many little-known texts; perhaps we should not attach too much weight to a work that does not seem to have circulated or exercised subsequent influence. Nevertheless, it seems to me to represent an example of a possible third line of development within Renaissance and early modern alchemy. In fact, one can retreat from the laboratory not only to the "oratory"[55] but also to the library and reach for the shelves of erudition. Here, in a doctrinal and erudite context, transmutation, wondrous remedies, therapeutic goals, and alchemico-religious intuitions could continue to be maintained in some unity; in any case, their textual basis could be preserved, even if they were transformed and mixed together in strange combinations. In this way, a textual and conceptual complex, which had already been organized and enriched, was transmitted. Such a complex could be elaborated even further, either leading to the pleasures of erudition or perhaps becoming the basis on which subsequent researchers could work according to different goals.

I believe that this third way of erudition was of considerable significance, especially in the fifteenth century, when the newly venerated classical authors met the venerable alchemists. Even in later alchemical projects—more engaged and coherent, based on more solid and better-thought-out philosophical grounds—erudition remains a feature and above all a mental attitude that is never entirely absent.

NOTES

I would like to thank Michela Pereira, who first drew my attention to Fabri's text, and Nancy Siraisi, who helped me in translating this paper.

1. Bologna, Biblioteca Universitaria (Fondo Caprara), MS lat. 104 (Frati, no. 138), fols. 245r–253v: "Incipit Liber Guylielmi De Dya de lapide philosophorum et de auro potabile ad summum pontificem. Gratulanti mihi dudum . . . " (hereafter cited as *De lapide*).

On this manuscript and Fabri's text, see Lodovico Frati, "Indice dei codici latini conservati nella R. Biblioteca Universitaria di Bologna," *Studi italiani di filologia classica* 16 (1908): 155–158; Lynn Thorndike, *A History of Magic and Experimental Science* (New York: Columbia University Press, 1923–1958), 4:342–344; on the Fondo Caprara, see Didier Kahn, "Le fonds Caprara de manuscrits alchimiques de la Bibliothèque universitaire de Bologne," *Scriptorium* 48, no. 1 (1994): 62–110. MS lat. 104 contains several other alchemical treatises; see notes 16 and 54, below.

2. *De lapide* seems to be the only work under Fabri's name. I do not deal here with problems concerning his identity and career, which would require too lengthy an analysis: see, however, Thorndike, *Magic and Experimental Science*, 4:342–344; Ernst Wickersheimer, *Dictionnaire biographique des médecins en France au Moyen Age,* new ed. with supplement by Danielle Jacquart (Geneva: Droz, 1979), 1:242; Elisa Mongiano, *La cancelleria di un antipapa: Il bollario di Felice V (Amedeo VIII di Savoia)* (Turin: Palazzo Carignano, 1988), p. 114; *Concilium Basiliense,* vol. 7, *Die Protokolle des Concils 1440–1443,* ed. Hermann Herre (Basel, 1910; reprint, Nendeln/Lichtenstein: Kraus Reprint, 1971), p. 201; Heribert Müller, *Die Franzosen, Frankreich und das Basler Konzil (1431–1449)* (Paderborn: F. Schöningh, 1990), 1:159, 2:603. See also the (not entirely reliable) notices and summary of Fabri's work provided by Giovanni Carbonelli, *Sulle fonti storiche della chimica e dell' alchimia in Italia* (Rome: Istituto Nazionale Medico-farmacologico, 1925), pp. 84–93.

3. Fabri's name is not recorded in medical documents related to the Savoy Court. Antonio Guaineri was the most famous of the physicians who officially worked at the court in that period. See discussion later in this paper; Danielle Jacquart, "De la science à la magie: Le cas d'Antonio Guainerio, médecin italien du XVe siècle," *Littérature, Médecine et Société* 9 (1988): 137–156; and eadem, "Theory, Everyday Practice, and Three Fifteenth-Century Physicians," in *Renaissance Medical Learning: Evolution of a Tradition,* ed. Michael R. McVaugh and Nancy G. Siraisi, Osiris, 2nd ser., 6 (Philadelphia: History of Science Society, 1990), pp. 140–160.

4. On the multifaceted personality and political attitude of Amadeus VIII/Felix V and the courtly culture that he promoted, see Bernard Andenmatten and Agostino Paravicini Bagliani, eds., *Amédée VIII-Félix V, premier duc de Savoie et pape (1383–1451),* Colloque International, Ripaille-Lausanne, 1990 (Lausanne: Bibliothèque Historique Vaudoise, 1992), especially the contributions of Jacques Chiffoleau, Catherine Santschi, Elisa Mongiano, Sheila Edmunds, Robert Bradley, and Terence Scully. See also Enea Silvio Piccolomini, *De viris illustribus,* ed. Adrianus Van Heck (Vatican City: Biblioteca apostolica vaticana, 1991), pp. 74–79; and idem, *Commentarii rerum memorabilium . . . ,* ed. Adrianus Van Heck (Vatican City: Biblioteca apostolica vaticana, 1984), 1:54, 435–441.

5. On aging and prolongevity, besides the classic introduction of Gerald J. Gruman, *A History of Ideas about the Prolongation of Life: The Evolution of Prolongevity Hypotheses to 1800,* Transactions of the American Philosophical Society, new series, 56, part 9 (Philadelphia: American Philosophical Society, 1966), see now Agostino Paravicini Bagliani, *Medicina e scienze della natura alla corte dei Papi del Duecento* (Spoleto: Centro Italiano di Studi sull'Alto Medioevo, 1991), esp. chaps. 6, 7, 9, and 10; Faye M. Getz, "To Prolong Life and Promote Health: Baconian Alchemy and Pharmacy in the English Learned Tradition," in *Health, Disease, and Healing in Medieval Culture,* ed. Sheila Campbell, Bert Hall, and David Klausner (New York: St. Martin's Press, 1991), pp. 135–145; Michela Pereira, "Un tesoro

inestimabile: Elixir e 'prolongatio vitae' nell'alchimia del Trecento," *Micrologus* 1 (1993): 161–187; Luke Demaitre, "The Care and Extension of Old Age in Medieval Medicine," in *Aging and the Aged in Medieval Europe,* ed. Michael M. Sheehan (Toronto: Pontifical Institute of Mediaeval Studies, 1990), pp. 3–22.

6. More or less contemporaneously, the Savoy court cook, Master Chiquart, was imperiously asked (as he repeatedly remarks) by Amadeus VIII to write a treatise on the courtly culinary art. Cookery is considered here as a "science et art, science de l'art de cuysinierie et de cuysine." In the prologue, Chiquart, too, lists "les quatre causes principales que doyvent estre en toute bonne oeuvre, c'est assavoir, cause efficient, material, formal et final"; see *Du fait de cuisine par Maistre Chiquart, 1420,* ed. Terence Scully, in *Vallesia* 40 (1985): 130, 127. Chiquart remarks that, in several cases, the good cook should follow the dietetic directions of the physician; for his part, Fabri stresses the theoretical value of the true science of cooking as opposed to the mere practice of "coquus mechanicus" (fol. 250r).

7. *De lapide,* fol. 245v: "Si peripateticorum solis auctoritatibus insistere opporteat et eorum conclusionibus uti, non est necesse ut degenerem ab eorum stillo et ideo fortassis sermo videbitur minus cultus. Sane malo cum philosophis sedere mensa inculta coloribus quam cum philologis gladium semper accuere et numquam percutere; sententie enim philosophie sui natura pulchre sunt et facunde." Fol. 246r: "Quia autem sanctitas vestra peripatetichorum scholam deligerit credo quia sola ista inter ceteras pauca protulit que non sint digna fide. Omnia enim que dixit Aristoteles eleganter probat ratione aut experimento."

8. In addition to the scientific and philosophical authors (mainly Aristotle and Avicenna) whom learned physicians were normally expected to master, Fabri also competently handled philosophical and patristic sources unusual in both medical and alchemical treatises: *De unitate et uno,* ascribed to Boethius; Boethius, *De consolatione philosophiae;* Isidore of Seville, *Etimologiae;* ps.-Dionysius, *Hierarchia angelica;* Apuleius, *De Deo Socratis;* Ambrose's commentary on Luke.

9. The basic text of this tradition is the pseudo-Lullian *Testamentum.* A critical edition of this work by Michela Pereira is forthcoming. For the time being I use the *Testamentum* edited in Jean Jacques Manget, *Bibliotheca Chemica Curiosa* (Geneva, 1702), 1:707–777; regarding the wonderful powers of the lapis/elixir, see pp. 776B–777A. Among the many studies she has devoted to this tradition, its theories, and development, see Michela Pereira, *L'oro dei filosofi: Saggio sulle idee di un alchimista del Trecento* (Spoleto: Centro Italiano di Studi sull'Alto Medioevo, 1992); eadem, "Teorie dell'elixir nell'alchimia latina medievale," *Micrologus* 3 (1995): 103–148; eadem, "'Mater medicinarum': English Physicians and the Alchemical Elixir in the Fifteenth Century," in *Medicine from the Black Death to the French Disease,* ed. Roger French et al. (Brookfield, Vt.: Ashgate, 1998), pp. 26–52; see also above, note 5.

10. See John of Rupescissa, *Liber de consideratione quintae essentiae* (Basel, 1561), liber primus, pp. 22–23.

11. See Chiara Crisciani, "Medici e alchimia nel secolo XIV: Dati e problemi di una ricerca," in *Atti del Congresso internazionale su medicina medievale e scuola medica salernitana* (Salerno: Centro Studi Medicina 'Civitas Hippocratica,' 1994), pp. 102–118; see also later discussion in this paper.

12. See *De lapide,* fol. 247r, and compare Albertus Magnus, *De mineralibus,* book 3, tractatus 1, in his *Opera omnia,* ed. P. Jammy (Lyon, 1651), 2:252a–b. On different forms of analogy between alchemy and medicine, see Chiara Crisciani, "Il corpo nella tradizione alchemica: Teorie, similitudini, immagini," *Micrologus* 1 (1993): 189–233, and Barbara Obrist, "Alchemie und Medizin im 13. Jahrhundert," *Archives internationales d'histoire des sciences* 43 (1993): 209–246.

13. See Pereira, "Teorie dell'elixir," p. 125.

14. *De lapide,* fol. 251v: "Quia [aurum] igitur est in complexione temperatum, ideo habet vim temperandi, et quia materia illius fuit summe depurata duobus humiditatibus superfluis, videlicet unctuosa et inflamabile et aquosa seu flematica evaporabili . . . ideo habet vim summe depurandi humores superfluos et confortandi suum simile videlicet humidum radicale, cum in eo sola humiditas que est radicalis remaneat." The remarks about the two dangerous humidities are virtually quoted from Albertus Magnus *De mineralibus,* book 4, tractatus unicus, chap. 7, "De natura et commixtione auri," in *Opera Omnia,* ed. Jammy, 2:264b–265a.

15. For the medical concept of radical moisture, see Michael R. McVaugh, "The 'Humidum Radicale' in Thirteenth-Century Medicine," *Traditio* 30 (1974): 259–283; see also Gad Freudenthal, "The Problem of Cohesion between Alchemy and Natural Philosophy: From Unctuous Moisture to Phlogiston," in *Alchemy Revisited,* ed. Zweder R. W M. von Martels (Leiden: Brill, 1990), pp. 107–116.

16. To give only a few examples, something of the kind can be found in Roger Bacon, *De retardatione accidentium senectutis,* ed. A. G. Little and E. Withington, in vol. 9 of *Opera hactenus inedita Rogeri Baconi* (Oxford: Clarendon, 1928), esp. pp. 43–46; idem, *De conservatione iuventutis,* in ibid., especially pp. 133–134, 139–140; ps.-Arnald of Villanova *De vita philosophorum,* ed. Antoine Calvet, *Chrysopoeia* 4 (1990–1991): 62, 68, 72–74; idem, *De conservanda iuventute,* in his *Opera omnia* (Basel, 1585), col. 818; ps.-Lull, *Liber Mercuriorum* (third part of *Testamentum*), chap. 18: "De aquis et medicinis pro humano corpore," Oxford, Corpus Christi College, MS 244, fol. 63va; John of Rupescissa, *De consideratione,* p. 23; *Tractatus de investigatione auri potabilis editus a quodam solemni medico,* Bologna, Biblioteca Universitaria (Fondo Caprara), MS lat. 104, fols. 271r–283v (a very interesting anonymous text in the same codex as Fabri's *De lapide*), at fol. 282v.

17. See, among the many studies he devoted to the "Alchimica" of ps.-Arnald, the contribution of Antoine Calvet, "Mutations de l'alchimie médicale au XVe siècle: A propos des textes authentiques et apocryphes d'Arnaud de Villeneuve," *Micrologus* 3 (1995): 185–209. See also Giuliana Camilli, "'Scientia mineralis' e 'prolongatio vitae' nel 'Rosarius philosophorum,'" *Micrologus* 3 (1995): 211–225; Michela Pereira, "Arnaldo da Villanova e l'alchimia: Un'indagine preliminare," in *Actes de la I trobada internacional d'estudis sobre Arnau de Vilanova* (Barcelona: Arxiu de Textos Catalans Antics, 1995), 2:95–174; and note 5 above. These and other recent studies on the relationship between alchemy and medicine in the fourteenth and fifteenth centuries point out the pivotal role of "Arnald" in the development of this network of medical-alchemical conceptions. Arnald, a famous and authoritative physician and alchemical *auctoritas,* involved in several texts on prolongevity (authentic, attributed, or with paternity as in the case of *De retardatione*—variously ascribed to Arnald or Bacon), became a more and more important and reliable reference, especially for those physicians interested in potable gold and alchemical remedies.

See, for some witnesses of this role, Chiara Crisciani and Michela Pereira, "Black Death and Golden Remedies: Some Remarks on Alchemy and the Plague" in *The Regulation of Evil: Social and Cultural Attitudes to Epidemics in the Late Middle Ages,* ed. Agostino Paravicini Bagliani and Francesco Santi (Florence: Società Internazionale per lo Studio del Medioevo Latino, 1998), pp. 7–39.

18. Rupescissa, *De consideratione,* pp. 22–24, 48–53; Arnald of Villanova, *De vinis* in his *Opera omnia,* cols. 586, 591. Outstanding praises of the two "suns" can also be found in the treatise of the *Solemnis medicus* (see note 16 above), fols. 279r–280v. Danielle Jacquart pointed out the relevance of *héliocentrisme* also in the sober ideas on alchemical remedies of Michele Savonarola: see "Médecine et alchimie chez Michel Savonarole (1385–1446)," in *Alchimie et philosophie à la Renaissance,* ed. Jean-Claude Margolin and Sylvain Matton (Paris: Vrin, 1993), pp. 109–122.

19. *De lapide,* fols. 252v–253r: "Non credo fore aliud telkchem permissum nisi verum yxir Hermetis quem Aristoteles vocat patrem omnis telchem vel thelesim"; "Est enim yxir recte factum fortuna fortunarum, quia divitias largitur ad plenas quadrigas, ut dicit Raymondus in Testamento. . . . Itaque nullum impedimentum morborum nec aliud potest prevalere adversus habentem."

20. The idea would later be developed, for instance, by Giovanni Bracesco in his *Lignum vitae,* edited in Manget, *Bibliotheca Chemica Curiosa,* 1:911–938; cf. Sylvain Matton, "Marsile Ficin et l'alchimie: Sa position, son influence," in Margolin and Matton, *Alchimie et philosophie,* pp. 155–156.

21. Petrus Bonus of Ferrara, *Pretiosa Margarita Novella,* edited in Manget, *Bibliotheca Chemica Curiosa,* 2:1–80; Nicolás Eymerico, *Contra alchimistas,* in "Le traité 'Contre les alchimistes' de Nicolas Eymerich," ed. and trans. Sylvain Matton, *Chrysopoeia* 1 (1987): 93–136. On the *quaestio de alchimia,* see Chiara Crisciani, "La 'questio de alchimia' fra Duecento e Trecento," *Medioevo* 2 (1976): 119–168; Barbara Obrist, "Die Alchemie in der mittelalterlichen Gesellschaft," in *Die Alchemie in der europäischen Kultur und Wissenschaftsgeschichte,* ed. Christoph Meinel (Wiesbaden: Harrassowitz, 1986), pp. 33–60; William Newman, "Technology and Alchemical Debate in the Late Middle Ages," *Isis* 80 (1989): 423–445.

22. The *quaestio* format would still be used later—for instance, by Benedetto Varchi, *Sulla verità, o falsità dell'archimia, questione,* ed. Domenico Moreni (Florence, 1827); see Alfredo Perifano, "Benedetto Varchi et l'alchimie," *Chrysopoeia* 1 (1987): 181–208.

23. Thaemo Judeus, *Quaestiones in quattuor libros Metheororum,* ed. George Lockert (Paris, 1515–1518), fols. CCIva–CCIIIra. Pomponazzi's *quaestio* is transmitted by three *reportationes:* Paris, Bibliothèque Nationale, MS lat. 6535, fols. 334r–350r; Osimo, Biblioteca del Collegio Campana, MS 45, fols. 122r–126r; and in his commentary on Aristotle, *Meteora,* in Milan, Biblioteca Ambrosiana, MS lat. R 96 sup., fols. 162r–241v, which I have not been able to consult. The last of these manuscripts has recently been edited. See Paola Zambelli, "Pomponazzi sull'alchimia: da Ermete a Paracelso?" in *Studi filologici e letterari in memoria di Danilo Aguzzi-Barbagli,* ed. Caniela Boccassini, Filibrary series, no. 13 (Stony Brook, N.Y.: Forum Italicum, 1997), pp. 100–122. See Bruno Nardi, *Studi su Pietro Pomponazzi* (Florence: Le Monnier, 1965), esp. pp. 79–84; Franco Graiff, "I prodigi e l'astrologia nei commenti di Pietro Pomponazzi al 'De caelo,' alla 'Meteora' e al 'De

generation,'" *Medioevo* 2 (1976): 331–361; and Amalia Perfetti, "Aristotélisme et alchimie dans l'anonyme 'Trilogio della trasmutatione dei metalli,'" in Margolin and Matton, *Alchimie et philosophie*, pp. 223–251.

24. See Crisciani, "La 'quaestio de alchimia'" and "Il corpo nella tradizione alchemica"; Newman, "Technology and Alchemical Debate"; Barbara Obrist, "Art et nature dans l'alchimie médiévale," *Revue d'histoire des sciences* 49.2–3 (1996): 215–286; Michela Pereira, "L'elixir alchemico fra 'artificium' e natura," in *Artificialia*, ed. Massimo Negrotti (Bologna: CLUEB, 1995), pp. 255–267.

25. *De lapide,* fol. 247r–v. Fire as light, flame, and glowing coal is introduced by Aristotle merely as an example of a possibly misleading attribution of species in *Topics* 5.5, 134b–135a. These three forms of "fire" appear later (with the correct reference to Aristotle) in Alexander Neckam, *De naturis rerum libri duo,* ed. Thomas Wright (London, 1863; reprint, Nendeln/Lichtenstein: Kraus Reprint, 1967), 1.17, "De igne," p. 57; they are also present in Bartholomaeus Anglicus, *De proprietatibus rerum* book 10, "De igne et eius proprietatibus" (chap. 4, "De igne"; chap. 5, "de flamma"; chap. 7, "de carbone"—I thank Jole Agrimi for this reference) and in Berthold Blumentrost, *Questiones disputatae circa tractatum Avicennae de generatione embryonis et librum meteorum Aristotelis,* ed. Rüdiger Krist, Würzburger medizinhistorische Forschungen 43 (Pattensen: Horst Wellm, 1987), p. 59; A very similar exemplum, based on the *ignea lux* that could be both *in carbo* and *in flamma,* is used by John of La Rochelle to explain the link between the rational soul and the body and the threefold nature of the soul; see Jean de la Rochelle, *Summa de Anima,* ed. Jacques Guy Bougerol (Paris: Librairie Philosophique J. Vrin, 1995), prima consideracio, VI (40), "De anima quantum ad corpus," pp. 125–130, esp. p. 127 (I owe this reference to Michela Pereira). However, John does not go on to develop the analogy between the threefold fiery soul and the Trinity.

26. For a survey of these moral qualities, see Chiara Crisciani, "Aspetti della trasmissione del sapere nell'alchimia latina: Un'immagine di formazione; uno stile di commento," *Micrologus* 3 (1995): esp. 158–162.

27. *De lapide,* fol. 249r. After having quoted the initial paragraph of *Tractatus aureus* ("Cum tanta etatis prolixitate experiri non desisterem . . . "), Fabri interprets it as follows: "In istis verbis Hermes unum deum confitetur et liberum arbitrium et diem judicii, resurrectionem et fidem."

28. *De lapide,* fol. 252v; see Bonus, *Pretiosa margarita,* p. 29A: "ipsa (alchimia) partim est naturalis et partim divina sive supra naturam."

29. *De lapide,* fol. 249r: "Venit [Deus] igitur multiplice prece placandus, ut apperiat artifici typos, figuras et analogias vatum et prophetarum antiquorum et ut auxilietur artificem in operatione sua." Besides Bonus, another outstanding interpretation of prophecies in alchemical terms is to be found in ps.-Arnald of Villanova, *Exempla in arte philosophorum.* See now the edition of Antoine Calvet, "Le 'tractatus parabolicus' du pseudo-Arnaud de Villeneuve," *Chrysopoeia* 5 (1992–1996): 145–47. See also Barbara Obrist, "Le rapport d'analogie entre philosophie et alchimie médiévales," in Margolin and Matton, *Alchimie et philosophie*, esp. pp. 56–64.

30. *De lapide,* fol. 249v: one of these cases (the philosopher who prefers to bite his own tongue rather than reveal the alchemical secret) is clearly a "modern" adaptation of

the political resistance of Zeno of Elea described by Diogenes Laertius, *De vitis philoso-phorum . . . libri decem* 10.5, 25–29.

31. See, among others, ps.-Albertus, *Libellus de alchimia,* in his *Opera omnia,* ed. Jammy, 21:3–4.

32. See John of Rupescissa, *De consideratione,* esp. the prologue; see also his *Liber lucis,* chap. 1, in Manget, *Bibliotheca Chemica Curiosa,* 2:84A. The cruel attitude of the oppres-sive tyrant against the brave, honest alchemist is also outlined in Thomas Norton, *Ordinall* (second half of fifteenth century), especially in the second part, where the virtuous deeds of the alchemist Thomas Daulton are told. He, although imprisoned, condemned, and hard-pressed by the king and other powerful personages, firmly kept secret the *magisterium* (in Manget, 2:293–294).

33. *De lapide,* fol. 249v: "Non debet igitur cogi philosophus directe vel indirecte nec male tractari, sed multipliciter honorari et verbis et factis dulciter attrahi."

34. *De lapide,* fol. 250r: "Et hic est verus modus inquirendi veritatem rerum, primo per theoricam cum vera doctrina philosophie, et non casualiter sicut empirici deceptores, vagi, omni prorsus lumine destituti vere theorice."

35. See Michela Pereira, *The Alchemical Corpus Attributed to Raymond Lull* (London: War-burg Institute, University of London, 1989): on the relevance of Fabri's contribution to the legend she notes, "Fabri is the first writer, so far as I know, to bring together the two parts of the legend—Arnald and the visit to England" (p. 43).

36. Fabri mentions Dastin, although he never quotes from his works; he uses both ps.-Lull, *Testamentum,* and Arnald (*De vinis, De sigillis*), and he explicitly attributes to the lat-ter the *De retardatione,* usually ascribed to Roger Bacon. Note that Fabri changes the title of the Arnaldian medical work "De vinis" into "De vinis seu elixir," thus clearly showing that in his own view, this text deals mainly with alchemy.

37. *De lapide,* fol. 249v: "Quot labores sumpserit rex Anglie Odoardus, qui in habitu heremite pro hac arte circuivit orbem terrarum"; cf. *Visio Edwardi,* edited in Pascale Barthélemy and Didier Kahn, "Les voyages d'une allégorie alchimique: De la 'Visio Ed-wardi' à l' 'Oeuvre royale de Charles VI,'" in *Comprendre et maîtriser la nature au Moyen Age: Mélanges d'histoire des sciences offerts à Guy Beaujouan* (Geneva: Droz, 1994), p. 519: "Cir-cuivi ego mundum ad ipsum inveniendum. . . . " The editors point out (pp. 495–496) that a version of the *Visio* with a prologue (in which Edward is represented as king, philoso-pher, and hermit) did indeed appear in the fifteenth century, when the legend of Lull in England also spread. The *Visio* and Fabri's *De lapide* are literally linked at least as regards the image of the king-hermit wandering in search of alchemy.

38. *De lapide,* fol. 249r: "Et utinam posses experimento rem sic probare uti probasti pery-pateticorum clarissima ratione. . . . Cui demum dixi philosophum contentari sola veritatis inquisitione indeque secutum est silentium et sic finis, pro tunc."

39. *De lapide,* fol. 253v; see Pereira, *L'oro dei filosofi,* esp. chap. 4.

40. *De lapide,* fol. 250v: "non ut es vestrum effundatis in experientia tantorum secreto-rum, sed ut videatis possibilitatem rei maxime fundatam in principiis nature."

41. See *De lapide*, for instance at fol. 247r, where some *experimenta* concerning heat and fire are mentioned; see above, note 7.

42. *De lapide*, fol. 251r: of the therapeutic virtues of gold Fabri, following Avicenna, maintains that in many cases "redducimus nos ad experimentum, cum magis conferat medico de talibus experiri quam ratiocinari"; fol. 252r: "Quibus omnibus contemplatis descendi ad experimenta et vidi aurum operari in corpore humano talia que si scriberem crederentur impossibilia."

43. *De lapide*, fol. 252r–v. In the thirteenth and fourteenth centuries, the possible link between alchemy and magic was addressed—either to deny or maintain it—in some texts by jurists evaluating the legitimacy of alchemy, in several decrees of religious orders prohibiting their members from its study and practice, and by some theologians (e.g., Thomas Aquinas and Giles of Rome). The link was strongly asserted by the inquisitor Eymeric, who accused all alchemists of making a pact with the Devil. See William Newman, introduction to *The "Summa perfectionis" of Pseudo-Geber*, ed. Newman (Leiden: Brill, 1991), pp. 30–40; Chiara Crisciani and Michela Pereira, *L'arte del sole e della luna: Alchimia e filosofia nel medioevo* (Spoleto: Centro Italiano di Studi sull' Alto Medioevo, 1996), pp. 45–53.

44. *De lapide*, fol. 252v: the forbidden goals are "fortuna et impedimentum, substantia et negotiatio, principatus et prelatio, coniunctio et separatio"; cf. Albertus Magnus, *Speculum astronomiae*, ed. Stefano Caroti, Michela Pereira, and Stefano Zamponi, under the direction of Paola Zambelli (Pisa: Domus Galilaeana, 1977), p. 33. This edition has been reproduced: see Paola Zambelli, *The "Speculum astronomiae" and Its Enigma: Astrology, Theology, and Science in Albertus Magnus and His Contemporaries*, Boston Studies in the Philosophy of Science 135 (Dordrecht: Kluwer, 1992).

45. *De lapide*, fol. 252v. Fabri here is commenting on the proem of "Bacon," *De retardatione*, p. 1, taken by him to be by Arnald; in quoting the passage, he radically alters its meaning: instead of "et inveni ibi vanitatem et temporis perditionem," he quotes and comments "et inveni veritatem quia per artem alkimie inveni yxir." Once more, he is thus endorsing the image of Arnald as an alchemist.

46. Even if we consider only Italian culture, it will suffice to mention the works of Augurelli, Bracesco, Quattrami, Percolla, Nazari. On alchemy and mythology, see H. J. Sheppard, "The Mythological Tradition and Seventeenth-Century Alchemy," in *Science, Medicine, and Society in the Renaissance: Essays to Honor Walter Pagel*, ed. Allen G. Debus (New York: Science History Publications, 1972), 1:47–59; François Secret, "Mythologie et alchimie à la Renaissance," in his "Notes sur quelques alchimistes italiens de la Renaissance," *Rinascimento*, 2nd ser., 13 (1973): 203–206; Joachim Telle, "Mythologie und Alchemie: Zum Fortleben der Antiken Götter in der frühneuzeitlichen Alchemieliteratur," in *Humanismus und Naturwissenschaften*, ed. Rudolf Schmitz and Fritz Krafft (Boppard: Boldt, 1980), pp. 135–154.

47. Petrus Bonus (first half of the fourteenth century) was perhaps one of the first authors who, in his coherent image of alchemy as both *scientia* and *donum Dei*, stressed its rational as well as its religious features and pointed out its links with prophecy, mythology, and poetry. See, for instance, *Pretiosa margarita*, pp. 29–30, 34, 42, and the titles of some

chapters— e.g., 6: " . . . quod haec Ars sit naturalis et sit divina, et quod per ipsam philosophi antiqui fuerunt vates de futuris miraculis divinis"; 9: "In quo ostendit . . . quod Philosophi hujus scientiae tetigerunt eam cum omnibus aliis scientiis." See Chiara Crisciani, "The Conception of Alchemy as Expressed in the 'Pretiosa Margarita Novella' of Petrus Bonus of Ferrara," *Ambix* 20 (1973): 165–181; Obrist, "Le rapports d'analogie." Although Fabri never refers explicitly to Bonus, it seems plausible to me that he knew *Pretiosa margarita.*

48. See Bonus, *Pretiosa margarita,* p. 4B: "cum ipsa (alchimia) omnes artes et scientias ad se trahat, et immisceat se eis"; p. 34A: "ita quod scientia ista nihil dimisit, quin a se detraheret, et sibi componeret."

49. That Fabri is here extolling, not explaining, the *lapis* was evidently clearly perceived by the pope, whom the text at this point represents as firmly interrupting the author's praises with "Satis est, de laudibus eius est superius" (fol. 253r).

50. See Getz, "To Prolong Life and Promote Health"; Pereira, " 'Mater medicinarum.'"

51. Michele Savonarola, *I trattati in volgare della peste e dell'acqua ardente,* ed. Luigi Belloni (Pel 59 Congresso nazionale della Società italiana di medicina interna, Roma, 12–14 ottobre 1953; Milan, 1953); see Jacquart, "Médecine et alchimie," and the article of Katharine Park in this volume.

52. Savonarola, *Trattati,* pp. 75, 80, 88.

53. Antonio Guaineri, *Opus preclarum ad praxim* (Pavia, 1518), fols. 219vb; see also fols. 28ra–b, 218–220, 238r, and above, note 3.

54. Karl Sudhoff, ed., "Eine Herstellungseinweisung für 'Aurum potabile' und 'Quinta essentia' von dem herzoglichen Leibartze Albini di Moncalieri (14. Jahrhundert)," *Archiv für die Geschichte der Naturwissenschaften und der Technik* 5 (1914): 198–201. Bologna, Biblioteca Universitaria (Fondo Caprara), MS lat. 104, the codex containing Fabri's *De lapide,* also contains three other texts, each different in scope and style, focused on potable gold, of which Albini's recipe (fol. 310r) is one. Besides the long treatise (fols. 271r–283v) of the *Solemnis medicus,* there is also a *Practica de auro potabili* (fols. 254r–255v), composed of a collection of short recipes plus one excerpt from ps.-Lull, *Liber Mercuriorum.* It is clear that the collector of these texts was very interested in this old remedy (the therapeutic virtues of gold were well known and easily to be found in practical medical texts), which came to be of outstanding importance precisely during the fifteenth century, perhaps because it was undergoing a new alchemical characterization and preparation: see Crisciani and Pereira, "Black Death and Golden Remedies," and Chiara Crisciani, "Oro potabile fra alchemia e medicina: due testi di peste," in *Atti del VII Convegno Nazionale di Storia e Fondamenti della Chimica, 1997,* published as *Rendiconti dell'Accademia Nazionale delle Scienze detta dei XL,* 21,2 (1997): pp. 83–93.

55. The reference is, of course, to the famous image, engraved by Hans Vredemann de Vries and inserted in Heinrich Khunrath, *Amphitheatrum sapientiae aeternae* (Hanover, 1609). The picture shows the adept working and praying in a dual purpose room: his laboratorium/oratorium.

10

The Homunculus and His Forebears:
Wonders of Art and Nature
William Newman

HOMUNCULI DUO

The intrepid reader of the *Chymical Wedding of Christian Rosencreutz* (1616) will encounter the following bizarre sequence of events. The hero of the romance, Rosencreutz, receives an anonymous invitation to the mysterious wedding of a king and queen, delivered to him by a beautiful, winged lady. After seeing a castle with invisible servants; a mysterious play featuring lions, unicorns, and doves; and a roomful of wondrous self-moving images, Rosencreutz finally meets the bride and groom. At the end of a sumptuous dinner accompanied by an elaborate comedy, the joyful couple, along with their royal retinue, are abruptly beheaded by a "very *cole-black* tall man."[1] After their blood is carefully collected, the bodies are then dissolved into another red liquor by Rosencreutz and a group of fellow alchemists. These laborants summarily congeal the fluid in a hollow globe, whereon it becomes an egg. The alchemists then incubate the egg, which hatches a savage black bird: the bird is fed the previously collected blood of the beheaded, whereupon it molts and turns white, and then iridescent. After a series of further operations, the bird, now grown too gentle for its own good, is itself deprived of its head and burned to ashes.[2]

This panoply of processes is an obvious recitation of the traditional regimens or color stages that were supposed to lead to the agent of metallic transmutation, the philosophers' stone.[3] Indeed, the philosophers' stone was often described as the end result of processes figuratively pictured in terms of copulating kings and queens who are murdered and reborn. But in the end, the bodies of *this* bride and groom are reassembled out of the ashes of the unfortunate bird by placing the moistened mass into two little molds. As they are heated, there appear "two beautiful bright and almost *Transparent little* Images . . . a Male and a Female, each of them only *four* inches long," which are then infused with life. These are identified in the margin as *Homunculi duo*.[4] The reader, having expected the end result to be the philosophers' stone, may be somewhat surprised at the outcome. This at least was the reaction of the alchemists who were employed at the court of the decapitated couple, for Rosencreutz informs us that they imagined the process to have been carried out "for the sake of *Gold,*" adding that "to *work in Gold* . . . is indeed a piece also of this art, but not the most *Principal,* most necessary, and best."[5] In short, according to the *Chymical Wedding of Christian Rosencreutz,* the real goal of alchemy is the artificial generation of human beings, and the manufacture of precious metals only a byline.

The author of the *Chymical Wedding* was Johann Valentin Andreae (1586–1654), a well-known Lutheran theologian and the composer of the utopian *Christianopolis.*[6] Perhaps it is unnecessary to say that for Andreae, the production of homunculi is largely an allegory of spiritual regeneration with the aim of charming the reader rather than teaching him to be a Frankenstein.[7] Andreae's reorientation of alchemy to the spiritual rebirth of man has a history as long and devious as the operations described by Rosencreutz. But even more tortuous is the path by which the homunculus reached Andreae, as there is nothing in the *Chymical Wedding* to suggest its origin. For Andreae's homunculus was born, oddly, from the confluence of two distinct traditions within the occult sciences; one of them a practical genre devoted to the artificial production of living beings, the other an apologetic literature whose goal was the defense of alchemy against its detractors. As I shall show, Andreae's source, Paracelsus von Hohenheim (1493–1541) and his followers, had already performed this fusion of traditions. Indeed, it is Paracelsus who made the generation of the *homunculus,* or artificial human, a theme sufficiently dear to Western civilization that Goethe's Faust carries one on his shoulder for much of *Faust* part 2. The following paper will have as its telos the analysis of Paracelsus' homuncular ruminations; but in order to achieve that end, we must first review the relation of Paracelsianism to late medieval alchemy.

Anyone who compares the alchemical writings of the thirteenth century, when the discipline was first being appropriated by the medieval West, with the output of Paracelsus and his followers will be struck by the vastly greater scope that the iatrochemists envision for their alchemy. The Paracelsian three principles—mercury, sulfur, and salt—are no longer simply the ingredients of metals and some minerals, as were the older mercury and sulfur inherited from the Arabs. Instead, the Paracelsians argue that they have discovered the components of the entire globe and its contents, even asserting that the heavens themselves are made up of their three principles.[8] A similar expansion of scope may be seen in the Paracelsians' claims for the medical role of the chemical art. The alchemy of Geber or Albertus Magnus limited itself to the replication and study of inanimate objects, while the expanded discipline of Paracelsus was above all a medical application of alchemical techniques, as well as a veritable chemical physiology, which used alchemy to explain a host of vital processes. Although Paracelsus was heavily indebted to earlier medical alchemists such as John of Rupescissa and pseudo-Lull, it is fair to say that there is nothing so comprehensive as his cosmological iatro-chemistry in the Middle Ages.[9]

In this paper I shall argue that a similar expansion took place in Paracelsus' view of the power of alchemy to replicate natural products, leading him and his followers to the position that human creative power was practically unlimited. The homunculus, as artificial human, was the crowning piece of human creative power, making its artificer a sort of demiurge on the level of a lesser god. It has long been acknowledged that the Renaissance tradition of natural magic promoted a view of "man the maker"; but as I shall also show, the use of the homunculus as a marvel of human art has its origin in a medieval debate focusing strictly on alchemy rather than magic, and this debate predates the Renaissance by several centuries.[10] Finally, I shall describe the sexual and religious ambiguities that the homunculus presented to Paracelsus, for these cast an oblique light on his mental landscape that throws several features into sharp relief. But before discussing these issues, we must first examine the philosophical background to Paracelsus' discussion of human creative power.

I. THE POWERS OF ART

It is a well-known feature of Aristotelian natural philosophy that the Stagirite made a point of distinguishing between natural and artificial products. In the *Physics* (2.1, 192b28–33) Aristotle distinguishes between a natural and artificial product by pointing out that the former will have "within itself the principle of its own making" (τὴν ἀρχὴν ἐν ἑαυτῷ τῆς ποιήσεως), whereas in the case of the latter, this principle resides "in some external agent" (ἐν ἄλλοις καὶ ἔξωθεν).[11] Whereas a seed naturally develops into a tree, a piece of wood does not naturally grow into a house. The artificial product requires a carpenter, who acts as the "external agent" by building the house out of wood. Hence natural and artificial products are essentially different. Nonetheless, Aristotle allows in the *Physics* (2.8, 199a16) that "the arts either, on the basis of Nature, carry things further than Nature can, or they imitate Nature" (ὅλως τε ἡ τέχνη τὰ μὲν ἐπιτελεῖ ἃ ἡ φύσις ἀδυνατεῖ ἀπερ-γάσασθαι, τὰ δὲ μιμεῖται), thus opening up an avenue for the argument that art can improve on nature.[12]

As I have documented elsewhere, the Latin alchemy of the Middle Ages began a sustained effort to erode the traditional Aristotelian dichotomy between natural and artificial products by building on Aristotle's admission that art can improve on nature.[13] This movement led numerous alchemists to a position that is strikingly prescient of the parallel claim made in Francis Bacon's *Descriptio globi intellectualis* four centuries later; the works of art and nature diverge "not in form or essence, but only in the efficient."[14] The theme had already become a *quaestio disputata* in alchemical writings of the High and late Middle Ages, reacting to the *De congelatione et conglutinatione* of the Persian philosopher Avicenna (980–1037), a section of Avicenna's *Kitāb al-shifā'* that was misleadingly attached to Aristotle's *Meteors* by Alfred of Sareshel at the beginning of the thirteenth century.[15] Involuntarily under the pseudonym of Aristotle, Avicenna there debunked alchemy by claiming generally that "art is inferior to nature, and cannot equal it, however much it may strive."[16] It is therefore imperative to realize that the Avicennian attack on alchemy is embedded in a strongly worded ne plus ultra concerning technology as a whole. As such, it became a locus classicus for the discussion of human art in a variety of realms, and remained so even in the seventeenth century.[17]

Whatever Avicenna may have said about human art in his other works, his position in the *De congelatione* was far more forceful than any taken by the Stagirite himself. Avicenna's attack, which terminates with the phrase "sciant artifices alkimie species metallorum non posse transmutari," came to be known in Latin simply by the incipit *sciant artifices*. The *sciant artifices* was a challenge to alchemy that could not be ignored: it was thus taken up by the army of subsequent alchemists and rebutted in many an alchemical *Theorica*. Let us consider one such example.

An influential *Book of Hermes* that was circulating already in the thirteenth century is organized around a succession of attacks on alchemy and their subsequent rebuttals.[18] One of these attacks takes up the cudgels of Avicenna, saying "Metallic bodies, inasmuch as they are works of nature, are natural, but human works are artificial, and not natural."[19]

The opponent of alchemy here merely states the distinction between natural works and human works: the implication, obviously, is that these are two radically different realms that cannot lead to the same products. "Hermes" replies with the following rebuttal:

> But human works are variously the same as natural ones, as we shall show in fire, air, water, earth, minerals, trees, and animals. For the fire of natural lightning and the fire thrown forth by a stone is the same fire. The natural ambient air and the artificial air produced by boiling are both air. The natural earth beneath our feet and the artificial earth produced by letting

water sit are both earth. Green salt, vitriol, tutia, and sal ammoniac are both artificial and natural. But the artificial are *even better than the natural,* which anyone who knows about minerals does not contradict. The natural wild tree and the artificially grafted one are both trees. Natural bees and artificial bees generated from a decomposing bull are both bees. Nor does art do all these things; rather it helps nature to do them. Therefore the assistance of this art does not alter the natures of things. Hence the works of man can be both natural *with regard to essence* and artificial *with regard to mode of production.*[20] (my emphases)

The response of Hermes begins with a set of empirical examples provided by the four elements; fire, air, water, and earth. The author wants to show that man can produce "artificial elements" that are fully identical to the naturally occurring forms. In like fashion he can make artificial forms of "green salt" (perhaps verdigris—copper acetate), vitriol (copper or iron sulfate), tutia (zinc carbonate), and sal ammoniac (ammonium chloride). These artificially produced minerals will not be simply equivalent to their natural counterparts—they will be better. Finally, new types of trees produced by grafting and bees "spontaneously produced" out of dead livestock are identical to their natural exemplars. Hermes concludes from this barrage of empirical evidence that art makes these multifarious products only by aiding nature. In a line that is astonishingly close to the viewpoint of Bacon, Hermes says that human works and natural works are identical as to essence (*secundum essentiam*), even if they differ according to their means of production (*secundum artificium*). This is effectively identical to Bacon's claim that the works of art and nature diverge "not in form or essence, but only in the efficient."

What is of particular significance at present is Hermes' insistence on the alchemist's ability to replicate not only inanimate matter but also lower lifeforms. Playing on the common belief that one could generate bees spontaneously merely by causing a cow's body to rot, Hermes claims that this provides yet another instance of man acting as the efficient cause to nature. The same idea occurs in the classic text of alchemy from the High Middle Ages, the *Summa perfectionis* of "Geber."[21] Here the author rebuts an antialchemical argument which claims that the art cannot succeed because it is dependent on the action of stars and planets whose exact positions cannot be determined by the inaccurate science of astrology.[22] To this Geber replies, "we see, when we want to lead a worm into being from a dog, or other putrescible animal, [that] we do not consider immediately the position of the stars, but rather the disposition of the ambient air, and other perfective causes of putrefaction other than that."[23] Like Hermes, Geber gives the alchemist an active role in generating insects from a rotting corpse, for he considers "the disposition of the ambient air." This is yet another instance of the power of human art. And like Hermes, Geber is intent on limiting this power to the production of lower life-forms. At another point the *Summa* even explicitly denies the ability of man to infuse a soul into matter.[24]

The position of "Hermes" and "Geber" was the one adopted by the majority of Latin alchemists in the Middle Ages. The alchemists make few references to attempts to produce what we would call "vertebrate" life-forms; and when these do occur, they are usually portrayed in negative terms. A good example of this tendency may be seen in the fourteenth-century *De essentiis essentiarum* spuriously ascribed to Thomas Aquinas. The pseudony-mous author reports an attempt made by Rāzī to create an artificial human being in a vessel, but he notes that even if this could happen, the creature would still probably lack a rational soul.[25] Let us therefore pass to a later thinker who seems to have harbored no such doubts.

II. PARACELSUS AND ARTIFICIAL LIFE

In 1572 the iatrochemical physician Adam von Bodenstein published a work supposedly written in 1537 by Paracelsus. This *De natura rerum,* which may be a reworking of a genuine Paracelsian text,[26] opens with a discussion of the art/nature dichotomy:

> The generation of all natural things is of two sorts, as [there is] one which happens by means of nature alone without any art, [while] the other hap-pens by means of art—namely alchemy. In general, however, one could say that all things are born from the earth by means of putrefaction. For pu-trefaction is the highest step, and the first beginning of generation, and putrefaction takes its origin and beginning from a moist warmth. For the continual moist warmth brings about putrefaction and transmutes all nat-ural things from their first form and essence, as also their powers and virtues. For just as the putrefaction in the stomach turns all food to dung and transmutes it, so also the putrefaction that occurs outside the stomach in a glass [i.e., a flask] transmutes all things from one form into another.[27]

The *De natura rerum* immediately places itself in the context of the al-chemical debate about the artificial and the natural by asserting that the gen-eration of all natural things occurs in two ways—either by means of nature without art or with the aid of art, that is, the art of alchemy. Although the au-thor is not overly concerned with philosophical niceties, he at once assimi-lates natural and artificial generations in saying that both come from "the earth" by means of warm, moist putrefaction. Thus the putrefaction that occurs in the stomach is not essentially different from that which occurs in a glass vessel: as pseudo-Hermes asserted, they differ only *secundum artificium.*

After a few words on the wonders of putrefaction, which allows one thing to be transmuted into another, the *De natura rerum* extends the foregoing logic to a discussion of eggs. In incubating her egg, the hen merely supplies the necessary heat for the "mucilaginous phlegm" (*mucilaginische phlegma*) within to rot and, in so doing, to become the living matter that will develop into a chick.[28] The key agent, once again, is putrefaction. But as is well known, this incubation and ensuing putrefaction can be performed artificially by means of warm ashes, without the brooding hen. More than this, if a living bird be burned to powder and ashes in a sealed vessel, and its remains left to rot into mucilaginous phlegm in "a horse's womb" (*venter equinus*—a technical term for hot, decaying dung), the same phlegm may again be incubated, to produce "a renovated and restored bird" (*ein renovirter und restaurirter vogel*). In this fashion, all birds may be killed and reborn, so that the alchemist becomes a sort of little god who brings about a miniature conflagration complete with a "rebirth and clarification" (*widergeburt und clarificirung*) of matter like that which will accompany the Last Judgment. This clarification of matter by the fire of the Day of Judgment is one of Paracelsus' habitual themes; he expounds it at length in his late *Astronomia magna,* the definitive statement of his philosophy.[29] We shall soon encounter another example of such quasi-incorporeal matter, though one that is clarified by a different means. The *De natura rerum* goes on to announce that the death and rebirth of birds forms "the highest and greatest *magnale* and mystery of God, the highest secret and wonderwork."[30]

Despite this categorical statement, the *De natura rerum* has even greater marvels to offer, as the author then says: "You must also know that men too may be born without natural fathers and mothers. That is, they are not born from the female body in natural fashion as other children are born, but a man may be born and raised by means of art and by the skill of an experienced spagyrist, as is shown hereafter."[31] Having introduced the homunculus, the text then digresses to discuss the unnatural union of man with animals, which can also produce offspring, though "not without heresy" (*so mag solches on kezerei nicht wol geschehen*). Still, one should not automatically treat a woman who gives birth to an animal as a heretic, "as if she has acted against nature" (*als ob sie wider die natur gehandelt hette*), for the monstrous offspring may only be a product of her disordered imagination.

Animals too can produce monsters, when their offspring do not belong to the same race as the parents. But the author of *De natura rerum* is more interested in the case of monsters which "are brought to pass by art, in a glass" (*durch kunst darzu gebracht werden in einem glas*). A good example of such artificial monsters is the basilisk, which is made from menstrual blood sealed up in a flask and subjected to the heat of the "horse's womb."[32] The basilisk is "a monster above all monsters" (*ein monstrum uber alle monstra*) for it can kill by its glance alone. Being made from menstrual blood, it is like a menstruating woman, "who also has a hidden poison in her eyes" (*die auch ein verborgenen gift in augen hat*) and can ruin mirrors and make wounds impossible to heal with her glance, or spoil wine with her breath. But the poison of the basilisk is much stronger than that of the woman per se, because it is the living and undiluted embodiment of her poisonous excrescence:

> Now I return to my subject, to explain why and for what reason the basilisk has the poison in its glance and eyes. It must be known, then, that it has such a characteristic and origin from impure [i.e, menstruating] women, as was said above. For the basilisk grows and is born out of and from the greatest impurity of women, from the menses and the blood of the sperm.[33]

One could therefore say that for the author of the *De natura rerum,* the basilisk is the epitome of the female itself, a valuation that does not seem to contradict the undisputed corpus of Paracelsus.[34]

Soon after this memorable account, the *De natura rerum* arrives at a lengthy description of the homunculus and its mode of generation. Coming directly after the discussion of the basilisk, which was made by a sort of artificial parthenogenesis, the homunculus seems to be its masculine twin. Just as the basilisk embodied the quintessence of feminine impurity, so the homunculus, created without any feminine matter, will serve as a magnification of the intellectual and heroic virtues of masculinity. But first let us relate its mode of production:

> We must now by no means forget the generation of homunculi. For there is something to it, although it has been kept in great secrecy and kept hidden up to now, and there was not a little doubt and question among the old philosophers, whether it even be possible to nature and art that a man can be born outside the female body and [without] a natural mother. I give this answer—that it is by no means opposed to the spagyric art and to nature, but that it is indeed possible. But how this should happen and proceed—its process is thus—that the sperm of a man be putrefied by itself in a sealed cucurbit for forty days with the highest degree of putrefaction in a horse's womb, or at least so long that it comes to life and moves itself, and stirs, which is easily observed. After this time, it will look somewhat like a

man, but transparent, without a body. If, after this, it be fed wisely with the arcanum of human blood, and be nourished for up to forty weeks, and be kept in the even heat of the horse's womb, a living human child grows therefrom, with all its members like another child, which is born of a woman, but much smaller.[35]

As we can see, the author of the *De natura rerum* introduces his homunculus within the framework of the traditional question of the limits of human art. Unlike the timid philosophers of old, the author says, he is willing to affirm the powers of human art in making a test-tube baby. And doubly marvelous will this creature be, having grown out of sperm alone, unpolluted by the poisonous matrix from which the basilisk took its origin. Because of its freedom from the gross materiality of the female, the homunculus is translucent and, as it were, bodiless. Like the "clarified" birds produced by alchemical techniques, the homunculus is almost incorporeal. Hence the author can use the homunculus as yet another excuse to vaunt the powers of human art, which he immediately sets out to do. The *De natura rerum* announces that from such homunculi, if they reach adulthood, arise further marvelous beings, such as giants and dwarves. These creatures have wonderful strength and powers, such as the ability to defeat their enemies with "great, forceful victory" (*grossen, gewaltigen sig*) and to know "all hidden and secret things" (*alle heimlichen und verborgne ding*). Why are they so gifted? Because "they receive their life from art, through art they receive their body, flesh, bone, and blood. Through art they are born, and therefore art is embodied and inborn in them, and they need learn it from no one."[36]

The reasoning here is straightforward. Because the homunculus is a product of art, in its mature state it has an automatic and intimate acquaintance *with* the arts, and consequently knows "all secret and hidden things." Hence the homunculus is not merely an artificial marvel in itself but a key to further marvels. It is the final expression of man's power over nature, as the author says, "a miracle . . . and a secret above all secrets."[37]

The *De natura rerum* and Earlier Tradition

At this point, it is fair to ask whether the composer of the *De natura rerum* has created this fantasy out of whole cloth or has drawn from earlier sources. As I have shown, the context of the *De natura rerum* is largely determined by the question of artificial and natural products. We have also seen that the creation of lower forms of animal made up one part of that debate, as in the *Book of Hermes* and the *Summa perfectionis* of "Geber." But the *De natura rerum* has gone far beyond those texts in its detailed and extravagant descriptions of artificial generation. Where is the author getting this material? Let us first consider the argument made by Gershom Scholem and affirmed by Walter Pagel that the homunculus finds its roots in medieval legends of the Jewish *golem*.[38] The golem was an artificial man created out of "virgin earth" by means of Cabalistic rituals involving Hebrew letters. As Moshe Idel has argued, however, there is little or no evidence that the golem was to be made of human sperm or sealed up in a flask.[39] For the origins of the homunculus we should seek a more proximate source.

In fact, the source of the *De natura rerum*'s homunculus can be found mainly in medieval Arabic literature on the generation of artificial animals, a tradition already described in Paul Kraus's famous book on Jābir ibn Ḥayyān, published in 1942. The corpus ascribed to the eighth-century Persian sage Jābir ibn Ḥayyān comprises over two thousand works, which were mostly written in the ninth and tenth centuries. Most of these works deal with alchemy and natural magic, and in them one finds instructions for the making of artificial humans. Jābir's *Kitāb al-tajmīʿ*, for example, advises that one take an undefined "element," "matter," "essence," "body," or "sperm" and seal it up in a mold with detachable parts.[40] One then inserts this into a perforated vessel, which is heated in a water bath to putrefy. By varying the shape of the mold, one can produce any sort of being, such as a young girl with a boy's face, or an adolescent with the intelligence of a man.[41]

Despite the similarity of Jābir's recipe to that of the *De natura rerum*, I have been unable to find any direct line of transmission from the Arabic of Jābir to our putative Swiss magus. But in the course of his description, Jābir mentions another tradition, attributed to Plato, which the Persian alchemist disavows.[42] I refer to the *Kitāb al-nawāmīs*, or *Book of Laws*, of pseudo-Plato, a work that was already known in Latin by the thirteenth-century bishop of Paris, William of Auvergne. William refers to this work as the *Liber neumich*, a corruption of the Arabic *nawāmīs*, but it is also called the *Liber vaccae* in honor of its first victim, a cow.[43] It is not unlikely that the author of the *De natura rerum* too may have known the *Liber vaccae*, and its bizarre prescriptions may be one source for his homunculus recipe.

Pseudo-Plato begins his book with directions for making a "rational animal," which I shall synopsize as follows: "Whoever wishes to make a rational animal should take his own water while warm, and let him mix (*conficiat*) it with an equal measure of the stone that is called stone of the sun. This is a stone that shines at night like a lamp until the place in which it is found is illuminated."[44] Then one must take a cow or ewe. Its vulva is cleansed with medicines and its womb made capable of receiving what is put therein. If a cow is used, the blood of a ewe is put on its vulva; if a ewe, the contrary. The orifice is then plugged with the stone of the sun. After this, the animal is put in a dark house, and every week it is given a pound of the other animal's blood to eat. One must then take some sunstone, as much sulfur, as much magnet, and as much green tuthia. One should grind them, mix with willow sap, and dry in the shadows. When the cow or sheep gives birth, one must "take that form and put it in that powder. For it will at once be clothed in human skin."[45] Then that form should be put "into a great glass or lead vessel." After three days it will be hungry and will move about. "Therefore feed it from that blood which has gone forth from the mother" for seven days. Then "the animal form which is agreeable to many miracles will be finished."[46] It can be used to change the progress of the moon, or to change one into a cow or sheep. "And if you take this form and feed and nourish it for forty days, and feed it with blood and milk, nothing else, and the sun does not see it," you may then vivisect it and use its fluids to anoint your feet, whereupon you can walk over water. Finally, "if a man has raised it and nourished it until a whole year passes, and left it in milk and rainwater, it will tell him all things that are absent."[47]

There are numerous parallels between pseudo-Plato's recipe for the rational animal and the *De natura rerum*'s homunculus, though there are also obvious differences. The choice of human sperm, the feeding with blood, the initial nourishing for forty days in a flask followed by a longer period of maturation, and finally the gift of preternatural intelligence are topoi shared by both texts. But there are multiple divergences as well, such as the complicated mixture of minerals that pseudo-Plato uses in order to clothe his rational animal with skin, or his advice that it should be eviscerated. Either the author of the *De natura rerum* has drawn on different sources or he has considerably toned down his primary source. At any rate, I think one must agree that there is sufficient resemblance between the *De natura rerum* and this Arabic literature of artificial generation to make a dependence on the tradition as a whole both plausible and necessary.

If we now pass from the *De natura rerum* to a work that belongs more definitely to the genuine Paracelsian corpus, it will be possible to cast our net a bit wider. In addition to the Arabic tradition of artificial generation, there is another source that Paracelsus may have used for his homuncular ruminations. I refer to the popular tradition of the *mandragora,* known even in Middle High German as *Alraun* or *Alraune.*[48] In his *Liber de imaginibus* of uncertain date, Paracelsus attacks dishonest apothecaries who carve roots to look like a man and sell them as *Alraun.* He denies categorically that any root shaped like a man really grows naturally.[49] Nonetheless, Paracelsus affirms in another passage that the mandrake can indeed be produced, even if the natural philosophers and physicians have enveloped it in error. In the *De vita longa* (1526/1527), after discussing the theory that pearls are generated from sperm, he says:

> the homunculus, which the necromancers falsely call "alreona" and the natural philosophers "mandragora," has become a topic of common error, on account of the chaos in which they have obscured its true use. Its origin is sperm, for through the very great digestion that occurs in a *venter equinus* the homunculus is generated, like [a man] in all things, with body and blood, with principal and lesser members.[50]

Here Paracelsus argues that the mandrake incorrectly described by necromancers and philosophers is really a homunculus, which they have misidentified. Paracelsus may be thinking here of the old German folklegend that the *Alraun* grew primarily beneath gallows, where it was generated from the sperm or urine of hanged criminals: in honor of its provenance, the *Alraun* was also called *Galgenmann*.[51] In order to understand his reasoning, one must realize that Paracelsus customarily employs the expression *venter equinus,* a technical term in alchemy for decaying dung used as a heat source, to mean any source of low, incubating heat. Thus it was easy for him to interpret the mandrake legend as a garbled recipe for the homunculus, in which the earth beneath the gallows acted as a *venter equinus.*

Implications of the Homunculus for Paracelsus

Having located the proximate sources of the Paracelsian homunculus, let us now pass to a discussion of its meaning for him. If we turn to Paracelsus' tract *De homunculis* (ca. 1529–1532), it becomes clear at once that the production of the artificial man, though an object of wonder and a means of advancing the power of human art, could also be a potent image of sin. Paracelsus begins *De homunculis* by observing that man has both a spiritual and an animal capacity; calling a man a wolf or dog is a matter not of simile but of identity. This refers to Paracelsus' theory of the microcosm, according to which man, who is made from the *limus* or dust of the earth, and not ex nihilo, contains all the powers and virtues of the creation within himself.[52] When someone acts in a bestial fashion, he therefore actualizes the beast within and literally becomes the animal whose behavior he imitates. It is the essence of a thing, not appearance, that determines its identity. The animal body of man exists independent of the soul, and it produces a defective, soulless sperm when one is possessed by it. It is from this defective, animal sperm, Paracelsus now tells us, that homunculi and monsters are produced: therefore they have no soul.

But this process can happen in different ways. First, as soon as a man experiences lust, sperm is generated within him. He has a choice at that point: either to act on his lust and let the semen pass out or to keep it within, where it will putrefy internally. If he should allow the semen to pass out of his body, it will proceed to generate as soon as it lands on a *Digestif*—that is, a warm, moist subject that can act as an incubator. This "polluted sperm" must produce a monster or homunculus when it is "digested."[53] Paracelsus remarks that this is also possible for women, though he adds that in their case it is more frequent for the seed to remain within, once generated by lust. It then putrefies internally and causes diseases such as a uterine mole, which mocks pregnancy but can lead only to a monstrous growth.[54] In the case of a male, the retention and putrefaction of sperm can lead to scrotal hernia (*Carnoeffel*) or another growth, for the diverted seed produces "flesh, decay, and lumps."[55] Interestingly, Paracelsus refers to this outcome as a "Sodomitic birth," for to him, even the internal production of seed without emission is a form of sodomy.[56]

The theme of sodomy occupies Paracelsus at some length. The logic of his argument leads him to conclude that intestinal worms and various rectal fauna are caused by the action of pederasts, and that the potential for producing intestinal homunculi is the real reason for Saint Paul's injunctions against the abusers of children. Similarly, the omnipotent generative power of sperm is used to explain the presence of horrible growths and even homunculi in the stomach and throat of sodomites who have ingested this dangerous fluid.[57]

At this point the reader may well wonder how one is to escape the destructive power of his or her own seed, given that soulless offspring are not produced only by such unorthodox sexual practices; they can result from mere seminal retention alone. The answer to this is as simple as it is shocking. Addressing himself to the reader as parent, Paracelsus tells us that either we must see to it that our sons get married or else we must castrate them, so that the root of this evil be dug up with all its branches.[58] In the case of women, there is simply no solution other than marriage. One is tempted at first to read this filial prescription as mere hyperbole, but some earlier remarks from *De homunculis* make it clear that Paracelsus is in deadly earnest. In a passage that begins abruptly from a fragment, he says:

> [God] has built his church on Peter, that is, on his chosen, so he will build his church on no other virgin (*jungfrau*). For one must not trust the same, [for] a reed in water is steadier. I announce this to you so that you understand that Christ does not want to have virgins (*jungfrauen*) whom he has

not chosen, because they are unsteady like the reed; rather he wants to have his own chosen, who remain faithful to him. But if man wants to hold himself chaste by force, from his own power, he should have himself castrated or castrate himself (*sol man beschneiden oder sich selbs beschneiden*), that is, remove the fountain where that lies of which I write. Therefore God has formed it—so that this may happen easily—not like the stomach or the liver, but outside the body. This is not given to women: therefore they are commanded by men. [If they are eunuchs,] they are either so by nature, or else God receives them with a sort of force, not according to their own will.[59]

Here Paracelsus expands on his notion that genuine chastity can only come with castration, since lust has the inevitable effect of generating seed. A self-professed virgin is not really such unless he has eliminated the very source of his own seed. From this statement Paracelsus arrives at a truly extraordinary conclusion: it is for the convenience of enacting their own self-mutilation that God has blessed men with external genitalia. Thus women, who have not the benefit of this option, must be placed under the rulership of men. To conclude this line of reasoning, men have a simple choice—they may either marry, in which case their semen is continually exhausted and used up properly, in producing ensouled children, or they should eliminate the production of further useless seed by self-castration. To do otherwise is to become the involuntary begetter of homunculi.

Even the most blasé of readers cannot fail to find Paracelsus' *De homunculis* an extraordinary document. The complex of ideas concerning sexual pollution, unnatural generation, disease, and religious purification by castration is, even by sixteenth-century standards, bizarre. No doubt some will be inclined to argue that the *De homunculis,* as one short tract among the huge literary output of Paracelsus, should be considered an aberration. But that is not the case. If we turn to other Paracelsian treatises, parts of the same complex emerge, though with some modifications. The fragmentary *De praedestinatione et libera voluntate* of about 1535 seems to argue that man has the freedom to choose whether he generate seed or not, saying that his free will consists partly "in the reception of the blood in the semen. . . . Thus you may live in purity, [or] in unchastity, whichever you wish."[60] Although this passage is more or less incomprehensible as it stands, Paracelsus seems to be saying that the generation of semen is a matter of choice, a message that he put in unforgettably draconian terms in *De homunculis.* In fact, the notion that seed is generated by choice receives much further expansion in Paracelsus' early *Buch von der Geberung der Empfintlichen Dingen in der Vernunft* (ca. 1520). Here Paracelsus says that men and women are born without seed.[61] Seed is only generated in a man or a woman by choice, in the following manner. The blood coexists in the body with a *liquor vitae,* which the fantasy (*speculatio*) can ignite just as fire ignites wood. When this ignition occurs, the seed separates from the *liquor vitae* dispersed throughout the body by a process that Paracelsus calls *egestio,* then passes into the *vasa spermatica.*[62] Whenever seed has been produced, Paracelsus says, the "light of nature is not, but is dead"; that is, the faculty of understanding has vanished. Consequently, he adds, it is necessary that the philosopher never generate seed. Indeed, God himself wants to have a "pure man, not a changed one"; that is, he desires a man unpolluted by the generation of semen.[63]

At another point in *Das Buch von der Geberung,* however, Paracelsus makes it clear that despite God's preference of the pure man over the impure, procreation is not a sin. His message is basically that of *De homunculis*—a good Christian has two choices, either to use his seed for the purpose of generation or to avoid its production altogether—although *Das Buch von der Geberung* lacks the overt injunction that we can achieve the latter goal only by means of self-mutilation. In essence, Paracelsus seems to be erecting two orders of men—a perfectly chaste philosophical elect, which never generates seed, and a progenerative plebs. He even goes so far as to suggest that the perfectly chaste man can experience physical rebirth through baptism, to have his Adamic, elemental body literally replaced by the flesh of the new birth. Such a regenerated man can become a *magus coelestis,* an *apostolus coelestis,* a *missus coelestis,* or a *medicus coelestis.*[64] The fate of the procreative man, however, is far less clear, for in many other places Paracelsus supports legitimate marriage.[65] It would exceed my scope here to try to resolve this vexed point in Paracelsus' philosophy. Let me merely reiterate that for *Das Buch von der Geberung,* at least, the message is that procreation, or even the generation of seed, eliminates the possibility of learning from the light of nature.

HABENT SUA FATA HOMUNCULI

As we have seen in the foregoing, Paracelsus has extremely ambivalent views on the matter of generating seed, at times passing into an almost Manichaean rejection of the "common man" who traffics in procreation. Yet one thing is clear. If one does in fact generate seed, he or she must look very carefully to its ultimate resting place. Once the sperm has been produced, neither abstinence nor emission per se is acceptable, since both can result in the generation of uncontrolled and dangerous monstrosities. According to Paracelsus' *De homunculis,* the only proper destination for male sperm is the female womb, the one environment guaranteed not to produce a homunculus. The

De natura rerum, on the other hand, whether genuine or not, has turned the pangenerative vice of human seed into a virtue. By means of the "alchemical" technique employed in incubating a flask at moderate heat, one can isolate the male seed from the female and thereby produce a transparent, "bodiless" homunculus. In this fashion, human art can generate a being unimpeded by the materiality of normal female birth, hence surpassing the artifice of nature.

Here we see the fruit of that confluence of traditions described above—the "rational animal" of the Arabic writers on spontaneous generation has combined with the Latin response to Avicenna to give birth to the Paracelsian homunculus. But this union was not without its dangers. Even in the Middle Ages, there was a powerful feeling that alchemy had transgressed on the creative powers of the godhead in its claim of mineral replication.[66] One version of the *Secret of Secrets* of pseudo-Aristotle contains the following relevant passage: "It must be known that it is impossible to know how to produce genuine silver and gold, since it is impossible to become the equal (*equipari Deo Altissimo*) of God the Highest in his own works."[67]

How much stronger would be the reaction to the homunculus! I shall cite but three examples from seventeenth-century England. Henry More, whose diatribes against "Eugenius Philalethes," or Thomas Vaughan, formed the pretext for writing his *Enthusiasmus Triumphatus,* saw Paracelsianism as the embodiment of philosophical enthusiasm. To More, Paracelsus was the "great boaster," whose "delirious Fancies" and "uncouth and supine inventions" found their epitome in the conceit that "there is an artificiall way of making an *Homunculus.*" Nor would More be pacified by such writers as Johann Valentin Andreae, who tried to allegorize the homunculus, being "ashamed of the grosse sense of it." More saw the artificial man as merely another instance of the Swiss boaster giving vent to "the wildest Philosophicall Enthusiasmes that ever was broached by any either Christian or Heathen."[68]

An equally unsympathetic view of the homunculus is found in an exact contemporary of More's, one not usually mentioned for her philosophical restraint. I refer to Margaret Cavendish, whose Epicurean *Poems and Fancies* appeared in 1653. Cavendish, despite her reputation for eccentricity, was consistently opposed to the claims of alchemy. Her comments on the homunculus are particularly enlightening, for unlike More, she treats the issue of artificial life within the context of the art/nature debate:

> The greatest Chymists are of a strong Opinion, that they can enforce Nature, as to make her go out of her Natural Pace, and to do that by Art in a

Furnace, as the Elixar, that Nature cannot in a hundred or a thousand Years; and that their Art can do as much as Nature, in making her Originals another way than she has made them; as *Paracelsus* little Man, which may be some Dregs gathered together in a Form, and then perswaded himself it was like the Shape of a Man, as Fancies will form, and liken the Vapours that are gathered into Clouds, to the Figures of several things.[69]

Like Henry More, Cavendish wants to see the homunculus as a son of Paracelsus' extravagant fancy, formed by free association from the residue in a flask. But she is unequivocal in her condemnation of the alchemical enterprise that the homunculus embodies—the surpassing of nature by art. Indeed, she is opposed even to the notion that art can equal nature, for as she continues to expostulate, this would make of man a little god.

Nay, they will pretend to do more than we ever saw Nature to do, as if they were the God of Nature, and not the Work of Nature, to return Life into that which is dead. . . . [F]or though the Arts of Men, and other Creatures, are very fine and profitable, yet they are nothing in comparison to Natures works, when they are compared. Besides, it seems impossible to imitate Nature, as to do as Nature doth, because her Waies and her Originals are utterly unknown: for Man can only guess at them, or indeed but at some of them. . . . [T]hough he can extract, yet he cannot make; for he may extract Fire out of a thing, but he cannot make the principle Element of Fire; so of Water and Earth; no more can he make the Elizar [i.e., Elixir] than he can make the Sun, Sea, or Earth. . . . But Nature hath given such a Presumptuous Self-love to Mankind, and filled him with that Credulity of Powerfull Art, that he thinks not only to learn Natures Waies, but to know her Means and Abilities, and become Lord of Nature, as to rule her, and bring her under his Subjection.[70]

It is fascinating to hear the resonances of Avicenna's *sciant artifices* in this passage and to witness Cavendish's denial of the very defense of art raised by the *Book of Hermes*—that man can "create" the four elements. Even the pious doubts of the *Secret of Secrets* commentator are echoed here, in Cavendish's complaint that the "greatest Chymists" confuse themselves with the "God of Nature." Yet the primary focus of Cavendish's attack is no longer the mere transmutation of metals, which she subjoins almost as a footnote, but the making of an artificial man. It is the mute witness of the homunculus, above all, that indicts the alchemist as an impious imposter. The sober natural philosopher must realize that "we scarce see the Shadow of Natures Works" but live in a twilight land at best, where we are apt to break our heads with errant wandering.

A final twist to the fate of the homunculus may be seen in the *Demonstration of the Existence and Providence of God* published by the Calvinist divine John Edwards in 1696. Edwards's book is above all a natural theology, and as such it expounds at length on the wondrous intricacy of the human body. The author finds particular support for his view in the fact that the symmetry and interconnectedness of the body's parts testify to the transcendence of their maker. This sets him apart from mere earthly workmen, who cannot create such organic perfection as to impart genuine life to their products. As Edwards says,

> This is no Workmanship of Humane Skill, here is no Automaton made by Art, no Daedalus's walking Venus, no Archytas's Dove, no Regiomontanus's Eagle and Fly. Here is none of Albertus magnus or Frier Bacon's speaking head, or Paracelsus's Artificial Homuncle. Here is nothing but what proceeds from a divine Principle and Art, and therefore cannot be reckoned among those mechanical Inventions which have an external Shew of Sensation and Life for a time, but are destitute of a vital Spring.[71]

Here Edwards ranks the homunculus among such famous mechanical automata as the brazen head of Roger Bacon and the dove of Archytas, in order to deny it any genuine self-moving principle. Even if the homunculus really can exist, it will only be a clever counterfeit of life and not a genuinely vital being. Remarkably, Edwards has managed to turn the argument of the Paracelsian *De natura rerum* on its head—where the author of that text used the homunculus as the final illustration of man's power over nature, Edwards employs it to demonstrate the feebleness of human art. It is nature alone, the living testament of the divine will, that can produce true life: the alchemist and mechanic can only fabricate a pallid imitation.

It is quite clear, then, that Paracelsus' readers in the seventeenth century were alert to the status of the homunculus as a hero of art, even when they rejected the artificial man as a fraud or a fancy. Few seem to have followed the path of Andreae in harnessing the homunculus to the yoke of Christian soteriology. And indeed, the homunculus as pictured either in the *De natura rerum* or *De homunculis* is an intractable vehicle of salvation. Neither the "bodiless" product of human artisanal mastery nor the obscene and tumorous growths of unbridled lust could serve the ministrations of the regenerate soul. In sum, by fusing together the traditions of artificial generation, alchemical debate, and an unorthodox Catholicism, Paracelsus and his epigones managed to create an image of the alchemist as a *magus coelestis,* approaching the creative powers of divinity itself. This holy magus held the keys of art and nature; in fabricating his homunculus, he could even mimic the supreme creative act of God, though on a smaller scale. Can anyone perceive this image without, like Margaret Cavendish, dimly hearing in the background the words of Genesis 3:5—"your eyes shall be opened, and ye shall be as gods, knowing good and evil"?

NOTES

1. Quoted in John Warwick Montgomery, *Cross and Crucible: Johann Valentin Andreae (1586–1654), Phoenix of the Theologians,* (The Hague: Martinus Nijhoff, 1973), 2:414.

2. Ibid., pp. 440–456.

3. For a similar and contemporary allegory involving death, resurrection, and color changes, see Basil Valentine, *Die zwoelf Schluessel,* in *Elucidatio secretorum, das ist, Erklaerung der Geheimnussen . . .* (Frankfurt: Nicolaus Steinius, 1602), pp. 398 ff. (entry taken from John Ferguson, *Bibliotheca Chemica* [Glasgow, 1906; reprint, Hildesheim: Olms, 1974], 1:239). For a discussion of such alchemical allegories, see William Newman, *Gehennical Fire: The Lives of George Starkey, an American Alchemist in the Scientific Revolution* (Cambridge, Mass.: Harvard University Press, 1994).

4. Quoted in Montgomery, *Cross and Crucible,* 2:458.

5. Ibid., p. 464.

6. Ibid., 1:122–131.

7. See Montgomery, *Cross and Crucible,* vol. 2, where he demonstrates this point exhaustively in his valuable commentary to the text.

8. Walter Pagel, *Paracelsus: An Introduction to Philosophical Medicine in the Era of the Renaissance* (Basel: Karger, 1958), pp. 82–104. For an assessment of Paracelsus in relation to medieval alchemy, see Wilhelm Ganzenmüller, "Paracelsus und die Alchemie des Mittelalters," in his *Beiträge zur Geschichte der Technologie und der Alchemie* (Weinheim: Verlag Chemie, 1956), pp. 300–314.

9. Pagel, *Paracelsus,* pp. 244, 258–259, 263–273. For the transmission of Rupescissa's work in German, see Udo Benzenhöfer, *Johannes de Rupescissa: Liber de consideratione quintae essentiae omnium rerum deutsch* (Stuttgart: Steiner, 1989).

10. Frances A. Yates, *Giordano Bruno and the Hermetic Tradition* (London: Routledge and Kegan Paul, 1964), pp. 144–156.

11. Aristotle, *Physics,* trans. Philip H. Wicksteed and Francis M. Cornford, Loeb Classical Library (1929; reprint, London: Heinemann, 1970), 1:109.

12. Ibid. p. 173.

13. William Newman, introduction to *The "Summa perfectionis" of pseudo-Geber,* ed. and trans. Newman (Leiden: Brill, 1991), pp. 1–40.

14. Francis Bacon, *Descriptio globi intellectualis,* vol. 5 of *The works of Francis Bacon,* ed. and trans. James Spedding et al. (London: Longmans, 1870), p. 506.

15. William Newman, "Technology and Alchemical Debate in the Late Middle Ages," *Isis* 80 (1989): 423–445, esp. 427.

16. Quoted in Newman, introduction to *The "Summa perfectionis,"* p. 49: "ars est debilior quam natura et non consequitur eam quamvis multum laboret."

17. Mary Richard Reif, "Natural Philosophy in Some Early Seventeenth-Century Scholastic Textbooks" (Ph.D. diss., St. Louis University, 1962), p. 238: "One final question briefly touched upon by several authors concerns the possibility of producing a truly natural product by means of human skill. The question is usually posed in this way: 'Can art effect certain works of nature?' The specific problem which they almost always have in mind is the transmutation of baser metals through the art of alchemy." See also Charles B. Schmitt, *John Case and Aristotelianism in Renaissance England* (Kingston, Ont.: McGill-Queen's University Press, 1983), pp. 193–205.

18. Quoted in Newman, introduction to *The "Summa perfectionis,"* pp. 6–15. For the Latin text, see pp. 52–56.

19. Ibid., p. 11.

20. Ibid., pp. 11–12.

21. See Newman, introduction to *The "Summa perfectionis,"* pp. 57–103, for a discussion of the indubitably Latin origin of the *Summa.*

22. Geber, *The "Summa perfectionis,"* pp. 643–644.

23. Ibid., p. 650.

24. Ibid., pp. 647–648.

25. Lynn Thorndike, *A History of Magic and Experimental Science* (New York: Columbia University Press, 1923–1958), 3:139. See Manchester, University of Manchester, John Rylands MS 65, fol. 205v, where the author ascribes the experiment to "Rasis in libro de proprietatibus membrorum animalium."

26. Karl Sudhoff rejected the authenticity of *De natura rerum* in its present form, though he suggested that it might contain "Hohenheimische Ausarbeitungen oder Entwürfe" (in Paracelsus, *Sämtliche Werke,* ed. Sudhoff [Munich: Oldenbourg, 1922–1933], 11:xxxiii), but Will-Erich Peuckert questions this rejection: see *Theophrastus Paracelsus: Werke,* ed. Peuckert (Basel: Schwabe, 1968), 5:ix. Kurt Goldammer also accepts the authenticity of *De natura rerum,* with reservations: "Der Gedanke der Substanzenseparierung hat dann auch die paracelsische Todesanschauung in jenen berühmt gewordenen Ausführungen der umstrittenen Schrift 'De natura rerum' geliefert, von der ich annehme, das sie in ihrer Grundidee echt ist, wenn auch eine Überarbeitung durch Schülerhände sich nicht ausschliessen lässt"; see "Paracelsische Eschatologie, Zum Verständnis der Anthropologie und Kosmologie Hohenheims I," *Nova Acta Paracelsica* 5: (1948): 52.

27. [pseudo?] Paracelsus, *De natura rerum,* in vol. 11 of *Sämtliche Werke,* ed. Sudhoff, p. 312: "Die generation aller natürlichen dingen ist zweierlei, als eine die von natur

geschicht on alle kunst, die ander geschicht durch kunst nemlich durch alchimiam. wiewol in gemein darvon zureden, möchte man sagen, das von natur alle ding würden aus der erden geboren mit hilf der putrefaction. dan die putrefaction ist der hoechst grad und auch der erst anfang zu der generation, und die putrefaction nimbt iren anfang und herkomen aus einer feuchten werme. dan die stete feuchte werme bringet putrefactionem und transmutirt alle natürliche ding von irer ersten gestalt und wesen, desgleichen auch an iren kreften und tugenden. dan zu gleicher weis wie die putrefaction im magen alle speis zu koz macht und transmutirts, also auch ausserhalb des magens die putrefactio so in einem glas beschicht, alle ding transmutirt von einer gestalt in die andere."

28. Ibid., p. 313.

29. See Paracelsus, *Astronomia magna,* in vol. 12 of *Sämtliche Werke,* ed. Sudhoff, p. 322. After the world has been consumed by fire in the final conflagration, everything will be as "ein eidotter ligt im clar." This will be a *perspicuum,* and this will be both a *chaos* and also "das wasser, von dem die geschrift sagt, auf welchem der geist gottes getragen wird."

30. Paracelsus, *De natura rerum,* p. 313: "das ist auch das höchst und grössest magnale und mysterium dei, das höchst geheimnus und wunderwerk."

31. Ibid.: "Es ist auch zu wissen, das also menschen mögen geboren werden one natürliche veter und mütter. das ist sie werden nit von weiblichem leib auf natürliche weis wie andere kinder geboren, sonder durch kunst und eines erfarnen spagirici geschiklikeit mag ein mensch wachsen und geboren werden, wie hernach wird angezeigt &c."

32. Ibid., pp. 315–316: "dan der basiliscus wechst und wird geboren aus und von der grössten unreinikeit der weiber, aus den menstruis und aus dem blut spermatis, so dasselbig in ein glas und cucurbit geton und in ventre equino putreficirt, in solcher putrefaction der basiliscus geboren wird."

33. Ibid., p. 315: "Nun aber damit ich widerumb auf mein fürnemen kom, von dem basilisco zuschreiben, warum und was ursach er doch das gift in seinem gesicht und augen habe. da ist nun zu wissen, das er solche eigenschaft und herkomen von den unreinen weibern hat, wie oben ist gemelt worden. dan der basiliscus wechst und wird geboren aus und von der grössten unreinikeit der weiber, aus den menstruis und aus dem blut spermatis."

34. See, e.g., Paracelsus, *De generatione hominis,* in vol. I of *Sämtliche Werke,* ed. Sudhoff, p. 305, where the female is viewed as the principle of all evil: "Das aber ein mensch vil lieber stilet als der ander, ist die ursach also, das alles erbars in Adam gewesen ist und das widerwertige der êrbarkeit, unêrbarkeit in Eva. solches ist auch also durch die wage herab gestigen in die samen nach dem ein ietlichs sein teil davon gebracht hat, nach dem ist er in seiner natur. denn etwan hat die diebisch art uberwunden, etwan die hurisch, etwan die spilerisch &c." Cf. the parallel locus in Paracelsus, *Das Buch von der Geberung der Empfindlichen Dinge in der Vernunft,* in ibid., pp. 278–281.

35. Paracelsus, *De natura rerum,* pp. 316–317: "Nun ist aber auch die generation der homunculi in keinen weg zu vergessen. dan etwas ist daran, wiewol solches bisher in grosser heimlikeit und gar verborgen ist gehalten worden und nit ein kleiner zweifel und frag under etlichen der alten philosophis gewesen, ob auch der natur und kunst möglich sei, dass ein mensch ausserthalben weiblichs leibs und einer natürlichen muter möge geboren

werden? darauf gib ich die antwort das es der kunst spagirica und der natur in keinem weg zuwider, sonder gar wol möglich sei. wie aber solches zugang und geschehen möge, ist nun sein process also, nemlich das der sperma eines mans in verschlossnen cucurbiten per se mit der höchsten putrefaction, ventre equino, putreficirt werde auf 40 tag oder so lang bis er lebendig werde und sich beweg und rege, welchs leichtlich zu sehen ist. nach solcher zeit wird es etlicher massen einem menschen gleich sehen, doch durchsichtig on ein corpus. so er nun nach disem teglich mit dem arcano sanguinis humani gar weislich gespeiset und erneret wird bis auf 40 wochen und in steter gleicher werme ventris equini erhalten, wird ein recht lebendig menschlich kint daraus mit allen glitmassen wie ein ander kint, das von einem weib geboren wird, doch viel kleiner."

36. Ibid., p. 317: "dan durch kunst uberkomen sie ir leben, durch kunst uberkomen sie leib, fleisch, bein und blut, durch kunst werden sie geboren, darumb so wirt inen die kunst eingeleibt und angeboren und dörfen es von niemants lernen."

37. Ibid.: "dan es ist ein mirakel und magnale dei und ein geheimnis uber alle geheimnus."

38. Gershom Scholem, "Die Vorstellung vom Golem in Ihren Tellurischen und Magischen Beziehungen," *Eranos-Jahrbuch* 22 (1953): 235–289; see p. 281. For Pagel's acceptance of Scholem's argument, see his *Paracelsus,* pp. 215–216.

39. Moshe Idel, *Golem: Jewish Magical and Mystical Traditions on the Artificial Anthropoid* (Albany: State University of New York Press, 1990), pp. 185–186. Idel rejects Scholem's hypothesis that the homunculus theory owed a debt to the literature on the golem.

40. Jābir ibn Hayyān, *Jābir ibn Ḥayyān, contribution à l'histoire des idées scientifiques dans l'Islam,* ed. Paul Kraus, Mémoires présentés à l'institut d'Egypte vol. 44–45 (Cairo: Institut français de l'archéologie orientale, 1942), 2:110.

41. Ibid., p. 111.

42. Ibid., pp. 104–105 n. 12.

43. Thorndike, *Magic and Experimental Science,* 2:735.

44. New Haven, Yale University, Codex Paneth, fol. 392vb: "Qui vult facere animal rationale accipiat aquam suam dum calidam[?] et conficiat eam cum equali mensura eius ex lapide qui nominatur lapis solis. et est lapis qui lucet in nocte sicut lucet lampas donec illuminatur ex eo locus in quo est."

45. Ibid., fol. 393ra: "accipe illam formam et pone eam in illo pulvere. ipsa enim statim vestietur cute humana."

46. Ibid.: "pone illam formam animalem in vas magnum vitreum vel plumbeum non aliud usque quo pretereant ei tres dies et pacietur famem et agittabitur. Ciba ergo ipsam ex illo sanguine qui exivit de matre. et non ergo cesses similiter donec pretereant septem dies. . . . ipsa complebitur forma animalis que convenit rebus multis mirabilibus."

47. Ibid., fol. 393rb: "Et si acceperis hanc formam et cibaveris et nutriveris ipsam usque quo pretereant ei .xl. dies et cibabis eam sanguine et lacte non alio et non viderit eam

sol. . . . Et si homo rexerit eam et nutriverit ipsam usque quo pertranseat ei annus integer, et dimiserit eam in lacte et aqua pluviali narrabit ei omnia absencia."

48. Friedrich Kluge, *Etymologisches Wörterbuch der Deutschen Sprache* (Berlin: de Gruyter, 1989), p. 22. See also Albert Lloyd and Otto Springer, *Etymologisches Wörterbuch des Althochdeutschen* (Göttingen: Vandenhoeck and Ruprecht, 1988), 1:168–170, and Johannes Hoops, *Reallexikon der Germanischen Altertumskunde,* 2nd ed. (Berlin: de Gruyter, 1973), 1:198.

49. Paracelsus, *De imaginibus,* in vol. 13 of *Sämtliche Werke,* ed Sudhoff, p. 378: "dem geb ich zur antwort und sag, es sei nicht war, das alraun die wurzel menschen gestalt hab, sonder es ist ein betrogne arbeit und bescheisserei von den landfarern, die dan die leut mer denn mit disem alein bescheissen. dan es ist gar kein wurzel die menschen gestalt hat, sie werden dan also geschnizlet und geformirt; von got ist keine also geschaffen oder die von natur also wechst, darumb ist weiter darvon nit zu reden &c."

50. Paracelsus, *De vita longa libri quinque,* in vol. 3 of *Sämtliche Werke,* ed. Sudhoff, p. 274: "homunculus, quem necromantici alreonam, philosophi naturales mandragoram falso appellant, tamen non nisi in communem errorem abiit propter chaos illud, quo isti obfuscaverunt verum homunculi usum. origo quidem spermatis est; per maximam enim digestionem, quae in ventre equino fit, generatur homunculus, similis ei per omnia, corpore et sanguine, principalibus et minus principalibus membris." A parallel passage is found in the German text of *De vita longa,* p. 304.

51. Will-Erich Peuckert, *Handwörterbuch der Sage* (Göttingen: Vandenhoeck and Ruprecht, 1961), 1:406.

52. Paracelsus, *Astronomia magna,* pp. 33–38.

53. Paracelsus, *De homunculis,* in vol. 14 of *Sämtliche Werke,* ed. Sudhoff, p. 331: "sonder das verstanden also, das also der polluirt sperma, so er sein digestion und erden begreift, on ein monstrum nicht fürgêt."

54. Retention of female seed was associated with the production of a mole by many physicians from the Middle Ages on. See Danielle Jacquart and Claude Thomasset, *Sexuality and Medicine in the Middle Ages,* trans. Matthew Adamson (Princeton: Princeton University Press, 1988), p. 153. They cite Albertus Magnus, *Quaestiones supra de animalibus,* ed. Filthaut, book 10, Q. 5.

55. Paracelsus, *De homunculis,* p. 332. "daraus wird nun fleisch, moder, truesen &c." Paracelsus was influenced, no doubt, by earlier medical concerns about the retention of seed. Galen comments on this problem and relates the case of Diogenes the Cynic, who supposedly masturbated openly as a means of prophylaxis; see Galen, *De locis affectis,* book 6, Kuehn, 8:417–420, as cited in Jean Stengers and Anne Van Neck, *Histoire d'une grande peur: La masturbation* (Brussels: Université de Bruxelles, 1984), p. 41. See also Jacquart and Thomasset, *Sexuality and Medicine,* p. 149.

56. Paracelsus, *De homunculis,* p. 332: "ein sodomitisch geburt."

57. Ibid., pp. 333, 334–335: "also wissen auch, das in den stercoribus humanis vilerlei tier gefunden werden und seltsam art, die da komen von den sodomiten, von welchen Paulus

schreibt, und sie nent knabenschender, wider die Römer &c. . . . dergleichen auch so wissen, das die sodomiten solch sperma in das maul fallen lassen &c, und also oftmals in magen kompt, gleich als in die matricem als dan so wechst im magen auch ein gewechs draus, homunculus oder monstrum oder was dergleichen ist, daraus dan vil entstehet und seltsam krankheiten sich erzeigen, bis zum lezten ausbricht."

58. Ibid., p. 336: "drumb ziehe und ordne ein ieglicher sein kint in ehelichen stant oder in das verschneiden, damit der graben der dingen abgraben werde, die wurz aus der erden gezogen, mit allen esten heraus gerissen."

59. Ibid., p. 331: ". . . Dan hat er auf Petrum sein kirchen gebauen, das ist auf den erwelten, so wird er auf kein ander jungfrau sein kirchen sezen. dan den selbigen ist nicht zu vertrauen, das ror im wasser ist bestendiger. das zeig ich euch dorumb an, auf das ir verstanden, das Christus nicht wil jungfrauen han, die er nicht erwelt hat, von wegen das sie wie das ror unbestendig seind, sonder wil sein erwelten han, die selbigen bleiben im bestendig. so aber der mensch sich selbs mit gewalt wil keusch halten, aus seinen kreften. so sol man beschneiden oder sich selbs beschneiden, das ist den brunnen abgraben, do das in ligt, darvon ich hie schreib. drumb hats got beschaffen, das wol mag beschehen, nicht wie den magen, nicht wie die lebern, sonder für den leib heraus. den frauen ist das nicht geben, drumb seind sie den mannen befolen, sie seient dan von der natur eunuchae, oder got erhalt sie mit zwangnus art, nicht nach irem fuergeben." Although *beschneiden* does not normally have the sense of "castrate," the parallel passage using *verschneiden* (*De homunculis,* p. 336) ensures that Paracelsus does have castration in mind. See Jacob Grimm and Wilhelm Grimm, *Deutsches Wörterbuch* (Leipzig: S. Hirzel, 1956), ser. 1, 12:1132–1133.

60. Paracelsus, *De praedestinatione et libera voluntate,* in his *Theologische und Religionsphilosophische Schriften,* ed. Kurt Goldammer (Wiesbaden: Steiner, 1965), 2:114: "unser freier will ist anderst denn der erst und scheidet sich vom ersten also: der erst steht in der nahrung des menschen, der ander steht in aufenthaltung des bluts im samen. . . . also du magst in reinigkeit leben, in unkeuschheit, welches du wilt."

61. Paracelsus, *Buch von der Geberung,* pp. 252–253.

62. Ibid., pp. 258–260.

63. Ibid., p. 253: "wo aber der same in der natur ligt, da ist das liecht der natur nit, sonder es ist tot"; p. 254: "denn er wil einen lautern menschen haben und nit ein verenderten, als der same tut so er in der natur ist."

64. Paracelsus, *Astronomia magna,* p. 315.

65. Gerhild Scholz-Williams, "The Woman/The Witch: Variations on a Sixteenth-Century Theme (Paracelsus, Wier, Bodin)," in *The Crannied Wall: Women, Religion, and the Arts in Early Modern Europe,* ed. Craig A. Monson (Ann Arbor: University of Michigan Press, 1992), pp. 119–137. See also Ute Gause, "Zum Frauenbild im Frühwerk des Paracelsus," in *Parerga Paracelsica,* ed. Joachim Telle (Stuttgart: Steiner, 1991), pp. 45–56.

66. Newman, "Technology and Alchemical Debate," pp. 439–442.

67. Roger Bacon, *Opera hactenus inedita Rogeri Baconi,* ed. Robert Steele (Oxford: Clarendon, 1920), 5:173: "Sciendum tamen quod scire producere argentum et aurum, verum

est impossibile: quoniam non est possibile equipari Deo Altissimo in operibus suis propriis."

68. Henry More, *Enthusiasmus Triumphatus* (London: J. Flesher, 1656), p. 46.

69. Margaret Cavendish, *Poems and Fancies* (London: F. Martin and F. Allestrye, 1653), p. 176.

70. Ibid., p. 177.

71. John Edwards, *A Demonstration of the Existence and Providence of God, from the Contemplation of the Visible Structure of the Greater and the Lesser World,* part 2 (London: Jonathan Robinson, 1696), p. 124.

11

NATURAL PARTICULARS: MEDICAL EPISTEMOLOGY,
PRACTICE, AND THE LITERATURE OF HEALING SPRINGS
Katharine Park

In the prologue of his *On the Causes of Wonders* (ca. 1370), Nicole Oresme drew a clear distinction between the causal knowledge of the philosopher or medical theorist and that of the practicing physician. "One thing I would note here is that we should properly assign to particular effects particular causes," he wrote, "but this is very difficult unless a person looks at effects one at a time and their particular circumstances. . . . Why Sortes is poor and Plato is rich, why an animal died at such a time, why pepper in small quantities is a laxative and a diuretic in large quantities, . . . why Sortes heard such a voice or saw such a marvel—how could we render their particular and direct causes and how could we know their particular circumstances? As I have said then," Oresme concluded, "I shall only show in a general manner that such things occur naturally, as do learned physicians who compose general rules in medicine and leave specific cases to practising physicians. For no physician would know how to say—if Sortes were ill—what kind of illness he has and how it will be cured, except by seeing him and considering the particulars."[1]

In this passage, Oresme recapitulated one of the truisms of Aristotelian natural philosophy: although particular natural effects undoubtedly had particular causes, the investigation of those effects and those causes lay outside the purview of the natural philosopher, who concerned himself only with necessary, certain, and universal knowledge (*scientia*). This exile of particulars from natural philosophical reflection marked the work not only of philosophers who subscribed to the traditional idea of demonstrative science found in thirteenth-century writers such as Albertus Magnus or Thomas Aquinas,[2] but also of later philosophers influenced by the particularist ontology and empirical epistemology developed by William of Occam. In fact, as John Murdoch has argued, that particularist ontology "did not mean that natural philosophy then proceeded by a dramatic increase in attention being paid to experience and observation (let alone anything like experiment) or was suddenly overwrought with concern about testing or matching its results with nature; in a very important way natural philosophy was not about nature."[3]

Thus insofar as natural processes or phenomena appeared at all in works of fourteenth-century natural philosophers, they tended to do so in a highly abstract manner: either as part of a speculative discourse, concerning (for instance) the imagined behavior of bars dropped through the center of the earth or wheels of ice rotating in ovens, or in the form of stock examples, such as the presumed behavior of the clepsydrae, bellows, and bottles of water that testified to nature's abhorrence of a void.[4] Neither of these kinds of appeals to experience had anything to do with actual, observed particulars, in the sense referred to by Oresme. Indeed, as Oresme noted, philosophers like himself tended to deal with problems of natural causation only in what he called "a general manner," just as professors of theoretical medicine left the management of particular cases to "practicing physicians." In this sense, as philosophers from Albertus Magnus to Jean Buridan had previously acknowledged, most actual natural phenomena lay—in practice if not in theory—outside the sphere of philosophical reflection: the result was what Murdoch has described as a "natural philosophy without nature."[5]

But does this mean that no one was interested in the causal study of particular natural phenomena? In this paper, I shall argue that the later fourteenth century in fact saw the appearance of a sustained tradition of inquiry and a coherent body of literature devoted to the causal analysis of individual phenomena based on meticulous and repeated sense experience. As Oresme suggested, the men who produced this literature were not philosophers but practicing physicians. My paper deals with a particular genre of medical writing: the monographic treatises on healing springs produced by Italian physicians in the period between about 1350 and 1450. Unlike many other contemporary works on medicine, these treatises did not reflect in the first instance a university context, though some of their authors did teach medicine at various northern Italian universities; they grew instead out of medical practice, and one kind of practice in particular: the attendance of physicians on noble and princely patrons. I shall propose that these treatises reflect, albeit hesitantly and defensively, one of the first attempts by philosophically trained European writers to develop a method of natural inquiry based on the study of particular natural phenomena: in this case, individual natural mineral springs.

This project was not an easy one. As their works testify, the physicians who pursued it were confronted by obstacles of very different kinds. In the first place, they had to develop new methods of empirical investigation—or borrow them from other contexts, such as medical diagnosis and prognosis, and from nonuniversity and nonphilosophical disciplines, such as alchemy. These methods involved not only experimental techniques such as distilla-

tion and alembification but also the habit and discipline of close sensory observation of natural phenomena—of what Oresme called "paying sufficient
attention" (*advertere satis*) to particular natural effects.[6] At the same time, these
physicians had also to wrestle with the traditional hierarchy of value that
privileged the demonstrative and certain knowledge of the natural philosopher over the probable knowledge of the *artifex,* which was branded with the
epistemological stigma of uncertainty and the sociological stigma of the mechanical arts. As Jole Agrimi and Chiara Crisciani have shown, this hierarchy
of value lay at the heart of the ideology that informed the medieval university and the intellectual world of what they have called *doctrina.*[7] In what follows, I shall argue that this new tradition of empirically based knowledge was
both motivated and legitimated by the patronage relationships of medical
practitioners to their aristocratic patients, and that it reflected a different cultural and institutional context from that of the university.[8]

   Mineral springs had long been a staple of Italian therapeutics, as we
know from Pietro da Eboli's early-thirteenth-century poem *On the Baths
of Pozzuoli,* dedicated to Emperor Frederick II.[9] But it was only in the
mid–fourteenth century that such springs began to capture the serious attention of professional medical writers in central and northern Italy.[10] This new
interest formed part of, and was clearly a response to, the dramatic and ac-
celerating revival of lay interest in thermal medicine, which had figured
prominently in Etruscan and ancient Roman therapeutics. The fourteenth
and fifteenth centuries saw the rapid development of a kind of spa culture,
organized around Italy's many thermal springs.[11] Located in the countryside
outside major urban centers, these springs became important sites not only
of medical pilgrimage, on the part of both rich and poor, but also of a kind
of social season (in May and then again in September) for Italy's ruling elites.
Extended visits to baths appear repeatedly in the correspondence of fifteenth-century Italian noble and patrician families such as the Medici, who
especially favored Bagno a Morba near Volterra, and the Gonzaga, who pa-
tronized Petriolo, near Siena.[12]

   Spurred by this growing interest in thermal medicine, physicians, municipal officials, and local entrepreneurs scoured the countryside looking for
promising new springs. As a result, though early-fourteenth-century medical
authors such as Pietro da Tossignano or Gentile da Foligno knew of relatively
few such sites—Porretta, near Bologna, and Abano, near Padua, were the
most famous northern Italian examples—early-fifteenth-century writers on
mineral springs, such as Ugolino da Montecatini or Michele Savonarola,
were familiar with literally hundreds of individual springs. Some of the newly
rediscovered ones were Roman, surrounded by classical ruins, which were

often rebuilt; others were previously completely unknown, except, sometimes, to locals.

Both municipalities and wealthy individuals—for example, Lorenzo
de' Medici and his mother Lucrezia Tornabuoni in Florence, and Pietro
Gambacorta, lord of Pisa—engaged in elaborate exercises in real estate development around these springs, constructing complexes of inns and palaces
around newly discovered sites or refurbishing the buildings of old ones.[13] In
preparation, they frequently sent their own physicians to investigate the
properties of the bath in question. For example, when Florence annexed the
territory of Volterra, it acquired a group of springs called Bagno a Morba; the
government saw the baths as a valuable asset, and on several occasions between 1388 and 1391 sent Cristofano di Giorgio, a prominent young Florentine physician, to analyze the water in order to determine its special
properties and to recommend how the area might best be developed.[14] In the
same way, Giovanni Dondi, a famous Paduan physician and author of one of
the earliest monographs on thermal medicine (in the 1370s), was instrumental in developing the nearby new bath of Casanova, near Abano, and may
well have had a financial interest in it.[15]

It was in this context that we see not only a revival of interest in Pietro
da Eboli's *Baths of Pozzuoli,* which produced a number of beautiful illuminated manuscripts of the work for various princely patrons,[16] but also the appearance, in mid- to late-fourteenth- and early-fifteenth-century Italy, of a
series of learned medical monographs on hot springs and natural baths. These
included Gentile da Foligno's *On Baths* (before 1348);[17] Tura di Giacomo da
Castello's *On the Baths of Porretta* (1351);[18] Jacopo Dondi's *On the Cause of the
Saltiness of Waters* (1355);[19] Francesco da Siena's *On Baths* (1399), dedicated
to Duke Galeazzo Visconti;[20] and, in the first half of the fifteenth century,
Antonio Guaineri's *On the Baths of the Very Ancient City of Aqua* and Bartolomeo da Montagnana's *On the Appearance, Location, Powers, and Operations
of the Baths Discovered in the Paduan Countryside,* composed for Lord Giovanni
of Pesaro.[21] Some of these were actually written in response to the explicit
requests of princely patrons, and all took as their principal frame of reference
the kind of elite medical practice to which the most ambitious physicians aspired. They bristle with specific and highly respectful references to individual noble patients—their illnesses, their travels to one spring or another, and
the outcomes of their treatment. This is particularly true of the three longest
and most interesting treatises that form the basis of this paper. Giovanni
Dondi's *On the Hot Springs of the Paduan Countryside* (ca. 1372) grew out of a
year's attendance on Duke Galeazzo Visconti.[22] Ugolino da Montecatini
compiled most of the information for his *On Baths* (1417, expanded in 1420)

while in the service of Pietro Gambacorta and Malatesta de' Malatesta, lords of Pisa and Pesaro respectively.[23] And Michele Savonarola dedicated his *On Baths and Natural Spas* (1448–1449) to Borso, son of the Marquis Niccolò d'Este, who had hired him as court physician for the impressive annual salary of four hundred florins.[24]

This medical literature on baths formed part of the flowering in Italy of a particular branch of medical learning called *practica,* which concerned the diagnosis, description, and treatment of individual diseases; *theorica,* in contrast, dealt with more general and abstract questions concerning physiology and the nature of health and illness.[25] In the thirteenth and early fourteenth centuries, medicine as an academic and intellectual discipline had modeled itself on natural philosophy—its unique claims to certainty located epistemologically in the logic of deduction and institutionally in the university.[26] By the middle of the fourteenth century, however, the Italian city-states were witnessing an astounding development of the marketplace for professional medical services.[27] This produced an explosion of opportunities for physicians (and to some degree also surgeons) for highly lucrative employment not only by a wide variety of large and wealthy institutions—hospitals, monasteries, confraternities—but also by the growing class of patrician families and small princely dynasties that monopolized power in the highly urbanized but politically fragmented world of early Renaissance Italy.

As a result, the later fourteenth and early fifteenth centuries saw intense interest and rapid development in the area of medical learning that had its roots outside the universities and looked for rewards that were not simply academic: large salaries, lucrative contracts, and high prestige. This intellectual arena was largely coextensive with that of the field called *practica,* which, though itself a regular part of the medical curriculum in northern Italian universities, was oriented not toward elaborating causal explanations but toward developing an effective diagnostics and therapeutics. It is in this large body of literature, produced by physicians trained at the university and sensitive to epistemological issues, that we begin to see the elaboration of a philosophically informed and experience-based study of natural phenomena grounded on the consideration of particulars—not only natural springs but also the plants, animals, and minerals that formed the mainstay of contemporary therapeutics.[28]

This process appears especially clearly in the treatises on healing springs, in large part because each individual natural spring had long been thought to have unique properties, stemming from its particular location and topography; this topographical uniqueness arose not only from the particular subterranean arrangement of mineral deposits and heat sources that was thought to give each bath its own composition and temperature, but also from the unique constellation of planetary influences that each place on earth received. That the rays from the heavenly bodies strike different latitudes at different angles and are received differently depending on the particular arrangement or responsiveness of the matter there was already a common-place idea in the thirteenth century: Albertus Magnus had stressed it in two very influential works, *On the Nature of Places* and *On Minerals,* where he related it specifically to the appearance of minerals and of springs with special properties.[29] Writing two hundred years later, Michele Savonarola reiterated this idea, noting that many effects in the natural world were purely local, like the appearance of gold and silver deposits in certain regions, or the fact that Tartars, with their wide, flat faces, were born in only one part of the world.[30]

The power of place and the consequent uniqueness of particular mineral springs had a number of important implications. From a commercial point of view, of course, they raised the crucial question of whether healing water could be bottled and marketed away from its place of origin, a controversy specifically addressed by Savonarola.[31] But the epistemological implications were even more serious. For if each spring was unique—so that even directly adjacent springs could have wildly different temperatures and properties—then the properties of each could not be deduced from first principles but had to be carefully derived from experience of the individual case. As a result, as Savonarola put it, referring to the mineral properties of the bath of Monte Grotto (first studied by Jacopo Dondi), "all these things are probable, lacking logical demonstration. But experience is the mistress of all these discords."[32]

Here, Savonarola was carefully and explicitly using the language of probability and opinion. He believed not that the properties of individual springs had no causes, only that the particularities of place meant that those causes could not be known with certainty and therefore those properties were not amenable to demonstrative or "scientific" knowledge. As a result, each spring had to be studied individually and with the utmost attention, using all the information available to the senses: the color of the water, its smell and taste, the nature of the illnesses it cured. Often this process required relatively elaborate experimentation; there was, for example, a standing debate over whether the minerals of a spring were best extracted by boiling, distillation, or evaporation (the method developed by Dondi at Monte Grotto).[33] Savonarola even described a dispute he had with one of his noble patients— the famous condottiere Francesco Busson, count of Carmagnola—over the relative heat of the baths of Abano and Sant'Elena. The matter was resolved by filling vials from both at exactly the same time, by the clock, and comparing the temperatures once the two samples were brought together, which showed them to be virtually the same—an experiment that vividly illustrates the difficulties of contemporary instrumentation. As Savonarola himself put it, "to measure (*mensurare*) the degree of heat with the degree of coldness is not easy."[34]

But how were learned physicians to identify promising new springs for testing, and how might their properties initially be known? Here Ugolino da Montecatini and Michele Savonarola both emphasized the utility of lay observation and the importance of nonelite informants. For example, Ugolino reported that local women used one of the baths of Abano to "clean out their uteruses" and another, very cold one in the Pisan countryside to treat infertility; following the example of the latter, his own wife, who had been unable to conceive in twenty years of marriage, was pregnant within the year.[35] Writing of another spring discovered recently near his home town of Montecatini, he noted that "it was frequented mainly by peasants suffering from pains in the joints . . . and they go there without following any rules [presumably of physicians]. They receive great benefit from its use. And they take certain plants and make a hollow and enter it, mixing the water with mud. And they say that water is more effective when mixed with mud."[36] Similarly, Michele Savonarola described a newly discovered spring near Carpi, "the [healing] power of which was first pointed out by animals," as he put it;[37] he explained that in 1448, when the local cattle became sick and began to urinate blood, they sought out the spring, drank from it, and were cured. The cowherds, who had observed this, notified the authorities in Carpi, who concluded that its water was generally good for disorders of the urinary tract.

Despite acknowledging the utility of lay experience, Ugolino and Michele both emphasized (presumably especially for the benefit of their noble patrons) that this was never sufficient and might in fact be misleading. Ugolino noted that mineral springs could be harmful if not used properly, and he recommended that a physician's advice be followed at all times. Savonarola also stressed that the indiscriminate use of baths, without expert attention to the patient's individual complexion, time of year, and proper mode of application, might be downright dangerous.[38] In this way, the singularity of the patient—which was constantly emphasized in elite practice—interacted with the singularity of the spring, producing a unique situation that was wholly unamenable to demonstrative analysis.

While repeatedly stressing the special competence of the learned physician, his deep theoretical knowledge supplemented by broad experience, fourteenth- and early-fifteenth-century Italian writers on thermal medicine expressed nonetheless a pervasive concern about the shaky epistemological status of their conclusions, in contrast to the demonstrative and deductive ideal of natural philosophy. We already find an explicit statement of this problem in the treatise of Giovanni Dondi, who commented, after giving his own explanation of the cause of heat of thermal baths, "I have not promised to demonstrate perfectly that this is the certain and proper cause, since it is difficult to promise any certainty concerning these things that are perceived by conjecture. . . . Whoever doesn't like [this cause] can seek out one more probable, because everyone possesses the free faculty of inquiring and forming opinions, as long as he supports his opinion with reasoning."[39] By the middle of the fifteenth century, Michele Savonarola appeared even more defensive: "I have described in positive terms this way of investigating the cause of the heat of thermal baths," he wrote, "[although] I judge that this material is not conducive to demonstration and cannot be defended from contradiction. But it has seemed to me the most expeditious [mode of investigation] and the most consonant with human minds. On account of this let no one criticize [literally, 'bite'] me, since I have thus [at least] supplied [the basis] for investigating another and perhaps truer cause."[40]

These defensive statements make sense in the context of the intense competition that characterized elite medical practice.[41] But they also illuminate the difficulty and insecurity of naturalists attempting to craft an alternative model of natural knowledge to that found in university-based natural philosophy, sacrificing the limpid certainty of *scientia* for the muddy waters of sensory experience and probable opinion. The epistemological status of medicine, part practical skill and part learned theory, had long presented itself as problematic. Late-thirteenth- and early-fourteenth-century scholastic writers such as Taddeo Alderotti, attempting to carve out for medicine a stable place in the map of academic knowledge, had struggled to relate their procedures to Aristotelian methodology and to stake their claims to scientific status. As Nancy Siraisi has shown for the University of Bologna, such questions were fundamental to the establishment of medicine as a university discipline, on the model of law, with all of the associated authority, prestige, and statutory protection.[42] The anxious remarks of Michele Savonarola, writing in the mid–fifteenth century from the point of view of practice, testify to the continuing effects of those epistemological and political struggles.

At the same time, however, fourteenth- and fifteenth-century medical writers also elaborated aspects of Aristotelian logic and epistemology that differentiated their work from that of contemporary natural philosophers. They explored the method of "resolution" (proceeding from phenomena to their first principles)[43] and emphasized the importance of cumulative experience and progress in the area of therapeutics. In this last connection, they also drew on Aristotelian methodology, which acknowledged the legitimacy of probable opinion and identified it with approval by a community of learned experts.[44]

This collaborative and cumulative process also informed the literature on mineral springs, where, for example, Giovanni Dondi cited the discoveries of his own father Jacopo in his treatise on the hot springs of Padua. Writing early in the 1370s, Dondi in fact relied primarily on textual evidence, although he occasionally referred to his own observations—for instance, concerning the small black particles found in those baths.[45] In contrast, fifteenth-century writers such as Ugolino da Montecatini and Michele Savonarola were much more likely to cite their own observations concerning newly discovered springs, or at least to collect information orally from local medical experts who had studied the phenomena themselves. Thus Ugolino noted that because he had never visited the springs at Siena, he was relying on the testimony of two Sienese physicians, Marco and Francesco; and when he went to Viterbo to check out the baths there, he wrote, "I wanted for the day that I was there to inform myself from the local doctors and others."[46]

Such remarks hint at the emergence of a nascent community of inquirers working together to accumulate and collate new information derived from the direct experience and observation of natural phenomena. This new development cannot be attributed to humanism or a self-conscious rejection of medieval methods: Savonarola invoked Dondi repeatedly and with great respect. Rather, it appears to have grown naturally out of both the earlier tradition of text-based *experimenta* and the professional demands of contemporary practice. Furthermore, as it appears in late-fourteenth- and fifteenth-century texts on thermal medicine, this process was a pale forerunner of the mid-sixteenth-century developments that Paula Findlen describes later in this volume: the references to collaboration are incidental and unsystematic, and the treatises reflect as clearly the bitterly competitive world of elite medical practice, so vividly described by Ugolino da Montecatini, who found himself locked in conflict with his implacable enemy Giovanni Baldi.[47] In this sense, we can hardly talk of a highly developed collaborative model for the kind of empirical inquiry sketched in the treatises on baths.

In addition to their colleagues, the authors of these treatises found an even more potent source of legitimation in their noble patients. It is striking how often fifteenth-century writers on baths stressed the impetus they received from their patrons and patients for the study of the properties of newly discovered springs. Ugolino noted that Malatesta de' Malatesta, lord of Pesaro, had pressed him to investigate the water of Bagno ad Aqua, near Siena.

Malatesta had already performed his own distillation of the water and, according to Ugolino, begged him to amplify those findings.[48] I have already mentioned Michele Savonarola's discussion with his patient the count of Carmagnola about the relative heat of the baths of Abano and Sant'Elena. Savonarola also cited another example of the involvement of local nobles in the study of new springs: when Galasio, a noble from Carpi, heard that the physician was writing a book about baths, he sent him a sample of a local spring—the one that had been so efficacious for the epidemic of bovine cystitis in 1448—and asked him to investigate it.[49]

Writers such as Michele Savonarola and Ugolino da Montecatini invoked their noble patients at every opportunity, hoping no doubt that their patrons' prestige and support would discourage other authors from "biting" them for venturing into the elusive and uncertain territory of particular phenomena and probable opinion. The result was not only a very different epistemological model of natural inquiry from the demonstrative ideal advocated by university-based natural philosophers; it was also informed by a very different sensibility from the impassive and distanced stance of the professor of *scientia,* engaged in transmitting to his students the certain causal knowledge he had received in turn from his own teachers. The language of the thermal treatises is autobiographical, at times confessional—this is particularly true of Ugolino's work—and shot through with the rhetoric of surprise and wonder. Giovanni Dondi's treatise is a case in point. "And so," he wrote,

> when I first saw these waters and considered their properties, . . . I wondered not a little and, not finding causes that were wholly satisfactory, I was for a long time in doubt on many points. But now I have learned . . . from long experience that there is nothing that is not marvelous, and that the saying of Aristotle in the first book of the *Parts of Animals* is true, that in every natural phenomenon there is something wonderful—indeed many wonders. Thus indeed it is, brother: among wonders are we born and placed and surrounded on all sides, so that to whatever object the eye first turns, the same is a wonder and full of wonder, if only we examine it for a little.[50]

This passage immediately placed Dondi outside the tradition of demonstrative natural philosophy, where wonder was a taboo emotion, the hallmark of the nonphilosopher, who was ignorant of causes and therefore marveled at unusual natural effects (*mirabilia*).[51] Instead, it relates Dondi's work to courtly writing and the literature of romance, the aim of which was to evoke wonder by the description of exotic and unfamiliar natural phenomena: petrifying springs, fountains of youth, city walls made of lodestones, castles lit by carbuncles.[52] In this way, Dondi and his fellows must have hoped, if they could not lay claim to the authority of the natural philosopher, cloaked in the certainty guaranteed by demonstrative *scientia,* they could at least cash in on the prestige of their courtly patrons and the associated glamor and charisma of romance.

In addition to clothing their epistemologically shaky enterprise in borrowed splendor and recalling their princely connections, the discourse of wonders also served another purpose: to focus the attention of observers on the phenomena at hand. It was all very well for Oresme to note that the causes of natural effects were divinable if one "paid sufficient attention"; but the habit of paying attention to natural phenomena, particularly relatively unprepossessing natural phenomena such as a pool of stinking, muddy water, required a special discipline of the senses and the mind.

To some degree, that habit of paying attention to the data of sense already informed the practice of the physician, who regularly inferred the illness and chances of his patients from minute and subtle changes in the color of their urine, the smell of their excrement, the sound of their breathing, the rhythm of their pulse.[53] From this point of view, all investigators such as Ugolino or Michele Savonarola had to do was transfer those techniques from the body of the patient to the physical phenomenon—from the urine to the water of a spring. Thus, in a chapter called "How to Investigate the Minerals of Natural Baths," Savonarola explained that you could determine the mineral contents of water by sight, taste, touch, and hearing: water with a predominance of *nitrum* was more transparent and much sharper in taste than water with a predominance of salt, while a predominately salty distillate could be identified because it was softer to the touch than *nitrum* and crackled when thrown into a fire. These kinds of determinations, Savonarola noted, were by their very nature subjective and required "exquisite familiarity with the forms of minerals, particularly concerning those parts that pertain to the senses. Whence," he concluded, "let them be silent who perhaps would wish to criticize (*mordere*) me."[54]

Savonarola's remarks here suggest another possible model for this habit of attention to particular sensory phenomena: the alchemical tradition, which in northern Italy had strong ties to fourteenth- and fifteenth-century medicine.[55] Like a number of contemporary physicians, including Guglielmo Fabri, whom Chiara Crisciani discusses elsewhere in this volume, Ugolino da Montecatini and Michele Savonarola both had alchemical interests. Savonarola had even written a treatise on *aqua ardens,* or ethyl alcohol, and he and Ugolino repeatedly invoked distillation and other techniques for determining the mineral composition of water—techniques that were also regularly used in the preparation of contemporary medicines.[56] Furthermore, as the example of Fabri underlines, alchemy had a long tradition of courtly associations, embodied in works like the influential *Secret of Secrets,* purportedly composed for Alexander the Great.[57] From an epistemological point of view, too, alchemy had much in common with medicine. Composed, like medicine, of both a theoretical and a practical part, and rooted in a vision of nature compatible with Aristotelian principles, it had nonetheless an uneasy relationship to the demonstrative ideal of philosophy, since an important part of alchemical knowledge, like medical knowledge, was constructed through contact with matter, mediated through the senses. Both Michela Pereira and Chiara Crisciani have called attention to its liminal status, suspended between art and science, mechanical and philosophical knowledge, the material and the spiritual world.[58] The tentative, autobiographical, even confessional tone of some of the passages in the medical treatises on baths, while certainly owing something to humanism and the influence of Dondi's friend Petrarch, also recalls the self-representation of the alchemist as apprentice or pilgrim, humbly making his way through the confusing world of sensory phenomena.[59]

But neither alchemy nor the discipline of diagnostics could offer much to the physician in the way of conventional intellectual legitimation. From an institutional point of view, alchemy was marginal to the world of university culture, unrepresented in the official curriculum; and it further labored under the stigma of late-thirteenth- and early-fourteenth-century ecclesiastical condemnations and a penumbra of associations with heterodoxy and fraud.[60] Practical medicine, in contrast, was well established as a university subject, but few claims could be made for the intellectual or social dignity of diagnostics, whose sensory discipline centered on examining the effluvia of the human body. As Petrarch had written in his *Invectives against a Doctor* (1350s), "you look into soiled basins, you examine the urine of the sick, and you think about gold. Why is it surprising that you, who have so much to do with things that are gloomy, dark, and yellow, should yourself be gloomy, dark, and yellow?"[61] In Petrarch's eyes, the practicing physician was mired in the realm of the senses and involved in what he characterized as a mechanical art.

Thus one of the functions of the discourse of wonder in treatises on thermal medicine was to elevate its objects, and therefore its investigators, by associating them with natural marvels; the aim, no doubt, was both to lay claim to their courtly associations and to confirm their remarkable healing powers. Dondi worked hard to redescribe the properties of the springs of Padua as "wonderful accidents" (*accidentia mirabilia*): their repellent taste, their unpleasant smell, the small worms that inhabited them, and the green and grey slime that accumulated at the bottom of their pools.[62] Such rhetorical strategies were at best stopgaps; they in no way lessened the difficulty and uncertainty of the enterprise in which Dondi was engaged. Anomalous, particularistic, and unassimilable to necessary and universal demonstrations, mineral springs demanded a new model of natural inquiry—one that did not turn its back when faced with novel and chance phenomena, but tried to craft a new set of procedures, rooted in experience, for untangling the complicated strings of causes involved. In contrast to natural philosophical teaching, this form of inquiry was progressive and open-ended; its practitioners did not envisage themselves as merely interpreting or transmitting a largely complete body of knowledge, but instead saw themselves as part of an ongoing enterprise of discovery. This enterprise was coordinated and validated by a new community of learned experts: physicians trained in academic natural philosophy but active in the contingent world of practice and legitimized by the approval of their noble clients. Such clients were familiar with the long history of wonders as the aristocracy of natural phenomena, rooted in the literature of romance and courtly recreation. Thus they accepted the proffered treatises not only as potential repositories of healing wisdom but also as offerings fit for a prince.

Although the world of the courts and of aristocratic practice was an increasingly important site for the production of medical learning in fourteenth- and fifteenth-century Italy, it is important to emphasize the continuities between it and the academic world. Most of the writers on thermal medicine taught in one or more Italian universities at some point in their careers: some, like Dondi and Savonarola, converted a notable academic reputation into princely patronage; others, like Ugolino, began in courtly and private practice, later moving into a university post. Thus their works on baths in no way repudiate fundamental scholastic assumptions concerning methodology, epistemology, or the structure of the physical world. But they did mobilize a clear set of rhetorical and literary strategies to underline their illustrious social connections and legitimize their immersion in the domain of contingency, sense perception, and particular effects. The treatises on thermal medicine reveal a fluid social and intellectual environment, marked by the interpenetration of medical, philosophical, and alchemical interests, as well as by easy communication between the worlds of university and court. Equally striking are the continuities with late-thirteenth- and early-fourteenth-century medical and pharmacological learning. Despite his friendship with Petrarch, Dondi's own medical work shows no sustained influence of humanism, and the same is true of Ugolino and Savonarola (though Savonarola's *On Baths* was translated into Greek by Theodore of Gaza).[63] Rather, their interests in contingent natural phenomena appear to have grown naturally out of the existing tradition of writing on therapeutics, as well as the exigencies of elite practice itself.

I would also stress the strong discontinuities between the empiricism of fourteenth- and fifteenth-century Italian balneologists and the modern idea of empiricism, which ultimately came to rest on phenomena that were replicable, classifiable, countable, and homogeneous. The healing springs studied by learned physicians and frequented by both peasants and princes were not phenomena of this sort but were natural wonders. They belonged to the world of the anomalous, the remarkable, and the bizarre, and their value resided in their singularity. Compared with contemporary natural philosophers, or even writers on theoretical medicine, Dondi, Ugolino, and Savonarola shied away from generalizations, even about the small class of thermal springs, preferring to focus on the properties of each bath: its temperature, its peculiar mix of dissolved minerals, and the diseases it was known to cure.

Unique phenomena of this sort resisted induction, much less deduction; they demanded rather what William Eamon, following Carlo Ginzburg, has called a "venatic epistemology," modeled on the hunt, which focused on tracing backward the complicated causal chains that produced particulars by reading the fragmentary evidence of natural signs.[64] Michele Savonarola specifically invoked the idea of sign when he wrote:

> The signs of doctors are never found to be definitive and infallible; rather they give knowledge approaching the truth. This is the source of judgment based on sense (*iudicium extimativum*), since different people judge differently concerning diseases and cases that occur. The same thing applies to untangling the minerals in baths, since because the signs are not altogether certain and definitive, it happens that people writing about those minerals may disagree. Thus you must flee doctors without good judgment, and thus one should not consider only one sign, but all signs or several and the most important ones, which should produce belief in matters of this sort.[65]

In this passage, as in others, Savonarola struggled to formulate explicitly the epistemological process that lay behind the judgments he and his fellow physicians made when they wrote about thermal springs, using both the medical language of sign, derived from the activities of diagnosis and prognosis, and the philosophical language of judgment and opinion. His efforts were at best halting, reflecting the rudimentary state of the enterprise. But they represent the aspirations of physicians writing on *practica* to forge an intellectually respectable study of particulars.

This endeavor came into its own in the next century, when natural history began to achieve disciplinary autonomy and when a whole range of fields related to the teaching of medicine, most notably anatomy and botany, began to develop a solid empirical base.[66] But it is wrong to see these developments as wholly novel, without roots in the earlier period. As Vivian Nutton has recently emphasized, the medical culture of the fifteenth century—the "missing century," as he calls it—was vigorous and innovative, marked by a growing rejection of traditional medical authorities and a strong emphasis on medical practice; the latter gave rise to an increasing clinical emphasis in teaching and a large and original body of writing on topics ranging from *materia medica* through surgical techniques and instruments to the nature and management of epidemic disease.[67] The early Italian medical literature on springs was part of this flowering, and it shows some of the first attempts to engage in the detailed study of individual natural phenomena, based on the carefully collated data of the senses. The dissectors and collectors of the sixteenth century developed these leads in important new directions, but they followed in the steps of their fourteenth- and fifteenth-century colleagues, who wrestled with the elusive and frustrating world of natural particulars, as well as with the difficult epistemological issues that such inquiry raised.

NOTES

1. Nicole Oresme, *De causis mirabilium,* prologue, in Bert Hansen, *Nicole Oresme and the Marvels of Nature: A Study of His "De causis mirabilium" with Critical Edition, Translation, and Commentary* (Toronto: Pontifical Institute of Mediaeval Studies, 1985), pp. 137–140.

2. See Benedict M. Ashley, "St. Albert and the Nature of Natural Sciences," in *Albertus Magnus and the Sciences: Commemorative Essays, 1980,* ed. James A. Weisheipl (Toronto: Pontifical Institute of Mediaeval Studies, 1980), pp. 85–94; William A. Wallace, "Albertus Magnus on Suppositional Necessity in the Natural Sciences," in ibid., pp. 102–126; and in general Eileen Serene, "Demonstrative Science," in *The Cambridge History of Late Medieval Philosophy,* ed. Norman Kretzmann, Anthony Kenny, and Jan Pinborg (Cambridge: Cambridge University Press, 1982), pp. 496–517.

3. John Murdoch, "The Analytic Character of Late Medieval Learning: Natural Philosophy without Nature," in *Approaches to Nature in the Middle Ages,* ed. Lawrence D. Roberts (Binghamton, N.Y.: Center for Medieval and Renaissance Studies, 1982), p. 174.

4. The example of the rod is from Swineshead and that of the wheel from Gaetano da Thiene: John E. Murdoch and Edith Sylla, in *Dictionary of Scientific Biography,* ed. Charles Coulston Gillespie (New York: Charles Scribner's Sons, 1981), s.v. "Swineshead, Richard"; and Murdoch, "Analytic Character," p. 200 n. 9. For the stock examples, see Edward Grant, *Much Ado about Nothing: Theories of Space and Vacuum from the Middle Ages to the Scientific Revolution* (Cambridge: Cambridge University Press, 1981), pp. 77–100.

5. Murdoch, "Analytic Character"; see in general the literature in note 2.

6. Oresme, *De causis* 3, pp. 270–272.

7. Jole Agrimi and Chiara Crisciani, "Per una ricerca su *experimentum-experimenta:* riflessione epistemologica e tradizione medica (secoli XIII–XV)," in *Presenza del lessico greco e latino nelle lingue contemporanee,* ed. Pietro Janni and Innocenzo Mazzini (Macerata: Università degli Studi di Macerata, 1990), pp. 9–49; eaedem, *Edocere medicos: Medicina scolastica nei secoli XIII–XV* (Naples: Guerini, 1988), esp. pp. 49–74, 137–156.

8. For a related set of reflections concerning the mechanical arts, see Pamela O. Long, "Power, Patronage, and the Authorship of *Ars:* From Mechanical Know-how to Mechanical Knowledge in the Last Scribal Age," *Isis* 88 (1997): 1–41.

9. Latin text and English and Italian translations are in Pietro da Eboli, *De balneis puteolanis,* trans. Carlo Marcora and Jane Dolman (Milan: Il Mondo Positivo, 1987).

10. See Domenico Barduzzi, *Ugolino da Montecatini* (Florence: Istituto Micrografico Italiano, 1915), p. 71.

11. See in general Federico Melis, "La frequenza alle terme nel basso medioevo," in *Congresso Italiano di Studi Storici Termali* (Salsomaggiore Terme, 1963), 1:38–49; Richard Palmer, "'In our Lightye and Learned Tyme': Italian Baths in the Era of the Renaissance," in *The Medical History of Waters and Spas,* ed. Roy Porter (London: Wellcome Institute for the History of Medicine, 1990), pp. 14–22; Ralph Jackson, "Waters and Spas in the Classical World," in ibid., pp. 1–13; and, especially, D. S. Chambers, "Spas in the Italian Renaissance," in *Reconsidering the Renaissance: Papers from the Twenty-First Annual Conference,* ed. Mario A. Di Cesare (Binghamton, N.Y.: Medieval and Renaissance Texts and Studies, 1992), pp. 3–27. This last is an excellent introduction to the subject.

12. See Yvonne Maguire, *The Women of the Medici* (London: George Routledge and Sons, 1927), esp. pp. 83–87, 101–109; Gaetano Pieraccini, *La stirpe de' Medici di Cafaggiolo,* 2nd ed. (Florence: Vallecchi, [1947]), 1:57–59, 63–68, 86–92, 128–133; Janet Ross, *The Lives of the Early Medici as Told in Their Correspondence* (London: Chatto and Windus, 1910), esp. pp. 112–116, 179–186; L. Guerra-Coppioli, *Il Bagno a Morba nel Volterrano e Maestro Pierleone Leoni da Spoleto, medico di Lorenzo il Magnifico* (Siena, 1915); and Attilio Portioli, *I Gonzaga ai bagni di Petriolo di Siena nel 1460 e 1461 (documenti inediti)* (Mantua: Eredi Segna, 1869).

13. Chamber, "Spas," pp. 18–20. For Lorenzo and Lucrezia, see also Maguire, *Women of the Medici,* pp. 106–109; Ross, *Lives of the Early Medici,* pp. 113–114. For Gambacorta, see Barduzzi, *Ugolino,* p. 32.

14. Florence, Archivio di Stato: Provvisioni-Registri 84, fol. 100r; 85, fol. 104v. At the same time, the councils elected six "officials of the baths" to administer the site. For evidence of municipal interest in baths as early as the later thirteenth and early fourteenth century, see Francesco Raspadori, "Legislazioni termali senesi nel Medioevo," in *Atti del XXI Congresso Internazionale di storia della medicina* (Siena: n.p., [1969?]), 1:35–39; Giulio Gentili, "I più antichi documenti sulle terme di Porretta (secoli XIII–XIV)," in *Congresso Europeo di Storia della Medicina* (Montecatini Terme: n.p., 1962), 1:236–244.

15. Michele Savonarola, *De balneis et thermis naturalibus omnibus Italiae* 1.3, in Tommaso Giunta, ed., *De balneis omnia quae extant apud Graecos, Latinos, et Arabas . . .* (Venice:

Giunta, 1553), fol. 17v. I base my discussion of Savonarola's treatise on the text in this edition, aware that it has been revised by Giunta.

16. See C. M. Kauffmann, *The Baths of Pozzuoli: A Study of the Medieval Illuminations of Peter of Eboli's Poem* (Oxford: Bruno Cassirer, 1959); Jonathan J. G. Alexander et al., *The Painted Page: Italian Renaissance Book Illumination, 1450–1550* (Munich: Prestel, 1994), p. 65.

17. Gentile da Foligno, *Tractatus de balneis,* in Antonio Cermisone, *Consilia Cermisoni* (Venice: [Bonetus Locatellus for Octavianus Scotus, ca. 1495–1497]); see Lynn Thorndike, *A History of Magic and Experimental Science* (New York: Columbia University Press, 1923–1958), 3:236. I have not seen this edition. A survey of the medical literature appears in Chambers, "Spas," 4–8.

18. Tura di Giacomo da Castello, *Trattato delle terme di Porretta* (Vicenza: Giovanni di Reno, 1473); see Barduzzi, *Ugolino,* p. 71, for additional bibliographical references. I have not seen this edition.

19. Jacopo Dondi, *Tractatus de causa salsedinis aquarum et modo conficiendi sal artificiale ex aquis thermalibus euganeis,* in Giunta, *De balneis,* fol. 109r–v. As in the case of Savonarola's treatise, I use Giunta's edition with some trepidation, not having access to earlier versions. Discussion and bibliographical details in Thorndike, *Magic and Experimental Science,* 3:386–387, 392–393; and Tiziana Pesenti, "Dondi dall'Orologio, Iacopo," in *Dizionario biografico degli italiani* (Rome: Istituto della Enciclopedia Italiana, 1960–), 41:104–111.

20. Francesco da Siena, *Tractatus de balneis,* Paris, Bibliothèque Nationale, MS 6979, fols. 1r–19v; cited in Thorndike, *Magic and Experimental Science,* 3:537–538. Francesco also served Urban V and Malatesta de' Malatesta, lord of Pesaro, another heavy user of mineral baths, for the six years between 1400 and 1406; see Thorndike, 3:537, and Ugolino da Montecatini, *Tractatus de balneis,* ed. and trans. Michele Giuseppe Nardi (Florence: Olschki, 1950), p. 47.

21. Antonio Guaineri, *De balneis Aquae civitatis antiquissimae commentariolus,* in Giunta, *De balneis,* fols. 43r–47r; see Thorndike, *Magic and Experimental Science,* 3:214–231. Bartolomeo da Montagnana, *De aspectu, situ, minera, virtutibus, et operationibus balneorum in comitatu patavino repertorum,* in Giunta, fols. 37r–43r.

22. Giovanni de' Dondi, *De fontibus calidis agri patavini consideratio,* in Giunta, *De balneis,* fols. 94r–108r. On Giovanni, see Tiziana Pesenti, "Dondi dall'Orologio, Giovanni," in *Dizionario biografico degli italiani,* 41:96–104; and, on this work, Thorndike, *Magic and Experimental Science,* 3:392–397.

23. Ugolino, *Tractatus,* based on the version of 1420. For the manuscript and early printed history of this work, see Barduzzi, *Ugolino,* pp. 72–74; Ugolino composed it in 1417, while practicing in Città di Castello, and augmented it with additions concerning Pietro da Eboli and the baths of Pozzuoli in 1420 (while lecturing on medicine at the University of Perugia). The version printed in Giunta, *De balneis,* fols. 47v–57v, is incomplete and modified by Pier Candido Decembrio, who restructured it and edited the Latin in order to offer it to Borso d'Este, who was also the dedicatee of Michele Savonarola's treatise on baths. On Ugolino's life and career, see Barduzzi, *Ugolino,* and Katharine Park,

Doctors and Medicine in Early Renaissance Florence (Princeton: Princeton University Press, 1985), ad indicem.

24. Savonarola, *De balneis;* for a discussion of this work, including problems in dating, see Thorndike, *Magic and Experimental Science,* 3:197–214. Biographical references in Danielle Jacquart, "Médecine et alchimie chez Michel Savonarole (1385–1446)," in *Alchimie et philosophie à la Renaissance,* ed. Jean-Claude Margolin and Sylvain Matton (Paris: Vrin, 1993), pp. 109–122; and Tiziana Pesenti Marangon, "Michele Savonarola a Padova: L'ambiente, le opere, la cultura medica," *Quaderni per la storia dell'Università de Padova* 9–10 (1976–1977): 45–102.

25. On the development of *practica,* see Nancy G. Siraisi, *Medieval and Early Renaissance Medicine: An Introduction to Knowledge and Practice* (Chicago: University of Chicago Press, 1990), p. 152; Luke Demaitre, "Theory and Practice in Medical Education at the University of Montpellier in the Thirteenth and Fourteenth Centuries," *Journal of the History of Medicine and Allied Sciences* 30 (1975): 103–123; and Danielle Jacquart, "Theory, Everyday Practice, and Three Fifteenth-Century Physicians," in *Renaissance Medical Learning: The Evolution of a Tradition,* ed. Michael McVaugh and Nancy G. Siraisi, Osiris 2nd ser., (Philadelphia: History of Science Society, 1990), pp. 140–141.

26. Michael McVaugh, "The Nature and Limits of Certitude at Early Fourteenth-Century Montpellier," in McVaugh and Siraisi, *Renaissance Medical Learning,* pp. 62–84; and Michael McVaugh, "The Development of Medieval Pharmaceutical Theory," in Arnald of Villanova, *Opera medica omnia,* vol. 2, *Aphorismi de gradibus,* ed. McVaugh (Granada: Seminarium Historiae Medicae Granatensis, 1975), esp. pp. 9–29, 89–120. See also Chiara Crisciani, "History, Novelty, and Progress in Scholastic Medicine," in McVaugh and Siraisi, *Renaissance Medical Learning,* pp. 118–139; Nancy G. Siraisi, *Taddeo Alderotti and His Pupils: Two Generations of Italian Medical Learning* (Princeton: Princeton University Press, 1981), pp. 118–146; and the literature in note 7 above.

27. See Park, *Doctors and Medicine,* pp. 85–116.

28. For sixteenth-century developments in this area, see Paula Findlen's essay in this volume, as well as Lorraine Daston and Katharine Park, *Wonders and the Order of Nature, 1150–1750* (New York: Zone Books, 1998), chap. 4.

29. Albertus Magnus, *De natura locorum* 1.4–5; *De mineralibus* 1.1.7–8. For a fuller account of this theory, see Katharine Park, "The Meanings of Diversity: Marco Polo on the 'Division' of the World," in *Texts and Contexts in Ancient and Medieval Science: Studies on the Occasion of John E. Murdoch's Seventieth Birthday,* ed. Edith Sylla and Michael McVaugh (Leiden: Brill, 1997), esp. pp. 140–142.

30. Savonarola, *De balneis* 2.1, fol. 11r.

31. Ibid. 2.3, fol. 16v; Savonarola concluded that the water was more efficacious on the spot, before it had been allowed to cool. By the sixteenth century, the interests of commerce over tourism seem to have decided the question in favor of commodification; see account book entries from 1507 in Florence, Archivio di Stato: Santa Maria Nuova 5806, fols. 4v–5r, 117v–118r (purchases of bottled water from the springs at Montecatini and Porretta); and Palmer, "Italian Baths in the Renaissance" p. 20.

32. Savonarola, *De balneis* 2.3, fol. 18v: "Haec omnia sic probabilia sunt, demonstratione logica carentia, sed experientia est omnium harum discordiarum magistra."

33. Ibid. 2.7, fol. 36v; see Thorndike, *Magic and Experimental Science*, 3:392–393.

34. Savonarola, *De balneis* 2.1, fol. 10v: "Nam videmus cavernarum aerem tempore aestatis per antiperistasim infrigidari, et tempore hyemis calefieri, sed mensurare gradum calidatatis cum gradu frigiditatis non est facile." For the temperature comparison described above, see 2.3, fol. 20r.

35. Ugolino, *Tractatus,* pp. 85 (quotation), 88–89.

36. Ibid., p. 95: "Adhuc ad istud accedunt maxime rustici qui vexantur a doloribus iuncturarum . . . et vadunt nullis regulis observatis. De cuius usu maximum sepe recipiunt iuvamentum. Et tollunt illas herbas et fatiunt ibi foveam et intrant fatiendo permixtionem luti et aquae et dicunt quod magis conferat illa aqua sic permixta cum luto."

37. Savonarola, *De balneis* 2.3, fol. 27r: "balneum hoc in tempore inventum est, brutis eius virtutem aliquam primo indicantibus."

38. Ibid. 2.3, fol. 19r–v; Ugolino, *Tractatus,* p. 92.

39. Dondi, *De fontibus* 4, fol. 99r: "Nec tamen promiserim perfecte demonstrare hanc esse certam et propriam illius causam, cum sit arduum de his, quae coniectura percipienda sunt, aliquid certi promittere. . . . Cui autem non placet, aliam ipse magis probabilem investiget, quoniam cuilibet est liber inquirendi facultas et opinandi, dummodo opinonem suam fulciat ratione."

40. Savonarola, *De balneis* 2.1, fol. 11v: "Hic autem modus investigandi caliditatis thermarum causam sic positive a me descriptus est, quoniam id iudico materia demonstrationem non patiente, a contradictione defendi non posse. Nam mihi expeditior et mentibus hominum consonantior visus est. Quare me quisquam non mordeat, cum sic ad investigandam aliam et forte veriorem causam dederim."

41. See Park, *Doctors and Medicine,* pp. 39–41, 114–117, 216–218, and note 47 below.

42. Siraisi, *Taddeo Alderotti,* esp. pp. 10–13, 121–137. See also McVaugh, "Nature and Limits of Certitude."

43. See Siraisi, *Taddeo Alderotti,* pp. 128–135; William A. Wallace, *Causality and Scientific Explanation* (Ann Arbor: University of Michigan Press, 1972–1974), 1:28–47, 65–86, 117–127.

44. See Ian Hacking, *The Emergence of Probability* (Cambridge: Cambridge University Press, 1975), pp. 21–23.

45. Dondi, *De fontibus* 7, fols. 105v–106r.

46. Ugolino, *Tractatus,* p. 123: "Et quia ego iam fui in Viterbio volui pro illa die qua steti cum illis medicis et aliis informari." The Francesco named by Ugolino (p. 49) was presumably the same Francesco da Siena who composed his own treatise on baths (see note 20 above). In his methods, Ugolino was following the example of Francesco da Siena

himself, who was sent by Urban V with seven other physicians to study the baths in Viterbo and who also consulted with local physicians concerning their properties; see Thorndike, *Magic and Experimental Science,* 3:537–538.

47. See Ugolino, *Tractatus,* pp. 46, 48, for descriptions of his competition with Giovanni Baldi to direct the medical care of Niccolò di Vieri de' Medici and Raimondo Massimo degli Albizzi.

48. Ibid., p. 111.

49. Savonarola, *De balneis* 2.3, fol. 27r.

50. Dondi, *De fontibus* 2, fol. 95v: "Ita et ego a principio videns has aquas et considerans praescripta accidentia que apparent in eis quae videntur extra naturam aliarum aquarum et aliorum fontium, non mediocriter admiratus sum, et non occurentibus [sic] causis illorum quae apparent quae plene satisfacerent, longo tempore in multis dubitavi, sed iam docentibus annis didici et experientia longa collegi nihil non esse mirabile verumque esse dictum Aristotelis primo de partibus animalium scribentis quod in unoquoque naturali inest aliquid mirabile, immo vero mirabilia multa. Sic profecto frater est: inter mirabilia nati et positi sumus et undequaque circundati adeo ut ad quodcunque primum oculos vertimus, id mirabile sit et mirabilibus plenum, si parumper profundimus intuitum." Cf. Aristotle, *De partibus animalium* 1.5, 645a16–23.

51. Daston and Park, *Wonders and the Order of Nature,* chap. 3.

52. On the literature of romance, see especially Edmond Faral, "Le merveilleux et ses sources dans les descriptions des romans français du XIIe siècle," in his *Recherches sur les sources latines des contes et romans courtois du Moyen Age* (Paris: Librairie Ancienne Honoré Champion, 1913), pp. 307–388; and Daston and Park, *Wonders and the Order of Nature,* chap. 1.

53. For some preliminary reflections on the sensory discipline of premodern European medicine, see the essays in W. F. Bynum and Roy Porter, eds., *Medicine and the Five Senses* (Cambridge: Cambridge University Press, 1993), especially Vivian Nutton, "Galen at the Bedside: The Methods of a Medical Detective" (pp. 7–16), and Jerome Bylebyl, "The Manifest and the Hidden in the Renaissance Clinic" (pp. 40–60).

54. Savonarola, *De balneis* 2.7, fol. 36v: "Quibus accipiatur quantum necessaria est indagatori minerarum, formarum mineralium exquisita notitia, praecipue ad eas partes, quae sensibus comprehenduntur. Ob quam rem taceant, qui fortassis me mordere voluerint." *Nitrum,* William Newman tells me, is either sodium carbonate or saltpeter.

55. Chiara Crisciani, "From the Laboratory to the Library: Alchemy According to Guglielmo Fabri," in this volume; Michela Pereira, *The Alchemical Corpus Attributed to Raymond Lull* (London: Warburg Institute, University of London, 1989), pp. 25–28.

56. Jacquart, "Médecine et alchimie"; Michela Pereira, *L'oro dei filosofi: Saggio sulle idee di un alchimista del Trecento* (Spoleto: Centro Italiano di Studi sull'Alto Medioevo, 1992), p. 12. Savonarola composed his *Libellus de aqua ardenti* in 1440 and produced a vernacular translation several years later, edited by Luigi Belloni in *I trattati in volgare della peste e dell' acqua ardente* (Milan: n.p., 1953); see Jacquart, "Médecine et alchimie," p. 111 n. 11.

57. See Agostino Paravicini Bagliani, *Medicina e scienza della natura alla corte dei papi nel Duecento* (Spoleto: Centro Italiano di Studi sull'Alto Medioevo, 1991), esp. pp. 34, 78, 228, 263–264, 351; Chiara Crisciani, "The Conception of Alchemy as Expressed in the *Pretiosa Margarita Novella* of Petrus Bonus of Ferrara," *Ambix* 20 (1973): 179–180.

58. Chiara Crisciani and Claude Gagnon, *Alchimie et philosophie au Moyen Age: Perspectives et problèmes* (Quebec: L'Aurore/Univers, 1980), pp. 64–74; Pereira, *L'oro dei filosofi*, pp. 136–140.

59. Crisciani and Gagnon, *Alchimie et philosophie*, pp. 73–74; Pereira, *L'oro dei filosofi*, pp. 91–92. For Dondi's friendship with Petrarch, see Park, *Doctors and Medicine*, p. 222.

60. Crisciani, "Conception of Alchemy," pp. 177–178.

61. Francesco Petrarca, *Invective contra medicum: Testo latino e volgarizzamento di Ser Domenico Silvestri,* ed. Pier Giorgio Ricci (Rome: Storia e Letteratura, 1950), p. 57; see also Coluccio Salutati, *De nobilitate legum et medicinae,* ed. Eugenio Garin (Florence: Vallecchi, 1947), 87.

62. Dondi, *De fontibus* 1, fol. 94v.

63. Jacquart, "Médecine et alchimie," pp. 114–115.

64. William C. Eamon, *Science and the Secrets of Nature: Books of Secrets in Medieval and Early Modern Culture* (Princeton: Princeton University Press, 1994), esp. 281–284; Carlo Ginzburg, "Clues: Roots of an Evidential Paradigm," in his *Clues, Myths, and the Historical Method,* trans. John and Anne C. Tedeschi (Baltimore: Johns Hopkins University Press, 1989), pp. 96–119. See also Hacking, *Emergence of Probability,* pp. 27–29.

65. Savonarola, *De balneis* 2.7, fol. 36r: "Nam cum signa medicorum sic efficacia et non fallentia non inveniantur, sed notitiam sunt dantia veritati propinqua, hinc surgit iudicium extimativum, quare diversi cum diversitate extimativae de occurrentibus aegritudinibus et casibus diversa deponunt iudicia. Ita et in explicanda minera thermarum fit, quoniam cum sic signa non habeantur omnino certa et efficacia, contingit in mineris thermarum scribentes variare. Quo loco accipias medicum non bonae extimativae fugiendum esse. Sic itaque non est de uno signo considerandum, sed omnia vel plura et potentiora illa sint, quae in consimilibus nobis fidem facere debent."

66. On these developments, see in general Harold J. Cook, "Physicians and Natural History," in *Cultures of Natural History,* ed. N. Jardine, J. A. Secord, and E. C. Spary (Cambridge: Cambridge University Press, 1996), pp. 91–105; Paula Findlen, *Possessing Nature: Museums, Collecting, and Scientific Culture in Early Modern Italy* (Berkeley: University of California Press, 1994); and Daston and Park, *Wonders and the Order of Nature,* chap. 4.

67. Vivian Nutton, "Medicine in Medieval Western Europe, 1000–1500," in *The Western Medical Tradition, 800 BC to AD 1800,* by Lawrence I. Conrad et al. (Cambridge: Cambridge University Press, 1995), esp. pp. 198–202. For a sense of some of the riches of this material, see Roger French et al., eds., *Medicine from the Black Death to the French Disease* (Brookfield, Vt.: Ashgate, 1998).

12

THE FORMATION OF A SCIENTIFIC COMMUNITY:
NATURAL HISTORY IN SIXTEENTH-CENTURY ITALY
Paula Findlen

I. "THIS PROFESSION OF SIMPLES"

In 1573 the archbishop of Bologna, Gabriele Paleotti, asked the Bolognese naturalist Ulisse Aldrovandi (1522–1605) to prepare a report on the state of knowledge in the Faculty of Medicine at the University of Bologna. Paleotti, one of the most prominent of the reforming bishops who defined the new religious and intellectual culture of post-Tridentine Italy, was eager to know what students were being taught, which lectures they preferred to attend, and finally what changes Aldrovandi himself recommended in the medical curriculum. After surveying what Aldrovandi termed "this Encyclopedia"—the length and breadth of the studies pertaining to medicine—he at last turned to his preferred subject, natural history, which he had taught for twenty years. Natural history, he wrote, was one of the "new classes that were not instituted in ancient times." Praising it as the first of the new subjects (*la più antica di queste Classi novamente introdotte*) to enter the Renaissance curriculum in medicine and natural philosophy, Aldrovandi noted that for fourteen years its teaching had been assigned to the third hour of the afternoon classes (in fact, it competed directly with the ecclesiastic historian Carlo Sigonio's lectures that year—tough competition indeed). Responding to Paleotti's question—"What should be changed?"—Aldrovandi's advice was simple: natural history should be taught all by itself, competing with no other discipline. Having clawed its way through the medical curriculum, out of the realm of practice and into the highest reaches of theoretical medicine (at least as Aldrovandi taught this subject), natural history ought to become the centerpiece of the academic study of nature, a fully independent field of knowledge that was required rather than optional for Bolognese students.[1]

Along with anatomy, natural history was perhaps the most widely discussed and hotly debated discipline among Renaissance natural philosophers. While fields such as astronomy and mathematics underwent important revisions during the same period, they did not change as quickly as natural history, nor did their curricular changes at the university level have as widespread ramifications for the way a generally learned public practiced and thought about science in the sixteenth century.[2] Between the 1530s and 1560s natural history experienced a remarkable resurgence in western Europe. Prior to that epoch, there had been only one short-lived attempt to reinvent the ancient discipline of natural history in the Italian universities, when Leo X assigned Gentile da Foligno to teach this subject at the University of Rome (La Sapienza) in 1513. However, the Sack of Rome in 1527 quickly aborted this initiative, leaving more established and less politically troubled centers of learning such as Bologna and Padua to take the lead in the 1530s.

During the 1540s newer universities such as Pisa and Ferrara had added natural history to the curriculum of their faculties of medicine; later in the century we can even find professors teaching natural history, or "medicinal simples" as the subject was initially called,[3] in Pavia, Parma, Siena, and Salerno. Those cities that did not have universities, such as Mantua, hired ducal botanists to tend their gardens (and guard the ingredients for antidotes in times of plague) and began to prefer town physicians who had benefited from this new form of training.[4] By the end of the century natural history was an established part of the medical curriculum, with special professorships, distinct locations in which to teach and work (botanical gardens and museums), a flourishing tradition of publications (ancient as well as modern), and a growing community of scholars who identified this way of studying nature as their primary goal.

The initial success of natural history in the university curriculum of late Renaissance Italy, as a largely medical subject that emphasized botany over other forms of natural knowledge, emboldened the second and third generations of naturalists to declare natural history central to the definition of natural philosophy itself. Such a move not only enhanced the intellectual status of natural history but began the process of gradually separating such fields as zoology, botany, and geology from medicine, as distinct but complementary forms of knowledge that could serve medical and nonmedical purposes. Of course such a change had not yet occurred in the sixteenth century, nor would Renaissance naturalists have discussed their goals in these terms. Instead they debated the extent to which natural history belonged to theory, focusing on questions of causality and classification, and to practice, concentrating on the uses of nature for humanity. Put simply, Renaissance naturalists wished to give an ancient discipline—the *historia* of Aristotle's animals and Theophrastus' plants, the *materia medica* of Dioscorides, Galen, and Avicenna, and to a lesser degree the *historia naturalis* of Pliny—a permanent place in the university curriculum by presenting it as the connective tissue that linked medicine to natural philosophy, making the study of terrestrial nature in the broad sense as important as the study of man.[5]

For Aldrovandi, arguably the most important and influential Italian naturalist of this period, natural history was the very thing that defined and completed the education of physicians and natural philosophers. Writing to the Bolognese nobleman Camillo Paleotti in 1585 about the necessity of convincing the Senate of Bologna to invest more money in the teaching of his subject, Aldrovandi declared: "Perhaps they do not know that by the maxim of Aristotle, Hippocrates, Galen, and all the other ancient writers, anyone who is deprived of this part of philosophy, so sensory and so useful, is not numbered among the philosophers and the physicians."[6] Paleotti, who had spent a good number of his leisure hours botanizing with Aldrovandi between Bologna and Trent when his brother the archbishop was at the Council of Trent, understood the significance of Aldrovandi's contention that natural history lay at the core of natural philosophy. Fourteen years later, on the eve of Aldrovandi's retirement from his professorship of natural history (*lectura philosophiae naturalis ordinaria de fossilibus, plantis et animalibus*) at the University of Bologna, Paleotti eulogized his friend for teaching "the philosophical history of natural things" for almost half a century.[7]

Retrospective assessments of a long and productive career such as Aldrovandi's inevitably omit a few details. In fact, Aldrovandi had held several different positions at the University of Bologna. For the early part of his career, Aldrovandi had been *lector de simplicibus,* a position that in no way specified the philosophical content of the natural history he subsequently taught upon his promotion to professor *ordinarius* in 1559. The terms of his initial position were quite typical for many Renaissance professors who were hired primarily to teach plant knowledge to aspiring physicians and little more; by contrast, as Charles Schmitt remarked in passing more than twenty years ago, the changing terms of Aldrovandi's later position signaled the move from medical botany to natural history.[8] In short, Aldrovandi, with his spectacular career at the University of Bologna, had been personally responsible for the intellectual elevation and expansion of natural history in the Italian universities.

By the end of the sixteenth century, natural history had been transformed from a form of writing, defined and shaped by the ancients, into an early modern discipline.[9] Its disciplinary status was marked by the use of such terms as "faculty" (*facoltà*) to describe its reappearance in the university curriculum and "profession" (*professione*) to identify the community of participants.[10] Increasingly the community of naturalists identified themselves through their shared commitment to the reform of the medical curriculum and their belief that studying nature was a calling in its own right; the former, in a medical culture that emphasized noninvasive herbal medicines, made botany the primary subject around which discipline formation occurred, while the latter placed botany in a continuum of studying nature that gave greater weight to the study of animals over plants and minerals because Aristotle had begun with fauna rather than flora. Naturalists also made room for topics we might nowadays label as mineralogy, geology, and paleontology on the presumption that inanimate objects (things dug up) were the least studied and most paradoxical parts of nature. Active efforts to reform medical botany and to expand the scope of and information constituting *historia naturalis* intertwined fruitfully, creating a self-consciousness among sixteenth-century scholars that they were on the verge of writing a new history of nature whose tangible results would be both intellectual and practical.

The disciplinary status of natural history did not rest solely on curricular innovations at a handful of universities. Often changes in curriculum made official activities that had already occurred outside of any specific institutional framework. The debates on Pliny's *Natural History* in Ferrara during the 1490s are an excellent example of how natural history became a topic of public debate and contention prior to the transformation of the university medical curriculum.[11] Such episodes shaped the discipline *through books:* the visibility and endurance of natural history during the Renaissance was due in no small part to the success of its publications and to the discussions that arose because of them as well as to the copious collections of natural objects, near and far, that demanded closer scrutiny in order to be known.

This essay will discuss the emergence of natural history as a discipline from the perspective of community formation. As I shall argue, the disciplinary status of natural history rested neither on a clearly defined topic nor on an agreed-upon set of procedures. Natural history continued to be an ungainly, encyclopedic enterprise that contained within it a cornucopia of diverse and often conflicting projects. If anything, its intellectual genealogy constantly called into question its disciplinary status; there are good reasons why botany enjoyed a sense of cohesion that natural history, or any other specific aspect of studying nature, never quite had until well into the eighteenth century. Despite a certain agreement that the empirical study of nature (what Aldrovandi, following Aristotle, called "sensory philosophy") was the primary goal, no one was really sure—at least prior to Linnaeus—what the point was of studying nature in this way, save to contextualize medical knowledge. In other words, the status of natural history as a discipline inhabited the unmarked terrain that included the more secure though intellectually contentious field of botany and the encyclopedic abyss of the project Pliny had defined as including just about everything. What factors contributed to the identification of botany and, from it, natural history as a distinctive subject—"this profession of simples" as Costanzo Felici called it in 1555, "our profession" as Francesco Calzolari termed it in the same year?[12] Given the important differences between, say, the collecting projects of Aldrovandi and the philosophical taxonomy of Andrea Cesalpino, what exactly bound the community of naturalists together?

In what follows, I examine one of the major episodes that contributed to the transformation of medical botany from a form of inquiry into a discipline that roughly approximated the one Aldrovandi had in mind by 1573: the publication by the Sienese physician Pier Andrea Mattioli (1501–1578) of a commentary on and translation of Dioscorides' *De materia medica*. It may seem strange to return to an ancient book as a means of describing an early modern community. Yet natural histories were, like encyclopedias, cosmographies, geographies, and other forms of compendia, collective projects whose success relied on the cooperation of many individuals. They were dynamic products whose constant revisions in terms of content reflected not only the fundamentally important developments in empirical knowledge but also the shifting parameters of an emerging scholarly community. In a world that still placed great weight on ancient authorities and that increasingly embraced printing as a means of communicating information, the editing and reediting of one text provides an unusual glimpse of the relationship between those with new knowledge and those who managed and controlled a form of knowledge that had its origins in a canonical text.[13] Mattioli's efforts to create the most authoritative version of Dioscorides' *De materia medica* provide a fascinating instance of how one especially influential naturalist portrayed the scholarly community of his day not only in light of the intellectual transformations it underwent in the mid–sixteenth century but also in light of his own ideas about how to define the very notion of a community.

II. THE TENTACLES OF MATTIOLI

Natural history emerged most clearly not through the accretion of university chairs in medicinal simples but through the activities surrounding what was probably the most well-read scientific book in the sixteenth century, Mattioli's Dioscorides. First published in Italian in 1544 and in Latin in 1554, it went into numerous editions in these languages as well as being the subject of singular translations in German, French, Spanish, and Czech. It comprised the vast majority of the seventy-eight editions of Dioscorides published in the sixteenth century and was republished through the middle of the eighteenth century, making it one of the few texts that spanned the entire early modern evolution of natural history.[14] Travelers reported back to Mattioli, not without amazement, that they had seen copies in Syria, Persia, Egypt, and among the Arabs. Allegedly there was even a Hebrew manuscript translation in Thessalonika.[15] In the 1560s Mattioli's publisher, the Venetian printer Vincenzo Valgrisi, estimated that he had printed 32,000 copies of the Italian editions alone, while Mattioli himself boasted in the preface to the 1568 Italian edition of sales of over 30,000 copies from the first ten editions.[16]

Such an astronomical level of success made Mattioli not only the most well-known naturalist of his generation but surely one of the most often read authors of his age. To put his publication record in perspective, Mattioli's book was less read than Erasmus' *Colloquies* and Luther's German Bible but better known than Thomas More's *Utopia,* many individual translations of Scripture, and of course the works of Copernicus and Vesalius.[17] Mattioli's self-consciousness about his success knew virtually no bounds, and he, not unlike his predecessor Erasmus, actively promoted his image as scholar with universal status in his field. Not without justification did the physician Girolamo Donzellino state: "there is no one in any corner of Europe, no man living today, to whom the name of Mattioli is not in some way known."[18]

Prior to the publication of his "Dioscorides," as contemporaries often called it, Mattioli had been one of many physicians collecting plants in his spare time. Neither his medical degree in Padua (1523), nor his short stint in the Roman hospitals (1523–1527), nor even his years as personal physician and political advisor to the bishop of Trent, Bernardo Clesio (1528–1539), had offered any hint of the important role he was about to play in the evolution of medical botany. Bereft of his patron with the death of Clesio in 1539, Mattioli accepted the post of town physician in Gorizia (Görz), a town north of the Friuli in the regions of Italy that still owed strong allegiance to the Holy Roman Emperor.[19] It was here, removed from any immediate contact with the activities of university professors at Bologna and Padua and in a somewhat remote but very interesting corner of the Holy Roman Empire, that Mattioli completed the first version of his Dioscorides.

Undoubtedly Mattioli himself did not fully anticipate the success of his work (though given what we know of his later ambitions, surely he had high hopes that it would eventually get him out of Gorizia). He was one of numerous humanists who, since the late fifteenth century, had aspired to translate and comment on the ancients. As Tiziana Pesenti demonstrates in her detailed study of Mattioli's publications, the earliest version of Mattioli's commentary was a fairly modest work, devoid of the Galenic additions, illustrations, extensive commentary, and lengthy pointed criticisms of the errors of others that emerged in the editions of the 1550s and 1560s.[20] At that stage, Mattioli was still a relatively obscure *protomedico,* perfecting his abilities to transform Jean Ruel's 1516 Latin edition of Dioscorides into a Tuscanized Italian that would do honor to the Greek physician's observations about the natural world. He was not yet the "author" of the text, actively competing with Dioscorides and Renaissance naturalists for the right to proclaim himself Europe's greatest naturalist. Instead he presented himself as a botanist who had entered a vast and long-neglected garden, filled with broken, tangled, and unappreciated plants, and had begun to weed it. Dioscorides, "the most faithful and diligent writer on simples," would be his guide.[21]

The initial response to the 1544 edition encouraged Mattioli to acquire a new publisher, Valgrisi, who added greater prestige to the project and placed it within a growing list of natural histories that he was printing when he brought forth the second edition in 1548, adding a sixth book—a treatise on poisons—to the original five. A pirated edition, with some rudimentary illustrations, appeared in Mantua the following year; both Mattioli and Valgrisi were so incensed that they not only mustered their political connections with the Gonzaga to have any subsequent printings halted but also set to work on a countertext: the edition of 1550. By then, Mattioli and his printer were already contemplating the possibility of a bigger audience. They translated Ruel back into Latin, thereby making the Latin version Mattioli's to sell to European medical students, and introduced illustrations into the book to conform to standards set by the German herbals of Otto Brunfels and Leonhart Fuchs; by 1554 they had created the "best" edition of Dioscorides, virtually cornering the market in *materia medica* textbooks.[22] In the process, they had begun to diminish the status of Dioscorides so that by 1565, Mattioli could state that he had outstripped Dioscorides in his knowledge of nature. So great was the contest between Mattioli and his chosen ancient that by the time his portrait appeared in the 1568 Italian edition (figure 12.1), it was accompanied by the following statement: "If the mind could be portrayed as the body, a portrait of Dioscorides will be of Mattioli."[23]

Mattioli's growing fame initially made him the definitive interpreter of Dioscorides, surpassing all other Renaissance physicians and philologists who had attempted to translate those words. It also secured his status as the foremost commentator on natural history texts of his generation. By 1549 Giovanni Odorico Melchiori called Mattioli's commentary "your Dioscorides." He elaborated: "I want to call it this because it seems to me that not only have you made it yours by having brought it into your native language . . . but by having made clear to all Italy, with most ample discourses, that which was

Georgij Handschij in Matthioli effigiem.
Si Mens, ut corpus, depingi posset, Imago   *Vna Dioscoridis, Matthioliq foret.*

Figure 12.1
"If the mind could be portrayed as the body, a portrait of Dioscorides would be of Mattioli." Source: Pier Andrea Mattioli, *I discorsi di M. Pietro Andrea Matthioli sanese* (Venice: Vicenzo Valgrisi, 1568). By permission of the Bancroft Library, University of California, Berkeley.

known here by few before."[24] Mattioli, never hesitant to heap praise upon himself, was quick to concur. By the 1552 edition, he described himself as the restorer of ancient learning, inheritor of the task set out by luminaries such as the Ferrarese physician Niccolò Leoniceno (great commentator on and critic of Pliny), who wrenched knowledge from barbarian hands. "I found myself among all others best-suited to bear upon my shoulders . . . the burden of interpreting in Italian Dioscorides's five books on the history and faculty of Simples." In his discourses, as his new version of commentary was called, Mattioli "with measured reasons and faithful authority" corrected the errors of the past and the present.[25] Soon naturalists gave up the pretext of referring to it as a work by a long-dead author and began to call it the "Mattioli." Writing to Aldrovandi in March 1561, the Veronese apothecary Francesco Calzolari noted, "Again a Mattioli has appeared."[26]

Between 1554 and 1577 Mattioli's Dioscorides emerged as the undisputed natural history of its day. "[A]nd when will his book appear?" queried Luca Ghini (1496–1556), the most distinguished naturalist in Italy before Aldrovandi, in December 1553.[27] Within a decade, it had become the talk of the papal and imperial courts, the one book that *any* aspiring naturalist ought to have read.[28] Naturalists recommended that princely patrons examine their hand-illustrated copies of Dioscorides to know more about the latest and most remarkable bits of nature, just as physicians, humanists, and learned apothecaries thumbed through more workaday versions designed more for use than for show. As the most published natural history next to Pliny's *Natural History* and the only ancient natural history that constantly changed to incorporate new information, Mattioli's Dioscorides became the book of record in which to revise scholarly knowledge of plants and a few select animals and minerals included in the final books. Scholars described objects that they sent to each other by referring to specific illustrations and descriptions in Mattioli's commentaries. "I saw that plant that Mattioli calls *androsaces* . . . ," wrote Ambrosio Mariano to Aldrovandi in May 1555. "I am certain that I did not see it among your plants when I was with you. This [specimen] conforms properly to Mattioli's picture."[29] Seeing nature and reading Mattioli's commentary became an ongoing, interactive process that epitomized in many ways the humanist ideal of a living text.

Mattioli's stature grew by leaps and bounds, and with it grew the science of simples. After sending the Holy Roman Emperor Ferdinand I a sumptuously decorated exemplar of his 1554 Latin edition—one that he bragged was more expensive than any book ever before sold or published in Venice[30]—he was rewarded with the post of imperial physician, caring for the emperor's second son of the same name. The Habsburgs so enjoyed Mattioli's presence in Prague that they bankrolled the addition of more notes and illustrations to the commentaries, allowing the pictures of Dioscorides' 600 plants and animals to grow to 1,200. At the same time, they ennobled Mattioli, his brother, and his eldest nephew and awarded him the post of imperial councillor. When Maximilian II became emperor in 1564, he made Mattioli his personal physician, further enhancing the stature of medical botany within the courtly medical hierarchy through the person of Dioscorides' most famous commentator. Thus from 1554 until 1571, when he retired to Trent, Mattioli participated in the Italian community of naturalists from afar. Nonetheless, it continued to be his primary point of reference. In April 1555, we find him asking to be remembered to "all his Italian friends."[31]

Throughout Italy, naturalists eagerly awaited the next edition—by this time they were appearing almost yearly—and circulated letters among themselves about the virtues and faults of the text. "Give me news of this other Mattioli," wrote the Riminese physician Costanzo Felici to Aldrovandi in 1557. "Will there be anything new when it has come out?"[32] The act of discussing and participating in the continued perfection of Mattioli's commentary became one of the most important communal activities engaging the attentions of Italian naturalists. Indeed, the ability to know, critique, and contribute to the book became a defining feature of this particular scholarly community, quickly elevating Mattioli's Dioscorides above such works as Brunfels's *Living Images of Plants* (1530–1536), Conrad Gesner's *History of Plants* (1541), and Leonhart Fuchs's *History of Plants* (1542) as *the* exemplary publication in natural history. In comparison to Mattioli's fast-paced revisions, all other texts seemed to have a static quality about them, since many were lucky to be published even once. Indeed the lack of news of Mattioli's latest additions to botanical knowledge essentially made one an intellectual outcast. "Mattioli's book with the addition of 135 herbs has not yet spread here," wrote Bartolomeo Maranta with some poignancy from Naples in 1558, "where beautiful things do not come until all the other famous cities have finished with them, which give them to us like the excrement of their wonderful concoctions."[33]

Mattioli himself encouraged these conversations when, in the 1550s, he began to acknowledge in print the assistance of fellow naturalists in the accumulation of material for his new and improved Dioscorides. At the same time, the choleric author sharpened his criticisms of other Renaissance naturalists—not only of previous commentators on Dioscorides such as Ruel, Fuchs, and the Portuguese physician Amatus Lusitanus (1511–1568) but eventually of virtually anyone who *dared* to disagree with him. Perusing the latest edition, one never knew where an acerbic comment—*lapsus Fuchsii, error Ruelli, calumniae Amathi*—might appear. Successive editions of Mattioli's

Dioscorides became a particularly important and visible location in which to define who really belonged to the emerging community of naturalists, at least as it was publicized in those writings, and what the terms of their relationships might be. His Dioscorides became the public document of record in which to inscribe one's presence in this particular sector of the learned world. Mattioli's own egotism about the role his work played in furthering knowledge and forging a sense of community was virtually boundless. When one friend did not thank him immediately for a copy of his book, Mattioli hypothesized that it was solely because he had neglected to include the recipient's name in the acknowledgments (never once entertaining the notion that the friend was mortally ill and therefore unconcerned about the contents of Mattioli's book!).[34]

Judging by the virtual absence of discussion of Mattioli's Dioscorides in the correspondence of naturalists prior to 1553, it seems reasonable to conclude that the preparation of the first Latin edition offered Mattioli the initial opportunity to define what he (and later Linnaeus) would call the botanical "republic." In July 1553 we find him thanking a youthful Aldrovandi for defending his writings "against those rabid dogs who seek to tear them apart."[35] Clearly Aldrovandi had been inscribed within the newly formed botanical republic as a loyal soldier in the battle against intellectual heretics, unchaste minds, and, worst of all, bad botany. Mattioli cultivated Aldrovandi's friendship with a level of craft that might have made even Machiavelli blush. Within months, he had persuaded Aldrovandi to part with two hundred of his most precious simples. Not only did Mattioli neglect to give Aldrovandi a written opinion of them, he also took the entire collection from Gorizia to Prague when he became an imperial physician in 1554, not hesitating to ask for more simples whenever his young colleague could spare them. In return, Mattioli offered Aldrovandi the prospect of being thanked in the preface to his commentary; with his name appearing in the dedicatory letter to the Holy Roman Emperor, Mattioli assured Aldrovandi, he would be "praised and celebrated throughout the world."[36] Surely immortality—and Mattioli had no doubt about the lasting value of his work—was ample compensation for the loss of a couple of hundred specimens?

In the early years of their relationship Aldrovandi was too much Mattioli's junior (and, in a sense, professional inferior) to complain. In 1553 Aldrovandi had not yet received his medical degree from Bologna; he was simply a particularly promising student who happened to possess a lot of natural objects. He shipped his simples and said very little, other than to complain privately to Mattioli's secretary, Giovan Odorico Melchiori, that such treatment did no honor to the meaning of friendship—the bedrock of the sort of *communitas* that defined intellectual relations among humanists. (To his credit, Melchiori responded that he "blushed at not being able to make good on Mattioli's promise.")[37] In the ensuing decades, as Aldrovandi's fame as a professor and collector grew, he began to insist that Mattioli offer him adequate recognition in the editions of Dioscorides and an occasional mineral or two from the rich deposits within reach of Prague.

By the late 1550s younger naturalists began to actively challenge Mattioli's erratic and imperious citation practices. After sending Mattioli an illustration of a sycamore in 1558, Aldrovandi observed that he had received no specific acknowledgment in the latest edition of the commentaries. Mattioli rushed to respond that he was not "eager to appropriate others' things"—the very accusation Paduan naturalists hurled against him in those same years—and suggested that he thought Aldrovandi was already famous enough to have no need of his praise. He assured his Bolognese colleague that this omission would be remedied in subsequent editions and began the process of moving Aldrovandi's name up through the ranks in his ever-lengthening letter of dedication, displacing other naturalists in turn. Thus by the 1568 edition, one could find praise of Aldrovandi appearing immediately after the encomium of the Habsburgs, botanical elders (Luca Ghini in Pisa and Gabriele Falloppia in Padua), and the first imperial physician (Giulio Alessandrini). Initially thanking Aldrovandi for sending him "hundreds and hundreds of plants"— belated recognition of much of the material that allowed the content of his Dioscorides to swell—Mattioli invoked his name again, a few pages later, when he specifically mentioned the sycamore illustration in question.[38] Mattioli's attentiveness to Aldrovandi's criticisms suggest that he understood well the emerging hierarchy of naturalists within the Italian universities. With Ghini and Falloppia dead, Aldrovandi had become the most important naturalist in Italy and deserved his rightful place in the "paper republic" Mattioli created every time he rewrote his acknowledgments.

III. THE BOTANICAL REPUBLIC

What was the character of Mattioli's botanical republic?[39] First and foremost, it was an *Italian* republic, created to restore glory to Italy in the spirit of Petrarch's and Machiavelli's famous statements to the same effect. While also placing his work in an international context that reflected the permeable borders of the republic of letters, Mattioli nonetheless privileged Italian contributors to the reformation of medical botany. In the preface to his 1557 edition, Mattioli expressed his pleasure at "having known that my lengthy efforts were appreciated by the Italians."[40] Knowledge may have known no boundaries—indeed, its parameters seemed to expand daily in an age of increased trade, travel, and conquest—yet Mattioli was quite sure where the best knowledge-makers could be found. Attacking the condescension of foreign scholars who honed their intellects in Italy without appreciating the talents of the Italians, he self-consciously echoed the long literary tradition of pitting the Italians against the barbarians. "These Barbarian traitors cannot suffer us Italians raising our heads. Nonetheless what good things they know, they learned in Italy, where they come as beasts and leave as men."[41] The Italy Mattioli envisioned was still the absolute measure of civilization, just as it had been for Pliny in his *Natural History* fifteen hundred years earlier.

Let us consider Mattioli's origins. A member of a leading Sienese family, he strongly identified with the lineage of intellectuals who saw writing in the *volgare* as a political and cultural statement. Hence his Dioscorides was first for Italians and next for the rest of the world. We might even argue that it was primarily for Tuscanized Italians who accepted the assumptions of the Sienese about the superiority of their language, culture, and nature. Although Mattioli spent, at most, two years of his life in Siena, it remained the center of his universe. "You pulled me out of my Tuscan nest, even though it was worth more than any other beautiful country," he wrote in his poem on Bernardo Clesio's palace. Mattioli sent his eldest son to Siena to be educated and boasted of meeting his famous compatriot, the papal banker Agostini Chigi, in Rome. When Giovanna de' Medici married Archduke Ferdinand, a happy fusion of Medici and Habsburg interests, Mattioli immediately rededicated the 1568 Italian edition of his Dioscorides to her.[42]

Coupled with these obvious statements of political allegiance to Tuscany was a strong sense of the natural superiority of this region of Italy. At various points, Mattioli praised "our most magnificent city of Siena" and the "sweet-smelling herbs of those delightful Tuscan hills." He also frequently invoked the natural knowledge of the Tuscans as a counterbalance to the faulty claims of foreigners. When Fuchs attempted to conflate "white thorn" (*Spina bianca*) with thistle (*Cardo*), for example, Mattioli howled at the idea that a plant that grew in the mountains could be the same plant found in the plains, "as all Tuscany bears witness."[43] Tuscany was the first model of nature, followed by the Trentino where Mattioli had spent the better part of his life. It was the source of the best plants, the best language, and the best citizens of the republic of letters. By contrast, foreigners had to earn their academic laurels by sitting at the feet of the Tuscans, most notably Mattioli himself. When the Salernitan physician Maranta dared to criticize Mattioli, Maranta was indicted for "Neapolitan arrogance" and immediately demoted from the status of colleague to disciple (as Mattioli moved his name down in the ranks of the acknowledged in the preface).[44] Clearly Maranta had not understood the place of the Neapolitans—Spanish subjects rather than free Tuscans—in the botanical republic.

Mattioli's political sensibilities about what it meant to be Italian emerged quite strongly in his description of his book as a garden, for it was not just any sort of garden, an *orto,* but the *orticello di Mattioli.* Given the politicized nature of Mattioli's other statements about what it meant to be Italian, it is not unreasonable to suppose that Mattioli envisioned his work as an improvement on Bernardo Rucellai's Florentine *orti oricellari,* those gardens in which everyone from Machiavelli to Benedetto Varchi had had interesting conversations about nature and politics in the first half of the sixteenth century. Mattioli borrowed liberally from the tradition of describing encyclopedias as gardens and forests when he described his version as "an expanded and amplified garden whose doors will stay perpetually open to anyone."[45] Indeed Mattioli may have tried to re-create this role during his years of service as Bishop Clesio's physician and advisor. We know that Mattioli traveled from Trent to Naples in 1536 to accompany the cardinal to an important meeting with Charles V. He also lived in the bishop's palace in Trent and in the cardinal's family home in Val di Non. Both settings provided Mattioli with the ideal ingredients for the creation of a pastoral enclave in which princes such as Clesio, the de facto ruler of Trent and the first Italian to win this appointment from the Habsburgs, could see studying nature as the perfect complement to the *vita activa.*[46]

Mattioli's community was also a Christian republic, tinged with the language of Tridentine reform that one might expect from a physician working in the vicinity of Trent precisely during those years when the Council (1545–1563) first met. Non-Catholics held no place in the interior of this world, existing instead on the margins to sharpen its definition. Here again Mattioli's personal circumstances undoubtedly had a great deal to do with the strength of his convictions. He had survived the Sack of Rome and arrived in Trent only shortly after the Peasants' Revolt of 1525 had brought rabble-rousing Lutherans to the doorstep of his beloved patron Clesio. Trent initially provided a haven for Mattioli from the troubles of Tuscany, which itself had suffered foreign invasion in the early sixteenth century. But it was not immune from the political and religious battles of the day. Cardinal Clesio's own concerns about the imminent invasion of the Protestants and his prominence in the efforts at church reform paved the way for the arrival of the Council in Trent. As it turned out, it was the Catholic Germans that the bishop-prince of Trent needed to fear. After the Holy Roman Emperor Ferdinand I died his son the Archduke Ferdinand began an aggressive policy of invasion, forcing

Clesio's successor Cristoforo Madruzzo to flee Trent for the safety of Rome between 1568 and 1578.[47] These were difficult times indeed in which to constitute an intellectual community whose boundaries changed with every shift of the political order. Mattioli found himself between two patrons who had once been allies and were now enemies.

What Mattioli's precise views on these events were is hard to say. But we can again discern a marked preference for the Italian side of things in ways that evoke Mattioli's other translation of an ancient text: Ptolemy's *Geography*. At some point during his stay in the Trentino, he sketched a watercolor of the Valli di Non and Sole that depicted the regions, beloved by Clesio, where he had botanized for many years. In it and in his Dioscorides commentaries, he clearly distinguished the Trentino from neighboring Tyrol, in explicit contradistinction to the very policy Archduke Ferdinand attempted to imposed in the 1560s and 1570s.[48] Once again, Mattioli's image of nature reveals his own perspective on the political and religious tensions that divided and shaped the republic of letters. Political experience had taught him to be suspicious of the Germans and other invading foreigners; religious circumstances intensified his distaste for anyone who threatened the integrity of the Catholic faith. One wonders how much of this contemporaries who traveled through Mattioli's chosen parts of Italy understood, since he offered one of the most powerful literary depictions of Italian nature in the sixteenth century. Long after his trip to Trent to visit Camillo Paleotti in 1562, Aldrovandi recalled how they had "visited all those places that Mattioli mentions in his histories."[49]

Drawing on his experiences, Mattioli envisioned the botanical republic as a godly community whose moral equilibrium he constantly assayed. Viewed from this perspective, Mattioli's commentary echoed the language of reform and renewal then popular among Catholic intellectuals such as Carlo Borromeo and Gabriele Paleotti. Describing himself in one of his later editions as someone with the "soul of a Christian physician," Mattioli presented his assessment of the status of various naturalists as a selfless act, done not out of personal spite and a desire for revenge against his detractors "but only to discover the truth for the benefit of the Republic."[50] This republic was implicitly a community of good citizens—worthy men who supported Mattioli's efforts through equally selfless acts. "[T]he chain of virtues and sciences [is] of such value that it binds the hearts so that even those who neither see nor know each other, love each other," he wrote in the preface to his posthumous 1581 edition of Dioscorides, thanking those anonymous donors who had helped Mattioli out of their love for his project and not for his person.[51]

Clarifying the behavior of good citizens in the botanical republic entailed identifying those unworthy souls who lived on the margins of this imagined community. Following such predecessors as Leoniceno, who had viciously attacked Arabic commentators on ancient natural histories in order to exalt the Greeks, Mattioli identified Muslims as a source of unruly and unreliable knowledge. To a certain degree, there was an intellectual basis for such criticisms, for Mattioli successfully corrected numerous errors and ambiguities that plagued medieval accounts of the content of Dioscorides. Yet it was often unclear, as with Pliny, how much the errors lay with Dioscorides, his medieval editors, and the presuppositions that all had brought to bear in assuming that a Mediterranean nature was equivalent to a more global conception of nature. Such ambiguity, however, did not serve Mattioli's own sense of the urgency of separating the damned from the saved. In this and many other instances, political and religious definitions of community strongly influenced the concept of an intellectual community.

Prior to his success with the Holy Roman Emperor in 1554, Mattioli dedicated several Italian versions of his Dioscorides to the bishop of Trent, Cristoforo Madruzzo. Describing how he had found natural knowledge, "from wild animals to the final remains of the last roots of those most noble plants, . . . in the dominion of Moors and Turks, men truly deprived of every gentility and politeness due to the coarseness of their nature," Mattioli neatly linked the barbarisms of foreigners to the barbarisms of the infidel.[52] As he argued, accepting simples from these pagans entailed nothing less than trusting the infidel with the truth of nature; instead, good Catholic naturalists needed to travel east in order to collect specimens themselves. In 1556 Mattioli heeded his own advice when he accompanied the Holy Roman Emperor to Hungary in the war against the Turks. Subsequently he relied on the Habsburg ambassador to the Ottoman court, Ogier Ghiselin De Busbecq, to provide a steady supply of plants and the famous sixth-century Codex Constantinopolitanus of Dioscorides that he consulted for his 1565 edition.[53] Implicitly, Mattioli acknowledged that valuable information about nature, as well as ancient exemplars of the text that he edited, lay in the East. But such materials needed to pass through Christian hands before they could become authoritative.

Mattioli reserved his most vicious criticisms for scholars nearby rather than for unnamed Moors in the distant East. Their disagreements tore more deeply at the fabric of the community he attempted to shape because they revealed the high level of internal dissent among European naturalists about the conclusions to be drawn from new specimens, as well as the confrontation between specimens and descriptions. So contentious was the field of natural history that shortly after Aldrovandi had accepted his promotion to teach the philosophy of natural history at Bologna, the famous Paduan anatomist Gabriele Falloppia wrote to Aldrovandi attempting to discourage him from moving away from other forms of medical teaching into the full-time teaching of natural history. "[Y]ou will enter a playing field in which there are nothing but angry people who contradict each other, who continuously upset each other and write against each other," he warned in 1561.[54] Mattioli was one of those who figured prominently on Falloppia's list of malcontents, though the surrounding names—essentially a who's who of zoology and botany at that time—suggested that the entire community was deeply divided over the intellectual and empirical outcomes of their work.

Taking Mattioli as our case study, let us examine how he dealt with detractors who enjoyed a similar degree of authority as learned commentators in possession of crucial empirical information. At every opportunity, he sought to discredit their intellectual conclusions by pointing to their flawed moral existence. Foreign naturalists such as Amatus Lusitanus (additionally suspicious as a Jewish *converso*) and the German prefect of the Paduan botanical garden, Melchior Wieland (1520–1589), both of whom publicly disagreed with Mattioli over various plant identifications in the 1550s and 1560s, were incontrovertibly barbaric, the very model of the "bad citizens" whom Mattioli wished to expunge from his botanical republic. So ferocious were their battles that even a century later, botanists still remembered them as a key episode in the history of their discipline. When the engraver chose to portray the history of botany on the frontispiece of Jean Bauhin's *Universal History of Plants* (1650), he included a vignette with the portraits of these three naturalists (figure 12.2). Below it lay one revealing word: *Dissentimus* (We disagree). By then few people read the works of Amatus and Wieland, leaving Mattioli the uncontested champion in the struggle for control of botanical knowledge.

In 1536 Amatus Lusitanus had published a Latin translation of Dioscorides, preceding both Ruel's 1546 translation and Mattioli's subsequent editions. This gave him the authority to comment on Mattioli's work, which he did in his *Enarrationes* (1553); there he noted some twenty mistakes of Mattioli among those of other naturalists. Mattioli now became the object of the sort of printed marginalia that he had used with great effect against others: "Mattioli contradicts himself," "Mattioli errs," and "Mattioli ineptly reproves Theophrastus." Mattioli's fury at Lusitanus, who publicly challenged his position among Renaissance commentators, seemed to increase with every passing year.[55] By the time he published his *Defense against Amatus Lusitanus* (1558), he had uncovered ten items to correct in the latter's work and

Figure 12.2

Mattioli, Wieland, and Amatus: "We disagree." Source: Jean Bauhin, *Historia plantarum universalis* (1650). By permission of the Research Library of the Getty Research Institute for the History of the Arts and Humanities.

one hundred mistakes to censure. Deliberately misspelling "Amathus" in order to rename Amatus "the Ignorant," Mattioli leveled the full weight of his authority in the medical circles of Catholic Italy against a Portuguese immigrant of dubious Christianity who had attempted to usurp Mattioli in defining the state of botanical knowledge.

In his *Defense,* Mattioli presented himself as working for the republic of letters, whose members demanded censure of Amatus—a judgment of a community rather than an individual. Accusing his Portuguese rival of everything from crypto-Judaism to heresy and apostasy, Mattioli linked criticism of himself to an abuse of the Catholic faith. Indeed, absence of faith became a precondition to dissent: "Just as there is no faith and no religion within you," he accused Amatus, "so in truth you are completely blind as to the medical art which you unworthily profess." He warned Amatus that further criticism would only elevate his, Mattioli's, standing among discerning readers: "Beware that your envy and calumnies do not still further increase and elevate my glory among learned readers and critics."[56] These words proved to be prophetic. Very quickly Amatus found himself out of a job and under suspicion of heresy. Unable to publish his own rebuttal in the face of Mattioli's strong influence among the Venetian printers and his extensive networks of allegiance, Amatus gradually faded from prominence in the botanical republic.

As the debate between Mattioli and Amatus concluded in the late 1550s, Mattioli found a new heretic on whom to lavish his attentions: Wieland. The Prussian botanist acquired knowledge at first vaunted by Mattioli in his earlier attacks against the infidels: he traveled in Greece, Syria, and Egypt during 1558 to 1560, returning to Padua laden with specimens for the university botanical garden. Such information, combined with Wieland's expertise in Greek, emboldened him to challenge Mattioli. The very year Mattioli attacked Amatus, Wieland published his own *Defense against Pier Andrea Mattioli* (1558), critiquing the Italian's translation of Dioscorides. Paper flew fast and furiously between the Venetian Republic and the Holy Roman Empire, as Mattioli attempted to demonstrate the superiority of his humanist skills. Linguistic competence soon became a matter of morality. Accusing Wieland of being nothing less than a "hermaphrodite," Mattioli implied that he had led his Paduan colleague, the anatomist Falloppia, down the path of moral decay through their alleged homosexual relations.[57] If Wieland could not be an apostate then he was surely a sodomite, banned from the botanical Eden of Mattioli's Dioscorides. So censorious was Mattioli of Wieland that when he heard the Prussian was off to Constantinople in search of plants from the East, he imagined this trip as a form of penitence that Wieland had undertaken for his sins against his botanical superior. Of course, when Wieland returned only to publish further rebuttals of Mattioli's views, redemption became out of the question.

Not all foreign naturalists could be dispensed with through vicious personal innuendo; such tactics were effective only with scholars who shared the same scholarly, publishing, and patronage networks as Mattioli. Mattioli's attitude toward northern European naturalists who did not involve themselves as deeply in the Italian scholarly community suggests how much his strategies for criticism relied on an understanding of the sociology of the republic of letters. For instance, he found no appropriate way to attack Conrad Gesner, the leading zoological authority of the mid–sixteenth century, simply muttering angrily to friends when the Swiss naturalist turned his attention to plants.[58] While more critical of the German Lutheran botanists Fuchs and Brunfels and the French naturalist Pierre Belon, Mattioli launched no personal attack against them. Instead he used his marginalia as a means of highlighting their errors, depicting them as perpetually untrustworthy. Given the success of his commentary, the gradual expansion of this sort of scholarly apparatus surely had its effect on the reputations of his northern colleagues among the Italians; it further reinforced the canonical status of Mattioli's opinion in this realm. At the height of his fame, Mattioli's comments were more copious, his illustrations better, and his criticisms sharper than those of any other rival. No doubt he would have been pleased to hear Sebastiano Soavi's request for "a Mattioli with figures" from Aldrovandi in 1567: "because we have a Fuchs, but it doesn't help very much."[59]

The assaults of other Italians presented Mattioli with some difficulties. If the ideal botanical republic were truly Italian, then how could he expel anyone from that birthright? The solution, it appeared, was a moral one. Italians need never be fully expelled; they had only to repent their sins to be reinstated. Imitating the Renaissance popes, Mattioli issued public sales of indulgences: homage to Mattioli in one's own publications, excellent specimens for his next edition, or a well-crafted letter of praise in Latin for his *Medical Letters* (1561)—a work Mattioli began to prepare in 1553 to immortalize reports of his fame and skill for all time—could earn a naturalist complete forgiveness if he had unwittingly sinned against the monarch of natural history. This, for instance, saved Maranta from expulsion from Mattioli's Eden when the two men could not agree on the correct interpretation of certain statements about the plant *Lonchite aspra*. To avoid the full wrath of Mattioli, Maranta composed a letter "to save his honor without blemishing mine." By the 1581 edition of Mattioli's Dioscorides, all had been resolved;

Maranta appeared between Falloppia and Aldrovandi, praised fulsomely as "the most diligent cultivator of the faculty of plants."[60]

The only unforgivable act an Italian naturalist could commit was to refuse to acknowledge the primacy of Mattioli's Dioscorides. In contrast to Maranta, who was fully redeemed, Aldrovandi's senior colleague in the teaching of natural history at Bologna, Cesare Odone, was placed in the circle of the damned. Describing Odone as his greatest enemy, Mattioli labeled him "piggish, disgraceful, vice-ridden, inhuman, envious, and pugnacious"—all for the alleged crime of refusing to introduce Mattioli's Dioscorides into the Bologna university curriculum (i.e., the crime of preferring other commentaries). On hearing that Odone planned to return to Puglia, Mattioli tartly remarked to Aldrovandi that he hoped Odone would find himself "teaching to locusts."[61] Aldrovandi did not make the same mistake, undoubtedly because he belonged to a younger generation that had begun their apprenticeship in natural history with early versions of Mattioli's commentary. During the 1560s through 1580s, as he lectured on various books in Dioscorides' *De materia medica,* he praised and used his friend's editions. As late as 1595, Aldrovandi was still searching for a hand-colored copy of Mattioli's commentary to add to his library.[62]

The case of Luigi Anguillara (ca. 1512–1570), first prefect to the Paduan botanical garden, is particularly instructive. By 1554 relations between Anguillara and Mattioli were visibly strained. Mattioli collected instances of Anguillara's ignorance and wondered aloud how one of Italy's greatest universities could allow such a man, the "eel-skinner" (a pun on Anguillara's name), to pretend to teach students botany. As far as Mattioli could tell, they didn't even know the difference between basil and lettuce after attending his demonstrations.[63] In retaliation for Anguillara's attempts to publicly correct some of his statements about the natural world, Mattioli removed Anguillara's name from the acknowledgments of the 1554 Latin edition. Confidently he wrote to Aldrovandi, "he will sin and repent since that is always the state of the envious."[64] For the next seven years Mattioli hounded Anguillara, delaying the publication of his *Simples* (1561) with the Venetian printer Valgrisi and impugning his reputation at every opportunity. These activities evidently produced the desired effect, at least temporarily. In June 1559, Mattioli wrote with great satisfaction to Aldrovandi regarding the publication of his latest Italian Dioscorides: "If you have not seen it, try to see it because it has turned out beautifully and, reading the Prologue to the readers, you will be able to see what I have newly said about you and *how I have restored Messer Aluigi to his place.*"[65] Mattioli's concept of community emerges

clearly in this final statement: the unfaithful had returned to the flock and order had been regained in the botanical republic. Of course Mattioli's anger
would be unleashed again with the appearance of Anguillara's criticisms in
print in 1561. But by then he was in the midst of fresh controversies with
other naturalists and no longer made the barbarians in Padua the focal point
of his attack.

IV. DEFINING A SCHOLARLY COMMUNITY

As we read through Mattioli's correspondence and successive editions of his
Dioscorides, an interesting and highly idiosyncratic portrait of the community of naturalists emerges. Mattioli's acknowledgments were forms of patronage, but they also forged an image of contemporary natural history that
reflected certain aspects of its structure at the middle of the sixteenth century.
Mattioli highlighted several types of naturalists in these informal histories of
his discipline and placed them in a well-defined relationship to each other.
First were the *commentators* on the ancients, humanists such as Ermolao Barbaro, Leoniceno, Antonio Musa Brasavola, Fuchs, Brunfels, and Euricus
Cordus—in other words, those scholars who had defined the genre that Mattioli now was perfecting. Even as he struck them down, he needed to maintain the dignity of the work they undertook in order to justify his own stature
in the field. Next came the "excellent and most experienced Preceptors of
Simples."[66] They were the *institutional* founders of natural history, most notably Ghini, who held the first chair in medicinal simples at both Bologna (1527–1544, 1555–1556) and Pisa (1544–1554), and subsequently
Aldrovandi. While Mattioli did not particularly praise prefects of botanical
gardens—workers rather than founders—he lauded the Great Council of Venice and the Barbaro family for having the foresight to imagine that the permanent acquisition and collection of plants would be important to the revival
of natural history.

Finally Mattioli thanked those *empirical* participants in the study of nature who shared their knowledge with him so that the garden of Mattioli
might soon resemble the garden of Eden, restored to perfection. This group
was a dynamic entity, ever changing as Mattioli modified his views about
who could best participate in the new natural history. In his earliest editions
Mattioli excluded apothecaries from this universe "for the most part, for not
understanding the Latin volumes of good authors."[67] By the mid-1550s he
identified the community of collectors as primarily composed of Italy's most
famous physicians (men capable of contributing to his *Medical Letters,* in other
words). It was undoubtedly because of the succession of controversies in

which he ensnared himself and his broadened appreciation for the empirical aspects of natural history that Mattioli later relaxed this standard. By 1568 we find him thanking the "noble and virtuous men of talent" who understood the all-consuming importance of Mattioli's project. They included prominent apothecaries such as Francesco Calzolari of Verona and Giulio Cesare Moderati of Rimini as well as the imperial ambassador in Constantinople, Busbecq, who had returned to Prague bearing a Greek codex of Dioscorides for Mattioli.[68]

As the order of acknowledgment suggests, despite the image of natural history as an empirical discipline, it was not quite one yet. Mattioli continued to privilege words over things throughout his entire career; for him, being a commentator and ultimately the new Dioscorides represented the highest position in the botanical republic of letters. Things were ephemeral; but a natural history made of words might last a millennium, creating a true empire of knowledge. His obsessive reworking of his commentaries on Dioscorides indicates his commitment to this way of viewing knowledge.

Contemporaries enamored of the material culture of nature noted and were often appalled by Mattioli's utter disregard for the tangible stuff of their profession. The Venetian patrician Pietro Antonio Michiel, creator of the most famous private botanical garden in Venice during this period, claimed that he did not even want to waste his time reading Mattioli's 1554 Latin commentary: "the Mattioli is not well made and . . . no wonder." Contrasting his enterprise with Mattioli's, he remarked, "It is no novelty to me that with an engraving or dry page and through discourse (*parlare per relatione*) one can do good things. But it needs a labor similar to mine who with diligence and even madness has suffered in raising, nourishing, and seeing plants from their beginning to end."[69] Elsewhere he remarked that what few bits of nature Mattioli deigned to send him were often old and not fresh. Mattioli, in other words, did not tend to the true garden of nature, being ever preoccupied with the paper garden he had created with the help of artists, engravers, and printers.

Aldrovandi, the recipient of this letter, shared Michiel's frustration. Despite his many requests for plants and minerals from Mattioli, who had full access to the diplomatic networks of the Habsburgs in acquiring fresh material for his own work, he rarely received anything but praise from his friend in Prague. At times Mattioli suggested that Aldrovandi surely had enough specimens in his museum in Bologna—how could one more matter? Yet Mattioli also denied Aldrovandi his share of the imperial spoils because he felt that a collector would not do anything with them that truly constituted knowledge. "[Y]ou know," he reminded Aldrovandi in 1566, "that I have never cared to observe dried plants, save for some of the rarest ones, having resolved that I do not wish to show them except in my commentaries where they are printed."[70] Such statements reflected not only Mattioli's optimism about his abilities to discern the truth of nature and record it in words and image, but also his sense that commenting on nature was surely a higher calling than collecting it.

From this perspective, Mattioli is seen to belong to a botanical republic that was in its twilight years as the sixteenth century came to a close. "[N]othing was ever greater in my prayers than to help the republic for posterity," he observed. Elsewhere he remarked that the costs of his project had been well worth it because it had been undertaken for the "profit of the entire community of the republic."[71] By the end of his long career Mattioli had probably acquired more detractors than admirers; few mourned his passing in 1577, as they had Luca Ghini's when he died in 1556.[72] By the time of Mattioli's death his project had run its course. Even Aldrovandi, who was Mattioli's logical successor in this enterprise, proclaimed himself a new Aristotle by writing his *own* natural history, based on the objects and books in his "theater of nature," rather than by commenting on Aristotle. In his philosophical history of nature—a project that Mattioli, with his preference for Dioscorides and for the *volgare,* surely did not share—Aldrovandi gave expression to a new vision of natural history as a discipline defined not by the world of one book but by the further institutionalization and expansion of an academic discipline. Nonetheless, Aldrovandi acknowledged the canonical status of Mattioli's publishing project when he instructed his own printer to produce his *Natural History* in the image of Mattioli's Dioscorides.[73]

Print culture, collecting practices, a new curriculum, humanist rhetoric, and epistolary consciousness all played important roles in the emergence of natural history in Renaissance Italy. No one activity defined what it meant for natural history to become a discipline, yet all of these factors shaped the process of interacting with new information. Contributing, reading, and reacting to Mattioli's Dioscorides was a formative experience for this early generation of naturalists. While others set the standard in the field and in the museum, Mattioli set the standard in print. And in print he stayed, throughout the sixteenth and much of the seventeenth century. His rapid and virtually unprecedented ascent in the world of courtly medicine did not transform natural history into natural philosophy; that was not his goal. But it surely provided a stronger foundation from which later naturalists such as Aldrovandi both made claims about the superiority of their chosen field and expanded that field from the narrow definition of medical botany to the more encyclopedic idea of natural history.[74]

), 1:269–279; Charles G. Nauert, "Humanists, Scientists, and Pliny: Chang-
aches to a Classical Author," *American Historical Review* 84 (1979): 72–85;
nna Ferrari, "Gli errori di Plinio: Fonti classiche e medicina nel conflitto tra
Benedetti e Niccolò Leoniceno," in *Sapere e/è potere: Discipline, dispute e pro-
università medioevale e moderna* (Bologna: Istituto per la storia di Bologna, 1990),
ristiani: 173–204. For a more general discussion of natural history as a human-
ne that situates these debates within a broader context, see Karen Reeds, "Re-
umanism and Botany," *Annals of Science* 33 (1976): 519–542

zo Felici, *Lettere ad Ulisse Aldrovandi,* ed. Giorgio Nonni (Urbino: Quattro
, p. 28 (Rimini, 8 September 1555); Cermenati, "Francesco Calzolari," p. 101
February 1555).

ocess of interacting with texts is also explored in Anthony Grafton, *Commerce
sics: Ancient Books and Renaissance Readers* (Ann Arbor: University of Michi-
997).

editions of Dioscorides, see John M. Riddle, "Dioscorides," in *Catalogus
u et Commentarium: Medieval and Renaissance Latin Translations and Commen-
Edward Cranz and Paul Oskar Kristeller (Washington, D.C.: Catholic Uni-
nerica Press, 1980), 4:1–144. See also Sara Ferri, "Pier Andrea Mattioli dai
materia medica," *Atti dell'Accademia delle Scienze di Siena detta de' Fisiocritici,*
978): xxii–xxxiv.

nnard, "Dioscorides and Renaissance Materia Medica," in *Materia Medica in
Century,* ed. Markus Florkin, Analecta Medico-Historica 1 (Oxford: Perga-
. 1; Giuseppe Fabiani, *La vita di Pietro Andrea Mattioli,* ed. Luciano Benchi
rgellini, 1872), p. 29.

rea Mattioli, *I discorsi d. M. Pietro Andrea Mattioli sanese* (Venice: Vincenzo
), vol. 1, sig. **r; Tiziana Pesenti, "Il 'Dioscoride' di Pier Andrea Mattioli
tanica," in *Trattati di prospettiva, architettura militare, idraulica e altre discipline
Pozza, 1985), p. 84. The following book appeared after this essay was com-
rri, ed., *Pietro Andrea Mattioli, Siena 1501–Trento 1578, la vita, le opere* (Pe-
emme, 1997).

points of comparison, see Lucien Febvre and Henri-Jean Martin, *The Com-
trans. David Gerard (London: Verso, 1984), pp. 218–219, 274–275; a more
w is offered in Elizabeth Eisenstein, *The Printing Press as an Agent of Change,*
idge: Cambridge University Press, 1979).

ellino's letter in Pier Andrea Mattioli, *Epistolarum medicinalium libri V
book 4. Quoted in Fabiani, *Vita,* p. 29 n. 1. On Erasmus' own strategies
arkably parallel to those of Mattioli, see Lisa Jardine, *Erasmus, Man of Let-
uction of Charisma in Print* (Princeton: Princeton University Press, 1993).

riod of Mattioli's life, see Franco Pedrotti, "Pietro Andrea Mattioli e il
iardini dei semplici e gli orti botanici della Toscana,* ed. Sara Ferri and Francesca
gia: Quattroemme, 1993), pp. 193–200; Giulio Cenci, "Nel IV cente-
ta di Pier Andrea Mattioli nel Trentino," *Studi trentini di scienze naturali* 9

Did Mattioli truly understand the botanical republic as a *disciplinary* community? That is a very hard question to answer. If we define discipline solely in terms of some sort of academic consciousness of the boundaries of knowledge, then we should probably begin our inquiry rather with Aldrovandi, who displayed a much greater self-consciousness about what it meant to be a *naturalist* and who actively created opportunities for himself to teach the whole of natural history. "I am pleased to hear of the high rank to which you have been elected to the chair in the natural philosophy of plants, animals, and fossils," wrote Maranta to Aldrovandi in 1561, "ordinarily teaching one and then the other intermittently, since they certainly are timely courses, worthy of every great and rare man." That same year others also marveled that Aldrovandi was teaching "Aristotle's history of animals, of metals, and who knows what else," when they heard about his latest promotion.[75] While Mattioli's expanded commentary on Dioscorides grew to include a few animals, it continued to emphasize botanical knowledge as the centerpiece of the Renaissance project in natural history. Yet in not excluding non-herbal materials, it created an important precedent for the work of later naturalists. Mattioli corresponded with and read a range of scholars who worked on many different parts of nature, so it is difficult to make hard-and-fast divisions between botanists and naturalists. He represented a crucial step in the process of transforming natural knowledge from an ancient form of learning into a kind of scientific inquiry that drew on the resources of an entire community to establish what was known.

Disciplines can emerge in many different ways. Certainly outsiders viewed Mattioli as one of the principal makers of a newly emerging field and identified it as a professional and intellectual pursuit that could be separated from other aspects of the medical profession. In his *Universal Piazza of All the Professions of the World* (1585), for example, Tommaso Garzoni included a chapter on "simples and herbalists" that described the revival of this ancient art in "modern times." Providing his readers with a list of its most distinguished participants, he highlighted "the work of Mattioli who, learning infinite things from Luca Ghini (in the science of simples undoubtedly the Prince), not many years ago commented in a most praiseworthy fashion on the work of Dioscorides, famous in this discipline."[76] In the eyes of Garzoni, an astute observer of the formation of identities, Mattioli belonged to an early modern profession that had brought an ancient discipline to its fruition.

The unique and unheralded success of Mattioli's Dioscorides as a publishing phenomenon created an important forum in which to think about the state of a field of knowledge and to define its participants with great precision. Its monopoly over the reading lives of other naturalists indicates that they

394

pondered carefully what Mattioli said about who and what counted in the republic of letters as well as how to view and record nature. Mattioli's artificial delimitation of his botanical republic as a community of Italian Catholics did not have lasting value, though it certainly reflected the tensions between the ideal of an international scholarly community and the realities of local intellectual networks; at a minimum, he helped implant the notion among Italian naturalists that they had preceded their northern European counterparts in offering a new and better interpretation of the natural world, even if this was not exactly true. Yet he also did much more than that: by identifying repeatedly in print the participants in the Italian project of reforming nature to each other, and by describing them as a community that collaborated around his book, Mattioli memorialized a set of relationships that existed not purely as a paper fantasy of one individual but also as a thriving intellectual network that self-consciously managed natural knowledge in the sixteenth century. In this way, the experience of reading Mattioli actively shaped and disciplined the community of naturalists, forcing them to contemplate the paradoxes of a "republic" ruled by an aging and autocratic monarch, whose own sense of community reflected the unique political and intellectual situation in which Italians found themselves at the end of the Renaissance.

NOTES

Thanks especially to Daniel Brownstein, Alix Cooper, Tom Kaufmann, and the anonymous referee for helping to improve this essay.

1. Bologna, Biblioteca Comunale dell'Archiginnasio, s. XIX. B3803, Ulisse Aldrovandi, *Informazione del rotulo del studio di Bologna de Ph[ilosoph]i et Medici all'Ill[ustrissi]mo Card[inale] Paleotti* (27 September 1573), esp. cc. 4r–6v. On Paleotti's educational reforms, see Paolo Prodi, *Il Cardinale Gabriele Paleotti (1522–1597),* 2 vols. (Rome: Edizioni di Storia e Letteratura, 1959–1967).

2. The locus classicus for this sort of inquiry is Robert S. Westman, "The Astronomer's Role in the Sixteenth Century," *History of Science* 18 (1980): 105–147. See also the more recent work of Mario Biagioli, in particular "The Social Status of Italian Renaissance Mathematicians, 1450–1600," *History of Science* 27 (1989): 41–95. For an overview of the relative status of different scientific disciplines in the Italian universities, see Charles Schmitt, "Science in the Italian Universities of the Sixteenth and Early Seventeenth Centuries," in *The Emergence of Science in Western Europe,* ed. Maurice Crosland (London: Macmillan, 1975), pp. 35–56; and idem, "Philosophy and Science in Sixteenth Century Italian Universities," in his *Aristotelian Tradition and the Renaissance Universities* (London: Variorum, 1984), pp. 297–336.

3. "Simples" referred to those natural objects that formed the basic components of premodern medicaments; they were the object of the ancient art of *materia medica,* as prac-

ticed by Galen and Dioscorides, from which natural presentation of natural history in the Renaissance u its medicinal value rather than its philosophical sig *naturalis.* Since *materia medica* was largely humoral, other treatments; this helps explain the dominance (tinctive form of disciplinary identity among Ren "simples" also included animals and minerals of m sons I will use the terms "medical botany" and "n the two subjects were closely intertwined.

4. I have discussed the genealogy of natural histo lecting in my *Possessing Nature: Museums, Collecti Italy* (Berkeley: University of California Press, 19

5. For background to the intellectual developme *Ancient Natural History* (London: Routledge, 19 medieval natural history, but see Jerry Stannard, *Ages,* ed. David C. Lindberg (Chicago: Universi and David C. Lindberg, "Natural History," in h *pean Scientific Tradition in Philosophical, Religious, 1450* (Chicago: University of Chicago Press, 1

6. Bologna, Archivio Isolani, *Fondo Paleotti* 59 Paleotti, Bologna, 11 December 1585.

7. Bologna, Archivio Isolani, *Fondo Paleotti* 5 *del Dottore Aldrovandi* (1599). I have dated this

8. Schmitt, "Science in the Italian Universiti

9. For further discussion of natural history's (ber, *Herbals, Their Origin and Evolution,* 3rd e 1986); Karen Reeds, *Botany in Medieval and* 1991); N. Jardine, J. A. Secord, and E. C. bridge: Cambridge University Press, 1996), and the Reform of Natural History in the S *plines: The Reclassification of Knowledge in* (Rochester, N.Y.: University of Rochester and emergence of scientific disciplines, Ru(plines: On the Genesis and Stability of the I *ence in Context* 5 (1992): 3–15, is particula historical emergence of disciplines differ s(

10. Bologna, Archivio Isolani, *Fondo* "Francesco Calzolari e le sue lettere all'Al

11. On the debates about Pliny, see Lynr *tory of Magic and Experimental Science* (New 4:593–610; Arturo Castiglioni, "The Sch *Science, Medicine, and History,* ed. E. Ash

Press, 195
ing Appro
and Giova
Alessandro
fessioni nell
2, ed. A. C
istic discipl
naissance H

12. Costar
Venti, 1982
(Verona, 23

13. This pr
with the Cla
gan Press, 1

14. On the
Translationur
taries, ed. F.
versity of A
discorsi sulla
ser. 16, 10 (1

15. Jerry Sta
the Sixteenth
mon, 1966),
(Siena: G. Ba

16. Pier And
Valgrisi, 1568
e l'editoria b
(Venice: Neri
pleted: Sara F
rugia: Quattr

17. For some
ing of the Book,
general overvi
2 vols. (Camb

18. See Donz
(Prague, 1561)
for success, ren
ters: The Const

19. For this p
Trentino," in *I*
Vannozzi (Peru
nario della venu

(1928): 34–58; and Franco Ottaviani, "Quattro generazioni di medici trentini (1539–1658)," in *I Madruzzo e l'Europa, 1539–1658: I principi vescovi di Trento tra Papato e Impero,* ed. Laura Dal Prà (Milan: Charta, 1993), pp. 673–677.

20. Pesenti, "Il Dioscoride," pp. 67–70; Jerry Stannard, "P. A. Mattioli: Sixteenth Century Commentator on Dioscorides," *Bibliographical Contributions, University of Kansas Libraries* 1 (1969): 68.

21. Pier Andrea Mattioli, *Di Pedacio Dioscoride Anazarbeo libri cinque dell'istoria e materia medicinale, tradotto in lingua volgare italiana da M. P. Andrea Matthioli sanese medico, con amplissime annotationi et censure* (Venice: Niccolò de' Bascarini da Pavone di Brescia, 1544), preface.

22. As Karen Reeds suggests, this was no small market; see her "Publishing Scholarly Books in the Sixteenth Century," *Scholarly Publishing,* April 1983, 259–272. Creating a definitive textbook for medical students (hence the necessity of a Latin edition) immediately made Mattioli's commentaries an international bestseller. Of course Mattioli had already boasted prior to 1554 that foreign scholars were attempting to read the Italian version because it contained such important information. Typically, Mattioli gave very little credit to Jean Ruel for creating the template for his version of Dioscorides. Ruel was acknowledged briefly in his preface of 1554, where Mattioli claimed to have "modernized" the French author's Latin. See Riddle, "Dioscorides," p. 33.

23. This transition offers interesting parallels with Erasmus' attitude toward Saint Jerome in the early sixteenth century: "Though why should it any longer look like something borrowed rather than my own?" he wrote. Quoted in Jardine, *Erasmus,* p. 68.

24. Giovanni Odorico Melchiori to Mattioli, Venice, 20 October 1549, in Mattioli, *Discorsi* (1568), vol. 1, sig. ★★★3v.

25. Pier Andrea Mattioli, *Il Dioscoride dell'Eccellente Dottor Medico M. P. Mattioli da Siena* (Venice: Vincenzo Valgrisi, 1552), sig. a. 2v.

26. References to the book as the "Mattioli" appear as early as 1553; see G. B. De Toni, *Cinque lettere inedite di Luca Ghini ad Ulisse Aldrovandi* (Padua: Tipografia Seminario, 1905), p. 13. Calzolari's letter is reproduced in Cermenati, "Francesco Calzolari," p. 115 (Verona, 3 March 1561).

27. De Toni, *Cinque lettere,* p. 12 (Ghini to Aldrovandi, Pisa, 14 December 1553).

28. "Today it is rumored in the papal palace by men who delight in minerals, plants, and every sort of animal, and almost everyone says that the Mattioli is rare." G. B. De Toni, "Spigolature aldrovandiane XIV: Cinque lettere inedite di Antonio Compagnoni di Macerata ad Ulisse Aldrovandi," *Rivista di storia critica delle scienze mediche e naturali* ser. 6, 3 (1915): 48 (Antonio Compagnoni to Aldrovandi, Rome, 17 April 1563). Similarly in his *studio* in Bologna in the 1570s, Antonio Giganti had only one book that he placed on the table in the midst of his objects—*Il Dioscoride;* see Gigliola Fragnito, *In museo e in villa: Saggi sul Rinascimento perduto* (Venice: Arsenale, 1988), p. 165.

29. Bologna, Biblioteca Universitaria (hereafter cited as BUB), Aldrovandi MS 382, I, c. 260 (Macerata, 25 May 1555)

30. C. Raimondi, "Lettere di P. A. Mattioli ad Ulisse Aldrovandi," *Bullettino senese di storia patria* 13, nos. 1–2 (1906): 19 (Mattioli to Aldrovandi, Gorizia, 17 February 1554).

31. G. B. De Toni, "Spigolature aldrovandiane XXI: Un pugilio di lettere di Giovanni Odorico Melchiori Trentino ad Ulisse Aldrovandi," *Atti del Reale Istituto Veneto di scienze, lettere ed arti* 84, part 2 (1924–1925): 612 (Melchiori to Aldrovandi, Gorizia, 1 April 1555).

32. Felici, *Lettere ad Ulisse Aldrovandi*, p. 37 (Rimini, 30 September 1557).

33. BUB, Aldrovandi MS 382, I, c. 90v (Maranta to Aldrovandi, Naples, 6 March 1558).

34. Raimondi, "Lettere," p. 22 (Gorizia, 20 May 1554).

35. Ibid., p. 13 (Gorizia, 12 July 1553).

36. Ibid., p. 18 (Gorizia, 27 December 1553).

37. De Toni, "Spigolature aldrovandiane XXI," p. 608 (Venice, 17 October 1554).

38. Mattioli, *Discorsi* (1568), vol. 1, sigs. ★★2v, ★★★4r–v. Compare this dedication to the one in the 1557 edition, in which Aldrovandi appears *after* Andrea Lacuna (a Galenic commentator) and Bartolomeo Maranta, without any special mention.

39. The events surrounding Mattioli's work offer a bird's-eye view into the workings of the sixteenth-century republic of letters. Recently this subject has received renewed attention. See particularly Jardine, *Erasmus;* Ann Goldgar, *Impolite Learning: Conduct and Community in the Republic of Letters, 1680–1750* (New Haven: Yale University Press, 1995); and Hans Bots and Françoise Waquet, *La république des lettres* (Paris: Belin–De Boeck, 1977).

40. Pier Andrea Mattioli, *I discorsi di M. Pietro Mattioli Medico sanese* (Venice: Vincenzo Valgrisi and Baldassar Costantini, 1557), sig. A.4v.

41. Raimondi, "Lettere," p. 40 (Prague, 29 January 1558); see also p. 42.

42. Pier Andrea Mattioli, *Il Magno Palazzo del Cardinale di Trento* (1539), in Ferri and Vannozzi, *I giardini dei semplici,* p. 193. Mattioli's 1581 and 1585 Italian editions also were dedicated to Giovanna de' Medici.

43. Ferri, "Pier Andrea Mattioli," pp. xxviii–xxix; Mattioli, *Discorsi* (1557), p. 323.

44. Raimondi, "Lettere," p. 57 (Prague, 29 October 1561).

45. Mattioli, *Discorsi* (1568), vol. 1, sig. ★★v; Felix Gilbert, "Bernardo Rucellai and the Orti Oricellari," *Journal of the Warburg and Courtauld Institutes* 12 (1949): 101–131.

46. Pedrotti, "Pier Andrea Mattioli," pp. 193–197. On Clesio's role in sixteenth-century political and religious life, see Paolo Prodi, ed., *Bernardo Clesio e il suo tempo,* 2 vols. (Rome: Bulzoni, 1987); and Dal Prà, *I Madruzzo e l'Europa,* pp. 15, 30.

47. Klaus Ganzer, "Clesio e la riforma protestante," in Prodi, *Bernardo Clesio,* 1:149–175; and Marco Bellabarba, "Il principato vescovile di Trento e i Madruzzo: L'Impero, la Chiesa, gli stati italiani e tedeschi," in Dal Prà, *I Madruzzo e L'Europa,* pp. 29–42.

48. Mattioli's *Le valli d'Annone e Sole* (ca. 1527–1542), is preserved in the Biblioteca Communale of Trent and discussed in Bellabarba, "Il principato vescovile di Trento e i Madruzzo," p. 137. On Mattioli's geography, see Cenci, "Nel IV Centenario della venuta di Pier Andrea Mattioli," p. 47.

49. BUB, Aldrovandi MS 21, IV, c. 66v (*Trattato della utilità et eccellenza delle lettura dell'hist[oria] natur[ale] sensata,* n.d.).

50. Pier Andrea Mattioli, *I discorsi di M. Pietro Andrea Mattioli* (Venice: Vincenzo Valgrisi, 1581), sig. ★★6r. For a theoretical discussion of the moral status of scientific endeavor, see Lorraine J. Daston, "The Moral Economy of Science," *Osiris,* n.s. 10 (1995): 2–24.

51. Mattioli, *Discorsi* (1581), sig. ★★2v.

52. Mattioli, *Il Dioscoride,* sig. a2r.

53. Riddle, "Dioscorides," p. 92, and Ferri, "Pier Andrea Mattioli," p. xxviii.

54. Falloppia is quoted in Giovanni Fantuzzi, *Memorie della vita di Ulisse Aldrovandi* (Bologna, 1774), p. 210 (Padua, 23 January 1561).

55. All the material in this passage, including quotations, is taken from Harry Friedenwald, "Amatus Lusitanus," *Bulletin of the History of Medicine* 5 (1937): 617–623. Both Amatus' *Enarrationes* and Mattioli's *Apologia* were extremely popular. The former went into six editions between 1563 and 1565 and the latter enjoyed an even longer life; it was reprinted as late as 1674.

56. Pier Andrea Mattioli, *Apologia adversus Amatum Lusitanum cum Censura in eiusdam enarrationes* (Venice: Vincenzo Valgrisi, 1559), p. 20. I have used Friedenwald's translation in "Amatus Lusitanus," p. 621.

57. Raimondi, "Lettere," p. 41 (Prague, 26 November 1558). On Wieland's publications, see Riddle, "Dioscorides," pp. 86–87, and Giorgio E. Ferrari, "Le opere a stampa del Guilandino: Per un paragrafo dell'editoria scientifica padovana del pieno Cinquecento," in *Libri e stampatori in Padova: Miscellanea di studi storici in onore di Mons. G. Bellini* (Padua: Antiniana, 1959), pp. 377–463. For further discussion of Wieland's polemics against other humanists, see Anthony Grafton, "Rhetoric, Philology, and Egyptomania in the 1570s: J. J. Scaliger's Invective against M. Guilandinus's *Papyrus,*" *Journal of the Warburg and Courtauld Institutes* 42 (1979): 167–194.

58. Mattioli wrote: "I hate that now he writes about plants." Raimondi, "Lettere," p. 4.

59. Alessandro Tosi, *Ulisse Aldrovandi e la Toscana: Carteggio inedito testimonianze documentarie* (Florence: Olschki, 1989), p. 180 (Camaldolo, 25 April 1567). On the issue of trustworthiness, see Steven Shapin's *Social History of Truth: Civility and Science in Seventeenth-Century England* (Chicago: University of Chicago Press, 1994).

60. For the letter, see G. B. De Toni, "Nuovi documenti sulla vita e sul carteggio di Bartolomeo Maranta," *Atti del Reale Istituto Veneto di scienze, lettere ed arti,* ser. 8, 14, part 2 (1911–1912): 1554 (Naples, 7 January 1561).

61. Raimondi, "Lettere," pp. 31–32 (Ratisbon, 19 January 1557).

62. Aldrovandi wrote, "Remember that Signor Nicolò Manossi in Venice writes that he has Mattioli on Dioscorides with the plants colored naturally." BUB, Aldrovandi MS 136, XXV, c. 87r.

63. Raimondi, "Lettere," p. 26 (19 September 1554).

64. Ibid., p. 24 (n.d., ca. June/July 1554). "From my book by now they have been cancelled . . ."

65. Ibid., p. 47 (Prague, 12 June 1559), emphasis mine; see Pesenti, "Il Dioscoride," p. 87.

66. Mattioli, *Il Dioscoride,* p. 7.

67. Ibid., sig. i.2r.

68. Mattioli, *Discorsi* (1568), vol. 1, sigs. ★★2r–v, ★★4v. On the role of physicians, see idem, *Discorsi* (1557), sig. Br. For a more detailed discussion of Mattioli's changing attitude toward collectors, see Paula Findlen, "Possedere la natura," in *Stanze delle meraviglie: I musei della natura tra storia e progetto,* ed. Luca Basso Peressut (Bologna: CLUEB, 1997), pp. 25–48.

69. Quoted in G. B. De Toni, "Contributo alla conoscenza delle relazioni del patrizio veneziano Pietro Andrea Michiel con Ulisse Aldrovandi," *Memorie della Reale Accademia di scienze, lettere ed arti in Modena,* ser. 3, 9 (1910): 41 (Venice, 10 April 1554). For another example of this critical view of Mattioli's work, see BUB, Aldrovandi MS 382, IV, c. 42v (Antonio Anguisciola to Aldrovandi, Piacenza, 15 December 1565).

70. Raimondi, "Lettere," p. 58 (Prague, 25 February 1566).

71. Mattioli to Aldrovandi, in Fabiani, *Vita,* p. 39 n. 2; see Pesenti, "Il Dioscoride," p. 89.

72. Space does not permit me to explore the circumstances surrounding Ghini's death, but this was surely the other formative moment for this same community of naturalists; it raised all sorts of issues about intellectual and institutional genealogies, the concept of natural knowledge as a form of patrimony, and the image of the botanical community as a well-defined family.

73. Pesenti, "Il Dioscoride," p. 103.

74. The essays in Vivian Nutton, ed., *Medicine at the Courts of Europe, 1500–1837* (London: Routledge, 1990), provide a useful introduction to the sort of position that Mattioli held.

75. Fantuzzi, *Memorie,* pp. 186–187 (Maranta to Aldrovandi, Naples, 20 April 1561); p. 210 (Falloppia to Aldrovandi, Padua, 23 January 1561).

76. Tommaso Garzoni, *La piazza universale di tutte le professioni del mondo* (Venice, 1651), p. 155.

13

EMPIRICISM AND COMMUNITY IN EARLY
MODERN SCIENCE AND ART: SOME COMMENTS
ON BATHS, PLANTS, AND COURTS
Thomas DaCosta Kaufmann

Stressing the nonquantitative aspects of science, this collection challenges a traditional historiography of science. History of science of the early modern period has tended to concentrate on physics and astronomy, most familiarly in accounts of the "Scientific Revolution." Emphasis on figures such as Bacon, Descartes, Kepler, and Galileo has also led to concentration on mathematics, on processes of reasoning, on induction, and on the role of experiment. Because of this focus on the quantitative, which is sometimes said to have marked the shift from earlier modes of "pre-" or "nonscientific" thinking, the role in the development of "science" of other aspects of investigation that might be called qualitative has been little noted. As a result, other sorts of observational processes have been understudied, other aspects of empiricism or an empirical approach ignored, and, finally, other sciences marginalized.[1]

The present paper responds to these issues as epitomized by Paula Findlen's and Katharine Park's essays in this volume. Park directly challenges some traditional emphases in her assertion of the importance and priority of medicine at the dawn of the Renaissance. For, with certain notable exceptions,[2] medicine has not usually received its due. Where its relation to the Scientific Revolution has not been simply neglected, the role of medicine has vexed rather than attracted historians of science.[3]

Findlen offers another, albeit implicit, challenge to an older "internal" historiography of science that tended to limit concerns to issues of method, theory building, or the accumulation of knowledge as an independent process. She follows the current of a few earlier treatments of the social or sociological parameters of scientific investigation that had also suggested that the definition of what constitutes science itself may change according to the societal circumstances in which its activities are pursued.[4] In this way the self-definition of science can be linked to the development of a scientific community. Findlen suggests how the change in conception of science may be related to the establishment of a scientific community in sixteenth-century Italy and to specific modes of community building.

The scope of both arguments can be expanded. While Park's and Findlen's papers may be situated in a more general historiographic context, they can of course first be considered in relation to the challenging work that both scholars have already produced. For example, in her book on early Renaissance medicine, Park establishes a distinction between the practice of empirics and the creation of medical theory by scholars at universities.[5] She characterizes the medical profession as one that contained both a small group of university-trained, and accordingly theoretically oriented, doctors and a whole collection of folk healers and other sorts of empirics. The world of early Renaissance medicine provided many contacts between these two disparate groups, and in this world practical therapeutics outside of the accepted academic medical tradition were demonstrably important for the historical development of medicine.

In her essay in this collection, Park extends her examinations further in place and time. Examining physicians not in Republican Florence but at the courts of northern Italy ca. 1400, she finds traces of activities and interests that, while not the same as those of seventeenth-century science, seem to anticipate them. These include most significantly an interest in the causal study of natural phenomena, and therefore in actual observed particulars. According to Park a new tradition of empirically based knowledge is to be found already in "one of the first attempts by philosophically trained European writers to develop a method of natural inquiry based on the study of particular natural phenomena"—in balneological treatises. The study of baths was undertaken for aristocratic patrons: some treatises were written in response to explicit request. These circumstances provide a much different social context for the empirical study of natural phenomena than does either Republican Florence or the Italian university and the *doctrina* promulgated therein.

Park's concern with this sort of question not only provides a link between some of her interests and Findlen's approach but also opens up a further line for inquiry that draws near to Findlen's topic. Like Findlen, Park mentions a nascent community of inquirers working together to accumulate and collate new information. It was this community that would determine the empirical validity of observation. But this community and its court connections may be related to another sort of investigation of nature, one that took place in many of the same places, at much the same time, and involved many of the same sorts of scholars and patrons who were concerned with balneology. Moreover, those engaged in this enterprise also had an obvious interest in particulars, as well as in an accumulation of observations.

This was the tradition of the herbal book, which is in fact also central to Findlen's concerns. Physicians would have been interested in both sorts of information, balneological and herbal. Both traditions represent and result from the approach of natural history, the study of matters that were not quantifiable but were, especially in the instance of herbals, nevertheless replicable, classifiable, and homologous. Both balneological and herbal sources provided information for medical practice that could have been derived from observation.

Most important, an ancient text that provided an authoritative source for the very *doctrina* on which that practice was based could have created a crucial connection between these interests. The book on plants (*Peri phytōn, De plantis*) associated with Aristotle in earlier times, though now usually attributed to the first-century-C.E. Greek philosopher Nicholas Damascenus, furnishes a link between the study of herbs and that of baths.[6] In a passage in 2.3, this Aristotelian treatise accounts for the composite constitution of wild herbs by reference to the origin of saltwater. The account entails an observation about the precipitation of salt and the condensation of vapors in baths.[7] Thus beyond the general empirical thrust that Aristotelian writings imparted to subsequent tradition, both balneological and herbal, this passage seems to establish a theoretical justification for combining an interest in baths and herbs. By calling attention to baths in the context of a discussion of plants, *De plantis* effectively creates a place for considerations of balneology within herbal doctrine that is founded in part on, as it were, Aristotelian as well as other ancient discussions of plants.

Although now regarded as pseudo-Aristotelian, *De plantis* belonged to the stock of traditional treatises of natural philosophy through the sixteenth century.[8] A translation from Arabic into Latin by Alfred of Sareshel was made probably before 1200, and it gained wide diffusion: at least 159 manuscripts containing *De plantis* have been identified.[9] Roger Bacon, Albertus Magnus, and Vincent of Beauvais all wrote commentaries on it. It was definitely used at the University of Paris in the thirteenth century, and we can assume that it would have been known at the northern Italian universities—especially at Padua, where instruction was given on the text of Dioscorides and where Aristotelian traditions were strong.[10]

In any event, associating an interest in plants with one in baths not only helps link the topic discussed by Park with that of simples, Findlen's topic, but has immediate pertinence for Renaissance science—and art. These realms are directly conjoined by Leonardo da Vinci, who also speaks of the importance of the use of pictorial images in the study of nature. Given what might be called the neo-Duhemian echoes of Park's arguments for the late medieval or early Renaissance origins of scientific empiricism, the absence of any mention of Leonardo in her paper, as in that of Findlen—as indeed anywhere else in this collection—is striking. Can it be that the alternative tradition Leonardo represents would mount too great a challenge to a unified view of the scientific tradition? All the more, then, do Leonardo's comments about the empirical utility of painting to those investigating nature, including water and plants, call for quotation and discussion here:

> Whoever disparages painting loves neither philosophy or nature. If you disparage painting which alone is the imitator of all the works to be seen in nature, you most surely will disparage an invention which with philosophic or subtle speculations, examines all qualities of forms: the sea, lands, animals, plants, flowers, which are surrounded by shadow and light. This is truly science and the legitimate daughter of nature, because painting is born of nature herself or, to put it more correctly, let us say granddaughter of nature because all things we sense are born of nature and painting is born of all those things.[11]

Among other allusions, these lines play on a passage in Dante's *Inferno* (11.97–105) where Dante places a discussion of *arte* in the mouth of Vergil.[12] It is known that Leonardo read Dante closely and responded to him in his own writings.[13] Here he takes the larger sense of *arte,* or art as understood more generally in Renaissance contexts,[14] and relates it more specifically to the art of painting. In Dante human *arte* is a *nepote* of God (as nature): painting is a *nipota* of nature in Leonardo. Leonardo's use of this trope has a specific purpose in his defense of painting.[15] It is deployed as part of an effort to redefine the meaning of art.[16]

Because of the context from which the reference to art is drawn, however, Leonardo's text also seems to imply a specific understanding of science, or *scientia*. Leonardo speaks of *scientia* where Dante refers to *Fisica*. Hence Leonardo seems to be speaking not just of science with the significance of knowledge in general but, like Dante, of the science of nature, as in Aristotle's *Physics*. And in Aristotle's *Physics* (199a) there is a passage that closely conjoins art (*technē*) and nature (*physis*): *technē* either completes (*epitelei*) things that nature has not accomplished, or it imitates (*mimetai*) nature. Consequently Leonardo seems to be deliberately tying painting to *scientia* in the sense not just of knowledge but of something akin to contemporary notions of science.

Moreover, the conjunction of terms in Dante allows for a reading of Leonardo's play on his text as applying not simply to science in general but to natural philosophy in particular. Dante begins the passage in question with the word *Filosofia* and explicates philosophy as referring to nature: that is, it is natural philosophy that he is discussing. Leonardo's first line also couples philosophy and nature. He goes on to describe just those sorts of subjects (including the sea, plants, animals) that might be associated with natural philosophy.

Citing the passage from Leonardo, in his own translation, James S. Ackerman thus seems justified in observing that Leonardo was able to refer to painting as a path to science or natural philosophy because of the dependence on visual evidence and observation implicit in nature studies. Ackerman claims that artists like Leonardo virtually preempted the fields of scientific investigation in many areas, including botany.[17] Although Leonardo is himself too late a figure to fall within Park's purview, both his comments reflecting Dante and his surviving drawings demonstrate that he continued earlier traditions. His interests in studies of plants and in studies of moving water, for example, are both abundantly represented in his drawings.[18] In them as in his anatomical studies, artistic techniques, such as the treatment of light and shadow that Leonardo specifically evokes in the passage quoted above, are developed to serve the investigation of nature, as a number of scholars have noted.[19] In turn, Leonardo's outspoken defense of the importance of painting for the study of natural phenomena establishes the visual arts as another manifestation of scientific learning.[20]

Leonardo thereby not only reminds us of the need to study visual sources as well as texts in the history of science, especially in its nonquantitative aspects, but he also leads us to reconsider another aspect of the implications of Park's paper for the study of herbals.[21] Herbals obviously involved an empirical concern with the growths found in nature and a focus on individual objects, as Leonardo's studies of individual plants and their particularities also suggest. Although herbal books did not necessarily impart their information in words, the collections of simples gathered together in the tradition of *materia medica* depended on illustration to disseminate their information. Hence the role of images depicting plants was central to the immediate—and empirical—concern of the herbal. The communication of the information they imparted depended on accuracy of representation: naturalistic verisimilitude came to be essential to establishing the identification of plants.

Herbals might seem to be about kinds, and their pharmacological properties were largely regarded as universal. But crucially, just as physicians studied individual springs, examining them as if they were individual patients, herbals also began to generalize from particular observations. While earlier, medieval artistic representations and illustrations of plants may have been copies of earlier pictures or composite idealizations, printed herbals of the sixteenth century, at least from the work of Brunfels on, in fact often depicted individual specimens, using them as the basis for illustrations.[22]

The development seen in printed herbaria had precedents, however, not only in the work of Leonardo, and of Albrecht Dürer,[23] but also in the work of illustrators of the period discussed by Park. While the dimensions and details of the change still require further investigation, art-historical researches have nevertheless long indicated that in the late fourteenth and early fifteen century, artistic practices changed from the use of the model book, in which previously existing types were recorded, to books with sketches, in which freely observed and executed drawings begin to be made.[24] And among these drawings there do exist studies of plants and animals made from observation of individual specimens.[25] These provided the basis for a transformation of the herbal, as they did for other sorts of natural historical compendia.

Already half a century ago Otto Pächt illuminated this development when he considered the importance of the relation of art to science in the herbal. Pächt thus pointed to another way in which the history of the visual arts is to be related to the concerns of the history of science. In an essay on early Italian nature studies and the origins of calendar landscapes in manuscript illumination, he demonstrated that some of the first new empirical studies of plants since antiquity presented in herbals, in the traditions of *materia medica* and of the *Tacuinum Sanitatis* in the late fourteenth and early fifteenth century, led to the creation of the new genre of landscape in painting. He argued specifically that the development of nature studies runs parallel to the growth of empirical science.[26]

Pächt's argument about these new creations has in turn implications for Park's thesis. The coincidences are striking. Pächt associated the new experiment, as he called it, in herbal illustrations with Padua, where there was present an important medical faculty at the university, where many other medical developments were made. But the exact context for the manufacture of the manuscripts discussed, and the origin of nature studies in model books, needs to be examined further. As in the instance of the work of Giovannino de' Grassi, workshops of illuminators of nature studies often seem in fact to have been associated with northern Italian courts.[27] The splendid illustrations turn the herbal books in question into luxury products that far surpass the practical needs (or means?) of a university physician; it is thus possible that these books, which Pächt first described as evincing an empirical turn, were related to, if not indeed created for, precisely the same milieus, and the same sorts of physicians, as those discussed by Park.

The herbal tradition also demands attention here, because herbals are central to Findlen's concerns as well. Like Park, Findlen has extended arguments from her earlier work, especially those in her book on scientific collections.[28] Making more particular the characterization of early modern science, she urges closer attention to the context in which science was carried out. Emphasizing the notion of culture, she argues for the importance in scientific activities of the formation of a scientific culture that depended on sensory engagement. Findlen has thereby broadened the category of activities that are to be studied under the rubric of science, taking scholarship into new arenas by asking what contexts made it possible for science to thrive in early modern Italy. She argues that late Renaissance natural history was defined by its audience as well as by the books that outlined its shifting parameters; the knowledge that it sought and imparted was therefore a matter of self-knowledge as well as of social knowledge.

In her paper Findlen proposes that natural history was a locus where theory and practice could be conjoined, because the practice of medicine could be linked with the study of natural history. She argues that in sixteenth-century Italy a scientific community for natural history was formed. This is a process that involved what she describes as the emergent book culture of the time. The dissemination of knowledge through and by books becomes part of what makes the new science. This argument raises a number of issues of broader importance that are worth further examination in relation to issues of science and art: in particular, the definition of what makes a community and its location.

Let us first question the definition of what Findlen describes as "Italy." For location cannot be simply equated with matters of late-twentieth-century language or geography. Although employing the vernacular may to a degree have defined speakers and writers in the sixteenth century, the issue is much more complicated than that. The definition (or self-definition) of "Italian" as applied to Mattioli in this regard cannot be taken to correspond strictly to the nation of Italy, because the use of the Italian language did not serve as a marker of Italian nationality at the time: other "nationals" outside of present-day Italy also spoke Italian.[29] The nation-state of Italy as a country of Italian-speakers did not yet exist; and because Italy as such did not yet appear on the map, the actual location of the place where Mattioli worked before going north also needs to be considered more carefully.

Mattioli worked in the city of Gorizia, now situated within Italy on its border with Slovenia. But in the later sixteenth century, Gorizia was within the bounds of the Holy Roman Empire; more specifically, it belonged to Carniola, a part of what then constituted *Innerösterreich* (Inner Austria), one of the Habsburg hereditary lands.[30] Trent, another place where Mattioli was active before he journeyed north, was a disputed territory. As a prince-bishopric, it was *Reichsunmittelbar:* it recognized as its suzerain only the Holy

Roman emperor in his person as ruler thereof. A counterclaim was made, however, that it was a Habsburg land, and rather than being ruled ultimately by the emperor (then a Habsburg), it thus belonged to the Habsburg dynasty as such. From 1564 onward, Mattioli's quondam student and patient Archduke Ferdinand II, regent of the Tyrol, tried to annex Trent to the territories over which he ruled, and in 1568 it was occupied. The year 1571 was a key moment in the struggle with the Habsburgs for control over the city.[31]

Can the dispute over Trent, and the difficulties for inhabitants that it created, have provided an impetus for Mattioli's departure from the city? Does not the Tridentine-Habsburg debate provide a more specific historical context for an understanding of Mattioli's remarks about Italians and others, especially his pro-Italian and related anti-German comments? And in general, do such remarks not reflect the continuing, long-lasting dispute and antagonism on the borderlands between the *welsch* and the *deutsch,* the conflict between Italian and German speakers?[32] It may be remembered that this conflict would have disastrous consequences, eventually leading to much loss of life on the battlefields of the First World War.

In any event, in the sixteenth century the extremely close relationships that existed between Trent and the imperial court belie an exclusively national or nationalistic interpretation. It is no accident that a whole sequence of scholars passed on from Trent to Vienna and Prague to serve at the imperial court. Not only Mattioli but also Giulio Alessandrino and Ippolito Guarinoni, to mention but the most famous, served the imperial court as botanists and physicians.[33]

In Vienna and in Prague these men would have become members of cosmopolitan courts, where they could have joined an intellectual community that consisted not only of Italian-speaking scholars but of natives of many lands; it was a polyglot group.[34] Among those with interests similar to their own, people like Mattioli could have encountered in Vienna or later in Prague scholars such as the Lusatian Paulus Fabricius, university professor of medicine and court mathematician; the Viennese Johann Aicholz, professor of medicine at the university there; the Silesian Crato von Craftheim; the Flemish botanist Rembert Dodoens; and later his fellow Netherlander Carolus Clusius (Charles de l'Écluse).[35] Several of these men are indeed known to have been friendly with Mattioli.[36]

Interaction between figures at court and in the city also suggests that the cosmopolitanism of the court affected the cities in which it was located. The Latinate, humanist culture of the Renaissance (as of the antecedent period) was an international culture. In one instance, Clusius lived on Aicholz's lands in Vienna, in the *Alsergrund* (now the ninth *Bezirk*), where his garden was located. The dual activity of Aicholz, who both owned a private garden and taught at the university, is typical of many in the Viennese group, such as Fabricius, who in addition to his court employment was, as remarked, also a university professor; they illustrate the multiple connections of this circle. Furthermore, the loci provided by the court for botanical activities in both Prague and Vienna would have attracted Mattioli as well as other figures in town. The imperial gardens that would have served botanists were located on Hradčany in Prague, in the castle precinct, as well as elsewhere in the city environs. At least as important were the gardens on the Prater in Vienna, about which Georg Tanner wrote in a poem; these also may have been helpful to Mattioli.[37]

Not only Clusius and Crato but many other figures in this group were writers on scientific matters, which pertained both to court and university concerns. Their publications, like those of writers who remained in the Italian peninsula, would also have disseminated their views. To follow Findlen, publications by authors at the imperial court would have helped make a community. Fabricius, for example, compiled a book, published in 1557, on plants found in the region of Vienna.[38] This was a work that would no doubt have been have interest to Mattioli when he came to Vienna, even if there is not yet evidence that he ever met Fabricius personally.

A further note about books and botanists in this community: the Netherlander Ogier Ghiselin De Busbecq, who brought the tulip to Europe, also brought back with him from Istanbul (Constantinople) a manuscript containing the text of Dioscorides—the famed Anicia Juliana Codex (also known as the Vienna Dioscorides), an ancient illustrated herbal that includes theriac illuminations.[39] Busbecq is probably that imperial ambassador to whom Findlen refers as having supplied material to Mattioli.[40] The Anicia Juliana Codex, perhaps one of those manuscripts, and in any event one that Mattioli knew, is of course a work of great interest to art history, the history of science, and history more generally. The time of contact between Busbecq and Mattioli, which fell between 1568 and 1569, suggests connections to other figures in a larger intellectual community. Because of the attention such a manuscript attracted, this community must also have included the imperial librarian Hugo Blotius (since not Prague but Vienna was then the seat of the imperial library).[41]

The Vienna Dioscorides was in fact but one of several manuscripts included in a trade in such materials. Manuscripts circulated back and forth from various centers. Italians such as Jacopo Ligozzi supplied nature studies for Rudolf II Habsburg as well as for the Medici and for private scholars such as Ulisse Aldrovandi.[42] Georg (Joris) Hoefnagel provided works for various courts, and his materials were also used by Aldrovandi.[43] Giorgio Liberale illustrated Mattioli's commentary to the *Materia Medica* of Dioscorides and supplied illuminated nature studies for Archduke Ferdinand of the Tyrol.[44] It can now be demonstrated that the imperial court artist Giuseppe Arcimboldo was regarded as something of an expert on natural history as well. Arcimboldo supplied specimens for the imperial collections, and he also executed nature studies for Aldrovandi. These were in turn used both for elements in his paintings of composite heads (thus for works of art) and for compendia of illustrations of creatures from the world of nature (thus for natural history, or science).[45]

Another point of practice would have helped form a community of informants. For many of the figures that have been mentioned—Alessandrino (Alexandrinus), Fabricius, Crato—held the position of personal physician to the emperor (*kaiserlicher Leibarzt*). This concern with medical matters gave them reason to have had commerce with Mattioli as well.[46]

Contrary to some of the evidence adduced by Findlen for the Italian situation, it is doubtful that this intellectual community was predominantly Catholic. At the imperial court Mattioli would have encountered a situation that was much more complicated; there in the 1550s the man who helped bring the Jesuits to Central Europe, St. Peter Canisius, could examine the Lutheran Fabricius for a position as court mathematician.[47] While by no means intentionally ecumenical, and also deeply involved in issues of church and throne, the rulers held beliefs of uncertain tenor; their courts remained open to men with various shades of opinion. Ferdinand I (emperor 1558–1564), though at times wavering in his religious attachments and in any case relatively tolerant, strongly supported the Roman church, but his successor Maximilian II, the patron of the Vienna group of humanists and scientists (1564–1576), was suspected of crypto-Lutheranism; Maximilian II was certainly open to scholars of all persuasions, including many Lutherans and some Calvinists.[48] The personal beliefs of Rudolf II (1576–1612), the great patron of art and science, are unclear; a wide variety of confessions were represented at his court, and ultimately he signed a Letter of Majesty granting freedom of religious practice in Bohemia.[49]

Regardless of the ruler's beliefs, many of the scholars in the Vienna and Prague circles of humanists were Protestants of various stripes. To mention just those with botanical interests, Dodoneus (Dodoens), Crato, and later Clusius were certainly Calvinists. Aicholz and Tanner were Lutherans.[50] Fabricius seems to have been a moderate Lutheran of Melancthonian persuasion, and he wrote a tract against the Gneseo-Lutheran Flaccists.[51]

While Rudolf II was also interested in a host of occult and scientific pursuits, including natural history, Mattioli's patron Emperor Maximilian II was highly interested in botany.[52] Maximilian's interest was part of his broader concerns with astronomy and other sciences.[53] We might therefore posit another definition of Mattioli's immediate scientific community. As noted, during Maximilian II's reign as well as during that of Rudolf II many connections existed between the imperial court and Italy; figures such as Aldrovandi employed court artists such as Hoefnagel and Arcimboldo along with Ligozzi, who in turn supplied materials to the imperial court. Moreover, Mattioli's son Ferdinando was one of the few individuals who were not dignitaries or members of the Habsburg family who were portrayed by Joseph Heintz, the court painter of Rudolf II.[54]

So, while Findlen suggests that Mattioli belonged to an Italian community from afar, we might focus instead on his connections with the community that he would have encountered more directly and immediately in Central Europe. This shift in attention applies even to the publication of books, for note must be taken of his work published in Prague by G. Melantrich. The republic of botanists was larger than the citizenry of any one nation. The community of scholars and scientists was much bigger, embracing all of Europe, as indeed the redaction and publication of editions of Mattioli's herbal book show. Several editions were published in Frankfurt am Main, and the Nuremberger Joachim Camerarius added illustrations and more text.[55]

Finally, a focus on court culture provides not only another link between Findlen's and Park's papers but also a further challenge to traditional historiography, and even to its revision as suggested in this collection. For the idea of the court servitor who is also a scientific innovator does not correspond to the current model of the scientific virtuoso found in England, or of his counterpart in Italy.[56] If however we examine the sites where scientific innovation actually occurred in the Renaissance, we may discover that they were located not only in the late trecento and early quattrocento duchies of northern Italy but also in Sforza Milan, and later in the northern and Central European courts—including most notably but by no means exclusively Rudolfine Prague. In both the period around 1400 and that around 1600, courts seem to have furnished important foyers for the sciences as well as the arts, and at both times an international court style may have existed in science as it did in art.

NOTES

1. See H. Floris Cohen, *The Scientific Revolution: A Historiographical Inquiry* (Chicago: University of Chicago Press, 1994). A broader statement about the importance of quantification has more recently been made by Alfred W. Crosby, *The Measure of Reality: Quantification and Western Society* (Cambridge: Cambridge University Press, 1997).

2. As in the work of Nancy G. Siraisi, e.g., *Medieval and Early Renaissance Medicine: An Introduction to Knowledge and Practice* (Chicago: University of Chicago Press, 1990).

3. See Harold J. Cook, "The New Philosophy and Medicine in Seventeenth-Century England," in *Reappraisals of the Scientific Revolution,* ed. David C. Lindberg and Robert S. Westman (Cambridge: Cambridge University Press, 1990), pp. 397–436.

4. Most notably R. K. Merton, *Science, Technology, and Society in Seventeenth-century England* (Bruges, 1938; reprint, New York: H. Fertig, 1970); see also the discussion in Cohen, *Scientific Revolution,* pp. 328 ff.

5. Katharine Park, *Doctors and Medicine in Early Renaissance Florence* (Princeton: Princeton University Press, 1985).

6. See Bernard G. Dod, "Aristoteles Latinus," in *The Cambridge History of Later Medieval Philosophy,* ed. Norman Kretzmann, Anthony Kenny, and Jan Pinborg (1982; reprint, Cambridge: Cambridge University Press, 1988), p. 47, for the current attribution. The issues of authorship and diffusion of this text are discussed in S. D. Wingate, *The Mediaeval Latin Versions of the Aristotelian Scientific Corpus, with Special Reference to the Biological Works* (London: Courier, 1931), pp. 55–56. Wingate makes the point that even if "pseudo-Aristotelian," *De plantis* contains genuine Aristotelian doctrine. Wingate also gives a different span, fl. 37–4 B.C.E. for the lifetime of Nicholas Damascenus (p. 55).

7. *De plantis* is most conveniently available in Aristotle, *Minor Works,* ed. W. S. Hett, Loeb Classical Library (Cambridge, Mass.: Harvard University Press, 1936), pp. 198 ff.

8. See William A. Wallace, "Traditional Natural Philosophy," in *The Cambridge History of Renaissance Philosophy,* ed. Charles B. Schmitt et al. (Cambridge: Cambridge University Press, 1988), pp. 211–212. It also belonged to the stock of works to be associated with psychology; see Katharine A. Park and Eckhard Kessler, "The Concept of Psychology," in ibid., p. 455.

9. See Dod, "Aristoteles Latinus," p. 79. For the possibility of the availability of other versions of the *De plantis* in the Latin West, see Wingate, *Mediaeval Latin Versions of the Aristotelian Scientific Corpus,* and further Charles Lohr, "Medieval Latin Aristotle Commentaries," *Traditio* 23 (1967): 313–413.

10. For the commentaries, see Wingate, *Mediaeval Latin Versions of the Aristotelian Scientific Corpus,* pp. 67 ff. Karen Meier Reeds, *Botany in Medieval and Renaissance Universities* (New York: Garland, 1991), pp. 7 ff., discusses the use of *De plantis* in medieval universities and mentions instruction in Dioscorides at Padua (p. 16). For Padua, known for its "Averroistic" tradition, it is also important to note that a commentary by Averroës might have existed on *De plantis;* see Wingate, pp. 71–72.

11. Leonardo da Vinci's *Treatise on Painting,* introduction, as translated by James S. Ackerman, "Early Renaissance 'Naturalism' and Scientific Illustration," reprinted in his *Distance Points: Essays in Theory and Renaissance Art and Architecture* (Cambridge, Mass.: MIT Press, 1991). 185.

The original text (Leonardo da Vinci, *Treatise on Painting* [*Codex Urbinas Latinus 1270*], facsimile [Princeton: Princeton University Press, 1956], pp. 4 ff.) reads: "Se tu sprezzarai la pittura la quale, e sola inuentrice de tutte le opere, euidenti de natura per certo tu sprezzarai una sottile inventione la quale con filosoficha, è sottile speculatione considera tutte le qualità delle forme, mare siti piante, animalie, herbe, fiori, le quali sono cinte d'ombra, e lumi et ueramente questa e scientia e legitima figlia de natura perche la pittura, è patorita da essa natura ma per dir più corretto diremo nipota de natura per che tutte le cose eviddenti sonno state partorita dalla natura delle quali cose, e' nata la natura, andonque nettamente la chiamaremo nipota d'essa natura, et prente d'Iddio."

12. Filosofia, mi disse, a chi la intende,
    Nota non pure in una sola parte
    Come natura lo suo corso prende
  Dal divino intelletto e da sua arte;
    E se tu ben la tua Fisica note,
    Tu troverai non dopo molte carte
  Che l'arte vostra quell, quanto puote
    Segue, come il maestro fa il discente,
    Sia che vostr'arte a Dio quasi è nepote.

Le Opere di Dante Alighieri, ed. Dr. E. Moore, rev. Paget Toynbee (Oxford: Oxford University Press, 1963), p.v.

13. See, e.g., Martin Kemp, *Leonardo da Vinci: The Marvellous Works of Nature and Man* (London: Dent, 1981), passim; on the *Divina Commedia,* see pp. 104, 162, 33.

14. See the paper by William Newman in this collection.

15. See most recently Claire J. Farago, *Leonardo da Vinci's "Paragone": A Critical Interpretation with a New Edition of the Text in the "Codex Urbinas"* (Leiden: Brill, 1991). Farago (p. 314), recognized that Leonardo's text in the *Trattato* might be a paraphrase of Dante's *Inferno,* canto 11, but made no further comment on the relation between the texts.

16. See the classic essay by Paul Oskar Kristeller, "The Modern System of the Arts," reprinted in his *Renaissance Thought II: Papers on Humanism and the Arts* (New York: Harper and Row, 1965), esp. pp. 176, 184, for Dante and Leonardo, and Anthony Blunt, *Artistic Theory in Italy, 1450–1600* (London: Oxford University Press, 1962), pp. 23 ff., for Leonardo.

17. Ackerman, "Early Renaissance 'Naturalism'."

18. These drawings have been frequently exhibited, and there is, as in all matters Leonardesque, an ever-increasing literature on them. A good starting point is the catalogue of an exhibition organized in London in 1989, *Leonardo da Vinci* (New Haven: Yale University Press in association with the South Bank Centre, 1989), with contributions by E. H. Gombrich, Martin Kemp, and Jane Roberts.

    For Leonardo's water studies, see the essay by E. H. Gombrich, "The Form of Movement in Water and Air," in *The Heritage of Apelles,* Studies in the Art of the Renaissance 3 (Ithaca, N.Y.: Cornell University Press; Oxford: Phaedon, 1976), pp. 39–56.

19. See Martin Kemp, "Il concetto dell'anima in Leonardo's Early Skull Studies," *Journal of the Warburg and Courtauld Institutes* 34 (1971): 115–134; Ackerman, "Early Renaissance 'Naturalism.'"

20. These relationships between the visual arts and science are well discussed by Kemp, *Leonardo da Vinci*. An earlier essay in a similar direction by a historian of science is V. P. Zubov, *Leonardo da Vinci*, trans. David H. Kraus (Cambridge, Mass.: Harvard University Press, 1968).

21. For the subject of herbal illustration, see in general A. R. Arber, *Herbals, Their Origin and Evolution: A Chapter in the History of Botany*, 3rd ed. (Cambridge: Cambridge University Press, 1986 [1st ed., 1912]); Wilfried Blunt and Sandra Raphael, *The Illustrated Herbal* (New York: Thames and Hudson, Metropolitan Museum of Art, 1979).

22. I am grateful to an anonymous reader of this article for suggesting I may have missed Park's point, thus leading me to elaborate these arguments and ensure that Ackerman's point, summarized here, not be missed.

23. See, e.g., Fritz Koreny, *Albrecht Dürer and the Animal and Plant Studies of the Renaissance*, trans. Pamela Marwood and Yehuda Shapiro (Boston: Little, Brown, 1988); and for the link with herbaria, see the review of this book by Karen Reeds in *Isis* 81 (1990): 768.

24. See for this issue most recently and fully Robert W. Scheller, *Exemplum: Model-Book Drawings and the Practice of Artistic Transmission in the Middle Ages (ca. 900–1450)*, trans. Michael Hoyle (Amsterdam: Amsterdam University Press, 1995), and Albert J. Elen, *Italian Late-Medieval and Renaissance Drawing-Books from Giovannino de'Grassi to Palma Giovane: A Codicological Approach* (Leiden: Elen, 1995), both with extensive bibliographies.

25. The relatively few examples on which judgments are based, as cited from the secondary art-historical literature by an author like Reeds, *Botany in Medieval and Renaissance Universities*, p. 185 n. 83, do not take into account the problem of survival of manuscripts and drawings from the period, and thus do not give an adequate picture of the problem. Already Otto Pächt (see note 26) considered the objects he was studying not as singular but as exemplary of a wider trend. This issue may be reconsidered in relation to information gathered in more recent studies and exhibitions, e.g., Marco Rossi, *Giovannino de Grassi: La corte e la cattedrale* (Milan: Silvana, 1995), and Paola Marini, ed., *Pisanello* (Milan: Electa, 1996); however, these works do not really tackle the question again.

26. Otto Pächt, "Early Italian Nature Studies and the Early Calendar Landscape," *Journal of the Warburg and Courtauld Institutes* 13 (1950): 13–47, especially pp. 25–31.

27. See ibid. and Rossi, *Giovannino de Grassi*.

28. Paula Findlen, *Possessing Nature: Museums, Collecting, and Scientific Culture in Early Modern Italy* (Berkeley: University of California Press, 1994).

29. The general issue of how language and nation relate to the definition of the Renaissance is reviewed in Peter Burke, "The Uses of Italy," in *The Renaissance in National Context*, ed. Roy Porter and Mikulás Teich (Cambridge: Cambridge University Press, 1992), pp. 6–20, esp. pp. 13–14; see further now Burke, *The European Renaissance: Centres and Peripheries* (Oxford: Blackwell, 1998). For the question of the relation of the imperial court to Italy and Italians in the time of Mattioli, see Heinz Noflatscher, "Sprache und Politik: Die Italienexperten Kaiser Maximilians II.," *Kaiser Maximilian II: Kultur und Politik im 16. Jahrhundert*, ed. Friedrich Edelmayer and Alfred Kohler, Wiener Beiträge zur Geschichte

der Neuzeit 19 (Vienna: Verlag für Geschichte und Politik; Munich: Oldenbourg, 1992), pp. 143–168. For the continuing importance of Italian as a spoken language (and cultural determinant) at the imperial court through the seventeenth century, see Thomas DaCosta Kaufmann, *Court, Cloister, and City: The Art and Culture of Central Europe, 1450–1800* (Chicago: University of Chicago Press, 1995), p. 270.

30. For a variety of essays on the social and cultural history of this region, see *Innerösterreich 1564–1619,* ed. Alexander Novotny and Berthold Sutter, Joannea 3 (Graz: Universitätsbuchdruckerei Styria, [1968]); *Graz als Residenz: Innerösterreich 1564–1619,* ex. cat. (Graz: Steirermarkischen Landesbibliothek am Joanneum, 1964).

31. The essays by Bruno Passamani, "Ragioni e struttura della mostra: Un principato per l'impero," and Marco Bellabarba, "Il Principato vescovile di Trento e i Madruzzo: L'Impero, la Chiesa, gli stati italiani e tedeschi," in *I Madruzzo e l'Europa, 1539–1638: I principi vescovi di Trento tra Papato e Impero,* ed. Laura Dal Prà, ex. cat. (Milan: Charta, 1993), pp. 11–28, 29–42, provide an introduction to the political vicissitudes of Trent in this era.

32. This conflict provides an introduction to some of the discussion of period eye in Michael Baxandall, *The Limewood Sculptors of Renaissance Germany* (New Haven: Yale University Press, 1980); a somewhat different interpretation of these issues is offered in Kaufmann, *Court, Cloister, and City.*

33. For Mattioli in this context of Trent physicians, see Franco Ottaviani, "Quattro generazioni di medici trentini (1539–1653)," in Dal Prà, *I Madruzzo e l'Europa,* pp. 673–679.

34. For the presence of these Italians in the midst of a host of other physicians, and the description of the courts as cosmopolitan, see R. J. W. Evans, *Rudolf II and His World: A Study in Intellectual History 1576–1612* (Oxford: Clarendon, 1973), pp. 203–204; idem, *The Making of the Habsburg Monarchy, 1550–1700* (Oxford: Clarendon; New York: Oxford University Press, 1979), p. 22. For Mattioli at the imperial court, see now Maria Ludovica Lenzi, "Dal 'Regno di Iatria' alla corte di Praga," in *Pietro Andrea Mattioli (Siena 1501–Trento 1578): La vita, le opere,* ed. Sara Ferri (Siena: Quattroemme, 1997), pp. 83–104.

35. For a recent treatment of Fabricius, see Thomas DaCosta Kaufmann, "Astronomy, Technology, Humanism, and Art at the Entry of Rudolf II into Vienna, 1577: The Role of Paulus Fabricius," in his *The Mastery of Nature: Aspects of Art, Science, and Humanism in the Renaissance* (Princeton: Princeton University Press, 1993), pp. 136–150; for Crato, see Howard Louthan, *Johannis* [sic] *Crato and the Austrian Habsburgs: Reforming a Counter-reform Court,* Studies in Reformed Theology and History 2, no. 3 (Princeton: Princeton Theological Seminary, 1994); for Dodoens, see *Luister en rempspoed van Mechelen ten tijde van Rembert Doedoens,* ex. cat. (Brussels, 1985); for Clusius, *Carolus Clusius und seine Zeit,* Wissenschaftliche Arbeiten aus dem Burgenland 54 (Eisenstadt: Amt der Burgenlandische Landesregierung, 1974); for Aicholz and for the whole group, see Joseph Ritter von Aschbach, *Die Wiener Universität und ihre Gelehrten 1520 bis 1565,* Geschichte der Wiener Universität 13 (Vienna: Verlagder K. K. Universität, 1888).

36. For Fabricius' friendship with Mattioli, see Kaufmann, *The Mastery of Nature,* pp. 144, 275 n. 20. For Busbecq and Mattioli, see note 40 below.

37. See Kurt Mühlberger, "Bildung und Wissenschaft: Kaiser Maximilian II. und die Universität Wien," in Edelmayer and Kohler, *Kaiser Maximilian II*, pp. 203–230, and in general Evans, *Rudolf II and His World*, pp. 119 ff. Also see still Aschbach, *Die Wiener Universität*, with further references.

38. Paulus Fabricius, *Catalogus Stirpium circa Viennam nascentium* (Vienna, 1557); see for further references Kaufmann, *Mastery of Nature*, pp. 146, 278 n. 40.

39. Vienna, Österreichische Nationalbibliothek, cod. med. gr. 1; for the literature on this manuscript, see *Der Wiener Dioscurides (Codex Vindobonensis Med. gr. 1)*, intro. H. Gerstinger (Graz: Akademische Drucke und Verlagsanstalt, 1965); Kurt Weitzmann, *Ancient Book Illumination* (Cambridge, Mass: Harvard University Press for Oberlin College and the Department of Art and Archaeology of Princeton University, 1959), pp. 12 ff., 15 ff. and *passim*.

40. For Busbecq and the court, see most recently Zweder von Martels, "On His Majesty's Service: Augurius Busbequius, Courtier and Diplomat of Maximilian II," in Edelmayer and Kohler, *Kaiser Maximilian II*, pp. 169–181. Busbecq's friendship with Mattioli is mentioned on p. 177. Findlen connects him with Mattioli on p. 384 above (with n. 53).

41. For Blotius and the imperial library, see Alphons Lhotsky *Die Geschichte der Sammlungen (Festschrift des Kunsthistorischen Museums 1891–1941)* (Vienna: F. Berger, 1941–1945), 1.1:164.

42. For Ligozzi and Rudolf II, see Thomas DaCosta Kaufmann, *The School of Prague: Painting at the Court of Rudolf II* (Chicago: University of Chicago Press, 1988), pp. 76, 95, 129 n. 28, and *Prag um 1600: Kunst und Kultur am Hofe Rudolfs II*, ex. cat., Kunsthistorisches Museum Wien (Freren: Luca, 1988) 2:138–143; for Ligozzi and the Medici, see *Palazzo Vecchio: Committenza e collezionismo medicei*, ex. cat. Firenze e la Toscana dei Medici nell'Europa del Cinquecento ([Milan]: Electa, [Florence]: Centro Di Alinari Scala, 1980), pp. 295 ff; for Ligozzi and Aldrovandi, see Giuseppe Olmi, "Osservazione della natura e raffigurazione in Ulisse Aldrovandi (1522–1605)," in his *L'inventario del mondo: Catalogazione della natura e luoghi del sapere nella prima età moderna* (Bologna: Il Mulino, 1992), pp. 61 ff.

43. An introduction to works done by Hoefnagel for the Habsburgs is found in Theadora Alida Gerarda Wilberg Vignau-Schuurman, *Die emblematischen Elemente im Werke Joris Hoefnagels*, 2 vols. (Leiden: Universitaire Pers, 1969), and Lee Hendrix and Thea Vignau-Wilberg, *Mira Calligraphiae Monumenta: A Sixteenth-Century Calligraphic Manuscript Inscribed by Georg Bocskay and Illuminated by Joris Hoefangel* (Malibu, Calif.: J. Paul Getty Museum, 1992); for Hoefnagel and Aldrovandi, see Olmi, *L'inventario del mondo*, p. 131 and note 25.

44. See most recently Alfred Auer and Eva Irblich, *Natur und Kunst: Handschriften und Alben aus der Ambraser Sammlung Erzherzog Ferdinands II. (1529–1595)*, ex. cat., Schloss Ambras, Innsbruck (Vienna: Kunsthistorisches Museum Wien, 1995), pp. 64–78, and Eva Irblich, "Naturstudien Erzherzog Ferdinands II (1529–1598): zur Kunstkammer auf Schloss Ambras bei Innsbruck," in *Thesaurus Austriacus: Europas Glanz im Spiegel der Buchkunst Handschriften und Kunstalben von 900 bis 1600*, ed. Irblich, ex. cat. (Vienna: Nationalbibliothek, 1996), pp. 209–225, both with further references.

45. On Arcimboldo, see further Irblich, *Thesaurus Austriacus,* pp. 233–260, and Thomas DaCosta Kaufmann, "Caprices of Art and Nature: Arcimboldo and the Monstrous," in *Kunstform Capriccio: Von der Groteske zur Spieltheorie der Moderne,* ed. Ekkehard Mai and Joachim Rees, Kunstwissenschaftliche Bibliothek 6 (Cologne: König, 1998), pp. 33–51.

46. On court physicians and their various origins, see Noflatscher, "Sprache und Politik," pp. 155–156.

47. For this incident involving Canisius and Fabricius, see Kaufmann, *Mastery of Nature,* p. 144.

48. See most recently Mühlberger, "Bildung und Wissenschaft," pp. 209 ff., 217 ff.; and now Howard Louthan, *The Quest for Compromise: Peacemakers in Counter-Reformation Vienna* (Cambridge: Cambridge University Press, 1997). Also see Otto Helmut Hopfen, *Kaiser Maximilian II. und der Kompromißkatholzismus* (diss., Munich, 1895).

49. The best treatment of the problem of the emperor's beliefs and of the question of religion at Rudolf's court remains Evans, *Rudolf II and His World,* pp. 84–115.

50. In addition to Mühlberger, "Bildung und Wissenschaft," see Grete Mecenseffy, *Evangelische Lehrer an der Universität Wien* (Vienna: n.p., 1967).

51. See Kaufmann, *Mastery of Nature,* pp. 144–145, with nn.

52. For Rudolf and the occult, see Evans, *Rudolf II and His World,* pp. 196–242; for Rudolf and flowers, a starting point is Kaufmann, *The School of Prague,* pp. 75, 76, 229, 244, 245; for Arcimboldo, Rudolf II, and natural history, see "Caprices of Art and Nature."

53. See Mühlberger, "Bildung und Wissenschaft"; Kaufmann, *Mastery of Nature,* pp. 140 ff.

54. See Jürgen Zimmer, *Joseph Heintz der Ältere als Maler* (Weißenhorn: Konrad, 1971), pp. 38–39, 143–144; Kaufmann, *The School of Prague,* p. 196, cat. no. 7.42.

55. The Melantrich edition is *New Kreuterbuch,* trans. G. Handsch (Prague, 1563). Editions by Camerarius include *De plantis epitome* (Frankfurt a. M., 1586) and *Kreutterbuch* (Frankfurt a. M., 1600). Mattioli's *Opera omnia,* ed. C. Bauhinus, was also published in Frankfurt in 1598.

56. See, however, the alternative picture created by the essays—including one by Paula Findlen ("The Economy of Scientific Exchange in Early Modern Italy," pp. 5–24) and one by Pamela H. Smith ("Curing the Body Politic: Chemistry and Commerce at Court, 1664–70," pp. 195–209)—collected in *Patronage and Institutions: Science, Technology, and Medicine at the European Court, 1500–1750,* ed. Bruce T. Moran (Rochester, N.Y.: Boydell, 1991). Other Italian examples are provided by the exhibition catalogues *La corte, il mare, i mercanti: La rinascita della scienza. Editoria e società; Astrologia, magia e alchimia; Firenze e la Toscana dei Medici nell'Europa del Cinquecento* (Florence: Electa, 1980), and Dario A. Franchini [et al.], *La Scienza a Corte: Collezionismo eclettico natura e immagine a Mantova fra Rinascimento e Manierismo* (Rome: Bulzoni, 1979), to mention just a few of a growing number of works on this theme. See, e.g., Kaufmann, *Mastery of Nature,* esp. pp. 188 ff.

INDEX